Natural Science in Archaeology

Series Editors
Bernd Herrmann
Günther A. Wagner

AF212724

For further volumes:
http://www.springer.com/series/3703

Rosemarie Klemm • Dietrich Klemm

Gold and Gold Mining in Ancient Egypt and Nubia

Geoarchaeology of the Ancient Gold Mining Sites in the Egyptian and Sudanese Eastern Deserts

Springer

Rosemarie Klemm
Institute of Egyptology
Ludwig-Maximilians-University
Munich
Germany

Dietrich Klemm
Dept. of Earth- and
Environmental Sciences
Ludwig-Maximilians-University
Munich
Germany

Translator
Paul Larsen
85, Bld. Mal. Leclerc
85000 La Roche sur Yon
France

All photographs, if not specifically marked, are taken by Rosemarie Klemm

ISSN 1613-9712
ISBN 978-3-662-52067-3 ISBN 978-3-642-22508-6 (eBook)
DOI 10.1007/978-3-642-22508-6
Springer Heidelberg New York Dordrecht London

© Springer-Verlag Berlin Heidelberg 2013
Softcover re-print of the Hardcover 1st edition 2013
This work is subject to copyright. All rights are reserved by the Publisher, whether the whole or part of the material is concerned, specifically the rights of translation, reprinting, reuse of illustrations, recitation, broadcasting, reproduction on microfilms or in any other physical way, and transmission or information storage and retrieval, electronic adaptation, computer software, or by similar or dissimilar methodology now known or hereafter developed. Exempted from this legal reservation are brief excerpts in connection with reviews or scholarly analysis or material supplied specifically for the purpose of being entered and executed on a computer system, for exclusive use by the purchaser of the work. Duplication of this publication or parts thereof is permitted only under the provisions of the Copyright Law of the Publisher's location, in its current version, and permission for use must always be obtained from Springer. Permissions for use may be obtained through RightsLink at the Copyright Clearance Center. Violations are liable to prosecution under the respective Copyright Law.
The use of general descriptive names, registered names, trademarks, service marks, etc. in this publication does not imply, even in the absence of a specific statement, that such names are exempt from the relevant protective laws and regulations and therefore free for general use.
While the advice and information in this book are believed to be true and accurate at the date of publication, neither the authors nor the editors nor the publisher can accept any legal responsibility for any errors or omissions that may be made. The publisher makes no warranty, express or implied, with respect to the material contained herein.

Cover design: deblik, Berlin
Cover figure: © crimson – Fotolia.com

Printed on acid-free paper

Springer is part of Springer Science+Business Media (www.springer.com)

We dedicate this book with gratitude to our daughters Judith, Katharina and Friederike who always have showed consideration for our lives as researchers and who superbly mastered our frequent periods of absence.

Preface

In the 1960s and 1970s, the Egyptian Geological Survey and Mining Authority (EGSMA) carried out a systematic survey of the gold deposits in the Egyptian Eastern Desert in collaboration with a team from the Soviet Technoexport Company. The work led to the compilation of detailed geologic maps and implementation of extensive geochemical analyses.

The project succeeded in identifying large numbers of partly economically viable gold mineralisations as well as unmistakable traces from ancient mining and prospecting activities. This unforeseen finding revealed that former populations had over centuries mined and prospected the region's gold ores while revealing almost if not just as efficient techniques as the ones used today.

Unfortunately though, a documentation of the archaeological heritage had not been included to the project, and even after the premature end of the Soviet-Egyptian cooperation, such issues continued to be completely ignored. In addition, reports on a number of already initiated subsidiary, geologic projects were not submitted. No convincing model for the genesis of the examined gold mineralisations was forwarded that might have contributed to the explanation of the most noteworthy consistencies regarding tectonics, lithology and formation processes that are necessary for the development of adequate prospecting methods.

In a joint project with the University of Assiut, Egypt, our attention was drawn to this aborted program. We consequently worked-out an interdisciplinary research method by which the following, yet remaining, issues needed to be addressed. These consisted of:

- Documenting and dating the hitherto virtually unknown archaeological heritage of ancient gold production sites
- Identifying and mapping former mining and prospecting sites and ascribing them to the chronologies of Ancient Egypt and Nubia
- Investigating the geology of ore deposits of formerly exploited gold occurrences with the aim of developing a genetic model
- Studying the development of prospecting methods through the course of history

In order to do so, three field campaigns in the Egyptian Eastern Desert (1989, 1990 and 1992/93) were initiated in a close cooperation between the Institutes of Geology and Egyptology at the University of Munich (LMU). Field work had been programmed to take place between the 28th and 22nd parallel N. Both already known as well as unknown sites were recorded within this territory during our survey. Our work consisted of documenting the sites' archaeological features at the surface as well as the geology of their respective surroundings which, whenever possible, was mapped.

Because the distribution of the ancient gold production sites turned out to spread far beyond the southern borders of Egypt and well into the Nubian territories of NE Sudan, it was only reasonable to extend our field work to regions further to the S.

Subsequently, three field campaigns were carried out in 1996, 1997 and 1999, and were limited to a zone between the 22nd and 18th parallels N, in Sudan.

In all, we managed to record approximately 250 gold production sites in six campaigns. They were dated by means of diagnostic surface finds and described with their respective geologic environments.

The term "gold production site" denotes not only sites exploiting primary gold deposits but also a large number of ones in which gold quartz ores previously collected from wadi sediments had been processed in the same way as the mined ores. Such areas were specified as "wadiworkings" as opposed to genuinely mined areas in which auriferous quartz veins had been extracted in either underground (mines) or opencast (trenches) processes. It turned out that at least from the New Kingdom onwards, and especially in the Early Arab Period, a large but ill-estimated portion of the produced gold had been retrieved from wadiworkings.

TM-satellite images (Landsat) were our most effective cartographic means with regard to specific issues pertaining to lithology. Aerial photographs, kindly provided by the Egyptian and Sudanese authorities, were also used. Later on, during evaluation work at home, for which there had unfortunately been too little time in the field, the higher resolution images of the Egyptian and Sudanese Eastern Deserts provided by Google-Earth proved of inestimable value and were subsequently also drawn upon.

In addition to the fundamental work by W. F. Hume (1936), localisation of the mining sites had become possible above all through the internal reports by the EGSMA (Egyptian Geological Survey and Mining Authority) and GRAS (Geological Research Authority of Sudan), essentially compiled as a result of the joint ventures with the mentioned Soviet exploration firm Technoexport. Internal reports by Robertson Research Ltd. and its subsidiaries, Minex-Egypt and Minex-Sudan, became accessible through archives in the UK and proved extremely helpful too, both during preparation work as well as in the field.

Evaluation of the gathered evidence in the field and of the photographic documents turned out to be very time-consuming. With the present, richly illustrated volume, we now hope to furnish an updated assessment as to the origin of the legendary Egyptian, Nubian, and no less remarkable Early Arab gold. The present study also intends to shed some light on the sheer scale of efforts and sacrifices the much admired gold from Ancient Egypt demanded from its manufacturers.

At this point, we would like to express our serious concern about the progressive destruction of many ancient mining traces almost everywhere–especially in the Egyptian Eastern Desert–by modern prospection activities, mining operations, and all-terrain vehicles. We very much hope that this book will raise the awareness of this valuable heritage of earliest human industries.

Munich

Rosemarie Klemm
Dietrich Klemm

Acknowledgements

The project was financed by the Volkswagen-Stiftung and to a lesser extent the German Research Counsil DFG (Deutsche Forschungsgemeinschaft). We would therefore first like to express our thanks to both institutions for their generosity and their positive stance during the entire duration of the project.

The Ludwig-Maximilians-University of Munich with its Institutes of Geoscience and Egyptology broad-minded facilitated this interdisciplinary work by providing access to laboratories and libraries, and not at least by supporting and encouraging us in every helpful manner, what we gratefully acknowledge.

Our field work in the Eastern Desert of Egypt would not have been possible without the close cooperation with EGSMA. We are therefore grateful to EGSMA for kindly having given us the permission to access its archives during the preparations of the field work and providing survey permits for the Eastern Desert, including ones issued by the military. We also thank EGSMA for generously having contributed to the project with its employees who escorted us in the terrain as friends and indispensable guides.

We are particularly indebted to the former Chairman of EGSMA, Dr. A. Dardir.

Furthermore, our thanks go out to the EGSMA geologist, Dr. M. Abu Bakr el-Hawari, who with much selflessness remained in our company throughout all field campaigns in Egypt, as well as his two colleagues, Ahmed Abdel Moniam, B.Sc., and Magdy Abdel Sattar el-Bery, B.Sc.

We also cordially acknowledge the efforts of our Munich collaborators, Dipl. Geol A. Kraus, Dipl. Geol., Dr. A. Murr, Dipl. Geol. G. Langwieder, Dipl. Geol. J. Spiech, and Dipl. Geol. A. Zwick, for their successful accomplishment of the difficult work in the Egyptian Eastern Desert.

Much is also gratefully owed to Dr. H.G. Kräutner for preparing numerous geologic maps around the vicinities of the gold deposits visited in Egypt. This work was based on lithologically processed TM-satellite images, our field observations and subsequent petrographic studies, and not least, the most informative maps and reports of the Technoexport-EGSMA joint venture, and here especially the report of Koshin and Bassyuni (1968).

We wish to express our thanks to the members of the German Archaeological Institute (DAI) in Cairo who supported us in every respect during our stays in Cairo.

In Sudan as well, we were very lucky to be able of benefitting from the generous assistance by the Geological Research Authority of Sudan (GRAS), in particular its director general, Dr. M. Omer Kheir, to whom we are very grateful. GRAS granted us insight into all documents relevant to our project.

Furthermore, we received much support from the National Board for Antiquities and Museums (NBAM), in particular its director general, Hassan Hussein Idris, to whom we are very thankful.

We also would like to mention the efforts of the late Mohamed Hassan, who took over the complex task of procuring the necessary field work permits. We are particularly indebted to him for this and for his continual help in Khartoum.

Our warmest thanks go to our Sudanese escorts of GRAS, the geologists Mohamed el-Hagh and Mustafa Khazim, for standing at our sides in every respect during the difficult field work in the Nubian Eastern Desert. We are equally grateful to the Munich geologists, first of all to Dr. Andreas Murr, who in an extremely helpful way contributed very much with his expertise to the success of all our campaigns in Sudan, but also to the graduate students in geology, Rupert Utz and Marcus Lang (in 1996), Christian Tichatschke and Florian Schmid (in 1999), for their assistance which by far exceeded their principal task of mapping the terrain.

We are also very grateful to the mining assessor, Torsten Prössdorf (†), for his constructive expertise during the 1997 campaign.

We also thank the numerous unnamed Bedouins from different tribes in Egypt and, above all, in Sudan who gave us their help during our repeated and arduous exploration excursions in the field. Luckily, our Sudanese colleague, Mustafa Khazim, knew their language, which proved of great benefit for us during recurring and prolonged negotiations with the desert inhabitants (Fig. 1). We were able to thank them in return with medical advice or small

Fig. 1 Our colleague Mustafa Khazim negotiating the right of way with a group of armed Bedja-Bedouins

Fig. 2 Bedouins waiting patiently for the opening of food boxes at Onib

gifts like matches, sugar, milk powder or tea, as these reticent people usually decline other gifts (Fig. 2). Only in one case of medical emergency was Rosemarie Klemm given access to a small tent reserved for women, where in faint light a number of women wearing gold jewellery were squatting around a fireplace from which toddlers had contracted burns that required treatment.

The field trips out to the Eastern Desert of Nubia, each lasting for several months through inhospitable terrain, could only be mastered through the assistance of technically competent personnel able to repair frequent breakdowns of vehicles and equipment. We therefore are particularly grateful to our drivers, Ahmed M. Nowar, Ibrahim Daleen, Mohamed Abdel Rasul, Abbas Mousa Sorro, Faolo Ato and Abdallah Osman. We also owe our thanks to H. Dietl, who as a qualified auto-mechanic took part in the expedition during the 1997 campaign.

Also much is owed to our cooks, Mohamed Khalifa, Abdalla Hamdan and Peter Bowl, who saw to that the team members remained physically fit to cope with the enormous strains.

Not least, the satellite images available on the Google-Earth virtual globe proved extremely helpful in locating and mapping many of the ancient sites, both in Egypt and Northern Sudan. We are most grateful for this and the general permission for using the images, which whenever displayed in this volume are quoted as obtained from Google-Earth.

We are much obliged for financial support of printing costs by Mr. and Mrs. Gert Wajsfelner, Mr. Klaus Wirth and Mr. Curt Engelhorn.

We are extremely grateful to Dr. Paul Larsen, who translated with expert knowledge the original German text.

Finally, we are very much indebted to the Springer-Verlag and here especially to Dr. Chris Bendall and Claus-Dieter Bachem.

Contents

Archaeological Chronology of Gold Mining in Ancient Egypt and Nubian Sudan

1

In the basement of the Eastern Deserts of Ancient Egypt and Nubian Sudan primary gold was usually mined in mineralised quartz veins and recovered from a previously ground ore meal through washing it out as a finely visible flitter (Fig. 1.1).

Gold-quartz mineralisations are formed through excretion from 150 to 350 °C hot, hydrothermal solutions originally circulating in open faults or shear zone systems. The widths of the auriferous quartz veins generally vary between few centimetres and 1.5 m. To extract them, a wide enough

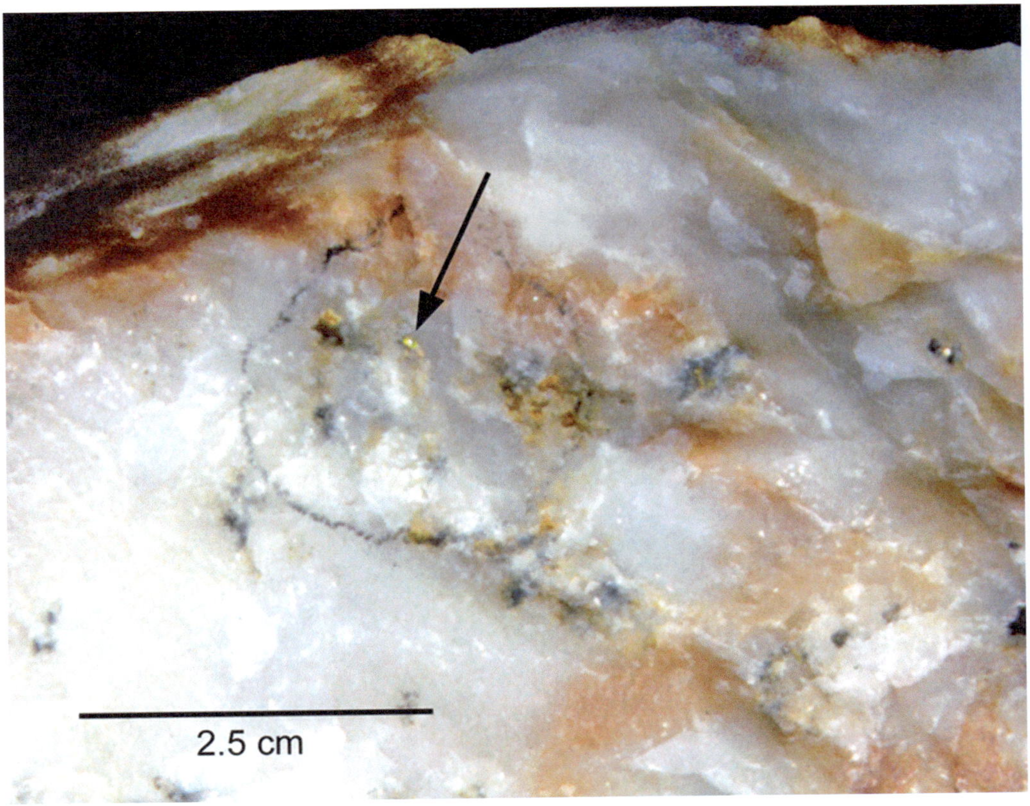

Fig. 1.1 Gold particle (*arrow*) contained in a quartz chunk from the El-Sid gold mine in Egypt

R. Klemm, D. Klemm, *Gold and Gold Mining in Ancient Egypt and Nubia*, Natural Science in Archaeology,
DOI 10.1007/978-3-642-22508-6_1, © Springer-Verlag Berlin Heidelberg 2013

trench was excavated to hold a labourer in order to assure that headway was made. Mining of very narrow veins therefore required painstaking, extra work for the removal of considerable amounts of barren wallrock.

To determine the actual gold concentrations extracted within the deposits throughout the different periods, different important variables need to be taken into account. First, gold contents may fluctuate considerably even within very small areas inside the same vein. Therefore preference was accordingly given to rich-ore-zones of which today, apart from occasionally occurring abutments, nothing is left. Now and then, information from recent mining records can be helpful in this respect. Together with and modern analyses they may permit rough estimates of the original gold contents. Today it is thought that the overall average gold content in many mining districts had been close to 1 ounce (31.5 g) per mined ton of quartz ores, although much higher grades can be assumed for certain ore-rich zones. The lower limit for gold extraction on the other hand, must have been near 10 g/t, considering that many ancient, leached-out heaps of quartz and sand ore residues (tailings) still contain gold grades of up to 5 g/t. At the turn of last century many of these tailings were therefore re-exploited by means of modern, alkaline cyanidation processes (Alford 1901; Schweinfurth 1903), which unfortunately led to an extensive destruction of the ancient extraction sites.

In addition to the mined gold quartz veins visible at the surface, gold was also extracted from quartz ores contained in the weathered rubble of the wadi beds and systematically processed since the New Kingdom.

Because of the arid climate, so-called placers or nuggets only played a secondary role in Egypt's ancient gold mining history. However, at the latest since the Early Arab Period, such placers had been dug-out usually in the form of very fine gold grains from alluvial sands in the wadi beds. Whereas such placer gold was separated from the sand by sifting and washing, auriferous quartz chunks extracted from the mines and recovered from the wadi beds needed first to be mechanically crushed-down to a powder fraction before the therein contained microscopic gold particles could be washed out.

In both cases gold dust ready for further processing, such as smelting, refining, smithing etc., was obtained. The evidence indicates that these processes had actually been carried out in specialised workshops in the Nile Valley apparently rather than at the extraction sites themselves. Because this study is only concentrated on the gold deposits, further metallurgical manufacturing procedures will have to be discussed elsewhere.

The archaeological evidence used for dating the human occupation in the mine districts relies mostly on the recorded mining and processing tools and the therewith associated findings on the processing techniques, be it in underground mines ore open workings in the wadi beds. Moreover, the mines were associated with nearby settlements, where special attention was given to features in domestic architecture, such as layout and settlement location in relation to the ore sources. In addition to the processing tools at the settlements, pottery proved whenever available to be extremely helpful for such approaches.

Epigraphic evidence that may add historical dimensions to individual mining sites is comparatively rare, and apart from a few exceptions is restricted to few sites already discussed in published reports (Gundlach 1977a, b).

Natural wadi erosion but particularly the recent soar in off-road, guided tours through the desert, have entailed that very sensitive features at the surface are becoming increasingly difficult to detect. The sites have also suffered severely at locations of modern exploitation and prospecting work. We therefore admittedly sometimes find our proposed dates to rely on relatively scanty archaeological indications. Such occasionally occurring insecurities, are however in a more general scope compensated by the high number of over 230 visited mining sites.

During field work we differentiated between following periods:

1. Earliest Hunters Period (fourth mill. BC).
2. Pre-and Early Dynastic Period ("two-hand-hammer" period) (~3000–2700 BC).
3. Old and Middle Kingdom (~2700–1800 BC).
4. New Kingdom to the third Intermediate Period (~1550–1070 BC).
5. Ptolemaic Period (~300–30 BC).

6. Kingdoms of Kush and Meroë in Nubia (~700 BC–100 AD).
7. Roman-Byzantine Period (~30 BC–641 AD).
8. Early Arab Period (Egypt: ninth to eleventh century AD; Eastern Nubian Desert: ninth to mid-fourteenth century AD).
9. Late nineteenth century – early 1960s.

In the following paragraphs a short discussion of the geo-archaeological specifics of each of the first eight periods is presented. For understandable reasons the last period was excluded from this discussion.

1.1 "Earliest Hunters" Period

In the beginning, the earliest discoveries of gold nuggets in the Eastern Desert of Egypt had probably been relatively sporadic and fortuitous. Such gold placers from the wadi beds are rare relics (secondary deposits) from the Pleistocene that had formed in the drainage system around auriferous quartz veins (primary deposits). Nuggets are typically much richer in gold than the gold from primary deposits. This is explained by the "forging" action provoked by mechanical transport of wadi rubble in flowing water, which also engenders the exsolution of their natural silver contents that are present in all Egyptian gold deposits.

Published analyses carried out on Egyptian gold artefacts (Hume 1936; Lucas and Harris 1962) have so far had relatively discouraging results as to the gold's origin in either primary or secondary deposits. Established gold-silver proportions seem to vary unsystematically within samples from various periods. Nevertheless, the results also show that artefacts often have gold percentages ranging between 78 and 85 % Au, which thereby are generally higher than those within primary deposits. This finding therefore seems to suggest the presence of placer gold, although this has so far not been substantiated by independent evidence.

The fact that gold artefacts generally remain comparatively seldom in the Pre-and Early Dynastic Periods is reflected by the fact that systematic gold mining still played a relatively minor role in the Eastern Desert of Egypt. This in fact, corroborates with the relatively low number of mines from this period recorded during our survey. Natural gold occurrences were probably not methodically sought for as opposed to later periods. It may rather have found its way into the Nile Valley as placer gold sporadically bartered by desert nomads. Especially golden beads recovered from Predynastic tombs had allegedly been smithed from nuggets and not cast (Petrie 1901; Quibell 1898). We may therefore hardly expect to find archaeological traces in the field from this type of gold "mining" in the Eastern Desert.

The high plain at Umm Eleiga, however, in some respect forms an exception. Through the site's unique deposit situation, it is conceivable that nuggets had been gathered here over lengthy periods, thus increasing the likelihood of archaeological traces in this area. Umm Eleiga is set in a plateau crossed by a fine web of auriferous quartz veins that had been reshaped in the recent geologic past (probably during the Pleistocene) by an intense fluvial activity. This resulted to the formation of up to several meters thick rubble layers in which substantial gold nuggets could develop under such formerly humid, climatic conditions. The gravel deposits were then relocated more recently by desert erosion, which lead to the exposure of the gold nuggets. Rock carvings displaying figurative and ornamental designs still witness the early presence of inhabitants in this desert plateau. Winkler (1938) ascribes this Predynastic group to the so-called "Earliest Hunters" of the Amratian culture and dates it to the middle of the fourth millennium BC. The up to 80 cm long, carved rocks show among others undulating, linear patterns that frequently end in spirals (Fig. 1.2). They also specifically depict wild animals, such as ostriches and elephants in loosely arranged compositions. So far, no representations of animal herds have been recorded (Klemm 1985). Between 80 and 100 decorated stones are still found in this approximately 50,000 m^2 large area. Most of the rock art has probably gone lost, chiefly through erosion. The high density of the rock art representations seems on any account to suggest a long period of occupation as well as to some extent an area quite wealthy in gold.

As in almost all other gold deposits in the Egyptian Eastern Desert, Umm Eleiga had also produced gold in later periods, notably the Early

Fig. 1.2 Stone decorated with wave-line and spiral motifs (signed by whiting) in the Eleiga gold mining district, Egypt. Pre-to Early dynastic Period

Arab Period, during which flat gold quartz veins striking just below the surface had bee exploited in trench-like pits.

1.2 Early Dynastic Period ("Two-Hand-Hammer" Period)

According to our observations, virtually all the earliest gold mines are set within similar geologic environments. They occur mainly in marginal areas rich in quartz-veins along highly weathered Proterozoic granodiorite intrusions, or in marginal areas of granodiorite, Neoproterozoic granites. Originally, these quartz vein systems had been enriched with primary copper sulphide mineralisations (mainly chalcopyrite and chalcocite next to galena, sphalerite and pyrite), which due to surface alteration processes had leached out and deposited in the form of typical, green malachite linings and other secondary copper minerals along the cleft structures of the surrounding wallrock. These are easily recognisable at the surface by green stains along the wallrock of the quartz vein systems.

In the area of study the primary, heavy metal sulphides of the quartz mineralisations are often associated with some gold. After decomposition and dissolution of the former, the latter remained inside the quartz in the form of fine flakes (often together with iron oxides). Therefore the green stains most certainly had functioned as dependable markers for ancient gold prospectors (Tawab et al. 1991).

However, it cannot be excluded that such delicate, green malachite linings at the wallrock clefts had originally been mined for their copper contents only, and that the link between the quartz veins and the gold was made at some later stage. This, on any account, led to the development of mining and processing methods specific to gold ores fundamentally different to the high-temperature smelting techniques used in the copper industry.

The studies conducted by the IFAO also involved excavations in settlements and mining areas in Wadi Dara and Gebel Mongul-South, in the N of the Eastern Desert (Tawab et al. 1990). One of the findings was consistent with our observation that the oldest mining and settling activities in this area go back to the Predynastic Period. The excavators also believe that exclusively copper ores had been mined in the beginning here and that gold processing had only started at a much later period (Early Arab Period, ninth century). The investigations at

Fig. 1.3 Heavy duty two-hand hammers used for crushing quartz ores during extraction. Higalig mine, Egypt

Wadi Dara also produced evidence for early copper metallurgy through the discovery of ancient furnaces (Castel et al. 1992, 1995).

Another characteristic feature of this period next to the malachite markers consists primarily of the use of unusually large and respectively heavy, calabash-shaped, stone hammers that because of their size (length up to 40 cm) and weight (up to 8 kg) could only be manipulated with both hands ("Beidhandschlägel" in German; Fig. 1.3). Their pounding surface is clearly rounded, and a slight depression at the far end was possibly meant to assure a better grip. It was interesting to note that several such mallets had been discarded just outside the mines, all with sharply edged chip marks at the blow surfaces. This had probably made them obsolete as none displays use wear at the chip marks.

The mines themselves, particularly in the case of perpendicularly plunging quartz veins, usually consist of relatively narrow shafts with noticeably smooth wall surfaces. This in fact, sheds some light on the actual extraction methods used in the early gold mines, in which the two-hand-mallets ("Beidhandschlägel") were used for making progress through the mine by directly crushing the quartz vein ores to a meal in situ. Weisgerber (1989) calls this the grinding ore extraction ("Zermalmende Erzgewinnung") which led to the smooth surfaces along the wallrock of the removed vein. The so obtained silica powder was subsequently collected and taken outside the mine for the following separation processes. For the lack of evidence it has yet to be established how and whether such processes were actually carried out at mine site itself.

The heavy, two-hand-mallets are occasionally associated with disc-shaped mallets that consist of round stone discs measuring between 15 and 25 cm in diameter. They are usually between 4 and 8 cm thick and have round, blunt edges used as pounding surfaces.

A common criterion for the oldest mines are usually single, narrow trench pits that rarely exceed depths of 5 m and lengths of 10 m. In mountainous terrain they are well discernible as elongated incisions in the surface. The mines provided just enough space for one to two labourers to work simultaneously at the far ends of the mine and approximately one labourer per meter in the central parts of the trench. The yield must therefore have been relatively low.

Fig. 1.4 Fist hammers (on the right) with originally pointed impact surfaces for chipping off ore fragments. Old Kingdom, Abu Mureiwat, Egypt

Typically, no settlement traces are found in the neighbourhoods of these mines. In addition, their geographic distribution is very thin with considerable distances between the mines, revealing no particular direction of prospecting. This rather compelling finding may suggest that the mines had been exploited by nomadic groups, and that only the striking, green malachite stains had led to their discovery. The absence of nearby stone huts would support the idea of transportable shelters made from organic material, though it is questionable whether this may ever be demonstrated.

1.3 Old and Middle Kingdoms

We recorded green malachite linings along the wallrock at all mines we were able to date to the Old and Middle Kingdom. They must therefore also have been the chief prospecting indicator for gold mineralisations in this period.

In general, the method of crushing the goldielding quartz at the extraction site itself was upheld during the Old Kingdom, although by now two new tools had been introduced. One is an oval mallet with grooves, weighing between 2 and 5 kg. Its carved groove served its fixing onto a bifurcated shaft. The other is an elongated fist hammer with a slightly ergonomical handle. Since their initial function had consisted of a crushing action, the efficient use of both tools had relied on their smooth and round impact surfaces. Therefore, many had been discarded immediately outside the mines after receiving flaws through unintentional chipping. On the other hand, the strike surfaces on many small fist hammers reveal that the miners by now had also begun to separate genuine ore chunks from the veins (Fig. 1.4). During the Middle Kingdom, the ores began to be processed outside the mine, which involved stone mortars. They too, are found near the mines and always occur together with the pointed hammers. The fact, though, that their number remains relatively low, seems to suggest that only little had changed in the current mining methods. One may also speculate whether over the time, many mortars had been gathered-up because of their versatility. In fact, similar objects are still being used today by the Bedouins. The stone mortars have a standardised appearance

Fig. 1.5 Mortar from red granite and pounder used for crushing quartz ores. Middle Kingdom, Daghbag, Egypt

and measure up to 35 cm in diameter. In their center is a hemispherical depression in which quartz gravel was crushed with an appropriate stone globe used as a pestle.

It is noteworthy that the Old and Middle Kingdom gold mines are grouped along the western mountain flanks of Egypt's Eastern Desert, with somewhat higher concentrations in the N. They are also located near the traditional passageways through the mountains to some extent reflecting a more systematic approach to gold prospecting in general with possibly master-planned expeditions starting-off from the Nile Valley. Indications to such expeditions are known from epigraphic evidence referring to officials' titles like the text quoting a "leader of the prospecting team" (Yoyotte 1975).

From this time onwards, domestic architecture within the mines' vicinities is also increasingly observable. As a rule, the dwellings are built in dry stone walls, whose construction material had been collected in the immediate neighbourhood. Their original heights seem to have been quite low. Virtually nothing is known about their roof covers.

The state in which the buildings are preserved is much dependent on the building material itself. Houses built from rounded, granite boulders have generally suffered more from weathering and are usually in a worse state than those built from flat schist slabs, or just rocks that are generally more resistant to weathering. On any account, architectural features are susceptible to alter very little over the millennia and on their own give only an approximated idea as to the dates of the sites. In this respect small finds are much more useful.

The findings seem to show that during the Old and Middle Kingdom the miners continued to be mostly constituted of desert inhabitants. One indication for this was the exclusive use of groove- and fist hammers with which the quartz rock was crushed directly inside the mines. By the Middle Kingdom though, there had been a noticeable transition to mortars for producing the ore meal required for further processing (Fig. 1.5).

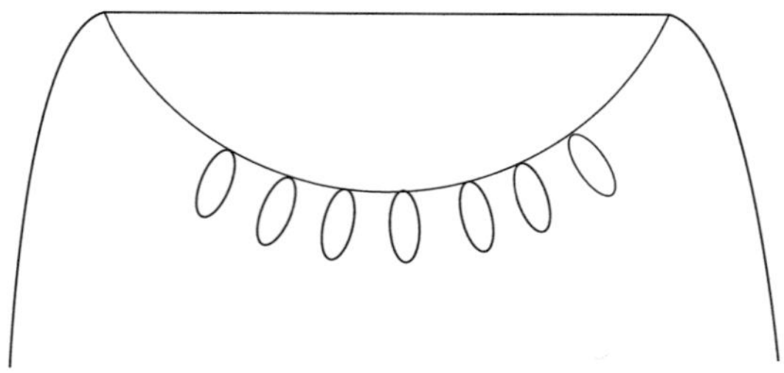

Fig. 1.6 Hieroglyph for gold with the phonetic value "nub", representing a pectoral or else a frame used for gold washing

Grinding techniques involving the mill, however, had already been applied in the Nile Valley for flour production since the Old Kingdom.

The lack of processing equipment and tailing sites in the mine- and settlement areas furthermore suggests that the processing tasks had been executed near water sources known to the local tribesmen, but which are extremely problematic to find today. Tailings were liable to form around the larger water holes already in this early period. Their traces, however, have often been erased by activities in later periods. Tailings can for example be observed at Barramiya and other gold mining districts, where wells have survived together with the remains from these early days of gold mining.

In assuming a primarily Bedouin tradition for the gold mining industry in the Eastern Desert of Egypt, the following reconstruction of the gold washing process may apply: The gold-yielding quartz slurry was first poured over an animal hide (sheep) on which the heavier gold dust got caught in the lower layers of the fur. In order to retrieve the gold dust, the fur was then set on fire, after which the melted gold was separated from the remaining ashes. The myth of the Golden Fleece finds in our view its origin in this process. In this respect, Aufrère (1990) took up an older interpretation for the hieroglyph for gold, which according to him may originally have depicted not a (golden) pectoral, as so far assumed, but in fact a device for gold processing. He thereby refers to a frame with a textile covering, in which the alleged beads hanging from a collar are interpreted as dripping water (Fig. 1.6).

As already mentioned, the Old and Middle Kingdom sites only reveal little surface pottery, in most cases fragments from so-called, red-burnished Medium bowls, which are also known from contemporary contexts at Wadi Dara (Tawab et al. 1990).

1.4 New Kingdom (Eighteenth to Twentieth Dynasty)

The upper time limit of the New Kingdom to the beginning of the Third Intermediate Period at around 1080 BC does not result from a conjectured lack of archaeological field data. From an archaeological perspective, there are no significant differences to the time of the preceding dynasties in the mining districts. The boundary is rather linked to history and to the political and particularly economic decline after the twentieth dynasty, which unquestionably affected the organisation of the gold mining industry (Helck 1975).

With the conquest of Nubia under the eighteenth dynasty, gold mining had been able to expand on a large scale into the southern territories. In Sudan, gold mining sites predating the New Kingdom are scarce and tend to concentrate around deposits closer to the Nile. The prospecting criteria had nevertheless been the same here as in the Egyptian motherland.

When speaking of New Kingdom time in Sudan it is noteworthy to emphasize that it means only the time span as documented for Egypt, which of course contains in Nubia different indigenous periods as shown in the Chronology of Nubia (see p. 611).

It was only with the exhaustion of the early mining districts associated to the copper mining industry at end of the Middle Kingdom that new prospecting methods had to be developed. The response to that need had resulted to a scenario of systematically targeted prospecting expeditions into the mountains with the intention to carry out experimental processing of selected quartz samples.

This apparently led to the identification of a new auriferous ore soon after. It consisted of a muted, grey vein quartz variety interfused with ore dust that thereby was added to the prospecting markers.

Consequently, the geologic environments in which auriferous quartz vein systems were exploited, multiplied considerably in the New Kingdom. By now the granodiorite complexes or the granodiorite margins of granite intrusions were

Fig. 1.7 Flat grinding mill and fist grinder from the New Kingdom for producing quartz powder. Marahig, Wadi Allaqi

seemingly no longer the exclusive gold mineral targets of the prospectors, but furthermore mafic and ultramafic (basaltic to serpentinite) rock sequences.

From this time onwards, districts dominated by clastic sedimentary units such as greywackes, siltstones, and conglomerates (mainly the so-called Hammamat series) were systematically included to the exploration work. Such rock sequences, however, are often only available in heavily altered, geologic surroundings exposed to metamorphism. Even in such terrain the prospectors had managed to identify the ores correctly. The fact that even modern geologic maps often fail to pinpoint such fine differentiations can only increase our esteem for the ancient prospectors' meanwhile reconstructible knowledge of geology.

With the introduction of the grinding mill to the mining industry in the New Kingdom, ore processing and prospecting methods improved significantly. Even without the green copper carbonate (malachite) stains it had now become much simpler to check the quartz veins for their gold contents, as sampling and preparation techniques had become more efficient. Prospecting was also extended to systematic inspection work on eroded ore material from the wadi sediments.

Tool marks in the mines reveal that at the beginning of New Kingdom ores were already extracted with metal chisels. The number of discarded stone mallets just outside the mines decreases to virtually nil. Outside the mine, the auriferous quartz chunks were pounded down to small, pea-sized fragments on flat stones serving as anvils. The gravel was then ground in special mills to the necessary grain size that permitted the release of the gold. This mill type had been used in the Nile Valley to produce cereal flour since the Predynastic Period (Roubet 1989) without beeing employed in the mining industry before the New Kingdom. Thus, it is conceivable that by that time gold mining had chiefly come under the direct control of the Egyptian heartland.

The grinding mill consists of a flat slab measuring about 30 × 60 cm, preferably of a hard rock, and a mobile grinding or runner stone (Fig. 1.7). The previously crushed ore was then ground down to a grain from which the gold spangles were separated in subsequent leaching processes.

Fig. 1.8 Milling gold-bearing quartz chunks in similar technique to that used during the New Kingdom in Egypt. Tanzania, 1991 (Photo credit: G. Borg)

Through use wear, the grinding stone and the flat slab end up by fitting perfectly together, and with continuing use wear a depression forms in the centre of the slab. As the depression deepens gradually down beyond a certain point, the efficiency of the grinding process considerably decreases. Instead of being discarded, the slabs were frequently reclaimed and reused as mortars in which ore lumps could be pounded.

Similar ore grinding techniques are still being used today in Tanzania (Fig. 1.8).

The stone anvil used for the first crushing process, by which the ore chunks were reduced to pea-size, consists of a stone block of a highly resistant rock, mostly andesite, dolerite, greywacke, or the like. Its surface is usually more or less square and measures between 30 and 50 cm, its thickness between 20 and 30 cm. Often, the surface displays a depression up to 5 cm deep as a result from the pounding action with a conical rock hammer (Fig. 1.9).

This new method subsequently led to a significant development in gold mining techniques in general. The rubble eroded down from the mountains into the wadis had been for the most received comparatively little transport, due to low hydrological activity. It therefore consists of relatively coarse blocks that also comprise weathering-resistant fragments from former quartz veins. With little experience, difference by selective testing could soon be made between auriferous and barren quartz varieties in the alluvium that were specific to each mining district. Quarrying the wadi alluvium for gold involved the extra advantage of the possibility to deploy workforces of virtually any size chosen.

Along the wadis where auriferous quartz rocks were being gathered and processed on vast scales, extensive settlement alignments began to form that still today often exhibit innumerable processing mills. According to our own estimates, the capacities of such sites could hold up to several hundred inhabitants working simultaneously in the wadi. This led to the production of enormous quantities of ores. In deep mines by contrast, the low capacities determined by the narrow shafts generally restricted the number of simultaneously working labourers to a range between two and five individuals at the utmost and accordingly the number of workers involved in the ore processing.

Fig. 1.9 Anvil slab and pestle for reducing ore chunks. Neguib, Wadi Allaqi

The auriferous quartz rocks are usually identified by a typically grey-brown colour of a muted-looking appearance, which in fact originates from fine ore inclusions and tiny gold sequins (Fig. 1.1). These rocks were then selected for further processing, whereas the barren, milky-white quartzes were stacked in small heaps to avoid work repetition. Such heaps are occasionally discerned in the wadis under low sunlight.

Associated to the quarries hereafter referred to as wadiworkings are heaps of leached quartz sand and even quartz dust that had accumulated around wells. Such tailings are often still preserved in the terrain (Fig. 1.10).

Next to the large surface wadiworkings the deep mines continued to be exploited. Many old mines still in operation received general enlargements. New mines were opened, too. However, copper oxide linings were no longer a stringent criterion for the prospectors. From now on as it seems, systematical testing of the quartz veins as to their gold contents became an integral part of the prospecting work.

In the New Kingdom, Egyptian gold mining spread to Wadi Allaqi in Lower Nubia. In Wadi Allaqi and its tributary valleys focus was set on wadiworkings, whereas deep mining received only little attention. This is connected to the context by which the markedly higher erosion activity in the mountains of lower Nubia led to substantially greater deposits of auriferous quartz vein fragments and even free gold dust in the alluvium. In the wider surroundings of a primary deposit one regularly comes across wadis with evident traces from wadiworkings, apparently resulting to significantly higher gold-yields than possible in primary deposits. With the decisive assimilation of Nubia into the Egyptian Empire at the beginning of the New Kingdom, this type of gold mining led to a massive increase of available gold, a situation highlighted by the amounts stated in the tribute lists in the Egyptian temples.

Fig. 1.10 Tailing consisting of processed sand residues. El-Sid mining district, Egypt

1.5 Ptolemaic Period (~300–30 BC)

To judge by the scale at which the mines run as well as by the settlement sizes, the gold mining industry in the Ptolemaic Period is also assumed to have had a relatively high output. By this time though, it had retracted mainly to the southern regions of the Central Eastern Desert. In Wadi Allaqi and Nubia by contrast, there are no traces from Ptolemaic gold mining found by us.

This is in contradiction to Castiglioni and Vercoutter (1998), who regard the site of Derahib as the Ptolemaic Berenice Panchrysios, which is actually an Early Arab settlement with two tower like fortified buildings.

New technologies continued to make their way into the mining sector of Egypt's Eastern Desert. Shafts were driven considerably deeper into the mountains. Especially in gently dipping vein systems, mines could from now on be vaulted and reinforced by abutments. The thereby significantly enhanced security inside the mines allowed for a deeper penetration into the rock. The depth limits, however, were also determined by the ventilation capacities inside the mines. Due to the high variability of a mine's internal architecture, aspects pertaining to ventilation also differed decisively. Generally, it is thought that the depths rarely exceeded 25–30 m.

Furthermore, a new method in ore processing was introduced. On a concave, 70–80 cm long and 30–40 cm wide millstone crushed gold quartz gravel was ground with an apron-shaped runner stone, weighing between 8 and 12 kg, while manipulated with both hands back and forth over the entire grinding surface of the millstone (Fig. 1.11). This milling method increased the efficiency of the grinding process considerably in comparison to that of the earlier method. In spite of the heavy weight of the runner stone, the grinding process was less strenuous, as the arc of the grinding surface permitted an ergonomic swinging motion based on the inertial mass of the runner stone. This method produced an even finer quartz grain, through which a higher rate of gold dust was released.

This mill type presumably represents a Ptolemaic import from the Aegean World. In the Minoan cities of Gournia and Phaistos on Crete, we were able to see such mills, although from contexts connected with flour production (Fig. 1.12).

Fig. 1.11 Concave Ptolemaic grinding mill with apron shaped runner stone. Hangaliya, Egypt

Fig. 1.12 Concave cereal mill from the Minoan city of Gournia, Crete

Fig. 1.13 Circular device for separating heavy metals, fourth century AD. Laurion, Attica

Another technology transfer from the Aegean World to the Ptolemaic mines in Egypt consists of a large, circular device used in the washing processes. With a diameter measuring between 6 and 10 m, it composes of trapezoidal, 80–120 cm long stone segments, in whose centre a groove channels the flow of ore meal slurries. The therein contained heavy mineral particles then deposit in small, perpendicular grooves in the bottom of the main groove. Next to the channel, the stone segments also display aligned depressions to gather the heavy mineral concentrates. Two such installations were found in the Ptolemaic quarters of Daghbagh. Two individual segments were recorded at Barramiya, and Bokari and probably an entire installation is preserved at Samut.

Similar installations at the ancient silver mines at Laurion date to the fifth and fourth centuries BC (Conophagos 1988) (Fig. 1.13).

The Ptolemaic Period miners generally continued the operations at the old mines, dwelled in the previously existing settlements, and used the old water wells. However, they extended the shafts and trenches considerably, whereby accumulated rock was moved as backfill to secure the mines. In the extended and above all, deeper shafts it had become possible to occupy more miners simultaneously. Metal chisels and mallets were used, as evidenced by the tool marks, which in analogy to the stone quarries of this period display more elongated shapes compared to those from the New Kingdom (Klemm and Klemm 2008a).

It may be conjectured that wadiworkings continued to function in this period, although this has not been confirmed by datable finds, like the Ptolemaic mill, in associated, earlier settlements along the wadis. By contrast, Ptolemaic processing tools tend to accumulate at the tailing sites, suggesting a fundamentally modified manner by which the ore processing was managed. The findings in fact suggest that processing had taken place at a central location, usually the well and no longer on an individual level around the houses.

Ore processing therefore no longer occurred in the yet occupied pharaonic settlements, but rather in specially arranged courtyards, near large tailing sites.

Pottery is sufficiently attested to from this period. Most of it consists of fragments from amphorae (handles and conical bases), cooking and domestic wares.

1.6 Kingdoms of Kush and Meroë in Nubia (~700 BC – ~100 AD)

In the gold mining regions of Nubia, between Nubt in the S, Wadi Allaqi in the N, and the Nile Valley a great many sites display New Kingdom type stone mills. However, they usually display evident traces from later reclaiming, usually consisting of a central depression in a formerly flat and oval grinding surface.

The spatial distribution of the mills is nevertheless undeniably limited to the New Kingdom sites of Nubia and especially the ones near the Nile Valley. Only approximate dates can be proposed for various reasons. By the secondary depressions in the originally flat grinding surfaces, the functional reclaim of the mills may all the same be clearly labeled as "post"-New Kingdom. Since the Ptolemaic concave mill is found neither in Wadi Allaqi nor at any other gold mining site in Nubia, its southern distribution is still ill-defined.

Through the introduction of the rotary quern (round mill) to Egypt by the Romans and through the close contact between Rome and the Meroë Kingdom, one may on the other hand expect that this by far more effective tool would have found its way into the Nubian gold mining sector, provided the latter still existed on a noteworthy scale. This has so far, however, not been confirmed by firm archaeological evidence.

The massive diffusion of the quern to Nubian gold mining sites first occurred with the Early Arab Period, between roughly the ninth and mid-fourteenth century AD. Its use, however, remained restricted mainly to the Eastern Desert with a documented western limit at Umm Nabardi, which is about 200 km to the E of the Nile.

Assuming that gold mining in the remote desert areas was linked to a strong central power, its resumption in Nubia in the era after the New Kingdom would not predate the twenty-fifth Kushite dynasty that had accessed the throne around 700 BC. A lower time limit would accordingly be concomitant with the rising influence from nomadic warriors of the Blemmyes tribe, who around 230 BC began to obstruct with increasing success the Meroë mining industry in the Nubian Eastern Desert.

Sofar, however, not much is known about gold mining under the kingdoms of Kush and Meroë in the Nile Valley and especially the Eastern Nubian Desert. Only detailed archaeological investigations may contribute to the reduction of the wide array of unanswered questions on this subject.

1.7 Roman-Byzantine and Early Arab Periods

During the Roman Period gold mining in Egypt descended into a phase of decline. In the Nubian Eastern Desert it partly even came to a standstill.

In the Egyptian Eastern Desert, it was restricted to some known productive districts near the desert routes and water supply stations (Hydreuma) that had already been established during the Ptolemaic Period and were now being developed under the Romans (Murray 1925; Sidebotham 1991, 2008).

The main mining activities during the Roman Period between the first and fourth centuries AD

Fig. 1.14 Bottom and top halves of a rotary mill (from two different mills). Early Arab Period mining district at Sagia, Egypt

concentrated in the Eastern Desert visibly on the well-evidenced stone quarries at Mons Claudius, Mons Porphyrites, Wadi Semna, Barud, and various smaller sites (Gnoli 1989; Harrell et al. 2002; Sidebotham et al. 2008). Even along the routes to the stone quarries though, gold extraction was discontinued at several, already existing mines. This for instance was the case at Fatira (Abu Zawal), a station on the route to Mons Claudius built on top of a Ptolemaic tailing site laid out around a well, near gold mines. No traces from gold mining in the Roman period were found here.

It was in this period that processing technology nevertheless underwent a final but fundamental improvement through the introduction of the rotary quern, which continued to remain widely in use for the coming centuries until the Early Arabic Period in both Egypt and Nubia. The earliest evidence from rotary mills in gold mines appears in the sixth century BC, in Gaulish Limousin, from where it spread throughout the Roman Empire (Cauuét 1991).

The rotary, or round mill consists of a flat, stationary stone with a diameter between 40 and 60 cm, and an upper, rotary disc with a central fill hole and a lateral perforation for the handle (Fig. 1.14). The rotor disc is slightly smaller than the stationary base-stone. With progression of wear from grinding the hard quartz ores the contact surface between the two superimposed stones gradually becomes perfectly even, thus producing an extremely fine powder from which the gold content is eventually extracted.

Compared to the traditional processing methods the increased effectiveness of the rotary motion resulted in conjunction with the finer quartz powder fraction to significantly higher proportions of released gold.

This mill type continued to be used throughout the Early Arab Period, and today is still occasionally seen in rural communities in the Nile Valley and the Eastern Desert for grinding cereals. Beyond the Egyptian borders it was widely used for grinding both cereals and ores in the

Fig. 1.15 Model of a domestic round mill in the Museo Municipal of Lagos, Portugal

entire Roman Empire and the British Isles until the Medieval Period (Fig. 1.15) (Childe 1943).

The Early Arab mines often date back to older complexes from the Ptolemaic and New Kingdom Periods. New mines though, were also opened. They concentrate in the southern sectors of the Eastern Desert and as in the Ptolemaic Period, tend to group along the coastal strip of the Red Sea.

The mines were enlarged and lowered down to greater depths. Ventilation limits for breathing and lighting were driven down to depths approaching 35 m. Existing mine shafts were therefore straightened-out and new ones were excavated as straight as possible, often in man-sized tunnels.

Whereas in earlier periods, ore lumps had been crushed by aid of relatively large anvil and grinding stones and even on the surfaces of runner stones, in the Ptolemaic Period more specialised, but much smaller stones were used for that purpose. They resemble small, rectangular ashlars with rounded corners whose measurements vary between 15 and 20 cm by 10 and 15 cm. They are mostly from dolerite with which the ore was crushed on stone slabs. Because of their size and weight, these hammer stones had to be manipulated with both hands. Many display wear-related depressions on two and sometimes (Fig. 1.16) several faces.

Furthermore, so-called inclined washing tables made from stone, often in a good state of preservation, were used in this period as sluices to separate the gold contained inside silica ore

Fig. 1.16 Globular pounders held with two hands and used for crushing quartz. Early Arab Period, Aradiya-East, Egypt

slurries (Fig. 1.17). Such tables were usually equipped with gutters and basins to assure that the water used in the washing process flowed back into a collecting basin where it was retrieved before being reused in another sluicing process. Such installations are often surrounded in a semicircular layout by heaps of leached-out quartz sands (tailings), which due to small hematite contents occasionally jut out in markedly red tones from the terrain. The inclined surfaces of the washing tables were probably covered with sheepskins, which as soon as they were saturated with the fine gold dust, were probably set alight in order to concentrate the gold.

At today's gold mines in Tanzania similar tables are covered with jute canvas sheets, a technique, which basically relies on the same principle as the ancient Egyptian gold washing method (Fig. 1.18).

Given the availability of dependable ore analyses, the ancient tailings have naturally prompted calculations as to the original quantities of extracted ores and gold. Estimations by Tawab et al. (1990) concerning a tailing of 30 m^3 resulted to a total of 150 t of ground and leached ores. At a grade of 10 g/t they estimated a total amount of 1.5 kg of obtained gold. Due to wadi erosion, such tailings, however, are usually no longer preserved to their original extents, and in most cases unknown amounts have been completely washed away, placing such calculations under a fully different light.

An essential archaeological feature concerns the settlement structures. As far as we were able to observe, the Early Arab settlers only in very few instances occupied the settlements of their predecessors. They therefore usually built their

Fig. 1.17 Remains of an inclined washing table including its backflow channel and collecting basin. Egait-South, Sudan

own, new houses. Exceptions are the gold mining settlements in Wadi Allaqi, where evidence seems to indicate that the Early Arab miners had moved into buildings dating to the New Kingdom. Usually though, the settlements are clearly separated from the older ones and often form independent complexes enclosed by walls, more or less similar to Roman road stations in the Eastern Desert. Early Arab houses may also scatter in open settlements, but then preferably in more hidden locations off the main wadis.

The houses and huts are generally built in the shell-facing technique with an inside filling of fine-grained gravel. Especially in Wadi Allaqi and the Nubian Eastern Desert, house walls may reach impressive thicknesses of up to 80 cm and are therefore often preserved to their original heights of about 2 m. Entrances are usually hidden away, facing the mountain slopes. Towards the wadi one sometimes notices walled screens. The settlements are often set in small and closed tributary wadis, which in addition, may be sealed-off by defensive structures at the inlets. The house plans are generally oval to round and rarely display right angles. Large prayer sites oriented to Mecca are relatively common outside the settlements.

A distinct, compact settlement type is encountered mainly in several areas of the Nubian Eastern Desert and near the Early Arab site of

Fig. 1.18 Gold concentrating on a wooden washing frame covered with jute bags to recover the gold particles, Tanzania, 1991. Basically the same method had been applied in Egypt at least since the New Kingdom Period (Photo credit: G. Borg)

Eleiga. It is characterised by orderly house alignments and storerooms and a central street running through its middle. It reflects a relatively high organisational structure of its building communities in contrast to the usually observed, loosely arranged settlements of that same period. We tentatively attributed these settlements to the Nubian campaign of the entrepreneur Al-Omari under the rule of A. Ibn Tulun (~ AD 880) and described by Al-Maqrizi (according to Urbain Bouriant 1895–1900).

In the Early Arab Period gold exploration in Egypt increasingly concentrated in the southern territories of the Eastern Desert, especially in Wadi Allaqi and the Nubian Eastern Desert.

Countless, small, but also large mining communities specialised in exploiting wadi alluviums, but also wadi trenches and even deep mines, witness the onset of something resembling a goldrush along the pilgrimage route through Wadi Allaqi to the Red Sea, in Sudan but also over large stretches of the Nubian Eastern Desert, including the Red Sea Hills.

According to historical sources, the apex of this phase occurred between the ninth and eleventh centuries AD, from the reign of A. Ibn Tulun to the middle of the Fatimid Caliphate (Alford 1901), and probably lasting to the middle of the fourteenth century AD in the Nubian Eastern Desert (Hasan 1967).

2 Value of Gold in Ancient Egypt

Since its earliest appearance, gold has fascinated virtually every culture on Earth. Its reflecting light, perpetual luster, unfading and invariable distinctiveness, and its connotation linked to the sun's eternity have until today been most likely at the quintessence of its symbolic value. "The sun itself, it is of pure gold" (Goethe, Faust II).

In Egypt, more than for its material value, gold played primarily a key role in the religious cult, as it was considered the "flesh of the gods". Statues of gods, well-guarded inside the temples' holiest of shrines were in fact made from pure gold. In the same vein the gilt peaks of high-standing obelisks warranted direct contact with the sun, as for instance illustrated by an inscription on an obelisk dedicated by queen Hatshepsut inside the temple at Karnak: "… then my heart incited me to fashion for him (the God Amun) two obelisks from electron, whose peaks reach up to the sky … thus I covered them … with electron, for my name to last eternally inside the temple, for always and forever."

Whether opulent death masks from pure gold like that of Tutankhamen, or the common private ones covered with leaf gold, the imperishable and insinuation associated with this metal reflected the determination of the deceased to safeguard his existence in the hereafter.

The earliest known gold objects from Predynastic Egypt come to us in the shape of small beads, which occasionally appear on necklaces, together with ones from other materials like carnelian and bone. A closer examination of their slightly amorphous shapes and structures reveals a manufacture by hammering rather than casting, thereby attesting that these are actually small nuggets formed by nature over a lengthy period through constant relocation of tiny gold particles contained in the debris of the wadi sediments. In such humid environments the silver contents tend to diminish as it is gradually washed out, thus increasing the relative gold content of the nugget. Therefore nugget gold is generally purer than that extracted from a mine.

In funerary contexts these early nugget beads were laden with spiritual significance and therefore rather associated with concepts of eternity than mere material values. Although the latter aspect became steadily more important with the progression of time, both aspects cannot be separated from each other in a social or political perspective.

From the Predynastic Period and the early Old Kingdom onwards, especially in northern parts of the Eastern Desert, acquisition and possession of gold had already been taking place through prospecting and mining rather than fortuitous discoveries of nuggets by nomadic populations. At first however, mainly copper ore deposits had been sought for, and it was probably only with the processing of these ores that inherent gold particles had been found, which at some later stage resulted to genuine prospecting for gold occurrences.

Expeditions were sent out to extract the gold from beforehand explored mining areas where under strenuous conditions and deprivation mining could then begin. Gold production soon became a matter of the pharaoh and his administration. This

R. Klemm, D. Klemm, *Gold and Gold Mining in Ancient Egypt and Nubia*, Natural Science in Archaeology,
DOI 10.1007/978-3-642-22508-6_2, © Springer-Verlag Berlin Heidelberg 2013

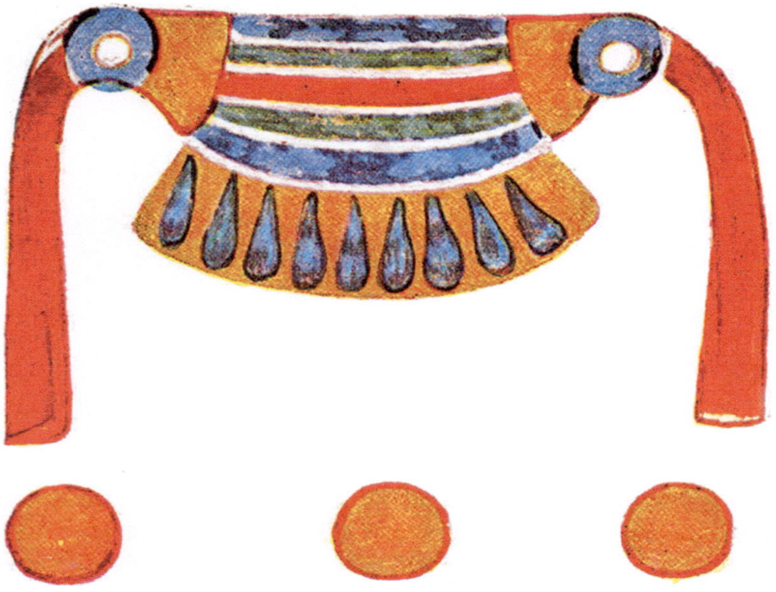

Fig. 2.1 Representation of a golden, bead-studded pectoral. The three-dot determinative for minerals below, denotes the sign as the hieroglyph for gold with the reading "nub" (after N. M. Davies, 1958)

same administration was also in charge of delivering, counting, and weighing the recovered gold as well as recording these actions in written documents. As a rule the delivered gold was kept in treasuries attached to the kingdom's most influential temples. Such lists, no matter their state of preservation, are known from all periods and state the amounts of gold delivered in the annual balances. In many cases they also contain information as to the origins and the quality of the gold.

Hence, any ruler was able to keep record over his kingdom's gold possessions and it was he who had the power to dispose over it.

An inscription on the so-called Palermo Stone (Schäfer 1902), an approx. 6 cm thick basalt slab of which other fragments are kept in Cairo, had probably been carved some time during the fifth dynasty as a copy of a much of older text. Reaching back to the Predynastic Period the inscription lists in chronological order the reigns from Aha (first dynasty, about 3000 BC) to Neferikare (fifth dynasty, about 2460 BC). In addition, the list gives important events attached to the respective reigns. For as early as the second dynasty it reads "the counting of gold and fields". With the beginning of the fourth dynasty (about 2630 BC) the list then continues with repeated donations dedicated to the gods and their land-possessing temples, but also large quantities of gold, which according to findings in the field may now be viewed as gold extracted from mines.

The hieroglyphic sign "nub" for gold depicts a gold collar (Fig. 2.1) and is encountered as an engraving as early as on prehistoric clay and stone tablets from the sites of Hierakonpolis, Nagade and Ombos, of which all are located in the mouth area of the ancient route leading to the Eastern Desert with its gold deposits (Vercoutter 1993).

Only approximate statements can be made as to the actual amounts of gold in circulation in Ancient Egypt. With regard to the quarried mines, one would need to measure them up accurately in order to get some idea about their respective yields. Beyond that, it seems however impossible to attempt any assessment concerning

Fig. 2.2 Scene from the tomb of Huy, viceroy of Kush, dated to the period of Tutankhamun (1333–1323 BC). It depicts Nubians delivering large quantities of ring-shaped gold ingots and gold dust in bags, each respectively marked with the gold hieroglyph

the yields of the countless mining pits, known wadiworking sites, let alone the unknown or partly-buried extraction sites, while keeping in mind that these date to diverging, yet undated periods in history.

In addition, especially for the New Kingdom we know of substantial gold imports and booty from conquered countries. From his campaigns in Syria for example, Thutmose III brought home the gilt battle chariot of the kings of Megiddo and Qadesh, and "ten gilt carriages whose shafts were of pure gold", and even a princess covered in gold. Inscriptions inside the tombs of the monarch's high officials report tribute payments consisting of cast gold ring ingots from Syrian rulers. This was the most common form of natural gold. In addition, there are illustrations of deliveries in bags containing gold dust (Fig. 2.2). The gold was weighed carefully with specific gold weights, generally figures in the shapes of cattle (Fig. 2.3). In the New Kingdom gold (and silver) and copper were weighed with among others, simpler shapes identified as weights by their inscriptions (Fig. 2.4).

In order to get some idea about Egyptian gold reserves, it seems adequate to consult the delivery lists from administrative and temple treasuries.

The most common Egyptian weight unit was the "dbn" equal to 91 g. According to Graefe (1999), however, this unit was divided by two for gold and silver, thus corresponding to 45.5 g. Because some documents cite "large dbn" at 45.5 g and "small dbn" at 13.1 g, the following calculation is based on Graefe's (1999) small dbn in order to avoid implausibly high figures resulting from a weight unit at 45.5 g per dbn. For reasons of clarity the following figures are given

Fig. 2.3 Weighing of gold, displayed as both ring ingots and dust (bags) above and below the weighing scene. The text to the left includes the hieroglyph for gold. Tomb of Paheri at El-Kab, eighteenth Dynasty (Photo credit: C. Hess)

Fig. 2.4 Gold weights of stone with different values. National Museum, Khartoum

kilograms. The large dbn values also appear in kg, in brackets behind the first figure:

Hatshepsut (1479–1458 BC)
Endowment to the temple at Karnak: 480.7 kg (1,669.5 kg)
Thutmose III (1479–1425 BC)
Gold entry from Nubia between the 32nd and 42nd year of reign: 150.5 kg (522.6 kg)
Booty from the battle at Megiddo in the 23rd year of reign: 39.3 kg (136.6 kg)
Income from a part of T's reign; raw gold in different qualities 332.8 kg (1,155.9 kg)
Total allowance for the temple at Karnak, from 46th year: 2,000.6 kg (6.948 kg)
Amenhotep III (1388–1351 BC)
Endowments pertaining to two different categories of gold: 412.4 kg (1,432.5 or 329.9 kg (1,145.8 kg)
Gold present to Niniveh and Hanigalbat: 1,320.0 kg (2 times 20 talents at 30.3 kg)
Ramesses III (1183–1152 BC)
Endowment to his temple in the temple district at Karnak: 655.0 or 54.0 kg (2.275 or 187.6 kg)
(according to the Harris Papyrus even for all temples only 17.8 kg)
Osorkon I (985–979 BC)
Endowment for the Heliopolis temple: 201.0 kg (698.2 kg)
For Amun gold and silver (?) 30,130 kg (104,650 kg)
Various 65.6 kg (228.0 kg)
Taharqa (690–664 BC)
For the temple in Kawa/Nubia over 9 years 15.7 kg (54.4 kg)

According to these randomly selected texts, gold production culminated under the reign of Thutmose III which corresponds to the time when Egypt was at the height of its expansion, imposing its rule over distant provinces throughout the Near East and Nubia up to the fourth cataract. In the 31st year of his reign, the annals of Thutmose III mention among others gold supplies from Syria contained in 41 protected, full bags, probably gold ore in the form of gold tinsels or nuggets (Helck 1965).

The texts disclose that the gold quantities decreased noticeably during the reign of the Ramessides. By the end of the twentieth dynasty, a period for which the epigraphic material remains virtually silent as to gold deliveries, a full-scale pillaging of private and royal graves was made public under Ramesses IX in the Theban necropolis and whose official investigations for the clarification of the incidents is recorded in the Abbott Papyrus (Kitchen 1983).

Nevertheless, according to the above list only a 100 years later Osorkon I disposed over the largest gold stocks ever recorded, and it is seems therefore tempting to connect this with the looted gold from the necropolis and its subsequent redistribution (Störk 1977).

Not included in the above list are the colossal amounts of gold cited in donation lists. Also, the numerous archaeological gold artefacts like the ~11 kg mask of Tutankhamun as well as his interior gold coffin weighing more than 110 kg to name but a few, make an impressive statement as to the scale of Ancient Egypt's gold prosperity (Fig. 2.5).

We may therefore assume that much larger amounts of gold than those suggested by the available texts had been in circulation, considering especially that over the time this non-perishable material has the tendency to accumulate rather than the contrary. Gold objects were melted down repeatedly and transformed to new artefacts, as observable by the heterogeneity of the data obtained from scientific analyses carried out on different samples from single artefacts (cf. Chap. 4). Even the gold jewelry sold today in Egyptian bazaars might well contain remnants from the country's ancient artefacts.

The available gold was essentially used for the manufacture of ornaments and jewelry of all sorts, cultic objects, amulets, and figurines of divinities. A comprehensive selection of such objects is found in the catalogue of the Vienna exhibition "Gold der Pharaonen" (Seipel 2001). In architecture gold was generally used for fittings in temple equipment like doors, columns and pillars, floors, and walls. In funerary contexts sheet gilding predominates on wooden coffins, gilt death masks and golden or gilt grave goods, like jewelry, accessories, and vessels.

Fig. 2.5 Inner sarcophagus of Tutankhamun from solid gold, weighing over 100 kg. Egyptian Museum, Kairo

Graefe (1999) gives a compilation of randomly gathered information on the weights of gold artefacts kept in various museums. They vary between a few grams to several hundred. The heaviest object in his list is a necklace from the "gold-rich" twenty-first dynasty attributed to king Psusennes.

The higher his rank, the more was an official able to supply his tomb lavishly with gold and thereby to lend expression to his aspiration for immortality. Deserving state servants were moreover rewarded by the king personally with the so-called gold of honor, often consisting of necklaces, bangles or other gold presents.

The wealthiest tombs however, were those of the kings. Unfortunately, only few have remained unscathed such as the most notorious one of Tutankhamun.

Considerable amounts of gold were required by the king for his foreign affairs and thus were too, contained in presents for distant rulers. Manufactured artefacts were not the only negotiated goods but raw gold as well. As revealed by the Amarna letters (Knudtzon 1964) however, presents could be demanded in counterpart by the pharaoh, as in the cases of the princesses for the harems of Amenhotep III and IV (Akhenaten).

Master planned and state-ordered pillaging of graves in Egypt took place during the first centuries following the Arab conquest 642 AD and ended only by the twelfth century (Rabie 1972). The subsequent beginning of the so-called goldrush in the South Egyptian and Nubian Deserts is most certainly to be seen in perspective with the dwindling gold resources from grave looting in the heartland. At the same time, new and lucrative profits were expected from the numerous gold producing sites rediscovered during earlier pilgrim voyages along the routes through Wadi Allaqi to the Red Sea.

Present-day gold producing sites in similar terrain situations (mostly wadiworkings) like those in the Bayuda Desert, in Sudan, may give some hint as to the yield of the exploited gold. During our stay at the site in 1999, about 500 people worked there, of which approximately 10 % were women and children. They worked at less productive locations and were mostly occupied with exploiting the sands.

According to the labourers, the day by day yield ranged between a minimum of 1 g (for most women) and a maximum of 15 g per person. The average yield for one male worker was given at 5 g/day. The annual work season varied between 3 and 4 months.

Because gold was plentiful like any other important commodity in Ancient Egypt, the country never developed into a trading nation. Absent luxury goods like ivory, exotic woods and rare animals were imported from the provinces or even more distant countries through diplomatic channels and barter economy.

The fact that gold was accurately weighed demonstrates that it was a precious material, although no specific reference value was attached to it as one might expect in a modern economy. Defining a realistic value for gold in Ancient Egypt proves therefore problematic, especially considering that if ever it had been fixed to some other reference value, it would have fluctuated considerably over its 3000 years of history.

The first although not minted coins eligible for contributing to an evaluation of a gold price appear sometime during the thirtieth dynasty, only shortly before the Ptolemaic conquest in 332 BC, in other words towards the end of what is often referred to as "Ancient" Egypt. Until then Egypt had no money economy comparable to that of today. The economy resembled more a barter economy and above all, it was guided by the state that redistributed all produced goods according to its inhabitants' requirements and its own will to respond to these needs.

3 On the Regional Geology and Genesis of Gold Deposits in Egypt and Nubian Sudan

Ore deposits are commonly viewed as geochemical anomalies whose origins are traced back to complex formational conditions in their geologic environments. Bearing in mind the geologic past of their immediate and distant backgrounds is therefore necessary for a comprehensive insight into the formational events pertaining to the ore deposits. Understanding such processes is fundamental for efficient prospecting with good forecasts for success.

The high accuracy of the ancient Egyptian prospectors, especially from the early New Kingdom onwards presupposes a thorough knowledge about the lithologic and tectonic implications surrounding the processes in deposit formation, at least as far as the Egyptian Eastern Deserts and Nubia are concerned. With their expertise the prospectors had managed to identify and exploit virtually all significant and outcropping gold deposits in Nubia and the Red Sea Hills in Sudan within a period of 150 years under the reigns of Thutmose III (1492–1479 BC) and Akhenaten (1351–1334 BC). Within at least the Egyptian Eastern Desert this successful prospection and exploitation however continued until the end of the Ramesside period because almost all the major gold deposits of this region display clear traces of mining from that New Kingdom periods, and all resumption of mining activities in later periods in essence foot these workings or their local follow-ups. It was not before the Early Arab Period (ninth to fourteenth centuries AD) that yet another numerous gold mining sites had been able to develop in the wadi floors (wadiworkings) of the Red Sea Hills and vast areas of the Nubian Desert, which indeed do not seem to reveal any linkage to the New Kingdom exploitations.

The situation was to a certain degree different with regard to the copper deposits. Prospectors had in fact been fully capable of identifying copper minerals through green malachite stains along the wallrock already in well earlier periods. It therefore hardly surprises if mining traces at copper deposits may date back to the earliest historic periods.

When copper mineralisations are associated to quartz veins in the Eastern Desert of Egypt and Sudan they also often contain gold. Consequently, malachite stains had therefore also become an important criterion in gold prospecting, particularly in the Old- and Middle Kingdoms.

For a better understanding of the development of local gold deposits and the appearance of ancient mines, it is necessary to give a short update of what is known concerning the geologic structure of NE Africa, and specifically the phase-rich, tectonic development of the so-called Arabian-Nubian Shield (ANS).

We thereby do not intend to recount the complex geologic history of the mountainous deserts E of the Nile, as this would exceed the archaeological interest of the present study.

Readers specifically interested in the geosciences of this part of the African continent may find more topic-related references in the appended bibliography, whereas the following paragraphs may only be taken as a first introduction.

R. Klemm, D. Klemm, *Gold and Gold Mining in Ancient Egypt and Nubia*, Natural Science in Archaeology,
DOI 10.1007/978-3-642-22508-6_3, © Springer-Verlag Berlin Heidelberg 2013

The African continent consists of several cratons* joined together by younger orogenic belts. Most geologists identify the African continent with four major cratons: the Kalahari Craton, the Congo Craton, the West African Craton, and the Central and East Sahara Craton. This, however, is a rough division, as there are in fact a number of smaller cratons that are "welded" together by orogenic zones. This is for instance the case for the Kalahari Craton, which actually consists of the Kaapvaal Craton in the territory of the Republic of South Africa as well as the Zimbabwe Craton, while both are connected by the tightly folded Limpopo Belt.

In all, seven orogenic events of varying ages are known to have taken place in Africa during Earth history (Cliffort 1970).

For the present purposes our attention is essentially concentrated on the East Sahara Craton (Fig. 3.1). However, Stern et al. (1994) doubt its real cratonic character as it is influenced in large parts by the Pan-African Neoproterozoic orogeny. Kröner (1979) too, views much of the East Sahara Craton in context with the Pan-African orogeny as a comprehensive isotopic rejuvenation and revitasation of the craton about 600 Ma (Megaannum = million years).

To the E of the West Sahara Craton is the Arabian-Nubian Shield (ANS) with its gold deposits Fig. 3.1. It was formed during the so-called Pan-African orogeny of NE Africa, which stretches down to the Mozambique Belt in the S (Vail 1988). According to geochemical isotope age analyses (compiled in Cahen et al. 1984) this orogeny dates to 550 ± 100 Ma.

According to Abdeen and Greiling (2005) the ANS is composed of a number of island arcs with parts of oceanic crust and some continental sequences. These individual series had been thrusted onto parts of the African Craton by major tectonic processes. They thereby collided at about 600 Ma and joined to so-called terranes on the western Gondwana supercontinent, which today corresponds more or less to Africa to the W of the Nile. These terranes partially compose of extended ophiolites with metamorphic sequences of mantle material. Today, they appear as serpentintes, metamorphic ocean basalts, and occasionally gabbros. The terranes furthermore contain (mostly metamorphic) marine sediments, mafic and acid volcanics, as well as continental residues, in other words, they consist of the characteristic geologic units of former island arcs and their marine basins.

During the submarine overthrust processes onto the African continent, these terranes had become intensely folded and thus forming basins taking up the detritus of the uplifting mountain units, which today at least in Egypt occur as greywackes, siltstones and conglomerates of the Hammamat sediments (partly quarried under the name of "bekhen" stone since antiquity).

In the entire Eastern Desert of Egypt and Sudan, these terranes became vigorously intruded by post-orogenic, granitoid magmas. Such events were particularly significant for the present context as the magmas furnished the thermal energy to set into motion their own gases and the water contained in the rocks. As these fluids moved towards cooler areas, they leached-out according to their dissolving capacities the rocks they passed.

Extremely low contents of gold and other metals were thus dissolved and on their thermally induced ascent precipitated predominantly with silicon dioxide in open fissures. In some cases though, gold was able to accumulate to economically worthwhile concentrations.

In order to reach such grades, the traversed rocks had to contain slightly above-average amounts of gold of at least 10 ppb (= parts per billion), which is mostly the case for ophiolite serpentinites and basalts as well as for their weathered products.

The combination of ophiolites (or their weathered erosion bi-products) and later granitoid magmas became an important geologic marker for gold prospectors. The discussion of the individual occurrences shows, however, that this seemingly simple principle is nevertheless constantly adapted by the geologic reality in the field.

Not least because of widespread volcanic activity, thick andesite to dacite sheets formed, which are referred to in Egypt as Dokhan-volcanics that among others contain deposits of "imperial porphyry" (Porfido rosso antico).

* (unusual terms and definitions see glossary)

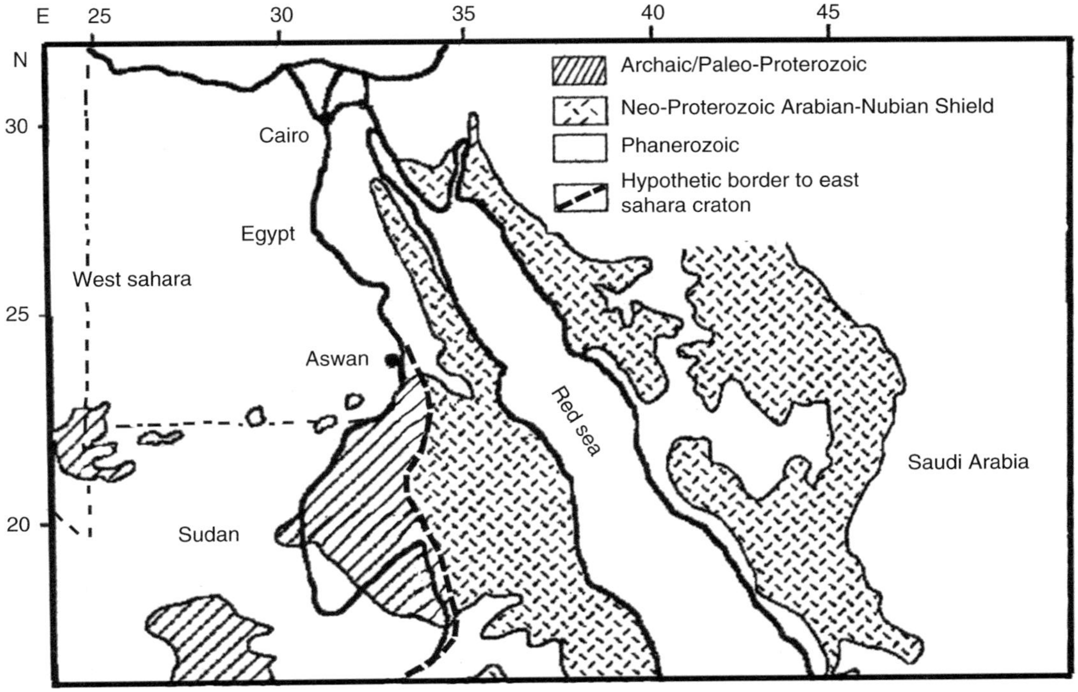

Fig. 3.1 Simplified distribution of the Palaeo-Proterozoic (Old African Basement) and Neo-Proterozoic (Arabian-Nubian Shield) and Phanerozoic (predominantly chalk, tertiary and alluvial sand) units in Egypt, Sinai, Arabia, and Sudan (Modified after Sultan et al. 1994)

The schematic map in Fig. 3.1 shows the distribution of the Pan-African components of the ANS in Egypt and Nubia. It reveals that the ANS partly overlies the East Sahara Craton. Furthermore, Figs. 5.1 and 6.1 reveal that virtually all ancient gold mines are located within units of the ANS.

In the Egyptian Eastern Desert the original, cratonic basement units are almost entirely veiled by later sandstone and limestone sequences. In Nubian NE Sudan, both in the area around Duweishat on the Nile as well as in the eastern Bayuda Desert (Vail 1988), rock units of the ANS are in superseding contact with more metamorphic, cratonic gneiss sequences. The fact that both sequences expose discordant tectonic stress patterns, and that especially the much diverging ANS-series are severely folded, demonstrates that the latter can only have been thrust onto the craton, where they are found today.

With the therefrom resulting plate tectonic models for the genesis of the ANS formulated among others by Al-Shanti and Mitchell (1976) and through the similarity of the ophiolites with the volcanic rocks to island arc volcanics, a so-called accretion model was proposed. Thereby, island arc rock-types were pushed together with the ophiolites i.e. accreted at the plate-tectonic collision zones, onto the continental margins. This process usually occurs in several stages, which are reflected by separate terranes in the ANS, especially in Sudan (Fig. 3.3). The terrane borders are often detected as ophiolite sequences in suture zones, and their ages are determinable by radio-chemical zircon analysis. Thereby, according to Kröner et al. (1992), the latest ophiolites in the W of the shield (Wadi Ghadir) are dated to 746 ± 19 Ma.

The Onib-Sol Hamed ophiolite complex in NE Sudan for instance stretches across the Red Sea to Saudi Arabia, and in parallel orientation further S is the Bir-Nakasib Umq suture (Fig. 3.3). The alignments of the suture zones are partly reported to indicate a terrane obduction from SE to NW (Shackleton 1994; Ries et al. 1983; Greiling et al. 1994), although Kröner (1985); El Gaby et al. (1984) diagnose an obduction from NE to SW in S Egypt (Fig. 3.3). As for the Eastern Desert, such compressed folds, which typically occur in nappe overthrusts, are oriented in a SE-NW direction (Greiling et al. 1994).

3.1 Stratigraphy of the Eastern Deserts of Egypt and Northern Sudan

The here discussed gold deposits exploited in antiquity occur almost exclusively in the Neoproterozoic sequences of the ANS in the Eastern Desert in Egypt and Northern Sudan. Their phanerozoic covers had been uplifted and mostly eroded in a balancing effect from the sinking of the Red Sea graben. The stratigraphy of the Egyptian Precambrian has received diverging assessments. Hume (1934) counts four sub-units, consisting of "fundamental gneiss" followed by "middle parashists" and "acid gneisses" and finally "upper parashists", the latter consisting of different granites. Akaad and Noweir (1969, 1978), on the other hand, differentiate between an older "Meatiq Group", a later "Abu Ziran Group", and a youngest "Hammamat Group." El-Ramly (1972) distinguish between two sequences also adopted by el-Gaby (1988, 1990), whereby the recumbent (earlier) part consists of a Pre-Pan-African "infrastructure" (= cratonic African Basement) and the hanging (younger) part of a Pan-African overthrust "suprastructure". Greiling et al. (1994) too, advocate a twofold division consisting of a "tier 1" (recumbent) and "tier 2" (hanging), though without proposing any dates. The twofold subdivision into an infrastructure and suprastructure seems in essence to be the more appropriate one, whereby the former (infrastructure) is well observed in Egypt but only sporadically near the Nile in the Eastern Desert of NE Sudan (Fig. 3.2). The following discussion of the subdivision makes much reference to the work carried out by Murr (1999), in addition to the authors quoted above.

3.2 Infrastructure

The age of the infrastructure is subject to controversy. El-Gaby et al. (1984, 1988, 1990); Hassan and Hashad (1990) view it as the eastern extension of the East Sahara Craton and therefore as part of the continental crust, which became overthrust by the suprastructure during the Pan-African orogeny.

This contends with the idea that the infrastructure is merely defined by a higher metamorphism of its sequences in Egypt and that it therefore is not part of the African crust (Ries et al. 1983; El-Ramly et al. 1984; Bentor 1985; Shackleton 1994). Greiling et al. (1994), who argues on behalf of published isotope analyses, draws the western border of the East Sahara Craton in Egypt more or less at the western side of the Nile. Near Aswan, however, the situation complicates. Especially in Wadi el-Hudi, locally occurring and tightly folded migmatite gneiss rather seems to belong to a primeval crust. This is even more the case for the anorthosite gneiss in the area of Wadi Allaqi and the garnet kyanite gneiss at Duweishat, in Northern Sudan (the hypothetical boundary of the East Sahara Craton in Fig. 3.3 is therefore much further E in relation to the units of the ANS as compared to Kröner et al. 1992). Evidence for an old source area in the Egyptian Eastern Desert, however, is furnished by a zircon date of 1.77 Ga obtained with granite boulders from clastic metasediments at Gebel Hafafit (Abdel-Monem and Hurley 1979). Nonetheless, other isotope samples from gneisses at Gebel Hafafit date to 680–700 Ma and can therefore hardly be attributed to an early African crust.

A similar situation is observed at the gneiss dome at Gebel Meatiq (Fig. 3.2), whose core consists of metamorphic rock and migmatite gneiss. Together with kyanite, staurolite and almandine parageneses they testify to a relatively high-grade metamorphism (Habib et al. 1985). To the N of the road, between Qena and Safaga, there are more areas with highly metamorphic gneisses. Apparently, the observed metamorphic deformations in the mentioned areas occurred through intensive overthrust tectonics. There

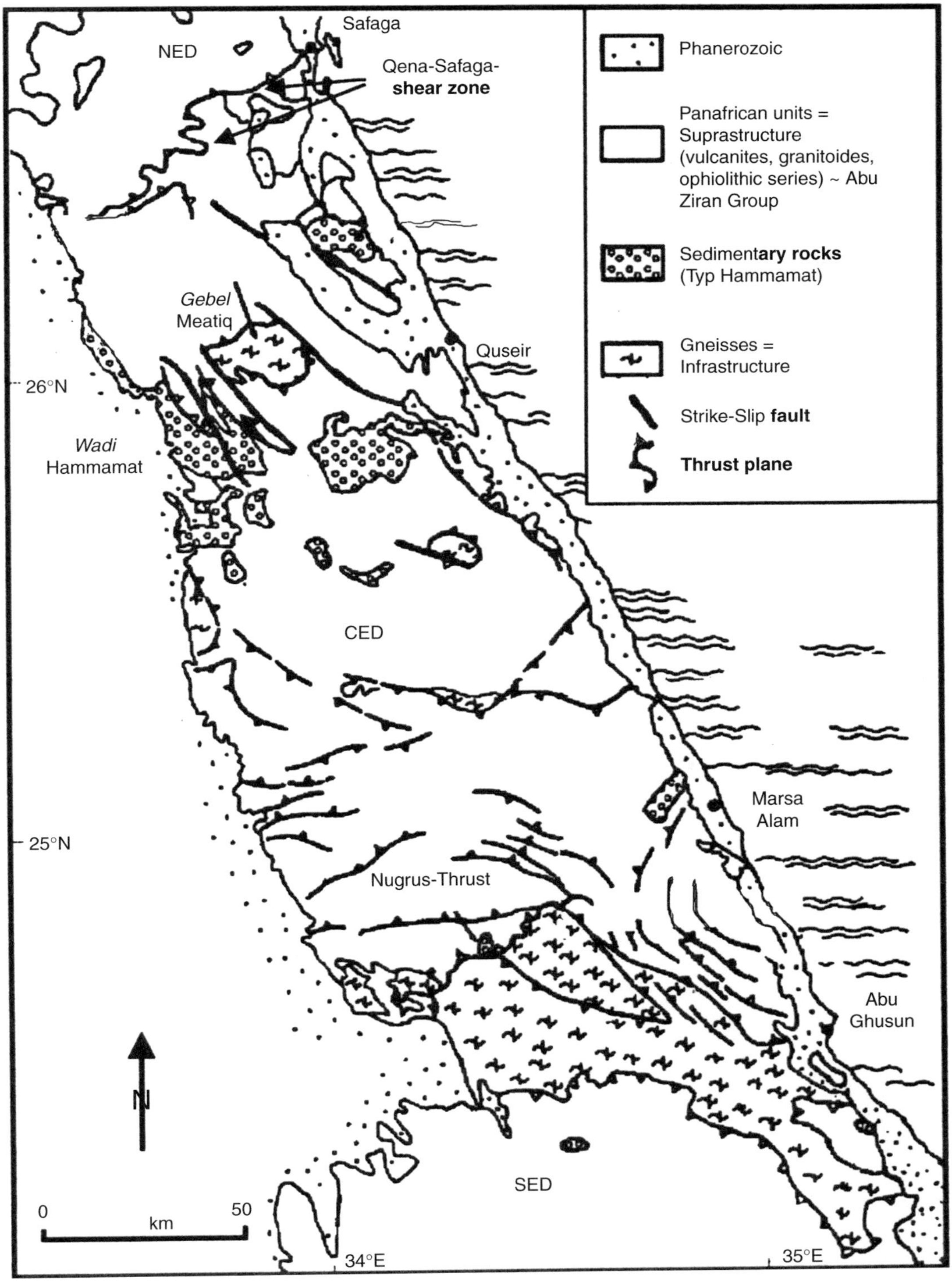

Fig. 3.2 Simplified geologic map of the Egyptian Eastern Desert. *NED* North-Eastern Desert, *CED* Central Eastern Desert, *SED* South-Eastern Desert (modified after Greiling et al. 1994)

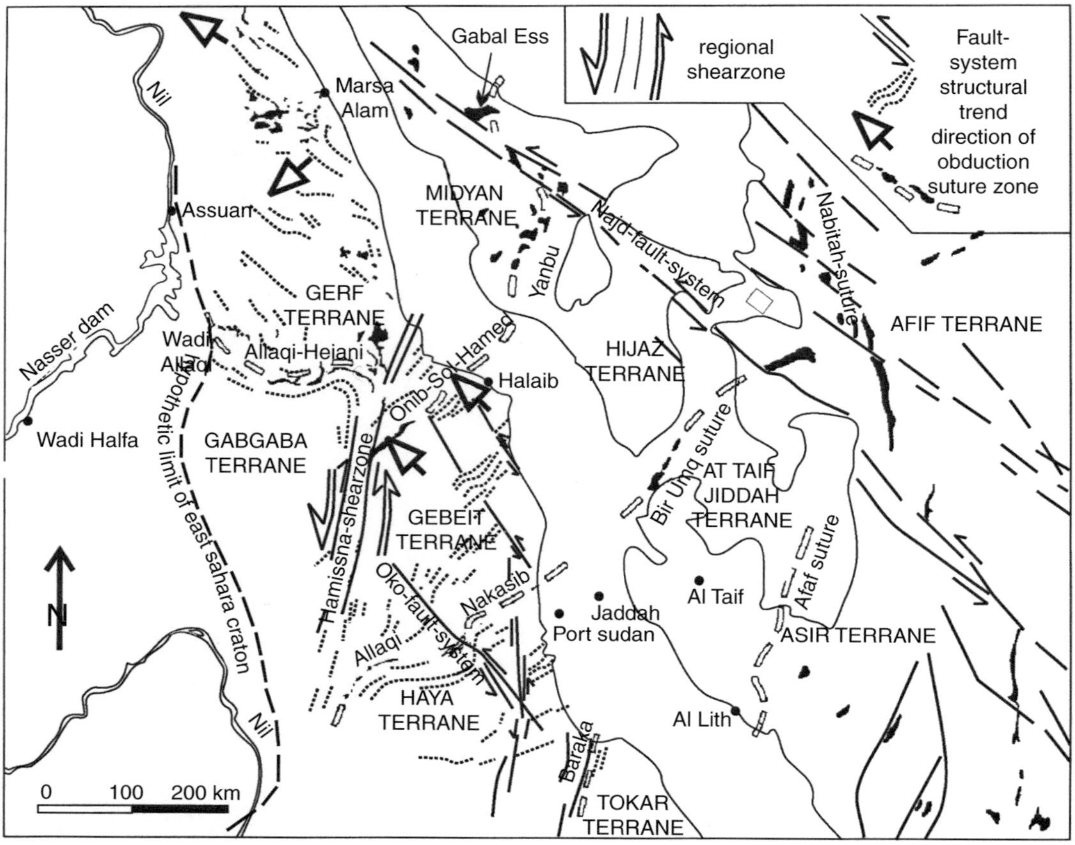

Fig. 3.3 Simplified structural map of the ANS displaying the suture zones (ophiolite ranges), terranes and the conjectured border to the East Sahara Craton (modified after Kröner et al. 1992)

seems to be no direct link between the ancient gold mines and the sequences of the infrastructure, if one sets aside the few deposits close to the Nile in Northern Sudan.

3.3 Suprastructure

Whereas according to Akaad and Noweir (1969, 1978), the suprastructure subdivides in two groups referred to as the Abu Ziran group and the Hammamat group, el-Gaby (1988) favours a threefold division, comprising an ophiolite-like sequence with volcanics of the island arc type, Hammamat sediments with associated volcanics, as well as post orogenic granites, which on the whole matches with the partition by Hume (1934).

The Eastern Desert of Egypt and Northern Sudan also preserve a number of almost complete ophiolite sequences. Garson and Shalaby (1976) and Langwieder (1993) describe a well-exposed ophiolite complex at Bir Umm el-Fawakhir in Egypt. El-Sharkawy and El-Bayoumi (1979) refer to another one in the vicinity of Wadi Ghadir. Other major ophiolite sequences are also found in Wadi Allaqi and in the Hamissana zone (Zoheir and Klemm 2007), in NE Sudan.

The ultramafic components of these ophiolites are mainly serpentinites derived from classic rocks of the upper mantle, such as harzburgite and more seldom of lherzolite and dunite (el-Gaby 1988), in which kneaded chromite lenses may occur. Along shear- and overthrust zones the serpentinites are often converted by CO_2-rich hydrothermal processes to talc and quartz carbonate.

Gabbros associated to these serpentinites are also generally affected by metamorphism and at their basis frequently contain pyroxenite cumulates. El-Gaby et al. (1984) and El-Sharkawy and El-Bayoumi (1979) describe massive diabase and sheeted dike complexes in Wadi Ghadir, which may reach thicknesses between 100 and 200 m. Pillow basalts of up to 100 m, have been described by Langwieder (1993) in the area of Umm el-Fawakhir. These are usually tholeiitic basalts with MORB affinity, but metamorphically transformed to now predominantly amphibolites.

Deep-sea sediments are rare, especially in northern and central parts of the Eastern Desert of Egypt. El-Sharkawy and El-Bayoumi (1979) describe up to 200 m thick sediments of clay and some chert in Wadi Ghadir. Stern (1981) views a back-arc in environment to be at the origin of the ophiolites, just as Abdel-Karim et al. (1996) do for the ones at Dungash.

Among the Egyptian ophiolites, particularly the serpentinites tend to display elevated gold values around 25 ppb (parts per billion). Even if this by far falls short of the lower limit for viable exploitation, it yet remains significantly higher than the usually observed levels around 5 ppb. Langwieder (1994) was able to establish grades of up to 180 ppb Au for serpentinites sampled at el-Sid.

It is usual to observe clastic metasediments in the ophiolite sequences. Sometimes they occur in melanges together with the serpentinites (Hassan and Hashad 1990). El-Bayoumi (1984) interprets them as deep-sea graben fills.

Kröner et al. (1992) obtained zricon dates of 746 ± 19 Ma from ophiolite plagiogranites, and Stern and Hedge (1985) date the Egyptian Eastern Desert island-arc volcanites to 720–770 Ma.

Reichmann and Kröner (1994) examined similar volcanosedimentary ophiolite units with island arc character in the Gebeit region of NE Sudan, which they dated with the Sm/Nd-method to 832 ± 26 Ma.

The actual ophiolite sequences are often superseded by slightly metamorphic volcanics with predominantly andesitic chemical compositions. Stern (1979, 1981), El-Gaby et al. (1984, 1988, 1990) and Hassan and Hashad (1990) describe them as "younger metavolcanics" as opposed to the basaltic "older metavolcanics". The younger series consist largely of andesite lava, but also of basalts and dacites-rhyolites, as well as volcanoclastics (Hassan and Hashad 1990). Such volcanics are commonly compared with today's island arc volcanics (Stern 1979, 1981).

Further up the stratigraphic sequence in the Egyptian Eastern Desert are the partially massive Hammamat sediments. They consist of clasitic to molasse-like sediments of greywackes and siltstones with intercalations of partly coarse conglomerates and little sorted fanglomerates. These rock series are particularly common in Wadi Hammamat, and have been quarried virtually throughout the entire Egyptian history as the so-called "bekhen" stone used for the production of finely crafted statues (e.g. the Menkaure triad), but above all, sarcophagi, cosmetic palettes, and bowls. As a matter of fact, broken sarcophagi that had been discarded at the quarries have become the eponyms of this wadi [Arabic Hammam = bath (tub)].

In contrast to the massive and large rock folds in Wadi Hammamat itself, the Hammamat series occur elsewhere usually in tight folds, which are considered unsuited for major works of art. All Hammamat sediments within the Pan-African orogeny are slightly metamorphic (greenschist facies) (Osman et al. 1993; Klemm and Klemm 1993, 2008b), which makes denotations like "metagreywackes" and "metasiltstones" etc. seem more appropriate. Because in terms of chemical composition, the Hammat series constitute only little altered sediment fills from the older island arc volcanics and ophiolite series, they also contain their trace elements, such as slightly enriched gold. Although it exclusively occurs in the lower ppb-ranges, it should nevertheless be regarded as a geochemical anomaly. This aspect is relevant concerning the genesis of the gold deposits and will therefore be returned to later in this discussion.

The Hammamat sediments also contain alternating strata of intermediary to acid volcanics as well as dike intrusions of such rocks. These are generally attributed to the Dokhan volcanics

(El-Gaby 1984), which are partially interlocked with the Hammamat sediments but mainly overlie the latter. Thereby, the Dokhan volcanics, which also comprise the imperial porphyry quarries much appreciated by the Romans, coincide at least partially in terms of chronology with the Hammamat sediments, even if most are younger. Genetically, the Dokhan volcanics represent subaerial, extruded dacites to rhyolites, whose later parts free of metamorphism. Auto-hydrothermal alterations, as described by Klemm and Klemm (1993, 2008b) for the imperial porphyry, are unaffected by this.

Thereby, in part massive Dokhan volcanics formed in the emerging areas of incipient orogeny, while Hammamat sediments including Dokhan pebbles accumulated in intra montane basins. In a subsequent compression phase, the molasse sediments were then folded (El-Gaby et al. 1988; Langwieder 1993; Murr 1993).

The age of the Hammamat sediments can only be determined indirectly. A recumbent Dokhan sample was dated with the Rb/Sr-method to 616±9 Ma (Ries et al. 1983), and an intruded granite inside the Hammamat series to 590±11 Ma (Dixon 1981). There seem to be two age groups for the Dokhan volcanics, one established to 625–610 Ma and the other to 600–575 Ma (Gillespie and Dixon 1983). Both groups also seem to reflect two separate magmatic events, which however, due to their close similarity to each other, are in terms of their petrography hard to distinguish from each other in the field.

The described rock spectrum of the island arc type in the Egyptian Eastern Desert also includes marbles, whose ratios increase further S. Locally restricted, enhanced levels of graphite occasionally give them a black appearance. Such marbles are found in the wider area around Baramiya at Gebel Rokhan, where Thutmose III is already known to have retrieved raw material for some of his statues. Massive and severely deformed marble banks are for the most found in the western Gabgaba terrane, in Nubian Sudan. There they occur in alternating sequences with slightly metamorphic rhyolites to dacites with tuffs and agglomerates.

These series compose more notably of acid to intermediate volcanics in Nubian Sudan and contain less sedimentary sequences than in the Egyptian Eastern Desert. Yet recurring petrographic records of slightly metamorphic greywackes, siltstones and conglomerates are particularly reminiscent of the Hammamat series though they are increasingly interlocked with marbles. These markedly volcanogenic series are commonly referred to as the so-called Nefirdeib formation. Equivalent to the Egyptian Eastern Desert, they exhibit affinities with island arc characteristics.

Compared to the Egyptian Eastern Desert where determination of the individual terranes is encumbered by intensive shearing and overthrust of the Pan-African orogeny (except the south-eastern Gerf-terranes; Fig. 3.3), Northern Sudan seems considerably less affected by compressional tectonics. Thus, the separation between the western Gabgaba and the eastern Gebeit-terrane by the so-called Hamisana suture is therefore much more convincing (Fig. 3.3).

3.4 Late to Post-Tectonic Intrusions

In order to understand the processes underlying the gold mineralisations and their prospecting in the Eastern Desert of Egypt and Northern Sudan, it is crucial to localise and discern the therewith connected, late to post-tectonic intrusions. Chemically, they are usually of granitic composition, but as a result from assimilation processes with their hostrocks they often change at their margins to granodiorite and even diorite. The recognition of such zones is essential, because they are areas in which auriferous quartz veins are repeatedly found. The ancient prospectors had undeniably been aware of this, especially in the New Kingdom.

El-Ramly and Akaad (1960) surveyed the granite intrusions in the Egyptian Eastern Desert and in doing so differentiated between "older" and "younger granites". The older ones revealed granodiorite to even tonalite compositions, and strictly speaking do not belong to the granites. In addition, at least some of these older granitoids (granite-like igneous rocks) were reported to be significantly deformed by tectonics. The younger, however, turned out as just slightly or not deformed at all and authentic granites, except for the ones in the marginal contamination zones mentioned above. Such categories are also equally identified in NE Sudan, but still await comparable examinations. In most but not all cases, the "younger granites" were reported to be reddish, probably as a result from fine hematite inclusions within potassium feldspar components.

In differentiating the granitoids in the Egyptian Eastern Desert, El-Gaby et al. (1988) proposed a syn- to post-orogenic group for the older granites and a late- to post-tectonic group for the younger ones. According to Harris (1982) the actual "younger granites" emplaced towards the end of the Pan-African orogeny. Stern and Hedges (1985) examined these granitoids with the Rb/Sr age-determination method and established a date of 709–614 Ma for the older, syn-orogenic ones and 596–544 Ma for the younger ones. These dates match approximately with those presented by Gillespie and Dixon (1983) obtained from U/Pb isotopes and ranging at 595–570 Ma, as well as those by Fullagar and Greenberg (1978) with the Rb/Sr method revealing an age of 586 ± 9 Ma (for the Fawakhir-granites) and 607 ± 8 Ma. Because of its importance for Egyptology, we here add the date of 565 Ma for the Aswan granite obtained by Abdel-Monem and Hurley (1978) with U/Pb method on zricon.

Another, by far more differentiated subdivision of granitoids was presented by Hussein et al. (1982); Hassan and Hashad (1990). The relevance of this subdivision, however, is comparatively low for the here treated topic regarding the gold mineralisations, and therefore is presently left aside discussion.

3.5 Tectonic Control of Quartz-Gold Mineralisation

Our survey of the ancient gold mining activities in the Egyptian and Nubian Eastern Deserts showed that the gold deposits as a rule are strictly bound to specific tectonic orientations. These consist partly of auriferous quartz veins mineralised inside gaps that had opened through extensional tectonics in the crust. Similar mineralisations could also occur within cavities that had been prompted through shearing. Since tectonic stress was the underlying factor for both opening processes, it turned out that a survey of the tectonic stress planes in the study area was of paramount importance in order to understand and therefore to prospect the occurrence of the gold deposits. During our field work in the footsteps of the ancient prospectors it soon became evident that this principle had been conceived and exploited, especially by the New Kingdom prospectors. They had manifestly conducted their sampling by concentrating almost exclusively on quartz vein mineralisations meeting certain local and tectonically induced opening orientations. Quartz veins that did not adhere to these orientations had been ignored. This conduct is only explained through a developed comprehension of such tectonic dependencies.

Being familiar with phenomena related to complex tectonic processes therefore proved essential as to the identification of the auriferous quartz vein mineralisations.

Repeated stress from tectonic compression and especially extension recurrently resulted to the opening of gaps in the crust of the Egyptian and Sudanese Eastern Deserts. The cavities subsequently filled either with magmatic melts or with quartz mineralisations from hydrothermal flow. Hydrothermal systems consist of hot aqueous solutions circulating in the crust trough joints and pore spaces whose capacities of holding different mineral solutions vary according to pressure, temperature, and pH level.

Due to the predominance of SiO_2 in the continental crust, this is generally also the main component in hydrothermal solutions, which once precipitated is referred to as quartz (chemical: SiO_2). Next to silicon dioxide, hydrothermal solutions also contain various salts, usually sodium chloride (NaCl), but also dissolved calcium ($Ca\,[HCO_3]_2$) and low concentrations of metal sulphides, such as iron, copper, lead, and zinc etc., and in few cases small amounts of gold in dissolved sulphide or chlorite complexes. Depending on the above mentioned parameters, some of the solution contents may precipitate to form separate minerals inside the quartz veins. For example, iron mineralises first together with sulphur to form pyrite (FeS_2), and if arsenic is available, to arsenopyrite (FeAsS), with copper it forms chalcopyrite ($CuFeS_2$). Lead together with sulphur form galena (PbS) and zinc with sulphur form sphalerite (ZnS). These are the most common metal minerals arranged according to frequency.

Whereas under normal circumstances, even in very low concentrations such metal sulphides are easily identified inside the quartz rock, gold usually occurs only in extremely fine particles no larger than a few microns (μm). Gold is also often contained in pyrite and arsenopyrite in such tiny particles that it remains invisible even under a high resolution optical microscope. The ancient miners had of course been unable to trace and extract such finely interwoven gold. Only in very few instances of severe weathering through dissolution of the sulphides a very fine, sponge-like lattice of gold was left behind and subsequently could become recovered. This type of gold always contains some traceable iron, no matter how thorough the processing, and is therefore safely identified with modern analytical methods. Such iron traces could actually therefore considered as reliable markers for substantiating the authenticity of gold artefacts.

3.6 Genesis of Gold Mineralisations in Egypt's Eastern Desert

This short outline of the complex geologic processes leading to the formation of the Eastern Desert of Egypt and Sudan and the emplacement of anciently exploited gold deposits, is now followed by a brief discussion of the genesis of the quartz veins themselves. It should however be kept in mind that each individual deposit has its own unique history, and that therefore are just as many deposit types as individual occurrences. They can nevertheless be grouped along the following, general criteria,:

1. Almost all gold mineralisations date at least to the Neoproterozoic era and the last stages of the Pan-African orogeny.
2. All gold mineralisations are located in the immediate proximity of late- or post-tectonic granitoid intrusions.
3. These intrusions penetrate among others into mostly folded and often imbricated sequences of molasse-like sediments and ophiolites, whereby their margins and roof areas may chemically change through assimilation of the wallrock.
4. Depending on the degree of assimilation and the chemical composition of the wallrock, this "hybridisation" of the rim and roof areas of the granitic intrusions leads to the formation of:
   (a) Granodiorites, through the assimilation of predominantly greywackes and siltstones and conglomerates.
   (b) Quartzdiorites to tonalites, through the assimilation of basalts and amphibolites.
   (c) Quartzgabbros to gabbros, through the assimilation of ultramafites (rocks from the upper mantle) and serpentinites.

Our geochemical examinations produced gold grades of up to 180 ppb for the ophiolite series, and yet 15–25 ppb for the molasse-like sediments of the Hammamat formation. We also observed that virtually all gold mineralisations stood in direct spatial context to such geologic units, so that according to this genesis model, they may be regarded as the primary source of gold pre-enrichment. Mainly in Sudan, another lithologic relationship to the gold mineralisations appears in the form of a genetic link to acid rhyolite to dacite volcanic rocks. There seems therefore to be a straight connection to former submarine, hydrothermal "black smoker" vents in analogy to ones observed today associated to acid volcanics in certain ocean beds.

In contrast to the aforementioned gold pre-enrichment in ophiolite and its bi-products Langwieder (1994); Murr (1999) were able to demonstrate that the original granite intrusions consistently displayed gold grades below the bottom detection limit of 5 ppb. The adjacent hybrid areas on the other hand displayed considerably higher, although fluctuating values of up to 50 ppb.

In describing the ancient gold exploitations it became apparent that many were bound to large, mostly NNW-SSE oriented shear zone systems, as seen in Fig. 5.91. Within this large shear zone system alone, are the sites at Hammama, Abu Had-North, Wadi Atalla el-Mur, Atalla, Umm el-Esh Sarga, Fawakhir, El-Sid, Umm Soleimat, and Hamuda. This example clearly demonstrates the importance of such shear zone systems as prospecting targets.

Actually, gold mineralisations often adhere to the shear directions, but others run almost perpendicularly to them. They are indicative of a tectonic control on the gold-quartz mineralisation caused by extension, often through the uplift from massive granitoid intrusions. The extensional faults occur either along cleavage planes previously formed in shear zones, otherwise along ac-planes, more or less in a perpendicular direction.

A common regime for the formation of gold quartz mineralisations in the Eastern Desert of Egypt and Nubia is therefore applicable:

Hydrothermal convection is set-off through the intrusion of large granitoid plutons into frequently, internally folded, metasedimentary and molasse-like sequences of the Hammamat type and its southern variants, containing imbricated ophiolites and island-arc volcanics. Low concentrated, but pre-enriched gold is thereby leached out together with silica and small amounts of sulphides from the wallrock and transported into zones of decreasing temperature and/or pressure. Mineral emplacement then takes places inside the extension gaps, for the most within the wallrock around the intrusion, but frequently also within its hybrid margins or roofs. The decisive factor here are the vein type cavities that offer ideal conditions for precipitation, as will be explained below.

Through frequently occurring carbonation within hydrothermal alteration environments relatively much CO_2 activity may be observed within the hydrothermal system, which is also confirmed by the results from the studies on the fluid inclusions. Furthermore, because of the values close to detection limit persistent occurrence of sulphide minerals in conjunction with the appearance of gold, its transport within the thermal circulation system probably also happened in connection with the sulphide complexes.

Botros (2002a, b) drew attention among others to the alteration zones generally observed in the wallrock along quartz veins as possible the indicators for gold mineralisations, though without developing the phenomenon to a criterion for prospecting work.

The investigations carried out by Murr (1999) and Zoheir (2004) revealed various chemo-physical mechanisms by which concentration of gold occurs. Furthermore, in a case study on the deposits of Fatira, Gidami, Atalla and Hangaliya, Murr (1999) was also able to describe the effect of granite intrusions on the mineralisations, which seems universally applicable to the entire Eastern Desert with only minor variations.

Similar, detailed studies have been carried out by Zoheir (2004) at the deposit Um el-Tuyor in eastern Wadi Allaqi. Much work has been completed in recent years on numerous anciently exploited gold deposits in Egypt's Eastern Desert, pinpointing crucial parameters concerning the formation of gold mineralisations. Thorough investigations by Hassan et al. (1999, 2004) on data from Hamash and Sukari, by Harraz (2002) on that from Atut, by Zoheir and Moritz (2007) on El-Sid, and Zoheir et al. (2008a, c), on Semna and on Umm Egat and Dungash (2008b) to name but a few, treating fluid inclusions from the auriferous quartzes have substantially contributed to our insight to the formation processes within each of these deposits. Next to the determination of the temperature ranges in which the quartz/gold mineralisation had taken place, the studies on the fluid inclusions also revealed important data on water salinity and gases, such as the most important one carbon dioxide but also less well represented ones like methane.

As already mentioned, there is a marked link between the occurrence of auriferous quartz veins and the marginal wallrock zones around granite intrusions. The granites themselves, however, apart from areas in the immediate vicinity of the mineralised gold quartzes turned out to have quite low gold values. The intrusions themselves may therefore be dismissed as genuine gold suppliers.

In many cases relatively high gold grades may be observed within about 1 m distance to the mineralised gold quartz. However, they decrease steeply from roughly several thousand ppb at the vein to less than 10 ppb about 70 cm away from it.

This anomaly is simply explained through reactions between hydrothermal fluids and the wallrock and occurs at almost any hydrothermal vein mineralisation. Nevertheless, in some cases exploitable gold grades may also form, especially when iron is released during these reactions. It then reacts with the gold-sulphide complexes and the free sulphur into pyrite or arsenopyrite in which the gold is contained. Only after the weathering of these sulphide ores, the insoluble gold remains visible at the vein boundary, and could thus be recovered by the ancient miners.

4 Analyses of Gold and Observations on Ancient Egyptian Gold Metallurgy

The here discussed gold had been almost exclusively extracted from quartz vein mineralisations. It is not pure gold but always contains some silver as well as traces of other elements.

Prior to the field work, we had initially hoped to be able to compare the data from our own analyses on gold samples retrieved from the deposits with those obtained from Ancient Egyptian gold artefacts, in order to get some idea as to the origin of the latter. Our hopes were soon let down though, because the investigation of the individual vein mineralisation and especially the tailing heaps showed that apart from few exceptions the annual gold yield of any given occurrence would have been too low for the production of any significant artefact in a separate metallurgical manufacturing process. Such objects therefore generally demanded a fusion of gold from different extraction sites, hence causing a chemical shuffle of the individual markers of each deposit.

Even if the yields of some important deposits, such as El-Sid, Barramiya, Umm Rus, or Sukkari had been large enough to allow for separate metallurgical processes, the chances of their isolation from the by far more common, mixed batches would have been extremely low. In addition, because of the high value of this material, the re-melting of gold artefacts was indisputably a commonplace practice throughout Egyptian history, at least until recently.

Yet chemical fingerprinting of ppb-scale (parts per billion) combinations of impurities specific to certain deposit categories with high-precision analytical methods have turned out to be of little value in this respect.

Another difficulty connected to the gold deposits is that many had undergone mineralisation in separate stages or generations, of which each displays diverging trace element characteristics. Furthermore, it turned out that non-negligible shares of ancient gold had previously been released in the deposit through the weathering of sulphide minerals in which they had been contained in small amounts (such as pyrite and arsenopyrite), which resulted among others to the loss of substantial silver contents. As demonstrated convincingly by Murr (1999) in the case of Fatira, the same gold deposit could therefore display contradictory gold compositions (Fig. 4.1).

In the discussion of the ancient gold mining sites in the Egyptian and Nubian Eastern Deserts, our observations are based on both our own and published data concerning gold contents. At first sight these, however, display extremely low values of usually only few grams of gold per ton of mined quartz and seldom reach or even exceed 30 g/t at an average. Although we cannot safely determine the actual gold grades of the ancient mines, which in most deposits are distributed very unevenly, in some mines it is nevertheless possible to take samples from abutments, resulting to reasonably safe average grade determinations. The latter varies between 10 and 30 g/t of mined rock. A supplementary constraint is that gold grades may be tied to certain zones within a quartz vein system and therefore dissipate completely with depth or just in a lateral direction. The ancient prospectors certainly only directed their attention first to such vein deposits whose outcrops showed at least

R. Klemm, D. Klemm, *Gold and Gold Mining in Ancient Egypt and Nubia*, Natural Science in Archaeology, DOI 10.1007/978-3-642-22508-6_4, © Springer-Verlag Berlin Heidelberg 2013

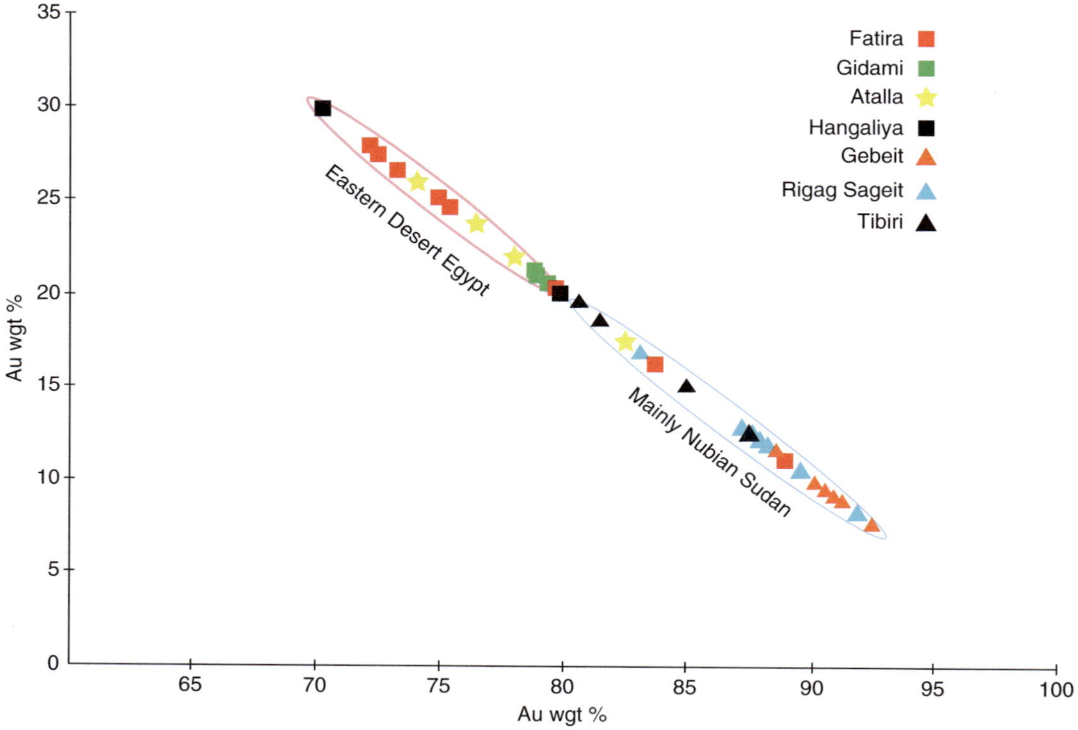

Fig. 4.1 Gold/silver content analysis in primary gold (native gold) samples from Egyptian and Nubian deposits (analyst: A. Murr) revealing generally higher silver contents in the Egyptian deposits compared to those in Nubia

some traces of visible gold. Rich-ore-zones were therefore obviously mined first, which means that these in general are no longer available today, leaving aside the mentioned abutments. Often, however, these have been removed as well, occasionally resulting to dangerous cave-ins (cf. Daghbag).

It is clear that a single quartz ore sample of only few kilograms may not be taken as representative for an entire vein deposit, but at best can only reveal the composition of the gold contained inside it.

A tendency towards decreasing silver contents in deposit gold from N (Egypt: Fatira and Gidami) to S (Sudan: Gebeit, Rigag-Sageit and Tibiri) reveals Fig. 4.1 (Goldanalysis). Currently, the explanation for this is merely based on assumptions relating to the different geologic settings. The gold contents in the Egyptian deposits are primarily based on hydrothermal exsolution of ophiolite series and their erosion sediments, in this case primarily the conglomerates and greywackes of the Hammamat series. The Nubian occurrences by contrast, derive mostly from so-called SEDEX deposits (sedimentary-exhalative deposits) bound to acid (rhyolite – dacite), submarine oxidised magmas. Due to extensive oxidation already on the sea floor of sulphide-rich, poly-metallic "black smoker" exhalation products that also contain small quantities of gold, slight amounts of iron oxide as well as insoluble gold with reduced silver contents remain. Through later hydrothermal leaching the gold was again dissolved, and re-deposited together with large amounts of quartz inside the vein cavities. This repeated dissolution of gold probably led to some loss of its silver contents, thereby explaining the observed lower silver grades in Nubian gold.

A large-scale and systematic, chemical survey of Ancient Egypt's gold artefacts unfortunately is still an awaiting implementation. The reasons for this for one the disinclination of the proprietors (museums and private collectors) to submit their collections to such investigations, even though modern analytical methods today are generally non-destructive. Moreover, high transport and

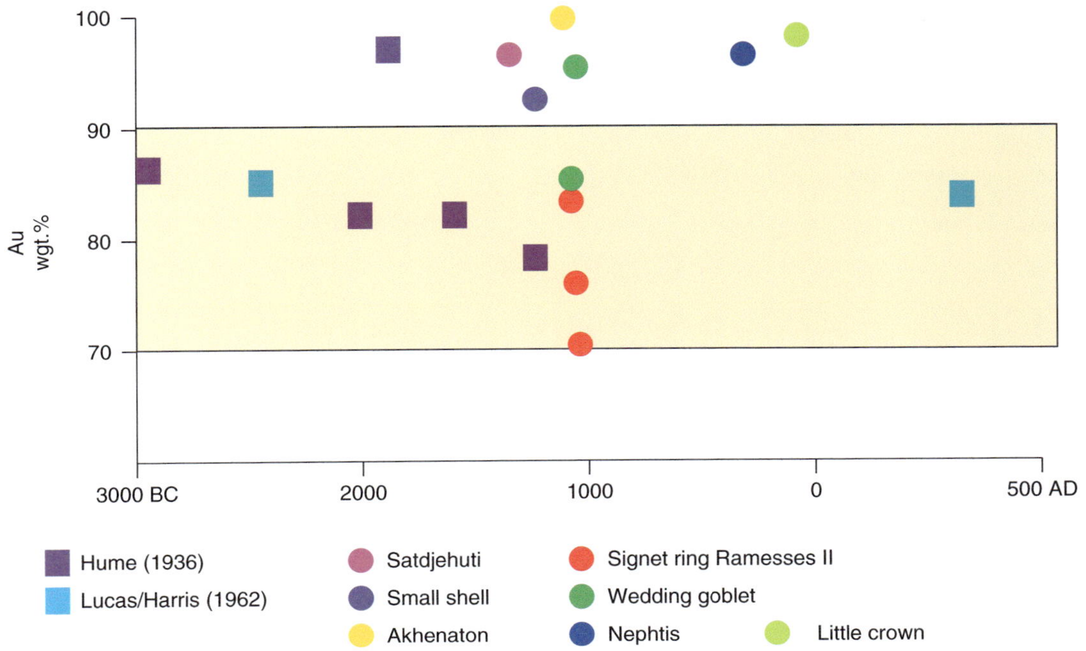

Fig. 4.2 Data from both, own and published analyses on ancient Egyptian gold artefacts. Hatched field represents average compositions of primary gold from Egyptian and Nubian deposits (analyst: A. Murr)

insurance costs brought about through the artefacts' wide dissemination would only add to the already considerable financial burden and time-consuming efforts such analyses entail.

Therefore, only preliminary details concerning the composition of Ancient Egyptian gold objects can be presented here as referred to in earlier reports as well as the data obtained from our own analyses.

Figure 4.2 (Gold analysis) displays a number of examined artefacts arranged in chronological order. Regrettably, virtually no exploitable data series are available from the Ptolemaic Period or later. In spite of their low number, it is nevertheless evident that the bulk of the data exhibit a relatively narrow range of chemical composition between 70 and 90 % wt. Au and 10–30 % wt. Ag. These values corroborate with the average proportions in the gold deposits of the Eastern Desert in both Egypt and Nubia. Apparently, the majority of the Egyptian artefacts were produced directly from the processed gold extracted from the gold deposits. The fact that some artefacts have low copper contents (gold analysis Figs. 4.3 and 4.8) is explained by the frequent copper sulphides or resulting malachite concentrations associated with the deposits, which remained as natural impurities inside the gold after the ore processing.

Nevertheless, some artefacts reveal significantly higher gold percentages than normally observed in primary, or so-called "mountain gold" deposits.

This may in essence, reflect one of the following three possibilities:

1. The artefact was made from placer gold like nuggets from river sediments. Through transport among the river boulders initially silver-rich gold is continuously "reforged", while the slightly more soluble silver decreases, especially near the surface of the nugget gold. The gold therefore becomes purer near the nugget surface. Indeed, the sections of nugget gold often reveal under the ore microscope slightly higher reflection intensities at the margins (gold-rich) than near its centre (silver-richer). Such nugget gold however, is only found in rapid flowing rivers with many boulders, which in the Egyptian Eastern Desert is inferred only for the relatively old Pleistocene wadi systems. This strikingly shimmering metal was already

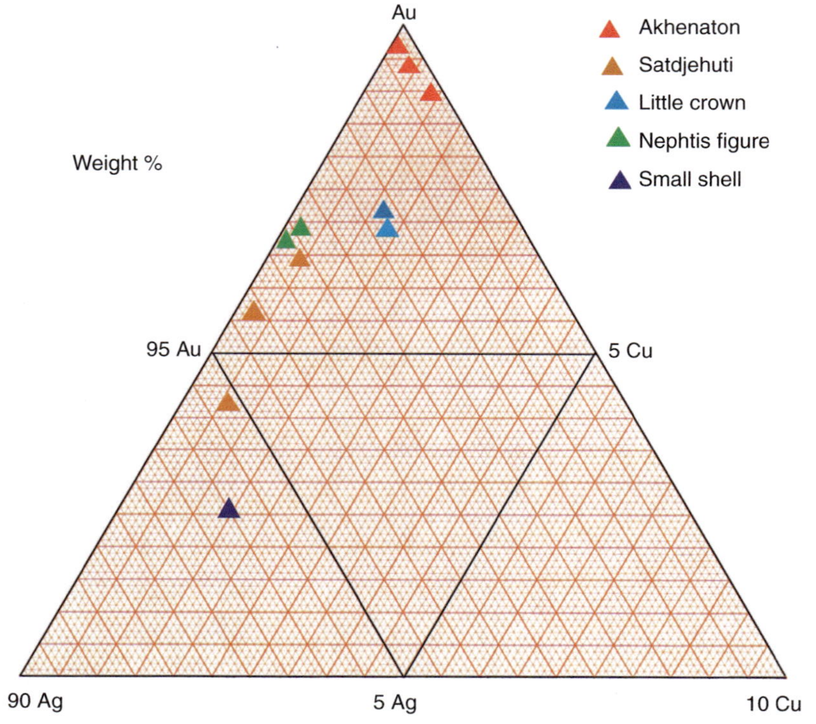

Fig. 4.3 Upper excerpt from a ternary plot showing the weight percentages obtained from microprobe analyses on the variables Au (gold)–Ag (silver)–Cu (copper) in artefacts with high gold percentages (analyst: A. Murr)

known to the earliest desert nomads, and probably even systematically collected. It therefore hardly surprises to find correspondingly high gold values in artefacts from these early periods. Characteristic for this type of gold are also frequently observed, fine platinum or platinoid inclusions in these objects, as for instance those reported by Ogden (1976). Such at first perplexing findings are nonetheless simply explained by the geologic environment with ophiolites in the catchment area of former nugget producing rivers in the Eastern Desert of Egypt and Nubia. Low contents of platinum may be contained in therein occurring serpentinite and similar ultramafic rocks. Through erosion they can end up in the river sediments where they are forged with the gravel transport into the much softer gold nuggets

2. In the discussions of the individual deposits we repeatedly refer to hydrothermal fluids containing gold-sulphides that migrate into fault systems. They first react with the wall-rock where iron contents are dissolved. Pyrite (FeS_2) and free gold then form in a subsequent reaction between the iron and the sulphur-enriched fluids and gold-sulphide complexes. Hence, small amounts of gold are incorporated to the growing pyrite lattice and later released again in oxidation and dissolution processes of the surface-near pyrite minerals. The gold is therefore set free, but in the meantime has lost much of its original silver contents in a so-to-speak natural refining process. Whereas with ancient mining technologies intrinsic gold could not be extracted from pyrite, it could on the other hand be relatively easily recovered at the oxidised surfaces of these alteration zones. In fact, many of the gold occurrences that had been mined in antiquity and that we were able to visit revealed extraction traces exclusively within these decomposed alteration zones, whereas the actual quartz vein had often remained untouched by oxidation. In the case that the gold from such occurrences was processed and melted apart, its higher levels of fineness

would accordingly have been preserved. Such high-quality gold, which excelled through a more "golden" hue and an improved ductility did of course not escape the attention of the Ancient Egyptian goldsmiths.

3. The third explanation is artificial gold refining. Such processes have repeatedly been described by ancient authors. The oldest (now lost) description of an applied gold refining process in Egypt dates to the late second century BC. It had been compiled by Agatharchides of Knidos and preserved as a copy in the Bibliotheca Historica by Diodorus about 100 years later. His descriptions which concern both, the work processes involved in mining operations and particularly that of gold refining, seem however so implausible that doubts may be raised as to whether he actually witnessed the processes, or whether this was only hearsay. In his own experiments, Notton (1974) was able to demonstrate that the ingredients lead, zinc, and barley bran added by Agatharchides had absolutely no effect on the refining of gold other than an undesirable one, especially as concerns the latter ingredient. Sodium chloride (NaCl) and earth, or crushed ceramics used for cementation processes on the other hand are essential. Accordingly, silver-containing gold flakes were heated in a sealed clay pot together with an amount of pottery shards and rock salt to about 800 °C for at least 24 h. In a first step sodium chloride (NaCl) reacts with traces of H_2O via HCI with silver to form gaseous silver chloride (AgCl). The silver chloride is then diffused out of the gold via the grain boundaries and fine channels to deposit at the vessel wall and the surface of the pottery fragments, which in another operation can be used for recovering the silver contents. Whether however, such processes had actually applied to extract silver in Ancient Egypt, is unknown. In his first cementation attempt Notton (1974) in any case obtained a grade of already 93 % Au from an electrum sample consisting initially of only 37.5 % Au. Of course, in order to obtain higher grades a repetition of the process is required.

Through a fortunate incidence, we were able to obtain a tiny sample of redundant gold foil after restoration work on a gold sheet applied to the lower part of Akhenaten's coffin. We decided to submit it to electron beam microanalysis, which produced gold values above 99 % (Fig. 4.3). Because of the unexpected result, it was additionally examined from all angles under the scanning electron microscope (SEM).

It turned out that the gold foil was composed of several forged, porous plates (Fig. 4.4). At the top side, dense twin lamellae were discovered, apparently resulting from the mechanical stress exerted during hammering of the sheet (Fig. 4.5).

At even higher magnifications one sees next to the twin lamellae characteristic, partially effaced holes between 1 and 3 μm in diameter (Fig. 4.6) representing nothing else than the diffusion channels through which the silver chloride was set free during the cementation process. In addition, a widening of the grain boundary is visible.

A comparison of this image with that of a Sardian (Lydia) gold foil from about 550 BC and refined according to the cementation technique (Fig. 4.7), in spite of the different magnification scales, hardly reveals any differences, especially considering that the Sardian plate had not been chased and therefore has no twin lamellae. Figure 4.7 represents a SEM micrograph of a gold sheet sample from Sardis refined in the cementation method. According to Geçkinli et al. (2000), it contains only 1.1 % Ag (sample 9A).

If one imagines the Sardian gold plate in Fig. 4.7 hammered out to 5 times its actual surface, then the holes and the expanded grain boundaries would of course become smaller, and the features in the image would become identical to those in Fig. 4.6. Since the described structures are characteristic for gold sheets refined in the cementation method, the technique is attested to in Ancient Egypt at the latest to the reign of Akhenaten (~1350 BC), i.e. about 750 years earlier than assumed for gold refining at Sardis (Ramage and Craddock 2000).

The Museum of Egyptian Art in Munich (Staatliches Museum Ägyptischer Kunst in

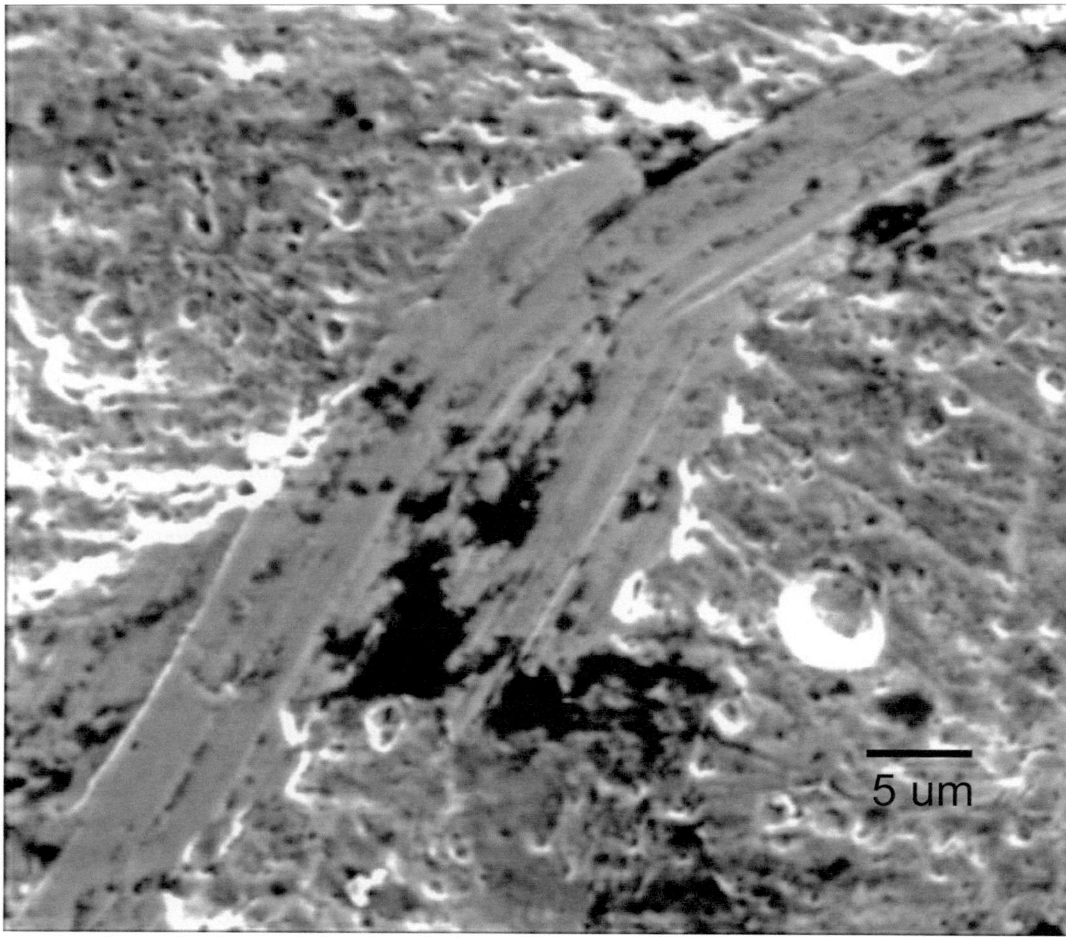

Fig. 4.4 SEM image (Scanning Electron Microscope) of a gold foil appliqué from the sarcophagus of Akhenaten. The analysis showed that it was composed of different gold sheets

München) generously provided additional, small gold sheet fragments from the death mask of princess Satdjehuti Satibu (~ 1550 BC) for micro-analysis. In this case, values between 94 and 96.5 % Au, 3.5 and 5.8 % Ag, and around 0.5 % Cu were established (Fig. 4.3). Although the microstructure revealed no clear evidence for the cementation method, its relative high purity at least suggests a refining process only for the production of the fine sheet gold for this high nobility death mask.

Because the malleability or ductility of gold increases with its purity, the Ancient Egyptian goldsmiths had been particularly interested in working with high purity gold, especially when it came to making very delicate objects like ultra-thin gold foils. Since gold hammering is attested to though iconography to around 2300 BC, e.g. at the Unas ramp and in the tomb of Mereruka (Fig. 4.9), one is tempted to believe that gold refining was in fact practiced already at this early stage, even if the mentioned representations are viewed as relative to the production gold foil rather than gold leaf (Drenkhahn 1976). Yet, additional data from analyses on contemporaneous gold foils have remained unavailable.

A shell pendant dated to the early eighteenth dynasty, also from the Egyptian Museum in Munich (ÄS 5301) revealed contents between 85 and 93 % Au and 1 % Cu (Fig. 4.3).

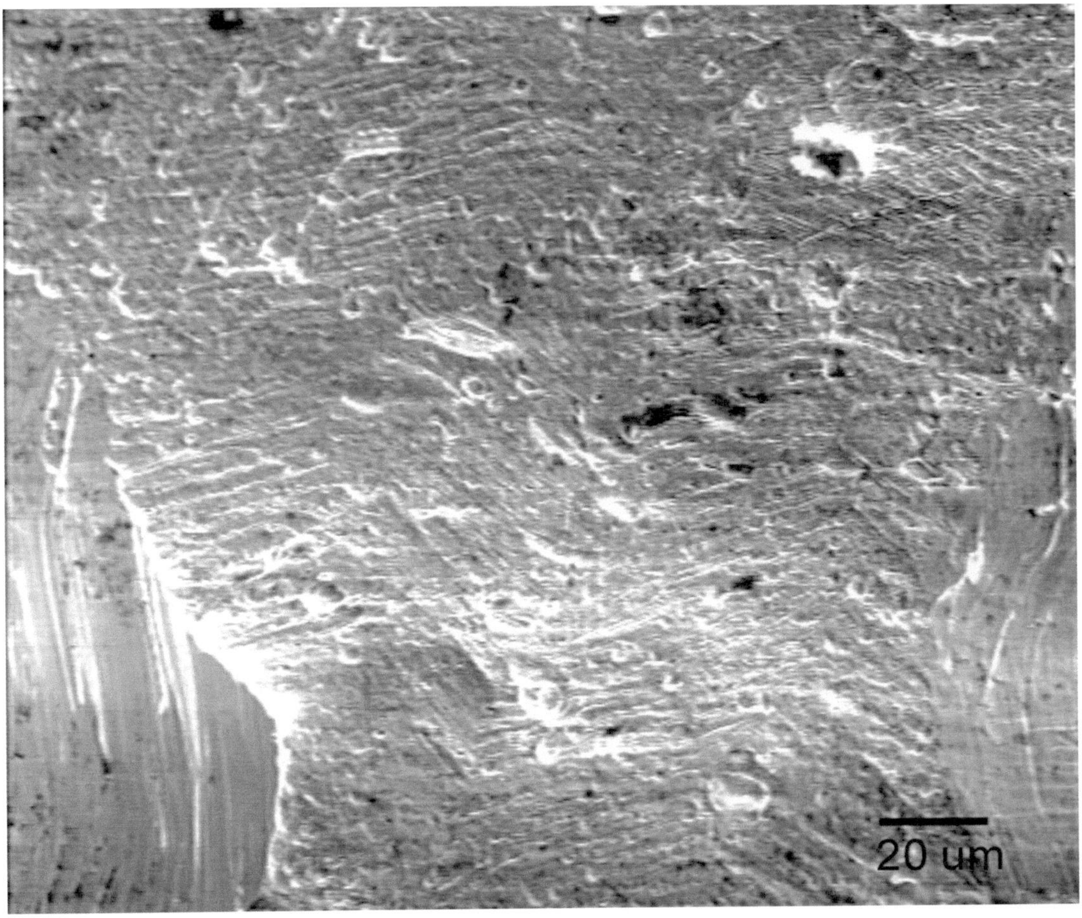

Fig. 4.5 SEM image of the same gold foil appliqué from the sarcophagus of Akhenaten. The image reveals an intensive twinning lamination of the sheet's singular gold grains, apparently resulting from mechanical stress through beating

As already discussed above, its high purity gold may also have been retrieved from the decomposed and hence copper-depleted alteration zone along a quartz vein.

This may also be the case for the mentioned death mask of Satdjehuti Satibu as well as for a small golden crown and a small nephtis- figurine with very low copper contents, which both date to the late periods.

Another remarkable example of probably refined gold from the time of Ramesses II (~1250 BC) are represented by the fragments of a wedding goblet with a bilingual inscription in Egyptian hieroglyphs and Hittite cuneiform for the occasion of the (erased) marriage between a Ramesside prince and a Hittite princess. One fragment consists of 95 % Au, 3–5 % Cu, and another of 85 % Au, 11.5 % Cu and 3.5 % Ag (Fig. 4.8). This object laden with high political symbolism was most likely made from refined gold.

Interesting data were obtained from a seal ring belonging to Ramesses II (ÄS 5851), whose different parts revealed proportions between 69 and 83.5 % Au, 3 to 5 % Cu, and 13.5 to 25.5 % Ag. Considering the importance of this royal object, the heterogeneity of its composition seems at first quite surprising (Fig. 4.8). On the other hand it is only understandable that the goldsmiths had preferred to use a lower grade gold alloy for this object, whose purely functional purpose after all remains evident. As also exemplified by the

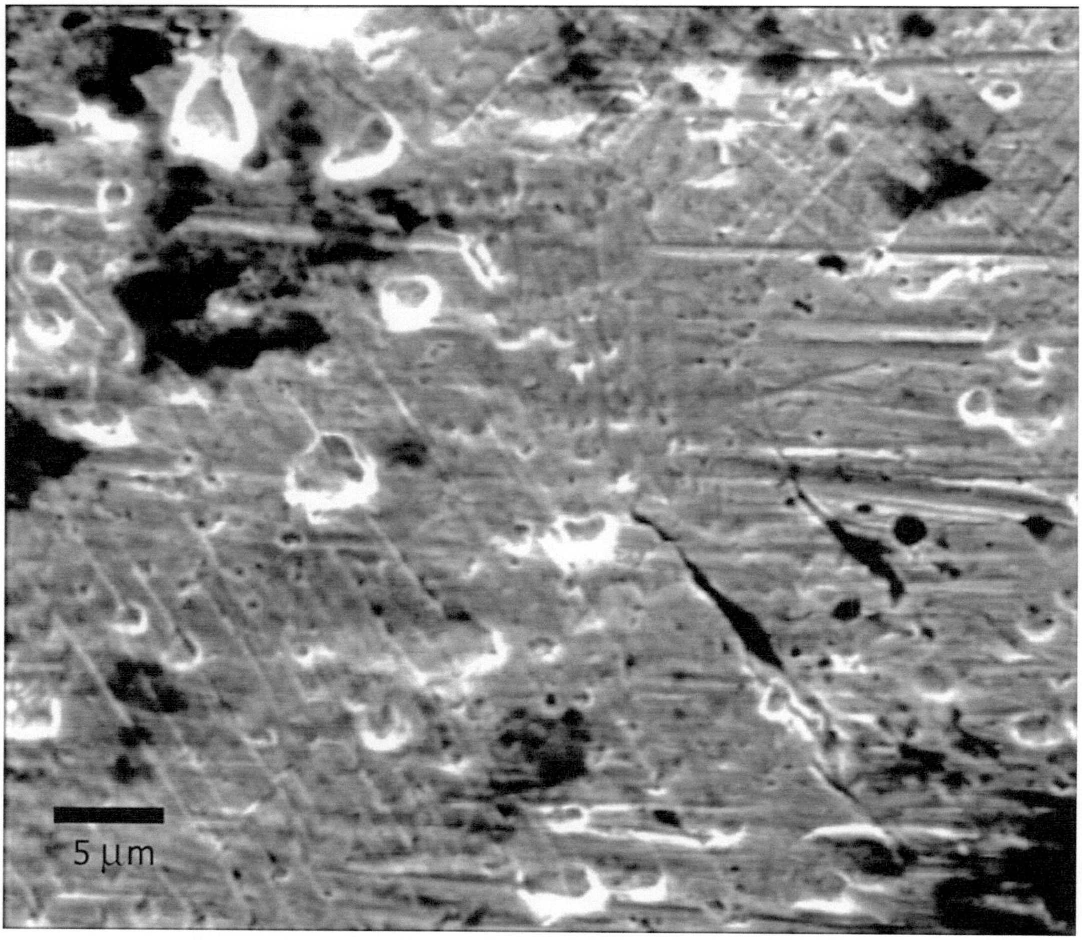

Fig. 4.6 Enlarged detail of the gold foil appliqué from the sarcophagus of Akhenaten. In addition to the twinning lamellae one also notices numerous 1–3 µm large perforations (SEM image)

diverging gold grades in the aforementioned wedding goblet, the goldsmiths had probably often preferred to use different gold components rather than melting them down to a homogeneous mass.

These data of course may hardly account for making any representative statement on the metallurgical methods applied in Ancient Egypt. The knowledge of gold refining technologies as early as the New Kingdom represents in any case a pioneering feat, predating by a millennium the equivalent technical knowhow by the Lydians, who so far have been viewed as the inventors of such processes (Ramage and Craddock 2000).

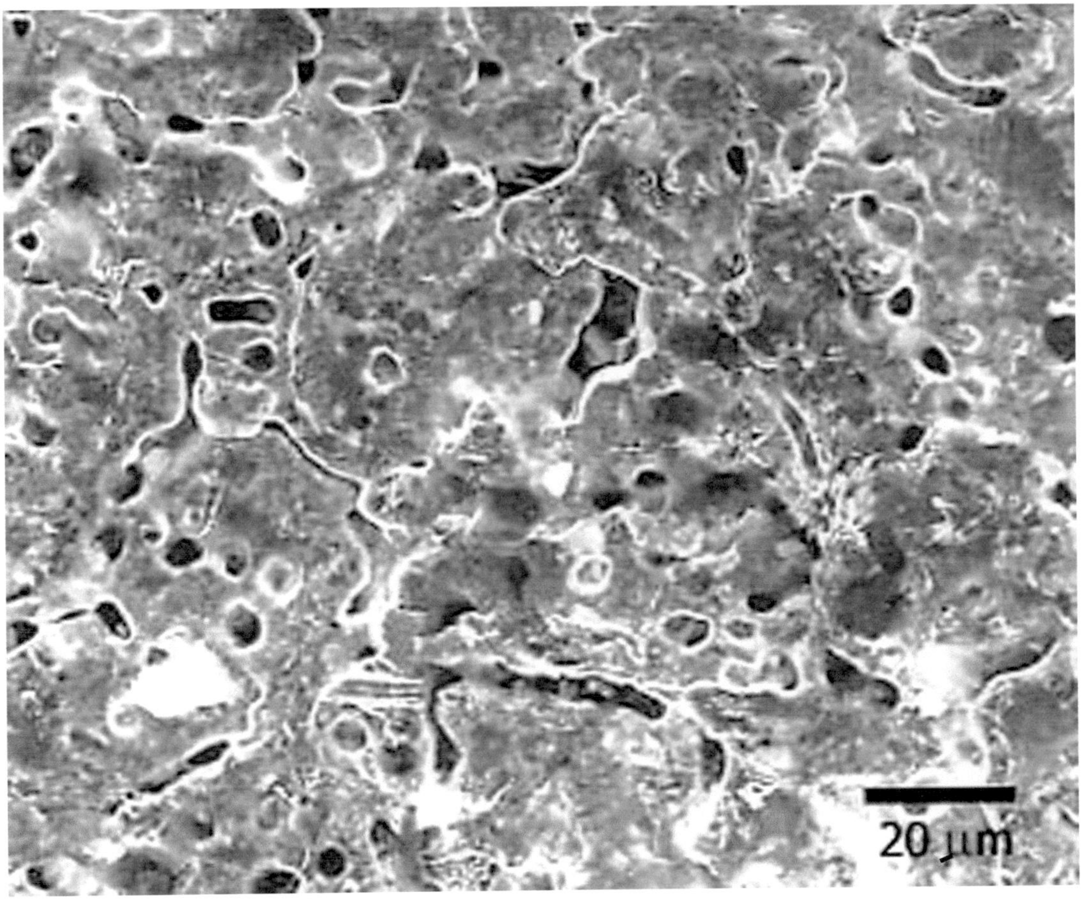

Fig. 4.7 Gold sheet from Sardis (Lydia) refined according to the cementation technique (550 BC, slightly modified after Geçkinli et al. 2000). Note the diverging scale to Fig. 4.6. (SEM-image)

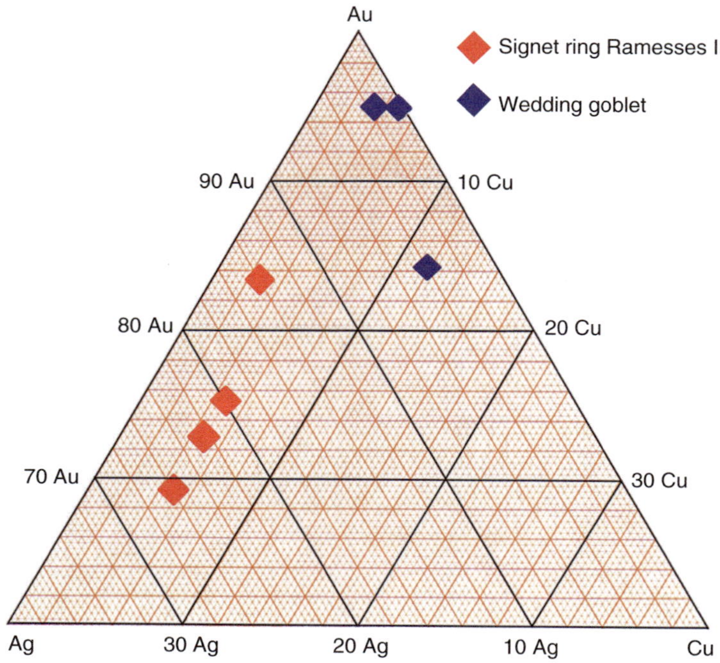

Fig. 4.8 Upper part of a ternary plot opposing the Au (gold)–Ag (silver)–Cu (copper) percentages in a signet ring of Ramesses II and an Egyptian-Hittite wedding goblet. Both artefacts reveal relatively heterogenous gold compositions which indicate different smelting phases (analyst: A. Murr)

Fig. 4.9 Relief inside the tomb of Mereruka (~2300 BC) depicting a scene (*to the right*) with an inscription mentioning the "beating of gold", thus hinting to the manufacture of sheet gold. It is not clear however, whether this implied the use of refined gold, especially as the hieroglyphic symbol is sometimes used to denote electrum, an argentiferous gold

5 Gold Production Sites and Gold Mining in Ancient Egypt

The here presented compilation of ancient gold production sites adheres to their geographic distribution from N to S, as rendered in the overview map in Fig. 5.1. It is divided up in seven groups, which by and large follow the arrangement by Hume (1936).

5.1 Most Northern Group

5.1.1 Wadi Dara (Old Kingdom Settlements and Mines)

Geographic position:	
Main settlement:	27°54′53′ N, 32°51′02″ E
Subordinate settlement:	27°55′11′ N, 32°51′05″ E ???
Smelting furnaces:	27°55′38″ N, 32°51′27″ E

After our visit to Wadi Dara, a French-Egyptian excavation team explored the district more closely in the early 1990s (Tawab et al. 1990; Castel et al. 1992; Grimal 1991, 1994). The chief observations made by this group are therefore taken into account in the following discussion.

The mining district of Dara 1 is divided into two main extraction zones on both sides of the here relatively wide Wadi Dara running from NE to SW (Fig. 5.2).

The basin-like situation at this site owes to the overlapping of two fault systems of which the older one is oriented in an ENE-WSW direction, overprinted by a later one oriented in a NW-SE direction.

The main settlement dating to the Early Dynastic and Old Kingdom stretches along the northwestern flank of the wadi (Fig. 5.3) and consists of three relatively large clusters of round huts as well as smaller ruins of isolated huts, about 600 m to the NW.

Two sites are located at the foot of the mountains in which several noticeable NW-SE oriented rock cuts and incsions revealing the former mines especially on the SE mountain slopes (Figs. 5.2 and 5.4). They follow mined mineralised quartz veins, mostly visible as open trenches and occasionally as deeper underground operations.

A closer inspection reveals green malachite layers along the wallrock of the small and parallel running pit trenches, which evidently points to primary copper contents in the quartz veins. Such features therefore at first suggest former copper mining activities, as in fact confirmed by the excavators (Castel et al. 1992, 1995). They were able to record ore extracting and processing through the discovery of tools (grooved stone axes and pounding slabs) and more importantly, of smelting sites and slags from settlement contexts. At a distance of about 1.5 km away from the settlement the excavators located the remains from furnaces on a slope, which forming a natural wind channel, unmistakably authenticated the evidence for early copper mining at Wadi Dara. The walls of the round huts cleared by Castel et al. (1992, 1995) revealed surprisingly well-executed masonry.

R. Klemm, D. Klemm, *Gold and Gold Mining in Ancient Egypt and Nubia*, Natural Science in Archaeology,
DOI 10.1007/978-3-642-22508-6_5, © Springer-Verlag Berlin Heidelberg 2013

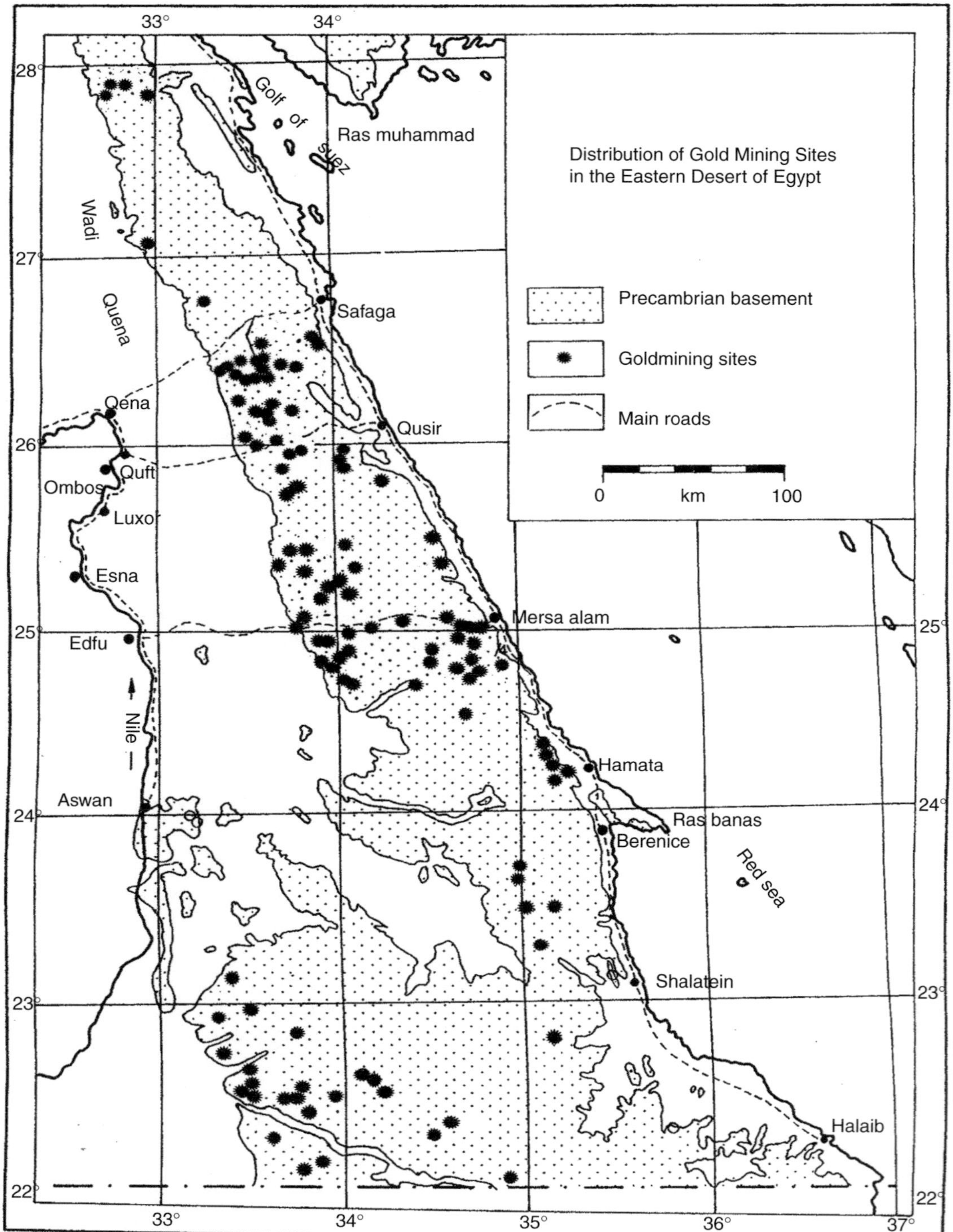

Fig. 5.1 Distribution map of anciently exploited gold occurrences in Egypt. After EGSMA (1986) and own fieldwork

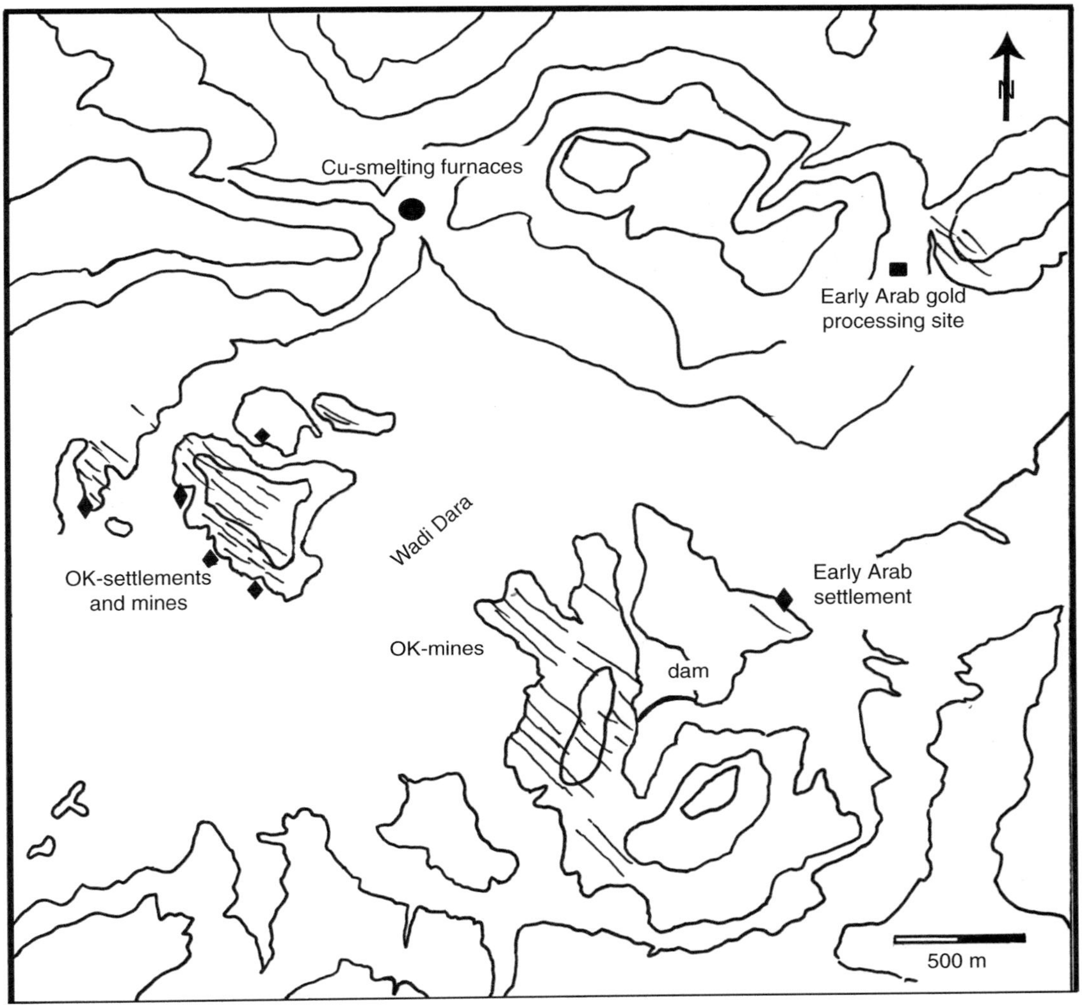

Fig. 5.2 Distribution of Old Kingdom and Early Arab Period settlements and copper/gold processing sites in Wadi Dara (R. Klemm)

In most cases, the huts had smooth floors with rectangular depressions, masoned beds, as well as small niches set into dry stone walls (Fig. 5.5).

Although there is no clear evidence for gold mining in addition to the Early Dynastic and Old Kingdom copper industry at Wadi Dara, it cannot entirely be ruled out either.

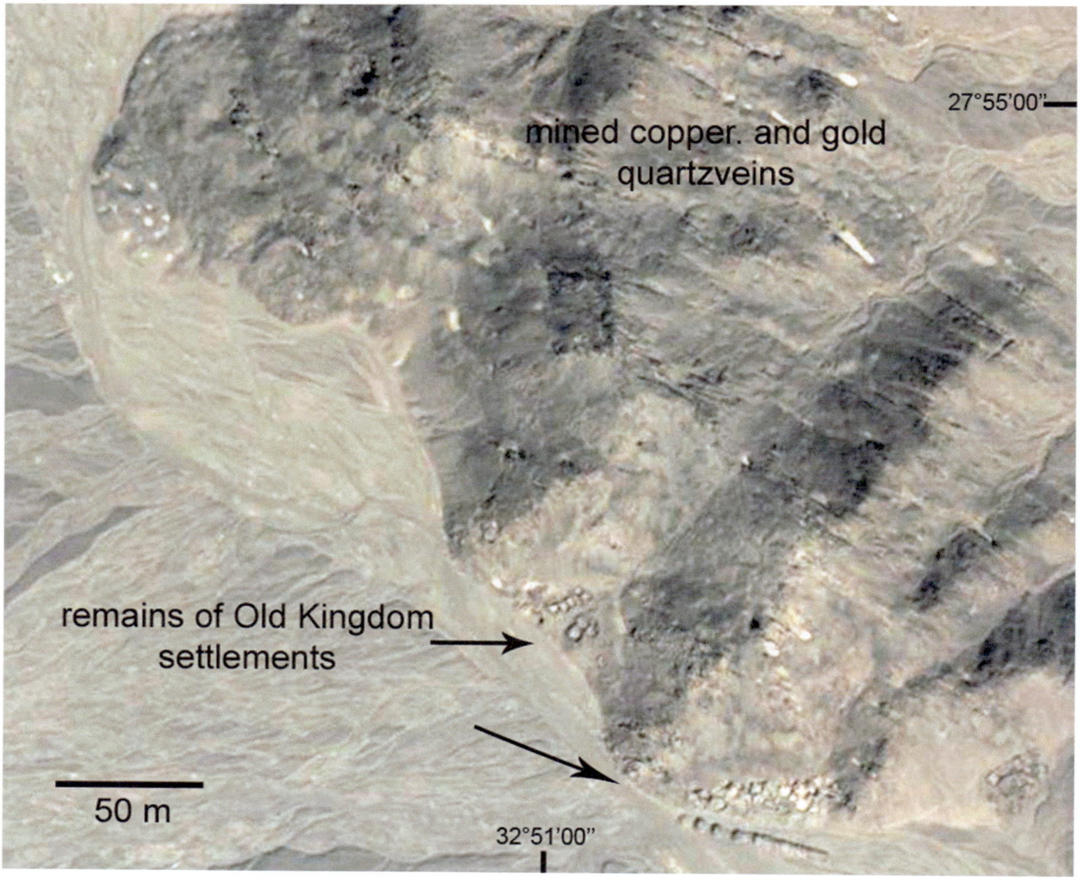

Fig. 5.3 Old Kingdom settlements and formerly exploited copper/gold quartz veins in Wadi Dara (modified Google-Earth image)

Fig. 5.4 During Old Kingdom exploited copper/gold quartz veins along the southern flanks of Wadi Dara

Fig. 5.5 Early Dynastic/Old Kingdom dwelling hut at Wadi Dara with carefully masoned walls. The upper, defensive wall is recent

5.1.2 Wadi Dara (Early Arab Period)

Geographic position (prayer site):	27°54′49″ N, 32°52′24″ E

About 1 km to the E of the copper extracting site 12–15 huts with partially well-preserved walls scatter over a distance of 300–400 m along the wadi edge. In contrast to the houses at Dara 1, these dwelling houses had probably also served as workshops, as demonstrated by evidence from various stages of the working process recorded within their walls. It comprised waste dumps containing ore chunks, quartz gravel crushed down to pea-size, powdered tailings, and slags from smithing, as well as tools (Fig. 5.2).

The abundant pottery and an open prayer site oriented SE and spreading-out over approximately 4 × 10 m in front of the site, date this settlement to a later occupation in the Early Arab Period. However, the whitish quartz lumps and the tailings unmistakably show that by now gold and no longer copper was being processed here.

Four rectangular buildings respectively measuring around 3 × 4 m with 60–80 cm wide face-shell walls are located in the northern part of the area. They exhibit stately, approximately 1 m wide entranceways, in one case with a monolithic door threshold. At some distance further N are the remains of another building, adjacent to which are a grave and a small prayer platform on an alluvial terrace.

This Early Arab Period occupation zone also includes a deep-mine forming an elongated, narrow cut, which follows a quartz vein leading towards NW above the settlement, along the eastern slope of the wadi (Fig. 5.6). At places it is 8–10 m deep, although much of it is buried under debris today. The mine was protected against collapse through several rock buttresses. The waste dump material as well as sections of the mine itself display a distinctive, green staining. Because such strikingly narrow mines are typical for very early mining and in addition, the clearly visible, green malachite linings at the wallrock had not only been key criteria for early copper prospectors but also for gold seekers, mining at first was possibly concentrated on copper. However, apart from a miner's dolerite hammer, no further evidence to support this was recorded at the mine's surface.

All remaining identifiable surface features in the area also date to the Early Arab Period. As it appears, already exploited quartz veins of earlier copper mines were checked and mined again in this period, but this time for gold. Nevertheless, gold was plausibly first and foremost retrieved from the alluvial rubble deposits in the wadi bed that had escaped the attention of the earlier miners.

The geologic surroundings of this copper and gold mining area consist of an intrusive quartz-diorite traversed by NW-SE striking shear zones exhibiting quartz mineralisations.

These distinctly visible quartz veins, whose thicknesses vary between 0.5 and 1 m, all strike NW-SE in an almost vertical dip. The mined quartz is milky white, and the centre of the vein contains calcite. Among the sulphide minerals, low contents of pyrite and to a lesser extent chalcopyrite could be observed.

A water retention complex totally silted-up with fine debris presents interesting features for the study the former processing technology. It is located in a small valley basin in the neighbourhood of the Arab settlement (Fig. 5.2) and had evidently served for collecting runoff after sporadically occurring rain showers. Such devices had been essential for both gold processing and the availability of drinking water. There are no clear indications as to the date of this small, curved dam, although other dams of this type, which seem exclusive to Early Arab gold producing districts have attracted our attention.

5.1.3 Umm Balad

About 10 km SW of the Wadi Dara mining district one arrives at another vast copper and gold mining area (Fig. 5.7).

After our survey in Umm Balad, the mentioned French-Egyptian team again carried out systematic geologic and archaeological investigations in the area, producing even more interesting results.

Fig. 5.6 Narrow trench pit at Wadi Dara from the Early Dynastic/Old Kingdom Period with clear copper linings (*bottom left*) and green stains at the wallrock

In the following we therefore also refer in addition to our own observations, to its work (Castel et al. 1998).

The mining area was formerly accessed through Wadi Qena, a wide wadi further to the W which is recognised from the mountains. Today one reaches the site by car in turning W towards the desert mountains at Ras Gharib as one arrives from the Red Sea road.

5.1.3.1 Umm Balad 1

Geographic position:	27°51′08″ N, 32°47′09″ E

The ruins of 50–60 simple, round huts measuring between 1.5 and 2.5 m in diameter, lie scattered, predominantly on the N side of the upper reaches of Wadi Umm Balad. In this area, the wadi is relatively narrow as it turns E, with its bed heavily

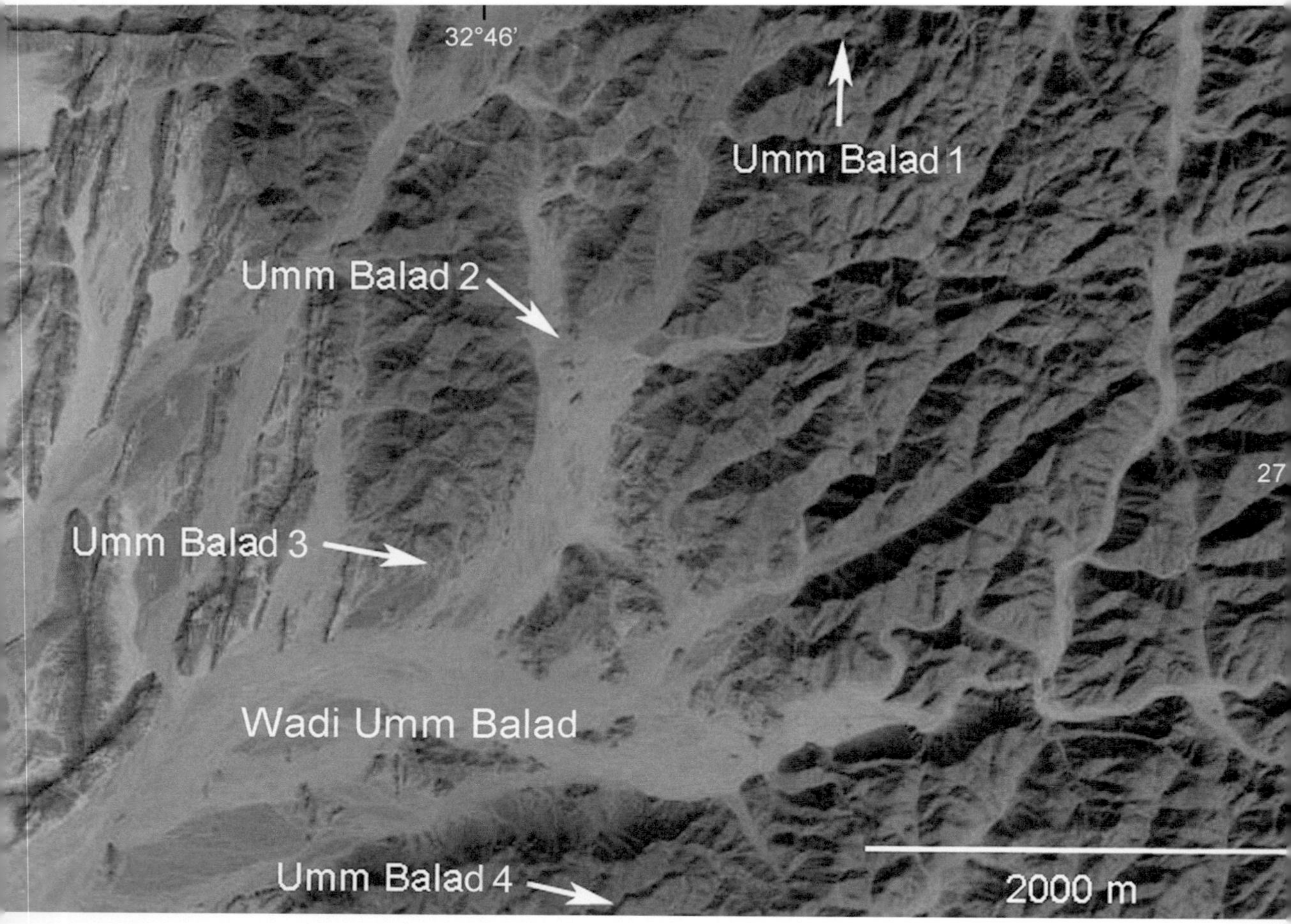

Fig. 5.7 Settlements and copper/gold deposits in Wadi Umm Balad (enlarged aerial photograph)

covered with river-rounded boulders. Round work platforms mingling in an irregular arrangement between the buildings are covered with fine sand and partly confined by simple stone alignments. From a distance, they are recognised as white patches in the landscape. Because of the narrowness of the wadi and especially the relatively high precipitation, as perceived from the gullies and a considerable rubble accumulation, the ancient miners had preferred to invest the slopes for both dwelling and working purposes (Fig. 5.8).

Tools lie scattered on the ground in- and outside the buildings. They were identified as mostly fragmented cube-shaped anvil stones and rotary mills, including the rotor discs.

Nearby, in the middle of the wadi is a very well-preserved installation for washing and processing stoneground gold ores (Fig. 5.9). It consists of an inclined stone socle in the centre of a largely intact water circulation system. Its function was to recycle water as it ran down from the inclined surface of the washing table. After being collected in a semi-circular basin it flowed through an approximately 6.5 m long, slightly curved channel back to the other end of the table where it was gathered in a 60 cm deep shaft.

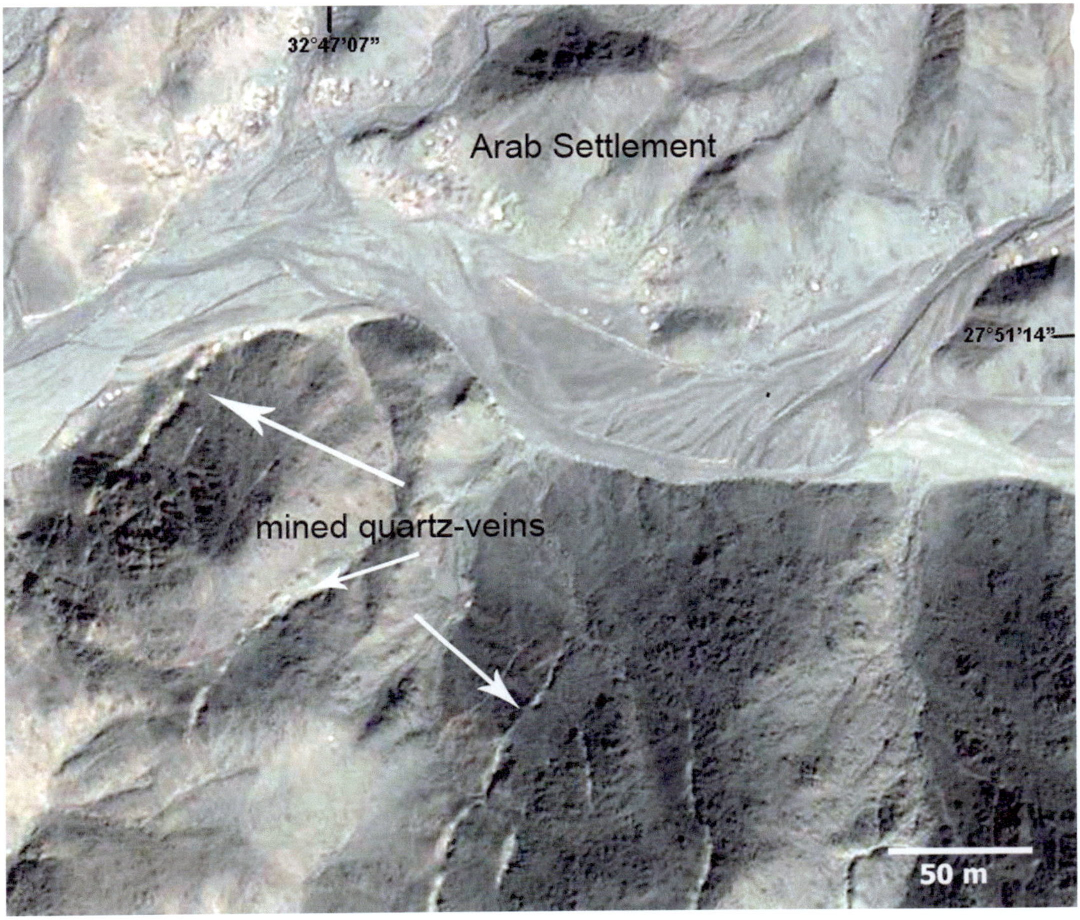

Fig. 5.8 Locations of the exploited auriferous quartz veins in relation to the Early Arab Period mining settlement at Umm Balad 1 (modified Google-Earth image)

A band of red tailing sands was observed around the entire installation.

A total number of four such washing tables were recorded within a distance of about 250 m. The westernmost one is characterised by two collecting shafts, which were probably connected to each other by an overflow system, through which gold quartz sands could accumulate at the bottom of the first shaft. Traces from recent soundings into these shafts indicate that samples had been taken for analytic gold checks, i.e. that even recently the shafts had been suspected to hold significant gold yields. Some round but no oval stone mills as known from the New Kingdom were found close to the installations. The few recorded pottery shards date without exception to the Early Arab Period. The entire zone has been disturbed to such an extent that most of the visible archaeological context is no longer preserved.

In a side wadi branching-off to the NE and at the eastern end of this zone is a large complex

Fig. 5.9 Washing table for gold processing in the Early Arab Period settlement at Umm Balad 1

with thick walls, consisting of two rooms and an added terrace on its eastern side. Nearby, there are several round millstone bases from light-grey granite and diorite and a small number of mostly fragmented rotor discs.

At the southern side of the wadi is an approx. 4 × 5 m large prayer site with a mihrab made of set stones. The niche points SSE, the direction towards Mecca (Fig. 5.10). At least three more, partly severely damaged prayer sites are still discernible in the surrounding area of Umm Balad 1.

In the Early Arab Period mining operations had concentrated around a deep mine following a gently sloping quartz vein ("filon K" in Castel et al. 1998). With an average content of 5 g/t (Castel et al. 1998) a total yield of 5–10 kg gold was estimated for this vein only. Four more quartz veins can be recognised in the area. Apparently, they had only been checked through shallow

Fig. 5.10 Muslim prayer site from the Early Arab Period at Umm Balad 1

scraping, and therefore not properly mined. The veins had been cleared in open trenches between 2 and 3 m deep. Some pounding stones and slabs are known from nearby areas, but no mills.

Some chalcopyrite was identified within the trial trenches. Apart from haematite, neither sulphide minerals nor traces from visible gold were observed in the mined seam itself.

The vein had been exploited over its entire width, although excluding the alteration zones next to the wallrock. At regular intervals of about 4 m, abutments had been left standing.

The mined vein quartz, which had formed in at least three generations, appears mainly in basic (formerly basaltic) metavolcanics. The fissure joints touching the quartz seam are intermittently filled-in with thin malachite layers. The vein quartz itself is to some extent clearly brecciated, while the breccia hollows tend to fill with haematite. The alteration zones of the wallrock concentrate around an average width of 15 cm (in few cases up to 25 cm) and are generally marked by epidote and sericite formations. In the surrounding area of the main quartz vein, the wallrock is sheared intensively, and the gaps are mostly filled with quartz. Lenticular and sheared metavolcanic rocks also appear in the quartz veins themselves. In such cases too, they display the same intensively altered minerals. A diorite (now affected by metaporhism) had penetrated into the metavolcanic sequence. Its grain size decreases clearly towards the wallrock, while up to 5 cm large hornblende and plagioclase crystals can be observed near its centre.

5.1.3.2 Umm Balad 2 (Fig. 5.7)

Geographic position:	27°50′19″ N, 32°46′16″ E

In the immediate vicinity of the four or five trial quarries, no clear-cut archaeological findings were made. This uncommon feature at ancient prospecting sites might suggest that this site is of a recent date.

These undatable trial mines follow andesite dikes, which develop in diorite environments. Close to the mostly heavily fractured andesite seams the diorite wallrock has altered to K-feldspar and is silicified. This occurs to such an extent that

Fig. 5.11 Ergonomically shaped one-hand hammers from dolerite at Umm Balad 3. Dated between the Early Dynastic to Old/Middle Kingdom Period

small quartz veins appear parallel to the andesite dikes, in which haematite and residues of chalcopyrite may appear. Primarily however, limonite inclusions with thin malachite interbeddings appear in such small quartz veins, suggesting a former presence of chalcopyrite minerals.

The presence of the malachite layers along the joints in the andesite and altered diorite wallrock had therefore probably been the reason for the mentioned trial mines.

5.1.3.3 Umm Balad 3 (Fig. 5.7)

Geographic position:	27°49′53″ N, 32°46′59″ E

Numerous stone hammers with ergonomically well-shaped handles from dolerite and local diorite lie scattered in an impressive debris spoil dump just below a carefully sealed adit at a mined quartz vein. They no doubt represent the former mining tools. The hammers date to a period between the Early Dynastic Period and the Old/Middle Kingdom (Fig. 5.11).

Unfortunately all shaft mouths measuring between 1 and 1.5 m are blocked by large rocks and therefore making any closer investigation of the mines difficult (Fig. 5.12).

The ruins of several roundish huts aligned side by side in a parallel arrangement to the former shaft were recorded directly opposite the collapsed openings on noticeably large debris heaps. The buildings divide into one or more rooms and cluster in two areas separated by a large spoil heap. The cluster on the western side (zone 2, B and C in Castel et al. 1998) consists of 50–60 huts, the eastern one (E and F) of about 40. In fact, according to the identification of the ceramics, the buildings date to a period between Nagade III and the fourth dynasty (Köhler in Castel et al. 1998). A small pottery shard was recovered with the engraved symbol of the Horus falcon above palace façade.

It probably reads "house of Horus" or Hathor. This makes sense in so far as Hathor is often associated with forces protecting mines (Köhler in Castel et al. 1998).

Fig. 5.12 Deep mine at Umm Balad 3

According to these findings, the main period of operations within the mostly blocked and silted up mines in this well-preserved mining area may therefore have taken place in the early part of the Old Kingdom. In those days, copper production had probably been the main purpose of the mines, especially as Castel and his team succeeded in recording several copper smelting furnaces in Wadi Umm Balad as well as in neighbouring Wadi Dara (zone 3, Castel et al. 1998).

The surrounding hills are speckled with old landmark towers, the so-called Alamat.

The mineral quartz vein, which had undoubtedly been exploited during the Old- and Middle Kingdoms and much later in the Early Arab Period, was bound to a shear zone system running into an andesite-basalt dike in a medium sized diorite, and altering the latter in an approximately 2–3 m thick zone. The andesite-basalt in its turn, is silicified and epidotisised along its shear clefts.

The quartz vein with its apparently primary copper-sulphide ore contents has completely oxidised to limonite through surface decomposition, while the shear structure of the wallrock is partially stained through thin malachite layers. The malachite content is nonetheless too low for reconstructing substantial copper mining for the periods concerned.

The cautious sealing of the numerous shaft openings are noteworthy. In six cases, the entrances are still distinguishable. Additional six to eight vein systems located further N, above the main vein had been thoroughly checked in antiquity, though without resulting to any systematic mining. These veins too, are located within shear zones directly overlying the diorite.

5.1.3.4 Umm Balad 4 (Fig. 5.7)

Geographic position:	27°48′45″ N, 32°48′24″ E

As opposed to the areas Umm Balad 1–3, this district has by and large remained unscathed.

The layouts of about five ruined buildings (5–6 m long and 3–4 m wide) are discernible at the northwestern wadi flank, but which are partly buried by slope rubble. In the vicinity of the mined pits, as well as within the house ruins there are numerous ergonomically-shaped fist hammers

Fig. 5.13 Grooved mallets (*arrow*) and ergonomically shaped fist mallets from Umm Balad 4

of stone (lengths 20–30 cm, diameters 5–8 cm) (Fig. 5.13). No diagnostic pottery could be recorded at the site's surface and especially, no remains from millstones. The findings nevertheless suffice to date the site to a period between the Old- and Middle Kingdom.

Three quartz veins show traces from mining. They measure between 0.3 and 1 m in width and are located in foliated shear zones in fine-grained granite. The main vein splits up forming a satellite vein, which however, was mined over a short distance only. Its very scanty ore mineralisation is primarily composed of chalcopyrite ($CuFeS_2$) and to a lesser extent of digenite (CuS_2). Nevertheless, in the mined parts close to the surface, the former sulphide minerals have oxidised almost completely to limonite [FeO (OH)], and the remaining copper contents appear in the form of green malachite $\{Cu_2[(OH)_2CO_3]\}$ in the shear clefts of the surrounding, fine-grained granite.

The extraction operations had followed these almost vertical quartz seams, which as they lead from the valley flanks form slit-shaped thrench pits, even today recognisable over lengths of 5–10 m and a maximum depth of 4 m. Even so, the total mining depth cannot be measured due to the debris fill. Because of the total absence of millstones, it may be assumed that mining had concentrated exclusively on copper ores, although the very low quantities of primary chalcopyrite tend to contradict this assumption. In spite of the early date, the undersized waste heaps suggest only little mining activity here.

Even though the Russian company SMW, which recently carried out thorough gold prospecting work in the Umm Balad area, was able to report gold grades of up to 232 g/t, (Gold 2009), it decided against initiating any further mining operations here.

5.1.4 Wadi el-Urf (South of Gebel Mongul)

The partly well-preserved mining district to the S of Gebel Mongul has been under exploitation ever since the Predynastic Period. As in the

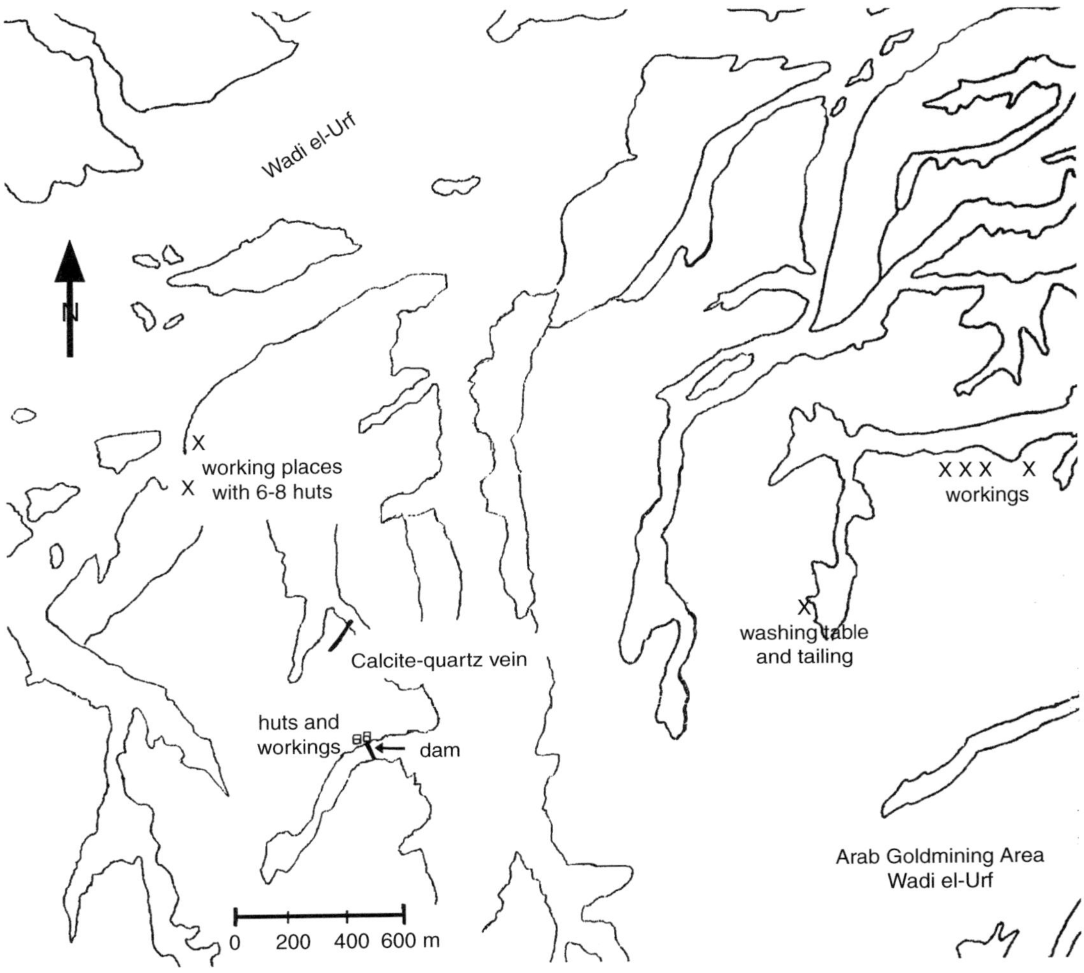

Fig. 5.14 Archaeological sketch map of the Wadi el-Urf-mining district (R. Klemm)

neighbouring districts of Umm Balad and Wadi Dara, Wadi el-Urf also occasionally delivers large, double-handled as well as normal fist hammers from the early extraction periods. Yet, the dating of the mines in the numerous quartz veins remains impossible without additional excavations due to the intensive mining activities in the Early Arab Period.

In this case, as well, the green malachite stains along the wallrock of the quartz veins had probably attracted the first Predynastic prospectors. Gold mining during the older periods doubtless continued until the Old Kingdom (Tawab et al. 1990). The second, clearly recognisable mining phase is the Early Arab Period, as is attested to by most of the visible archaeological remains at the surface. The only, yet disputable evidence from the New Kingdom is represented by the find of a singular, oval grinding mill.

Tawab et al. (1990) presented a thorough investigation of the Wadi el-Urf district and included their findings in detailed area sketches.

5.1.4.1 Central Processing Site (Fig. 5.14)

Geographic position:	27°48′41″ N, 33°04′52″ E

Three washing tables have been recorded at the central processing site in the Wadi el-Urf mining district (also referred to as "South of Gebel

Fig. 5.15 Well-preserved Early Arab period washing table and tailing heap in Wadi el-Urf

Mongul"). Two of them are located about 100 m away from each other and are in an excellent state of preservation. Both are surrounded by horseshoe-shaped heaps of reddish, ferrous quartz meal (tailing). Koshin and Bassyuni (1968) reported gold grades of 3.87 g/t obtained from tailing samples and thereby indicating substantially higher values for the primary gold ores. Their results also contradicted the findings by Riad et al. (1977) who reported uneconomic gold grades for this site. The southern washing table is best preserved. As it stands today, it is 0,90 m wide, 4.10 m long, and 0.90 m high. The northern one is 3.90 m long, 1.10 m wide and 0.90 m high. Scattered around it, there was a number of round mills and mill fragments (Fig. 5.15).

The third one is located approximately 30 m to the NW from here. In this case too, it is associated with a waste dump of fine quartz sands, and though it has badly suffered, one still clearly recognises the washing table through its inclined surface.

In the area between the three devices are numerous round huts and simple, flat work platforms. Round mills including rotor discs as well as pounding and grinding slabs, and a number of typical, cubic anvil stones with depressions on each of their six sides are found in the washing site's immediate vicinity. Additional equipment of the same sort is found scattered between a house alignment along the southern slope and the flats with the washing tables.

Eastern huts:	27°48′54″ N, 33°04′47″ E

A small number of huts and a more important architectural complex are located hidden away in a small side wadi. To the W, in the valley basin in front of the buildings one makes out a work platform surrounded by tailing heaps.

Western huts:	27°48′55″ N, 33°03′37″ E

This site consists of approximately six simple huts on the western side of a small wadi and a large house with several rooms, at its eastern side. Its walls are masoned in the shell-facing technique and are filled with gravel in the middle. The associated mines are visible as open pits along the crests of the surrounding mountains.

No processing devices have been identified at this site.

Beyond the N-S oriented wadi, additional ruined huts cling to the slope at the edge of a scree gully. The huts had been provided with retaining walls as a protection against the slope rubble.

The mines are located just above the huts, half way up the slope. Neither tools nor washing tables have been observed within the site's area.

The mines are generally slightly larger on the W side of the wadi and are deeper than the ones on the E side. Many mine openings had been carefully sealed after their abandonment.

The few collected ceramic shards consist essentially of amphora fragments from the Early Arab Period (about ninth to eleventh centuries).

No water wells are known from the entire district. The only still viable one is located approximately 5 km away in a straight line, at the foot of Gebel Mongul, the region's most important mountain.

5.1.4.2 Huts and Workings

Geographic position (wadi centre):	27°48′34″ N, 33°03′45″ E

Numerous huts flank the small wadi along whose delimiting ridges countless, ancient trench pits seam the parallel running quartz veins running from NE to SW.

In this area too, scattered pottery fragments from amphorae unquestionably date the site to the Early Arab Period.

Further S there is a small artificial dam used for retaining runoff water, which necessary both for livelihood and gold ore processing. The dam bars-off a narrow tributary wadi, which during rainfall forms a water retention basin of an approximate surface of 500 m^2. During our visit, the basin was dry, although relatively thick sludge deposits clearly reveal its former role as a water reservoir (Fig. 5.14).

Several, so-called Alamat towers were noticed in the surrounding mountains.

In terms of geology, the mineral ore deposit area S of Gebel Mongul and S of Wadi el-Urf is located in a very complex environment.

The area is marked by a dome-like granite magma intrusion into a rhyolite-dacite lava cover of Neoproterozoic Dokhan volcanics. More granodioritic features intermittently mark this magma and towards the centre of the intrusion stock, it is a biotite-granite. The entire sequence is intruded by unusually dark andesite and pinkish-red, felsitic dikes.

Eliwa et al. (2010) recently presented a detailed analysis of the apparently terrestrial volcano-sedimentary sequences in the el-Urf area. The authors identified an alternation of conglomerates and greywacke with mainly ignimbrite lavas, whose age they were able to determine at 615 Ma.

The mined quartz mineralisations show intermittent compositions in the uppermost zones, because next to quartz, repeated high occurrences of barite and specularitic haematite are noticeable. Both point to a high oxidation potential often observed in the upper parts of hydrothermal systems where they interfere with oxygen-rich surface waters. Whenever this is the case, no appreciable gold contents if at all are expected in such areas. They then only occur in deeper parts of the vein system. According to Riad et al. (1977), the gold grades of the exploited quartz veins are in fact rather low. They vary between 0.02 and 6 g/t, and only in few samples reach values between 20 and 50 g/t.

Often, thin malachite linings occur within the wallrock of such veins, pointing to former primary copper sulphide contents. Such phenomena must certainly have attracted prospectors as early as the Predynastic Period, as massive two-hand pounders occasionally occur in spite of intensive re-occupation in the Early Arab Period.

Hydrothermal fluids have to various degrees partly kaolinised, sericised, and epidotised the wallrock next to the quartz veins over several metres.

At the "huts and workings" location (cf. Fig. 5.14) several small quartz veins are found with thicknesses ranging between 5 and 10 cm while containing clearly less barite and haematite. Conversely they contain pyrite and to a lesser extent chalcopyrite. The gold contents of these ores are known to reach 10 g/t. Noteworthy is the occurrence of powellite ($CaMoO_4$) in one of these small veins. It probably derived from the oxidation of primary molybdenite (MOS_2).

Hamimi et al. (1994), Wetait (1997), and Botros and Wetait (1997) advanced an interesting thought as to the genesis of these volcano-sedimentary sequences. They see the alteration halos around the quartz veins at Umm Balad, Wadi el-Urf, and above all, the Dokhan-volcanic sequences as indicators for a characteristic porphyry-copper genesis and explain this by the region's relatively high copper anomalies, jointly with the gold mineralisations. Although this possibility has received little attention among geologists active in Egypt, it principally cannot be entirely ruled out either.

Koshin and Bassyuni (1968) refer to this mining district as Wadi Dib.

5.2 Northern Central Group

A map published by Abd el-Nabi et al. (1977) gives a series of smaller gold deposits, supposedly exploited in antiquity. Although we were unable to visit these sites, their geographic locations are nevertheless included to this overview:

Wadi Qena: 27°09′44″ N, 32°49′27″ E
Wadi Hammad West: 27°09′10″ N, 32°54′47″ E
Wadi Hammad: 27°06′43″ N, 32°57′20″ E
Wadi el-Atrash: 27°10′41″ N, 33°08′48″ E

5.2.1 Ghozza (Fig. 5.16)

Geographic position:	26°52′08″ N, 33°06′29″ E

Several decades ago, Meredith (1952) already furnished a report on a Roman post located at an older gold washing site.

Indeed, approximately 7.5 km W of the Roman water supply station at Bab Mughenigh there is a building complex from the Roman Period into whose walls were masond a large number of apron-shaped runner stones considered typical for the Ptolemaic Period. To the E of the complex is an extensive Ptolemaic gold production settlement surrounded by a wall in the northern third of Wadi el-Ghuz. The Roman camp itself was established on the former tailing spoils. The former well near the tailing site was integrated to the camp. The well and the camp were protected by fortress-like ramparts. Within the walls in the large courtyard are next to the lower parts of mills dating to the Ptolemaic Period two additional grinding mills from the New Kingdom. We here are therefore dealing with a mining district originally dating to the New Kingdom, which then was reactivated in the Ptolemaic Period, and finally the Roman protection camp was established with the released material (Klemm and Klemm, 2005).

This camp consists of a large building with a series of rooms along the eastern and western walls around a central courtyard. The same principle also applies to its NE extension which again consists of an open courtyard surrounded by rooms within a wall enclosure (Fig. 5.16).

The walls of the main camp were built from local rocks including Ptolemaic runner stones which apparently had been gathered in large quantities near the well and the tailings. The handles of the runner stones are usually broken-off, and their surfaces display distinct traces from pounding.

The walled mining settlement from the Ptolemaic Period has a NE-SW orientation. It is approximately 150 m long and 50 m wide and is partly very well-preserved. The site is nonetheless threatened by a gully running straight through it, cutting it in two (Fig. 5.16). A large central building is recognised in its south-western part.

About 250 m further S into the wadi, one comes across a severely eroded site whose irregular fortification walls enclose an area approximately the same size as the Ptolemaic settlement. It possibly dates to the New Kingdom, whereas its function may simply have been the protection of a large, free space reserved for gold washing. However, for the lack of surface evidence this assumption yet requires verification from excavated material.

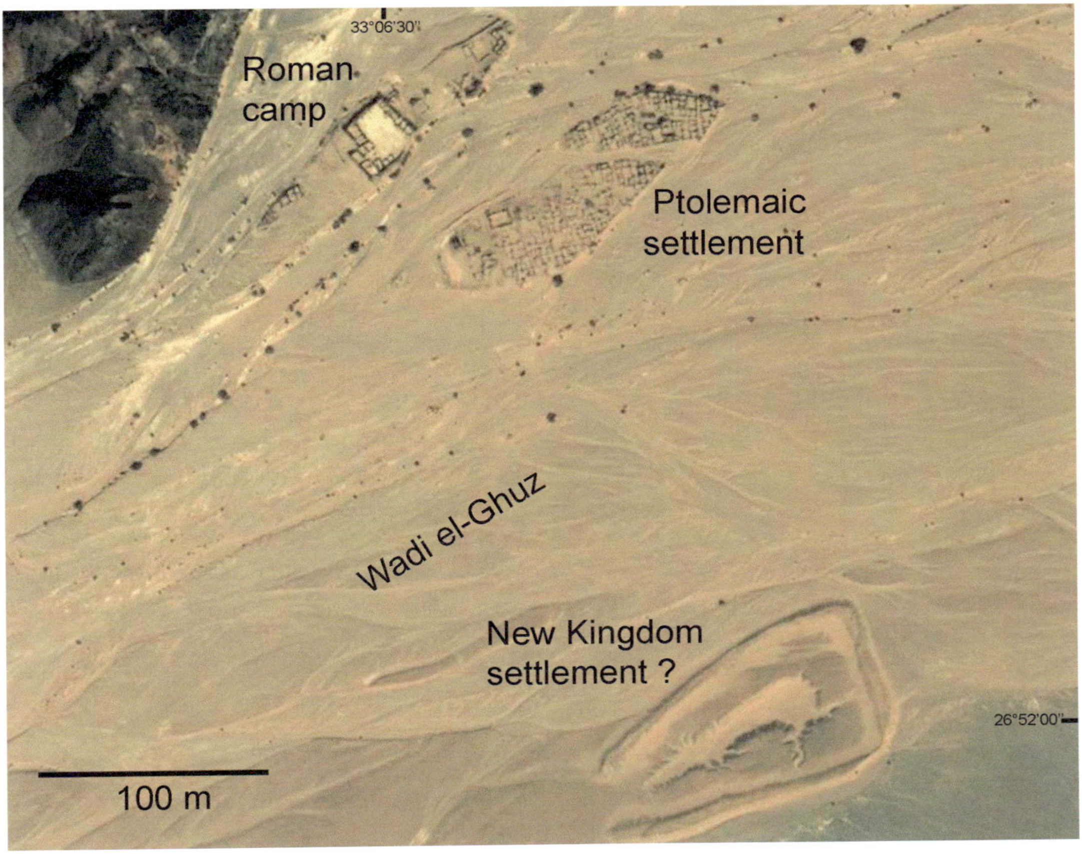

Fig. 5.16 The New Kingdom/Ptolemaic/Roman gold mining sites at Ghozza in Wadi el-Ghuz (modified Google-Earth image)

The entire district reveals no evidence from gold mining in quartz veins. To judge by the settlement and the wall-protected well, the ores at this site had been obtained through the systematic gathering of auriferous quartz fragments in the wadi alluvium (wadiworkings). The quartz chunks could be collected from depths between 0.5 and 1 m. The decisive advantage of this method in comparison to deep mining was that it permitted the hire of large labour forces. With comparatively little training the labourers would have been thought to distinguish between barren and auriferous quartzes. Furthermore, this method did not involve the arduous extraction in hard quartz veins. The sheer size of the Ptolemaic settlement and the conjectured gold washing site to its NW strongly suggest activities connected to such wadiworkings. Systematically carried out sampling operation by Abd el-Nabi et al. (1977) on well-distributed quartz chunks from the shear zones in the surrounding hills on the other hand, revealed low gold grades ranging between 0.02 and 1 g/t, in few cases 5 g/t.

The geologic surroundings of the wadi catchment area are affected by contact-metamorphism from a nearby intrusion of granitic magmas into a series of former greywackes, siltstones, and conglomerates of the Hammamat-formation, making this area an typical environment for gold quartz mineralisations. Granitic intrusions are furthermore evidenced through numerous granite apophyses of different sizes in the immediate surroundings. Through decomposition and erosion of likely auriferous quartz veins in the surrounding mountains quartz fragments have most probably accumulated in the wadi sediments over the time.

Because of the good state of preservation of the ancient sites in Wadi el-Ghus, including the enigmatic, walled site in the S of the wadi, the benefits from carrying out systematic archaeological investigations in the area would seem rather promising.

5.2.2 Fatira (Abu Zawal)

Geographic position (Roman fort):	26°40′32″ N, 33°14`33′ E

A. Murr (1994, 1999) plotted the Fatira district in a geologic map, and R. Klemm in an archaeological one, using aerial photographs and lithologically processed satellite images (Fig. 5.17).

The most noteworthy mines and settlement sites in this important district are located in Wadi Abu Zawal, an eastern tributary of Wadi Fatira, which in its upper course leads in north-easterly direction to Gebel Fatira and the Roman stone quarries at Mons Claudianus.

Surface finds date the earliest occupation to the New Kingdom. In this period preference was apparently given to mining the wadi alluvium, as indicated by the extensive settlements stretching along the wadi courses. They display the usual processing inventory comprising elliptical grinding mills, fist-sized grinding stones, cylindrical pestles, and anvil slabs with hemispheric depressions corresponding to the sizes of the pestles.

Fatira I (Fig. 5.17) exemplifies the situation at such barely discernible New Kingdom settlements in parallel alignment to the wadi. The expected small heaps consisting of waste material from former wadiworkings, however, have been eroded away beyond recognition. In the case of Fatira II on the other hand, such heaps are relatively well-visible. In addition, several small mines in the mountains testify to the activity of deep mining. Traces from wadiworkings are best observed at the site Fatira III, in the form of small spoil dumps as well as countless grinding mills, grinding stones, and millstone fragments. At the adjacent settlement near the western wadi rim too, numerous mills and tool fragments are still located in situ. Because of intensive mining during the Ptolemaic Period, it cannot be ascertained for sure whether or not the mines located to the E of Fatira III had been mined during the New Kingdom as well.

In the Ptolemaic Period, earlier mining activities from the New Kingdom had been resuscitated at the site. Old mine shafts had been enlarged and new ones had opened. Prospection work for yet unknown deposits had also been carried out to the N of Wadi Abu Zawal. Yet, most extraction sites from this period are found in the mountains between Fatira II and III. They have typically vaulted, partly very sinuous galleries descending to average depths between 8 and 12 m, although only few of them are accessible today. Many have been re-examined under modern prospecting work. To the E of this district there is an unusually large, U-shaped tailing, which though lacking the otherwise expected washing table in its central part, contains Ptolemaic and New Kingdom millstone fragments.

The main settlement from the Ptolemaic Period is located at the southern limit of Wadi Abu Zawal, opposite a wall-lined water well, in the centre of the wadi (station Abu Zawal; Fig. 5.18). In spite of the damage inflicted by modern mining, ancient walls are still visible at the settlement site. The buildings seem to consist of room alignments with spacious courtyards in-between. Occasional debris heaps contain among others pottery from the Ptolemaic and Roman Periods (Fig. 5.19).

The site was for the most part re-occupied, or continued to be occupied during the Roman Period, which is partly attested to by inscribed pottery shards recovered from the upper parts of the settlement's debris heaps. George Nachtergael (Brussels) kindly furnished a translation based on a photograph of an ostracon in Greek language from one of these mounds. It reads the following words: "Psentaesis to Ammonius, his brother,

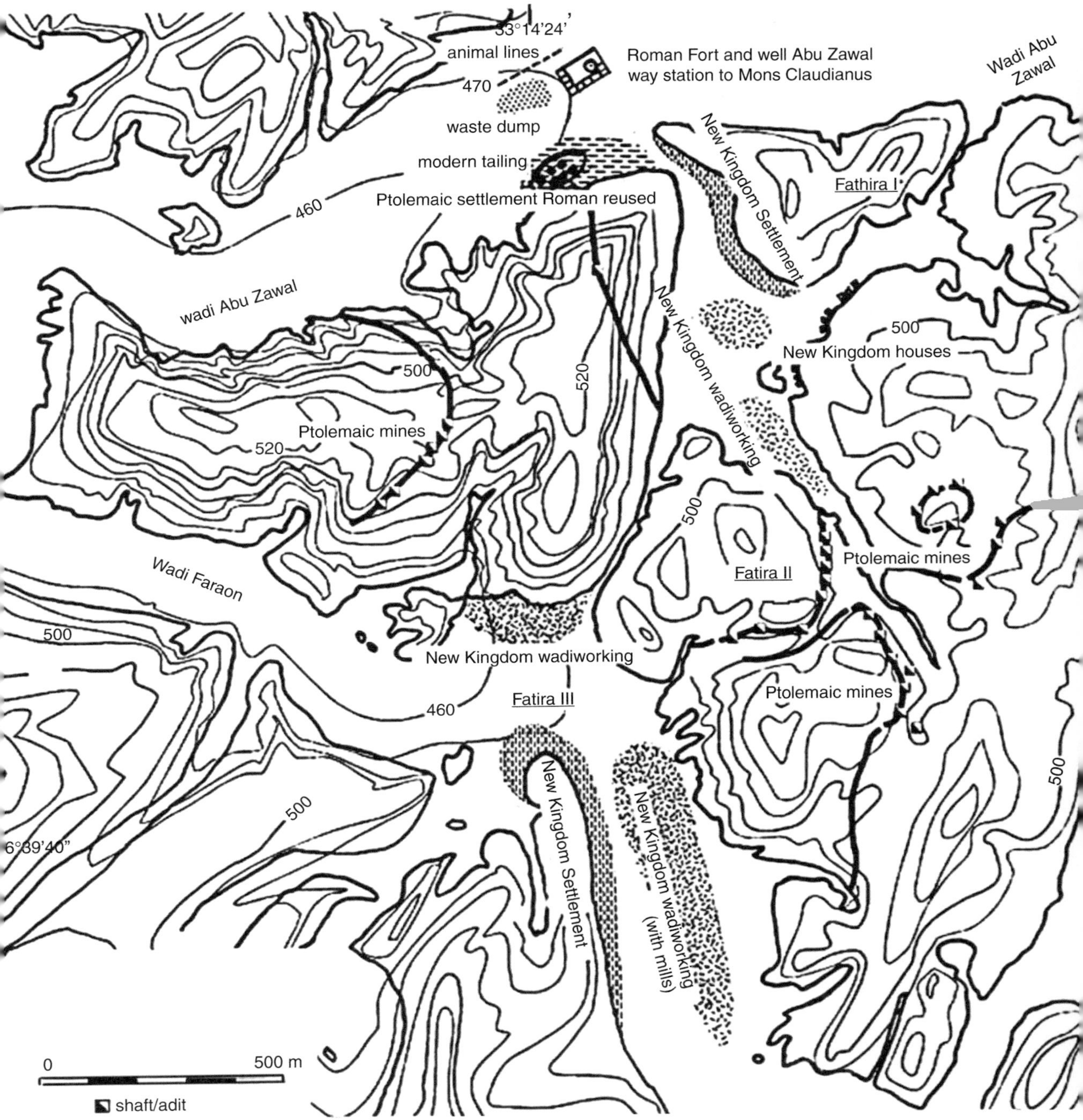

Fig. 5.17 Archaeological sketch map of the Fatira mining district (R. Klemm, modified after Murr (1994))

I greet you. Accept the delivery of a bag with pork dung. Greetings, Kasulla. I hope you are well." ("Psentaèsis à Ammônius son frère, salut. Prends livraison d'un sac d'excréments de porc. Salue Kasullas. Je souhaite que tu te portes bien.") (Fig. 5.20).

Nachtergael dates the letter to the first half of the second century AD, which corresponds to the Roman Period in Egypt.

The expedient had apparently asked his brother, whose place of residence is unknown but conceivably was located in the Nile Valley, to

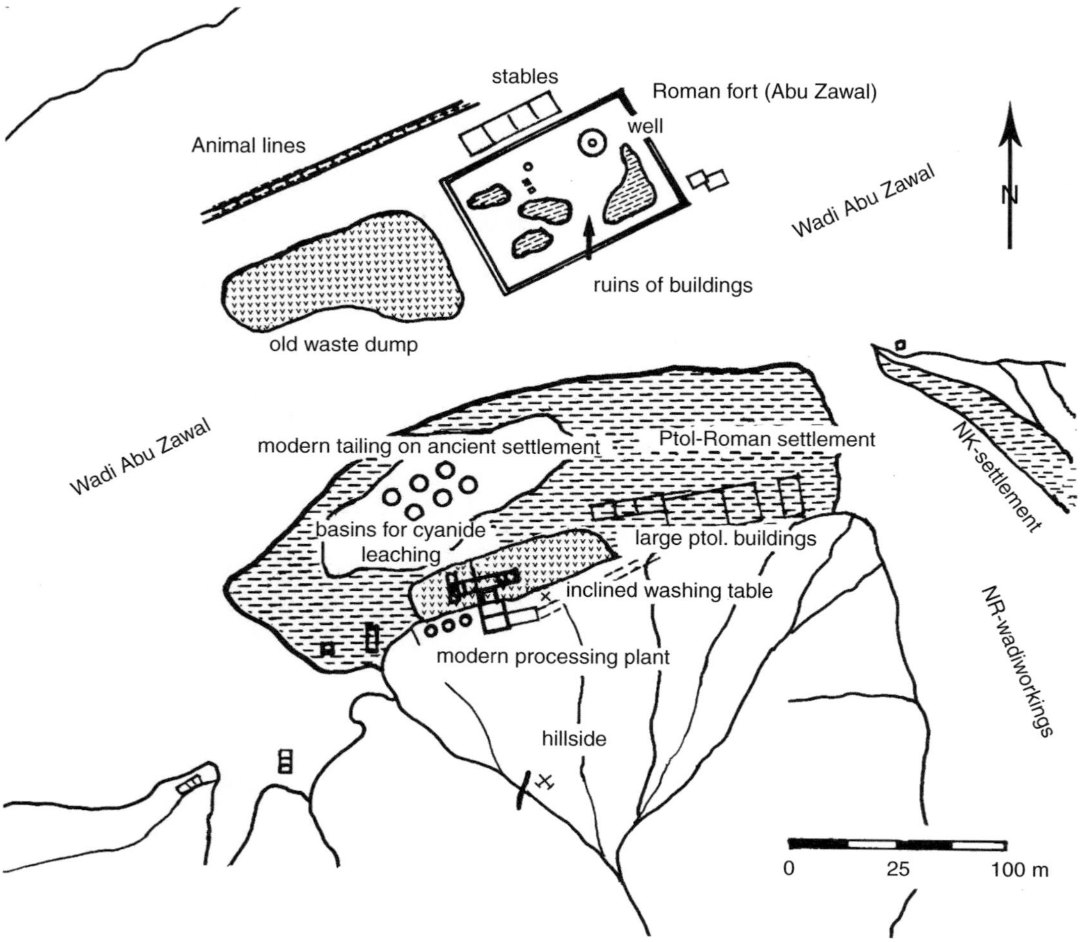

Fig. 5.18 Archaeological sketch of the Roman Fort at Abu Zawal and large Ptolemaic-Roman settlement (R. Klemm)

send a bag of (dry) manure with the next shipment to the camp.

Interestingly, next to a new word, probably denoting a bag made from tapes or the like and a rare word for manure or dung, the text seems to reflect a situation by which people staying in the desert held small gardens for their livelihood. During their prolonged stay in Wadi Heimur/Allaqi geologists from the EGSMA-project too, are reported to have kept small gardens next to their tents.

A series of caves, which after close examination seem to have functioned as storage cellars, stretches along the slopes of the settlement. Since they had been lowered only a few metres into the rock, it seems unlikely that they represent initial stages of mining.

The Roman Period occupation inside the fort as well as the settlement had no apparent link to mining activities, but were exclusively connected to the Abu Zawal fortress built around an important well (Fig. 5.19). According to the evidence (millstones) it dates back to the Ptolemaic Period, probably even to the New Kingdom. No round millstones are known from the sites, which seems to indicate that even in the Early Arab Period no

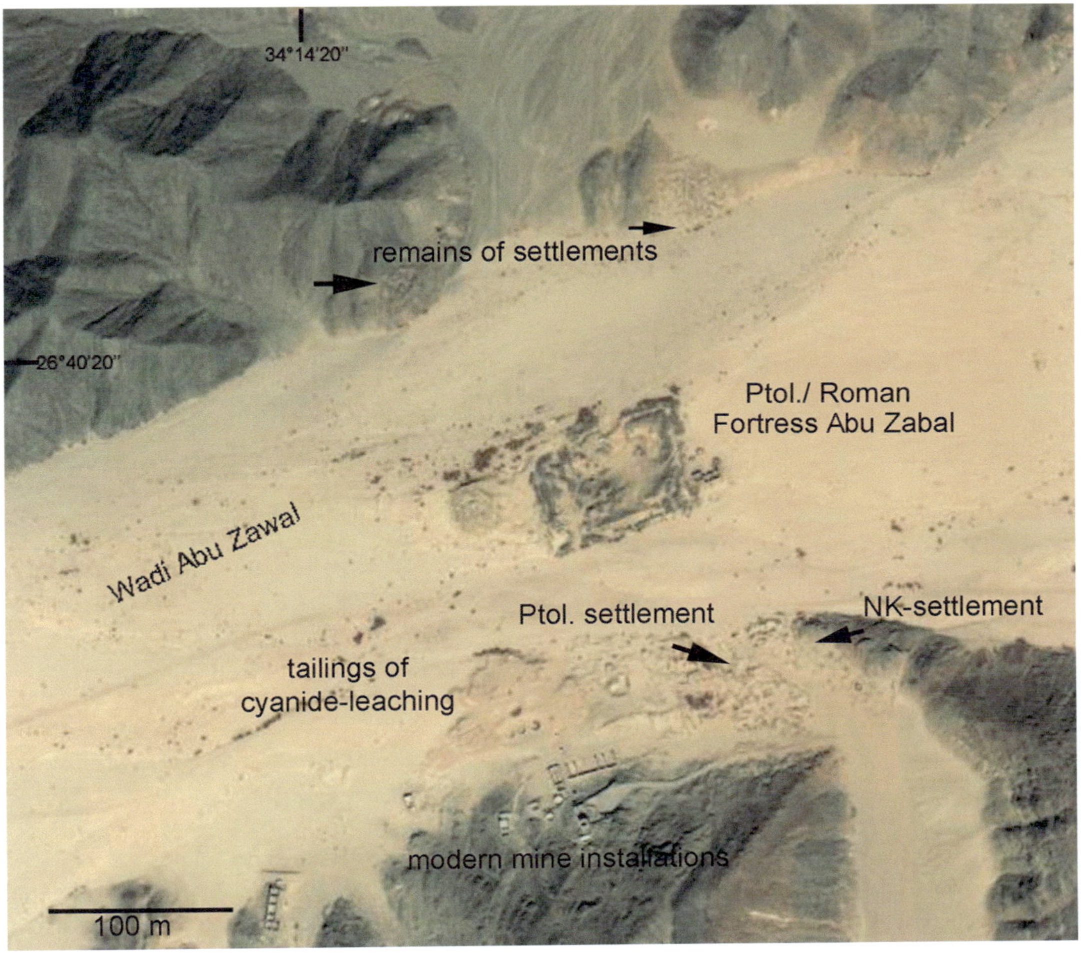

Fig. 5.19 Recent mining complex at Wadi Abu Zawal (first half twentieth century) and New Kingdom, Ptolemaic, and Roman sites (modified Google-Earth image)

mining activities had been pursued at the site. The main extraction periods must therefore have taken place during the New Kingdom and the Ptolemaic Period.

The large quantities of ceramics from the fort are mostly Roman and consist of dark-brown amphorae with burnished tops and grooved bases ending in elongated and solid cones.

At the western side of the fort's E wall a combustion layer was recorded, consisting of a white deposit from burnt limestone. It presumably served for producing the mortar whose traces are partly seen at the basin near the western corner of the fort, as well as in what is interpreted as water troughs. A completely sooted and burnt Ptolemaic, apron-shaped runner stone had probably once been part of this local limekiln.

By the Roman Period the well had probably already silted up and had needed further excavation at the moment of the building of its protective wall. The wall's masonry includes among others discarded millstones as well as ceramics probably recovered from the well's near surroundings. Building material (stones and pottery)

Fig. 5.20 Ostracon dating to the first half of the second century AD requesting the delivery of manure (translated by G. Nachtergael)

retrieved from the neighbouring settlement had probably also been used for erecting the fort's walls as their components much resemble the debris found at the settlement.

The well had probably been the major water source for the entire mining district until its fortification in the Roman Period, when it gained the rank of a road station on the route to the stone quarries at Mons Claudianus. The room alignments along the inside of the walls surrounding a well almost in its centre meet the typical construction standards encountered at Roman forts controlling Egypt's Eastern Desert. On its outside, to the NW are so-called "animal lines", or elongated drinking troughs which had been sealed with a pinkish mortar. Within the severely disturbed contexts at the room alignments, the masonry of the fortification wall, and the adjoining, outside debris heap one notices chippings and larger fragments of the bright-coloured tonalite originating from the nearby Mons Claudianus stone quarries. Raw half-products of small stone vessels are also found, which proves that small objects from this appreciated material had also been produced at the site. Immediately to the W of the fort is a waste dump consisting of debris material from the excavated well. Prior to any archaeological excavation at this dump it can only be speculated whether it exclusively composes of spoil material from well or whether it too, may contain tailing sands. Ptolemaic mills from this dump of which one had been transformed from a former New Kingdom grinding stone would at least hint in this direction.

With the modern era, the well served for the supply of pumping installations in the mining industry. Today, the same site is disfigured by its ruineous buildings and concrete pillars, as well as resulting wide-spread damage to the archaeological layers.

A carefully filled track running parallel to the slope leads southwards from the westernmost house at the Ptolemaic settlement of Wadi Abu Zawal to the Ptolemaic mines. Further away it runs

Fig. 5.21 Wadiworkings and New Kingdom settlement remains in Wadi Faraon

over a saddle to Wadi Faraon, the location of the pharaoh's mining operations and wadiworkings (Fig. 5.21).

As in the case of El-Sid, where New Kingdom settlements and mines are virtually devoid of amphora pottery (plates and bowls only), in Wadi Faraon too, one finds only few plates and bowls. Water was therefore probably transported and kept in animal hides in both wadis. Similar to El-Sid, where a large tailing is seen near several mines in the mountains, Fatira as well, reveals a tailing heap close to a small mine in the middle of the mountains (Fig. 5.22). In both cases, the water required for ore processing was probably transported from the well through the wadi to the mine, and vice versa, ores to the well.

Within the framework of this investigation project Murr (1994, 1999) surveyed and thoroughly described the geology of the Fatira area (Fig. 5.23). The following paragraphs summarise his observations:

The oldest units in the immediate surroundings of the ore deposits are sedimentary rocks of the Hammamat formation in an isoclinal syncline. The foot wall substrata are ill-sorted conglomerates with poorly rounded components. Toward the hanging wall siltstones become increasingly frequent as they alternate with the conglomerates. They partially show a graded stratification. Inside the fold core, there are siltstones only. A granitic pluton is intruded into this sediment sequence. Through partial melting of the wall-rock, the periphery of the pluton is contaminated to granodiorite. Towards the centre follow syenogranite, monzogranite, and finally, normal biotite granite. The contact zone with granodiorite generated metamorphism of the siltstones, which display hornfels-pavement-structures and poikilitic biotites. To the NE, granite and sedimentary rocks are bordered by a gneiss-migmatite-granite fault zone. According to El-Gaby et al. (1988) this zone represents a deeper tectonic level. In the NW of the area, there is an outcrop of porphyric alkaline rhyolite. It overlies the Hammamat sediments in angular discordance. Rhyolitic and to a lesser extent basaltic dikes cut-off the sedimentary rocks, granitoids, and the gneiss-migmatite-granite area in a NE-SW direction and testify to a tectonic crust extension in a NE-SW direction. The fault between granite/

Fig. 5.22 Large Ptolemaic tailing residues at Fatira II

sediment rocks and the gneiss-migmatite-granite area along the dike orientation is also displaced to the NE.

Since the granitoid complex is intruded into the folded molasse sediments and at the same time remains undeformed, it is evidently post-tectonic. Zircon and feldspars with rounded cores were identified in thin sections. It was therefore assumed that crust anatexis had generated the granite.

Productive quartz mineralisation usually follows an intermediary, 1 m thick vein that lead through granodiorites and contact-metamorphic siltstones. The vein strikes NNE and dips 15° SW. At contact with the vein, the granodiorite distorts and fractures. In addition, the mineralisation is located in a brittle shear zone of the vein. In the W of the main deposit area, the vein alters slightly along the shear zone and quartz mineralisations are absent. In the centre of the main deposit area, the wallrock of the mineralisation is altered to a high extent. The shear zone is mineralised with small quartz veins containing the ore. Primary sulphide ores have turned into limonitic iron ores. In the E of the main deposit area, a massive quartz vein has formed displaying widths of up to 1.5 m with a strong silicification of the wallrock. Ancient and modern extracting activities have concentrated on the central part of the main deposit. Only test galleries were found in the W. Further to the NW, there is another ancient mining district located within-contact metamorphic siltstones. The mineralisation too, is in an andesite dike with the same orientation as the one mentioned first. Small quartz veins displaying haematite mineralisations develop to the NNE into a large quartz vein. Another, anciently exploited quartz vein to the NE of this extraction area strikes NNW and dips at an angle of 63° SW. This quartz vein is only few centimetres wide and situated within the granodiorite.

Investigations of the petrography and mineralogy of the ores show that their genesis had occurred in the quartz vein as well as in the wallrock (granodiorite, andesite dike). Murr (1999) identified two different quartz generations. The first is a milky-opaque quartz in contact with the wallrock and regarded as the main pyrite-gold ore. In syeno- and monzogranites as well as in the granodiorite located at some distance from the gold deposits at Fatira, gold grades are normally below the detection limit of 5 ppb. Therefore, the

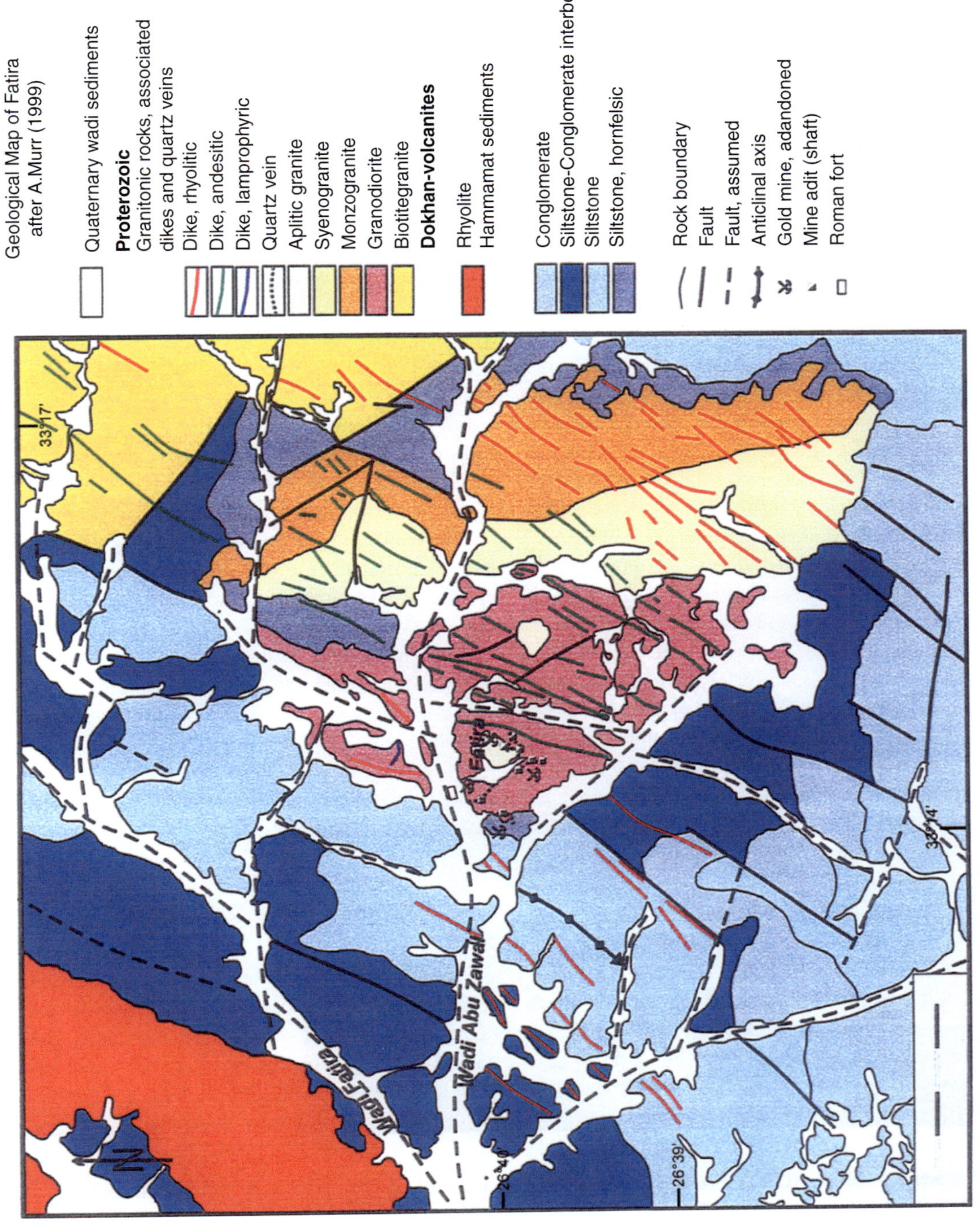

Fig. 5.23 Geological map of the Wadi Fatira surroundings (Murr 1999)

granites and granodiorite disqualify as primary gold sources.

The granodiorite is thoroughly altered within the immediate surroundings of the vein. Plagioclases and alkali feldspars are intensively sericitised. Biotites and iron-chlorite are no longer present. One recognises zones of former biotite by a preserved rutile and its typical sagenite-lattice, as well as zircons. A colourless chlorite with normal interference colours (low Fe contents) replaces biotite. In some areas, this chlorite changes to muscovite. Along fissures, the sericite transforms to stilpnomelane. Magnetite no longer occurs. Because of its brown translucence in thin sections and sharply edged, square to lozenge-shaped profiles, the ore was described as a lepidocrocite ore. These are pseudomorphs after pyrite. Ore formation takes place on a subordinate level in the entire rock. Ideomorhphic ores (pseudomorphic after pyrite) are sporadically found inside the quartz. Most ores are inside the colourless chlorite or the secondary muscovite. Other ores occur together with stilpnomelane along fissures.

From these observations, a precipitation mechanism is derived for gold contained in granodiorite: biotite and magnetite are converted to low-ferrous chlorite or pyrite. Fe from magnetite, biotite, or originally ferrous chlorite reacts with sulphide to form pyrite. A potential sulphide source is the same mineralising fluid that also transports gold. In a reaction between the gold-sulphide complexes and the wallrock iron, the gold-sulphide complexes decompose, thus causing precipitation of gold. The same reactions are also observable in granodiorite-wallrocks of numerous other gold mineralisations in the Egyptian Nubian Eastern Deserts.

The vein that is followed over the entire deposit area is least altered at the W side of the main extraction area. During mineralisation, the vein undergoes hydrothermal alteration. Plagioclase is transformed quantitatively to sericite while the latter take in the outer contours of the former. The rock consists largely of grossly arranged sericite aggregates with interceding quartz particles. Chlorite no longer occurs; former mafites are completely seriticised. Another relic mineral from the original rock is zircon.

According to detailed investigations of the liquid inclusions of the quartz generations involved in the mineralisation process, these were possibly formed in temperatures between 180 and 410 °C within a low saline environment.

Two essential factors can be reasoned for the concentration of gold in the Fatira deposits by means of the results obtained from petrography, geochemistry, and micro-thermometry (Murr 1999). Through the firm bond between the gold and sulphide minerals (pyrite) as well as the relatively low salinities of mineralised fluids, it may be inferred that gold was transported as a gold-sulphide complex in the fluid. Precipitation of the ore minerals from the fluid is engaged on the one hand by a reaction of sulphide with Fe from the wallrock, thus leading to the formation of pyrite and concurrent gold precipitation. On the other hand, the gold-sulphide complex can also disintegrate through ex-solution of the CO_2-H_2O phase. However, through the predominant occurrence of pyrite in the wallrock, the former mechanism seems more significant in provoking the gold precipitation. While gold remained undetectable in pyrite using the available petrographic methods, an increase of its grain size coincides with the decomposition of pyrite to limonite (c.f. Mumin et al. 1994). Thereby, the gold becomes visible under the microscope.

5.2.3 Abu Shehat

Geographic position:	
Ptolemaic processing plant (?):	26°35′09″ N, 33°11′55″ E

About 8 km (beeline) E of the military airport at Wadi Qena one notices a peculiar, rectangular construction in Wadi Abu Shehat associated to a round installation, measuring about 20 m in diameter. It closely resembles the Ptolemaic gold processing plants known from Daghbag, Barramiya, and Samut.

Yet, its entire structure is covered by sand to such a degree that it just barely is recognised in a much enlarged Google-Earth satellite image (Fig. 5.24).

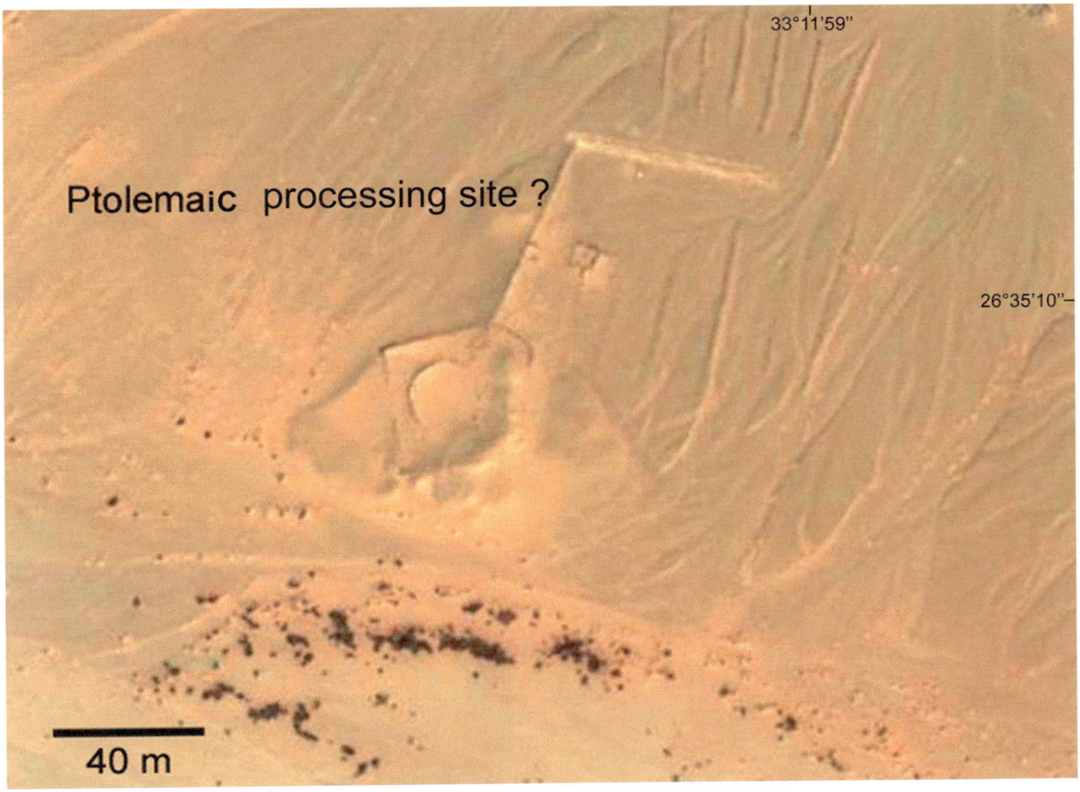

Fig. 5.24 Abu Shehat: Ptolemaic processing plant (?) in Wadi Shehat (modified Google-Earth image)

Because the closer surroundings reveal no other traces from either mining or settlement, which would hint to former wadiworkings, the function of this unit will remain obscure as long as systematic archaeological investigations are not carried out. Undoubtedly, additional archaeological remains are covered by the area's thick alluvial deposits.

5.2.4 Wadi Abiyad

Geographic position:	26°33′16″ N, 33°54′29″ E

In the northern part of Wadi Umm al-Huwaytat, about 2 km N of the modern phosphate mine and S of the tomb of Sheikh Awwad, in a small tributary valley, are the ruins of an elongated settlement in an approximate, parallel orientation to the main wadi. For the lack of archaeological investigations, its date is unknown.

Due to the complete absence of artefacts related to mining, this settlement must have had some other function. An ancient pilgrim's settlement on the route to Safaga may seem plausible. For that account, however, this raises the question as to why no other comparable settlement on the pilgrim route to Safaga is found in the Eastern Desert.

5.2.5 Abu Mureiwat (Fig. 5.25)

Geographic position (Old Kingdom mine):	26° 31′02″ N, 33°38′41″ E

Abu Mureiwat is to date one of the rare sites that contain clear evidence for mining activities

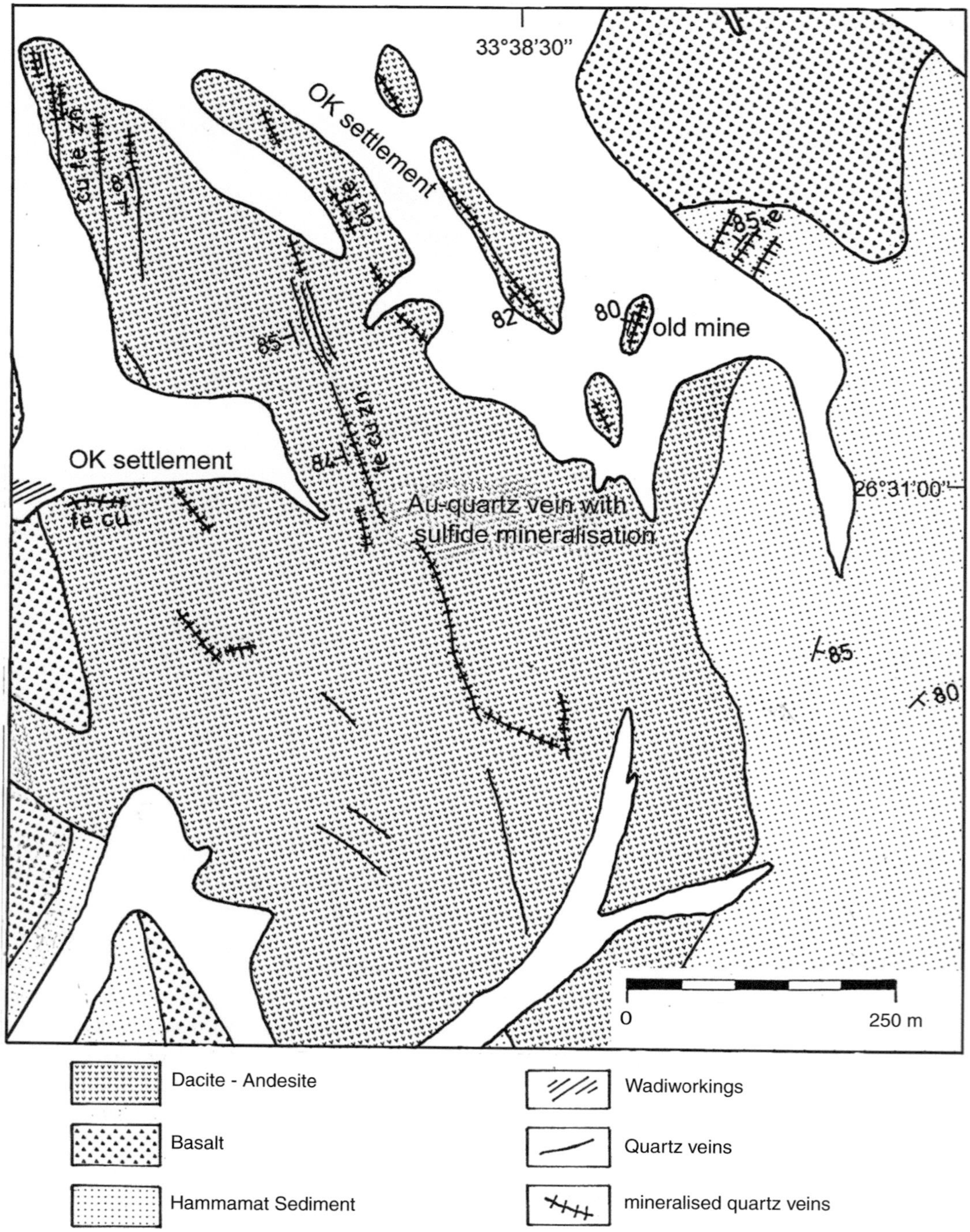

Fig. 5.25 Geological sketch map of Abu Mureiwat (modified after Armanious et al., 1971)

Fig. 5.26 Early Dynastic "Red Mine" at Abu Mureiwat with red spoil heaps. The greenish heaps in the foreground originate from the wallrock and contain numerous stone tools

during the Pre- and Early Dynastic Periods. This evidence concentrates on one hand in the so-called "red mine" situated in the valley basin (Fig. 5.26). Here, a clearly visible feature at a hillock in the valley plain displaying traces from ancient mining is represented by a rib-shaped outcrop from quartz. A red-coloured debris heap from ferric oxides (from former sulphide minerals) and a green-stained wallrock through malachite indicate together with otherwise good exploitation conditions to gold mining operations in a very early period.

The layout of the mine itself is somewhat disorderly. Though it follows the quartz vein, it also branches out into singular ore pockets, thereby leaving the impression of a confused maze of small corridors and caves. Moreover, the mine had probably been exploited repeatedly in later periods.

As in the case of Wadi Dara, consideration needs to be given to the possibility that mining had aimed to begin with at copper (malachite), as suggested by the eye-catching, green stains, and that the ore's gold contents had only been discovered later on in the Old Kingdom, leading subsequently to a shift from copper to gold extraction.

Stone tools at various stages of wear litter the waste dumps at the red mine. Most consist of grooved stone mallets often displaying blow chips at their surfaces, but also fist hammers, stone globes and disc-shaped stone hammers, which according to the findings, are sofar among the oldest mining tools known from Egypt (Fig. 5.27).

The second outstanding quartz vein crosses the plain at the western edge of the valley basin. It too, exhibits numerous locations of ancient extraction. Its date is more recent than the "red mine" and belongs according to the pottery to the Old Kingdom. Stone tools are found in the rubble along the entire length of the mostly buried mining sites. Their dates either match or contemporary with the Old Kingdom, as is supported by the ceramic evidence in the neighbouring settlement. Indeed, thanks to the prospecting trenches

Fig. 5.27 Early Dynastic/Old Kingdom stone tools from the "Red Mine" at Abu Mureiwat

by EGSMA some years ago in the mining area itself as well as in the badly preserved settlement layers at the northern end of the vein, much ceramic material has been recovered. It essentially composes of so-called fine, red "Meidum bowls". In addition, strikingly large, two-hand stone hammers utilised in the settlement's wall masonry are comparable to ones from Higalig, the hitherto oldest mine known from the Predynastic Period. This suggests a slightly later date for the settlement. Nonetheless, numerous malachite-covered ore chunks lie scattered within the settlement area, which could indicate that copper mining had still been practiced here in the Old Kingdom.

Furthermore noteworthy in Abu Mureiwat are very small anvil stones with tennisball-sized spherical hammer stones indicative of more cautious pounding processes, possibly comparable to those known from Egyptian emerald mines at Sikait, where they have served in the careful extraction of emeralds contained in quartz chunks. This more careful method is even perhaps connected to the very beginnings of gold processing, by which singular gold flakes were separated from the surrounding rock.

The houses in the settlement are built from loose rocks wedged against each other. Their dilapidated walls are preserved in one masonry layer only and display round as well as rectangular ground plans.

Next to the mentioned Meidum bowls, a second pottery series is found outside the settlement. It consists of dark-red to brownish amphorae manufactured in a very coarse, brushed ware. The brushed décor had been irregularly executed over the vessels' bodies in an oblique orientation. The same pottery is also known from other nearby sites, where it appears in Early Arab Period contexts. Presumably, only small-scale wadiworkings had been carried out here. Scanty rests from a gold washing table dating to this period are still distinguishable beyond the western mountain ridges bordering the valley.

Finally, a third, slightly sinuous vein stretches out along the western side of the valley, about half way up the mountains slopes. This vein had been exploited over several hundred metres and reveals a number of deep galleries of which most are buried by debris. To judge by the fist hammers found in the slope rubble, mining must have taken place during the Old Kingdom.

Additional ancient mines as well as the gold washing place from the afore-mentioned Early Arab Period are located at the western slopes of the ridge.

No archaeological traces from the New Kingdom were recorded from the entire Abu Mureiwat district, nor are there any tailing sites in the area.

The deposit area is geologically located in a sequence of acid and mafic metavolcanics, which are covered by metasediments and intercalations of banded ironstones (Fig. 5.25). The geologic sketch map is based primarily on an older map by Armanious et al. (1971) and has only been slightly modified by our team. The mafic metavolcanics are essentially andesite sequences and reveal neither clearly determinable lava covers nor tuff- or pillow layers in the terrain.

The dacitic units are comparatively bulky. Noteworthy for all rock types are a distinctive surface weathering and an intensive limonitic staining, which presumably owes to the decomposition of pyrites (cubic cavities) formerly contained in these rocks.

The deposit area of this metavolcanic-sedimentary mountain environment is marked by two, well-developed shear zone systems in an approximately NNW-SSE or NNE-SSW direction. They display different degrees of quartz mineralisations and are the main bearers of the gold- and copper minerals. Especially around the ancient mines, the quartz veins expose areas of pronounced, brown colouring through limonite, being the residue of sulphide minerals that have oxidised at the surface. This has also been confirmed by the drill core samples taken by the MINEX/Egypt Company in about 50 m depth. The samples were intensely mineralised by pyrite and to a lesser extent by chalcopyrite, pyrrhotite, sphalerite, and galena. Visible gold was not recognised inside the cores. In areas of broader openings of shear zones these are mostly filled-in with milky-white, bulky quartz. This quartz apparently contains no gold and as exemplified in Fig. 5.26, where the central, bulky quartz vein was left untouched, did not attract the interest of the ancient miners. Eldougdoug (1990a, b) gives more information on the geologic surroundings as well as a more detailed account on the gold anomalies of the volcano-sedimentary sequences at Abu Mureiwat.

Zoheir and Akawy (2009) determined temperatures between 260 and 380 °C for the genesis of the vein quartz, which approximately coincides with the temperatures of the ore genesis. They further argued that sulphur and carbon dioxide yielding fluids had transported small quantities of gold-sulphide complexes, which when reacting with iron had formed pyrite with traces of gold. With the oxidation of pyrite, gold was subsequently released. This explains why the ancient mining zones concentrate exclusively within the oxidised periphery of the quartz veins.

According to the extensive analyses by the MINEX Company, the gold contents of the sulphide-rich drill cores vary within a 5–12 g/t bracket.

The considerable amounts of gold contained in banded iron stones as determined by Botros (2004) had not been identified by the ancient prospectors and had accordingly been left untouched. It seems however that such rock plays an important role in the primary pre-enrichment of gold and the supply of auriferous quartz veins, as often observed in connection with the metasediments in Egypt and Nubia.

The most ancient quarry site with evidence from Predynastic, disc-shaped stone hammers as well as Old - Middle Kingdom ceramics is located at the SE end of the wide valley of Abu Mureiwat. The area is marked by an intensively sheared and thoroughly oxidised vein system, striking NNE-SSW with an almost perpendicular dipping. The vein system is clearly distinguishable by two protruding ribs. It was apparently only after the exhaustion of this easily exploitable vein system from the Old Kingdom onwards, that the NNW-SSE striking flatter veins in the wadi bed and the ones at the northeasterly sloping flanks had been

Fig. 5.28 Archaeological sites in Wadi Gasus (modified Google-Earth image)

tackled for ore extraction. In ancient times in general, only the well-oxidised zones of the ore-yielding veins had been mined, whereas the primary sulphide zones had not been reached. According to the analyses by MINEX, such zones begin at an approximate depth of 40 m below the terrain's surface. An unsheared vein with a very thin, oxidised crust located further W had apparently been too hard for the early miners and had therefore been left unexploited in spite of its high gold contents.

Smaller quartz veins, often oriented in a perpendicular direction to the main veins are all totally devoid of gold.

Recurrently, the surface metavolcanics display short apophyses and small intrusive stocks composed of tonalites and to a lesser extent of granites. They originate apparently from a slightly deeper lying granite body, which had superficially penetrated the series from below. They seem to have furnished the thermal energy for the hydrothermal processes that subsequently lead to ore formation.

5.2.6 Wadi Gasus (Fig. 5.28)

Geographic position:	
Early Arab settlement with tailing-site:	26°30′15″ N, 33°51′05″ E

For long, Wadi Gasus has been an important route between the Nile Valley and the harbour at Philoteras (mod: Safaga). To reach the namesake

Fig. 5.29 Early Arab Period gold processing site with tailing and broken round mills in Wadi Gasus

gold mining settlement, one leaves the asphalt road in Wadi Gasus shortly after the rock inscribed with the cartouches of Amenirdais and Schepenwope (Schweinfurth and Ermann 1885) and then turns-off towards SW (Fig. 5.28). After approximately 1 km one arrives at the settlement located at the western side of a deep valley basin.

The cartouches had been hewn into the northern wadi flank on smooth surfaces from Hammamat rock (greywacke). The route leading down Wadi Gasus in south-westerly direction is also called Wadi Gowa, which is why the site with the inscriptions is sometimes also referred to as Wadi (or Bir) Gowa.

The Early Arab mining settlement, well-protected by a gorge-like incision into the Hammamat series, is located in a side valley of Wadi Gasus amidst coarse grained and pink granites. At the periphery the granite turns by assimilation into granodiorite, while still containing numerous xenoliths from the intruded metasediments.

In the core area of this granite, one clearly recognises the roof of the intrusion with a contact-metamorphic overprint of the Hammamat series.

Near the settlement, numerous small mining sites are located in the surrounding granite hills. They generally do not penetrate deep into the mountain. Numerous footpaths connecting the single mines to each other traverse the hills. Natural depressions at the bottom of scree gullies along the northern and eastern mountain flanks had no doubt been used as water reservoirs in former times. In addition, through the construction of a dam at the end of a wadi chasm, significant amounts of rainwater could be harvested. Such water management techniques are well-known from Early Arab contexts and in particular observed at sites with evidence from so-called wadiworkings. At many locations in the settlement of Wadi Gasus as well, remains from intensive processing activities of mostly alluvial quartz chunks are distinguishable even today. They originate from small quartz veins, which in a dense pattern cut through the neighbouring granite mountains. The small mine shafts observed in the mountains may hardly have sufficed to generate the three large tailing piles preserved in the plain (Fig. 5.29).

Especially around these tailings, numerous round stone mills, cubic anvil stones and pounding stones were observed in situ. Inclined washing tables are otherwise only roughly discernable. Round mills are also found in great numbers inside the hut ruins.

The dwellings themselves are mostly badly damaged. This is partly due to the fact that they had been built from rounded rocks of a very coarse-grained granodiorite, which is highly vulnerable to decomposition. On any account, the spheroid shapes warranted only for little wall stability. The architecture consists mainly of round huts, each comprising only of one room. Like the additional, three to four-roomed buildings, they scatter thinly along the foot of the mountains or in the valley plain. Because of their severe weathering, they are not easily distinguished from a distance, as they tend to blend-in with the background scenery composed of similar rocks shapes.

According to the entire archaeological assemblage, including the pottery and above all, an abundant green-glazed pottery, the site's occupation period is exclusively restricted to the Early Arab Period.

Other former quarry sites line-up along both slope sides of Wadi Gasus as they follow the easily visible quartz veins. The veins belong to a roughly NE-SW oriented shear zone system in siltstone-greywacke sequences of the Hammamat series. These rocks are still slightly affected by contact metamorphism, whereby the hornfelsic overprint increases markedly at the contact with granodiorite.

The geologic map (Fig. 5.30) furthermore shows that folded ryolites occur in the Hammamat series. In the marginal areas further W bordered by a graben, Eocene limestone sequences appear and potentially yield extractable galena.

5.2.7 Gold Processing Site at Bir Semna (Fig. 5.31)

Geographic position:	26°28′13″ N, 33°35′46″ E

The gold processing site "Bir Semna" is about 1 km to the N of the actual locality at Bir Semna, located in Wadi Semna, through long stretches of which runs a single-track railway line along a pilgrim route. The tool evidence at the site designates it as a gold processing site from the New Kingdom and the Ptolemaic Period. The site is fortified with a partly preserved, heaped-up earth levee protecting a central well, which today is buried below debris (Fig. 5.31). The structure measures about 100 × 80 m, in a parallel orientation to the valley near its eastern flank of the Wadi Semna. Only the N and E walls have survived. On the roughly 6 m wide wall crest in the N are the ruins of several huts. A relatively large house complex leans on the E defences. More ruins from huts are located on a wadi terrace outside the retaining wall.

In spite of intensive erosion, distinct traces from reddish tailings are clearly visible inside the walled zone. In the NW part of the site is a rectangular, approximately 80 cm wide, 40 cm deep, and 5 m long basin. Its inside was sealed with a reddish plaster and had evidently served for washing quartz ores as part of the plant. Similar washbasins have been recorded at the Ptolemaic processing sites at Gidami, Daghbag, and Barramiya.

Numerous Ptolemaic concave stone mills including the apron-shaped runner stones from reddish granite porphyry are found at the site. Mills from greywacke too, lie scattered in the surrounding terrain. In one case, an oval stone mill from the New Kingdom had been reshaped to a runner stone some time during the Ptolemaic Period, a fine piece of evidence testifying to the development of the tool sequence we propose (Fig. 5.32).

The numerous potsherds recorded at the eastern house essentially stem from Ptolemaic Period amphorae.

In spite of our intensive explorations within the catchment area of this installation, we found no traces from ancient mining here. Conceivably, the operations had formerly consisted of an extensive alluvial mining system requiring an important central processing plant like the one at Wadi Semna. Because of the complete absence of tailings and other related findings in the mining district of Semna located approximately 3 km away, it seems conceivable that the ore extracted in this area was actually processed at Bir Semna. According to the local Bedouins, modern mining attempts had failed at Semna mine itself because of the adverse conditions concerning the water supply. Similar conditions may be conjectured for past periods, under which the miners had been forced relocate the ore processing to some other, more distant site.

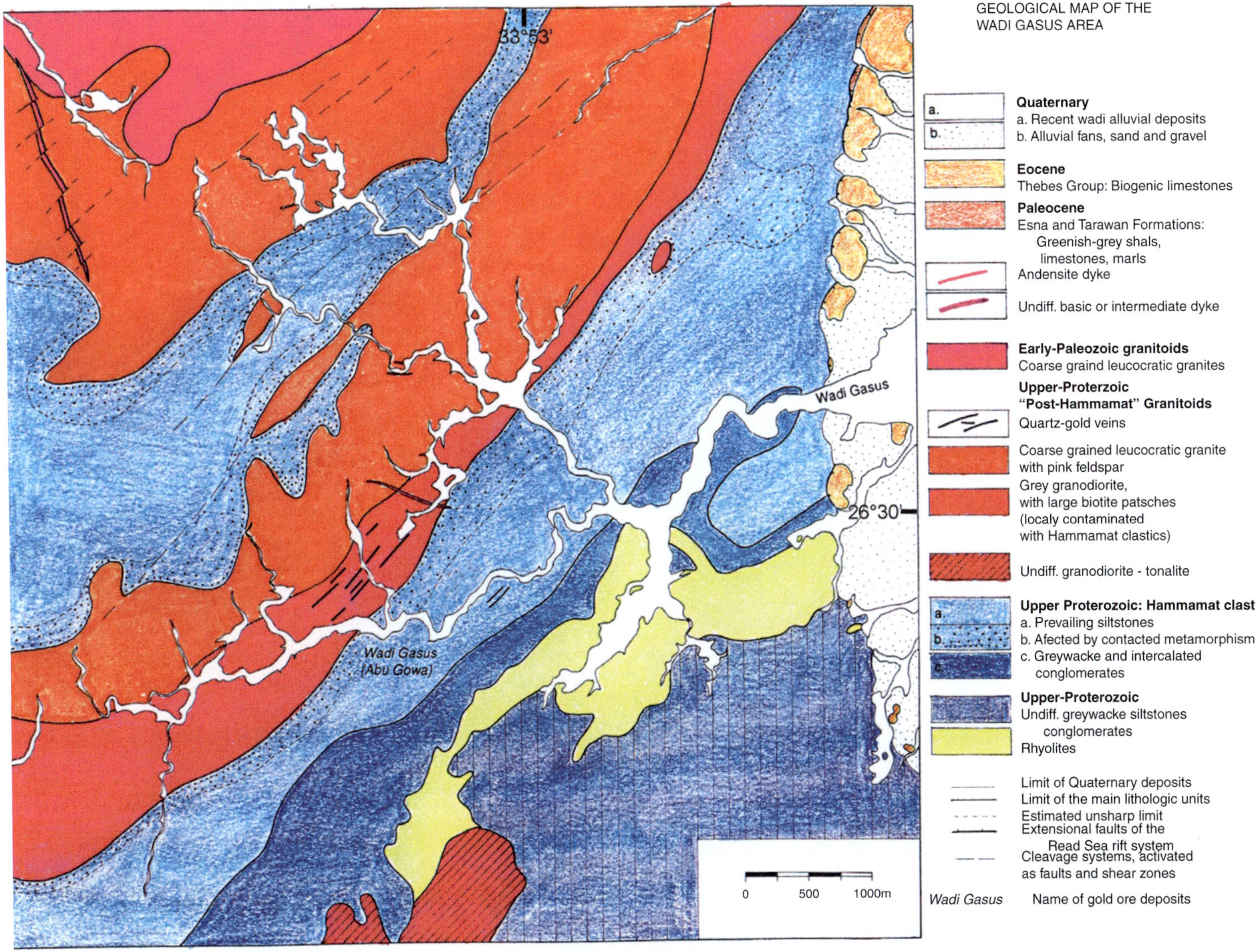

Fig. 5.30 Geological sketch map of the Wadi Gasus surroundings (compiled by H. Kräutner)

Fig. 5.31 Gold processing site at Bir Semna. The finds date the site at least to the New Kingdom. The rectangular fortification walls probably date to the Ptolemaic Period

Fig. 5.32 Double handed apron shape runner stone belonging to a concave, Ptolemaic ore mill made from an oval New Kingdom grinding mill

Fig. 5.33 Central mining district at Semna with ancient and modern complexes (modified Google-Earth image)

5.2.8 Semna (Fig. 5.33)

Geographic position (main vein):	26°26′31″ N, 33°35′21″ E

The Semna deposits lie in a small, tributary wadi, which roughly runs in parallel direction bout of 3.2 km to the S of Wadi Semna.

Unfortunately, the ancient mining district at Semna had been badly destroyed during mining operations at the beginning of the twentieth century, and later in the 1950s. The surface finds are therefore no longer fit for the reconstruction of an even approximate picture of the ancient mining activities there.

The district is accessed through a roughly 2 km long and relatively narrow trail beginning in actual Wadi Semna and leading to Gebel Semna on whose slopes the ancient mines are found (Fig. 5.33).

These essentially consist of two parallel vein shafts oriented perpendicularly to the mountain range across which they lead. Judging from the sinuous and narrow appearance of the galleries, mining activity presumably goes back to at least the New Kingdom, whereas the fist hammers observed in the slope rubble even suggest a date in the Old or Middle Kingdom. Two vertical service shafts and ruins from modern company buildings attest to recent mining activities at the site.

In the main wadi near the foot of the exploited mountain is a rectangular feature partially surrounded by eroded walls, especially at the side facing the wadi (Figs. 5.33 and 5.35). As it seems,

Fig. 5.34 Semna, mined quartz vein showing light-coloured, barren quartzes from an earlier quartz generation. Only its auriferous, central part has been removed

the walls may originally have served for retaining water, as water gullies lead directly into the walled area. The wall itself consists of wadi rubble. At the slope side, the wall crest is 2 m wide. The preserved part of the eastern wall is 44 m long. In this area an apron-shaped, Ptolemaic runner stone was recorded.

The installation very much resembles the processing site at Bir Semna.

It is remarkable that generally only few architectural remains have survived in this mining district, although this is possibly due to destruction from recent mining. They include two alignments of six, well-preserved houses with round ground plans at the entry of the district. Their walls built in shell-facing masonry filled-in with finely grained materials stand in preserved heights of up to 1.60 m. The houses date to the first mining phase in the early years of twentieth century and are known from similar arrangements and buildings at the mining site at Atalla.

On either side of an approximately 500 m long, tributary wadi leading to the southern mining areas, one can make out approximately ten more, badly ruined houses of an older date. The wadi itself is marked by numerous, rather deep excavation holes, which according to our accompanying Bedouin had been dug out by the British mining directorate in its search for water.

A large hole had been dug approximately 200 m further to the E, in the middle of the main wadi for the same purpose. It is completely dry today, and it probably never generated large quantities of water.

Apart from the mentioned fist hammers and the Ptolemaic runner stone, no other tools were found at the surface in Semna. Therefore, ore processing must for the most part have taken place at some other site, considering that significant problems connected to water shortages had probably been encountered here in ancient times, as testified also by the large water reservoir.

The ceramics in the mine area consist mainly of dark-red, coarse, and obliquely brushed amphora shards from the Arab Period. In addition, there are also some light-coloured Ptolemaic

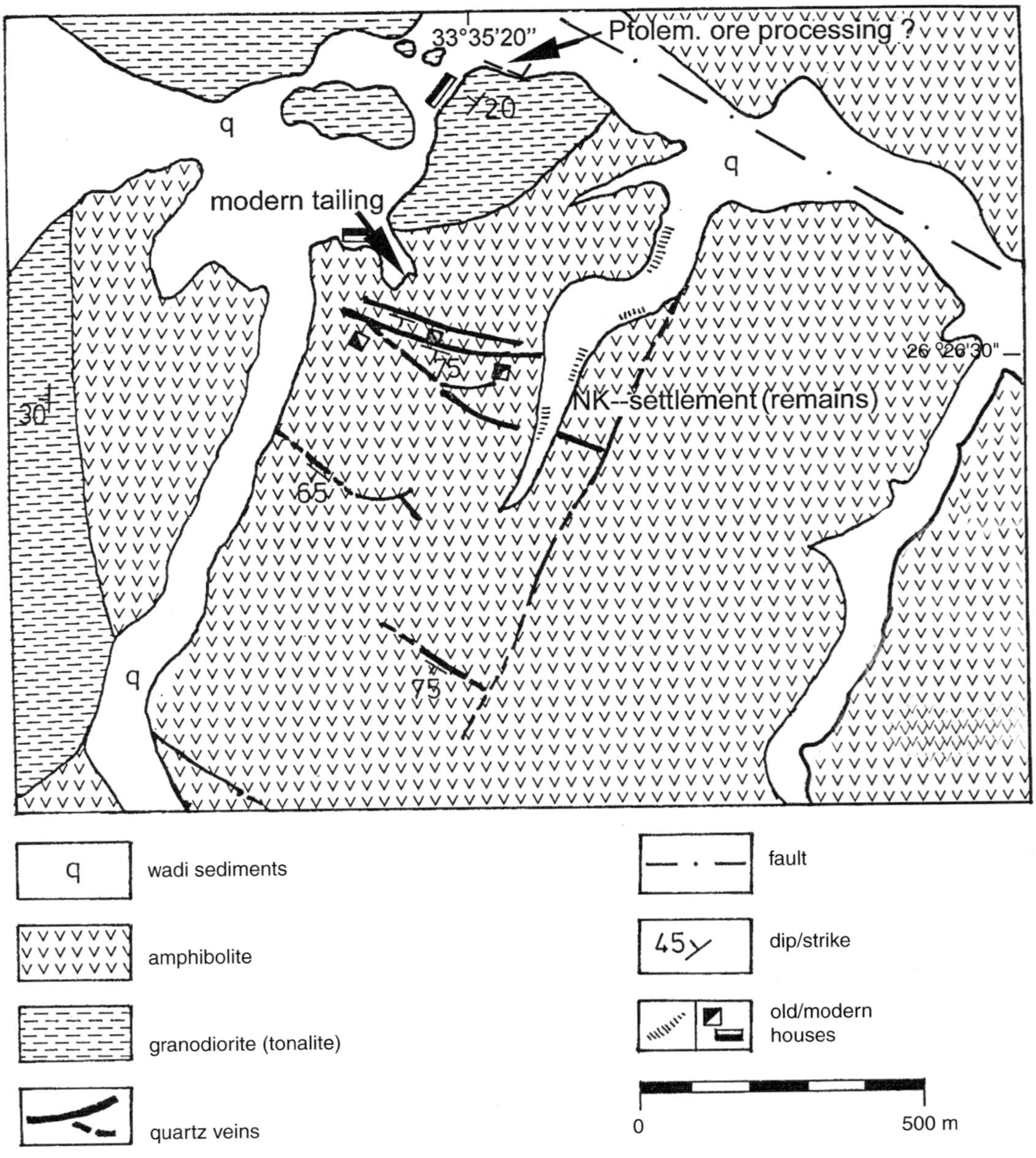

Fig. 5.35 Geologic sketch map of the close surroundings of the Semna mining district (modified after Zoheir, 2008c)

amphora shards and very few shards from the New Kingdom.

The gold quartz mineralisations of the Semna deposits are situated in zones of strong shearing in a mafic amphibolite sequences. Due to contact metamorphism by intruded granodiorites, they acquire a dioritic, at times even gabbroid appearance and are discerned as schist-like amphibolite from intensive shearing. Through assimilation and direct contact to the mafic wall-rock, the granodiorite may develop tonalites.

To the N and in the W, these mafic sequences are associated with metarhyolites which are transformed to quartz-rich sericite-chlorite schists.

The ancient quarry sites follow three quartz veins, which for their part are the product of several mineralisation generations. Noteworthy are an intensive shearing in the same direction as the quartz vein mineralisations and an iron-rich (brown stains), hydrothermal alteration of up to 3–4 m around the quartz veins.

The main extraction vein whose width varies between 1 and 6 m, strikes ESE-WNW (110°) and dips at angle between 70° and 50° S. These variations result in a lenticular vein structure, which probably originates from syntectonic mineralisation processes. As a rule, the widest quartz pockets had not been exploited. The mined gold mineralisations are restricted to the peripheries of finer shear zone mineralisations of substantially smaller thicknesses. At regular intervals of 10–15 m, pillars had been left behind to assure safety, which today furnish excellent sampling conditions. To the E the extraction zones almost reach down to the surface level of the small side wadi (Fig. 5.34). This corresponds to a maximum extraction depth of 45 m. As a result, the exhausted mines appear as large, hollow, and vertically standing negative discs. To the W, the vein disappears below the wadi sediments, whereas the ancient extraction zone already ends before the last hill.

A substantial part of the ancient dwelling huts was located close this mine. A detailed investigation by Zoheir et al. (2008c) confirms the interaction between wallrock alteration and sulphide ore precipitation. Systematic analyses of the fluid inclusions inside the quartz from these zones lead to the conclusion that their formation and the precipitation of the ores had taken place within temperatures between 220 and 270 °C.

Under modern mining operations, three shafts had been sunk into the vein system. The main shaft reaches down to about 15 m below the ancient extractions (Koshin and Bassyuni 1968).

Two anciently worked quartz veins located to the S of the main vein can only be followed at the surface over short distances. One of them is visible close to the main vein and is limited to the easternmost hill part. At the same time, the vein located further to the S (southern mine) is also the one highest up (Figs. 5.33 and 5.35). It stands out by a large quartz (ore?) waste dump (arranged according to block sizes), which seemingly grew with the accumulation of debris from later, apparently unsuccessful mining attempts.

In a strict sense, apart from the invisible gold, ore mineralisation in this case too, is limited to pyrite and to a lesser degree chalcopyrite, whereby keeping in mind that both minerals are only recognisable in the upper, oxidised zones by the typical brown stains or malachite layers on the wallrock. In addition to these main ores, there also was a subordinate mineralisation of arsenopyrite, galena and sphalerite.

At the peripheries, the ore veins are accompanied by mostly intensively sheared, aphyric andesite dikes cut at acute angles by later, additional andesite dikes.

5.2.9 **Wadi Margh-West** (Fig. 5.36)

Geographic position:	26°26′00″ N, 33°31′28″ E

Extensive ruins of houses dating to the New Kingdom are distinguished along Wadi Margh's W and E sides. This wide, sandy valley forms the northern prolongation of Wadi Baghlog. In spite of the fact that in some cases the house walls are preserved to only few stone layers, more than 50 intact oval stone mills and countless runner stones from the western side only, date the complexes to the New Kingdom (Fig. 5.37).

Particularly on the west side, ten well-preserved and about 15 more, severely eroded houses oriented parallel to the wadi reveal that mining had taken took place exclusively within the wadi alluvium (wadiworkings). Accordingly, large numbers of recorded oval stone mills support this observation. Most had been re-used later as pounding slabs as indicated by hollow dents resulting from hammering inside their central elliptical depression (Fig. 5.38). This phenomenon is known from other sites as well, and is best explained as follows: Whenever an oval depression resulting from wear reaches a certain depth and considering that the length of a grinding surface was determined by that of an arm and its width by the hand-held runner

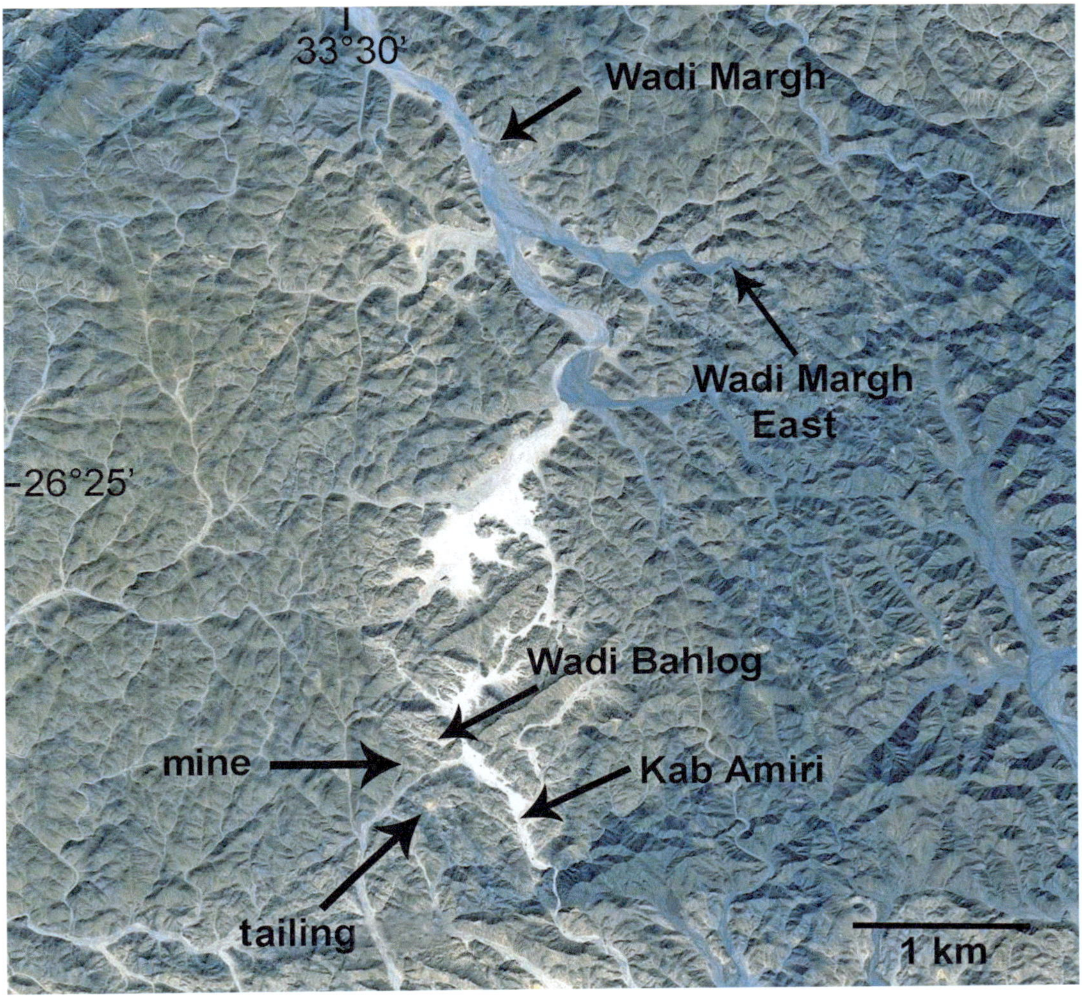

Fig. 5.36 Ancient housing and mining remains in Wadi Margh and Wadi Bahlog (modified Google-Earth image)

stone, and that therefore both parameters were not subjectively alterable, the grinding surface of the worn oval stone mill becomes too small and the milling process in its turn, ineffective. Such mills were then used as pounding slabs for crushing larger ore lumps down to pea-size. This procedure seems also to indicate that mining operations had lasted for a lengthy period. The pestles used for crushing the ores are usually cylindrical stones which, because of their partly large diameters, needed to be manipulated with two hands.

No mining traces were detected in the surrounding mountains. In other words, gold production had probably been limited to wadiworkings only, whose traces have unfortunately been wiped-out by occasionally occurring wadi floods.

Similar house alignments are also known from the eastern side of the wadi. Here they cluster in individual groups on flat elevations along the wadi edge. Here too, several New Kingdom oval stone mills had been converted to later pounding stones. A short-lasting, second occupation seems to have taken place on this side of the valley

Fig. 5.37 Typical New Kingdom row house in Wadi Margh-West

during the Arab Period, as revealed by a relatively well-preserved house, whose masonry comprises a number of New Kingdom mills. Moreover, a nearby, small hearth-like feature had been delimited on three sides by vertically standing mills from the New Kingdom.

With regard to its geology, the lower part of Wadi Margh is composed of andesite metavolcanics into which granodiorite is intruded. It is visible at the surface along the upper reaches of the wadi and may be regarded as part of an assimilation margin of a massive granite intrusion few kilometres further W. Accordingly, in the catchment area of the wadi sediments the productive mineral quartz veins correspond to those in Wadi Margh-East.

5.2.10 Wadi Margh-East

This small, tributary valley branches-off to the SE from Wadi Margh and then turns E after about 1.5 km. Many, heavily eroded house ruins are noticeable all along this narrow wadi. The surface finds include the typical New Kingdom oval mills and anvil stones from the wadi bed. The exploited gold quartz veins are located within the upper third of this small valley. They strike and dip respectively at 80°/55° S, 62°/80° S, 115°/72° N, 98°/70°S and 102°/90° Their thicknesses vary between 0.2 and 0.8 m (Fig. 5.39).

The surrounding hills are composed of coarse-grained granodiorite to tonalite. Here and there, they contain large amounts of xenoliths from

Fig. 5.38 New Kingdom grinding mill later reclaimed as a pounding stone (or mortar) in Wadi Margh-West

hornfelsic metasediments, indicating a short distance to the intrusion roof. Predominantly NE-SW striking andesitic quartz veins traverse the entire area.

The mineralised quartz veins are all associated to well-distinguishable shear zones. The sheared parts display well-visible, oxidised bands along the exploited veins. This is probably due to surface decomposition of only former pyrite, as the contact surfaces are not lined with malachite. This mineral rather tends to occur along pure, lens-shaped quartz veins with thicknesses varying between 0.8 and 0.2 m. As it seems, the latter must have promised higher gold yields, since they had evidently been given preference in the extraction process.

Fig. 5.39 Blocked up mine in Wadi Margh-East

5.2.11 Wadi Bahlog

Geographic position:	
Early Arab settlement:	26°24′09″ N, 33°30′12″ E
Mine:	26°24′05″ N, 33°30′06″ E
Tailing site:	26°23′58″ N, 33°30′13″ E

5.2.11.1 Early Arab Settlement (Fig. 5.36)

This settlement is located to the W of Wadi Bahlog (Fig. 5.40). It is somewhat hidden away in a narrow valley, running parallel behind the first mountain ridge. Here again, one comes across round huts within a weathered granite

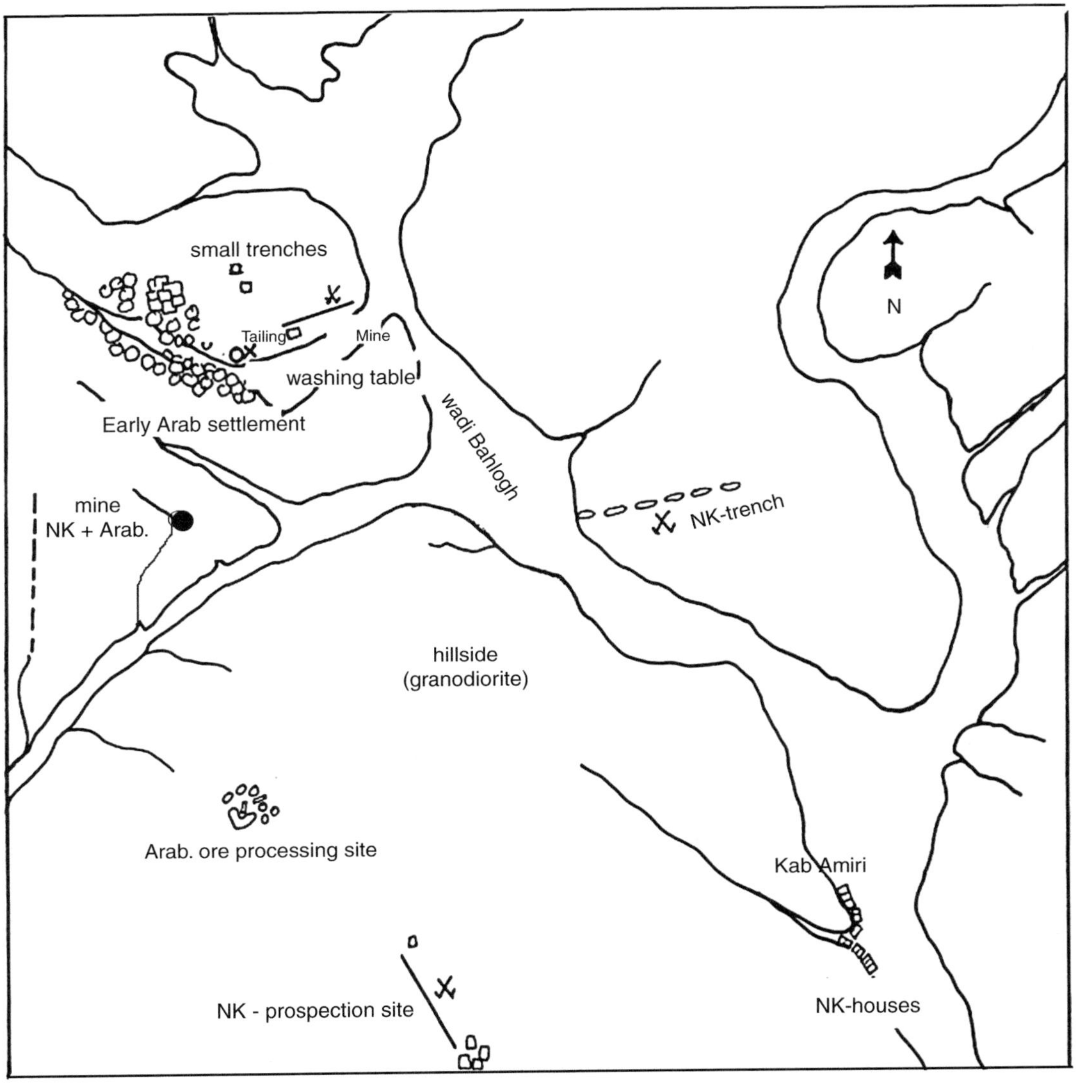

Fig. 5.40 Archaeological sketch map of the gold mining district in Wadi Bahlog (R. Klemm)

scenery bordered by the mountain flanks. Large, natural granite nodes had often been integrated to the wall masonry, especially in positions facing the valley. Both, inside and outside the huts, of which most consist of single rooms, large amounts of ceramics were observed. They include three different types of amphorae. The most common one has a hastily added spout to its shoulder (Fig. 5.41). The second type has a narrow to wide neck with only one handle at the shoulder. The third one displays a wide, cylindrical neck and two thick handles attached to its shoulder.

Furthermore, large cooking pots had been lowered into the ground outside the huts, as indicated by associated remnants of charcoal and scorch marks.

An inclined washing table amidst reddish tailing sands arranged in the shape of a horseshoe next to a large heap of unprocessed quartz lumps is located at the southern end of the wadi's eastern side. Two small, parallel quartz veins had been

Fig. 5.41 Early Arab Period domestic pottery shards from the relatively well-preserved settlement at Wadi Bahlog

mined on the mountain slope, directly above the processing site. Here and within the entire area to the N lie scattered round mills and small cubic anvil stones dressed on all six sides.

The majority of the huts concentrate on the south-western side of the valley. Some are in a very good state of preservation (with up to 1.20 m high walls). A large house with several rooms is located further to the NW flank (Fig. 5.42).

As it seems, the site had been suddenly abandoned before depletion of the ore stocks. This is partly indicated by the intact ceramics and above all, the unprocessed quartz ore heap near the tailing site.

Underground mining had been pursued approximately 250 m further SW, on the mountain crest. Outside, near the mine entrance relatively fresh chisel marks clearly reflect an attempt in the Arab Period to widen and

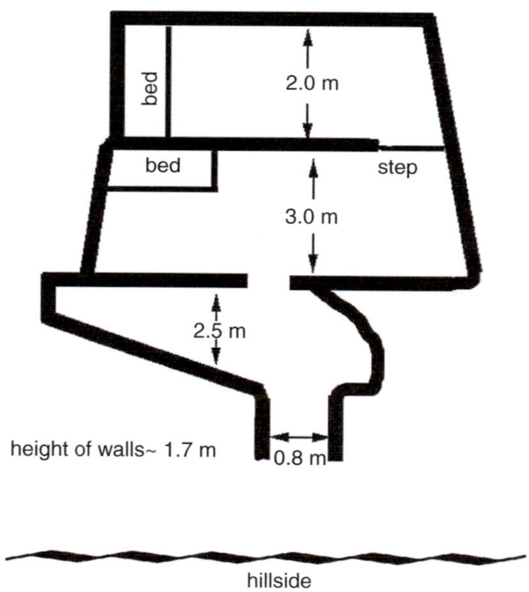

Fig. 5.42 Unusual house ruin within the Early Arab Period settlement in Wadi Bahlog

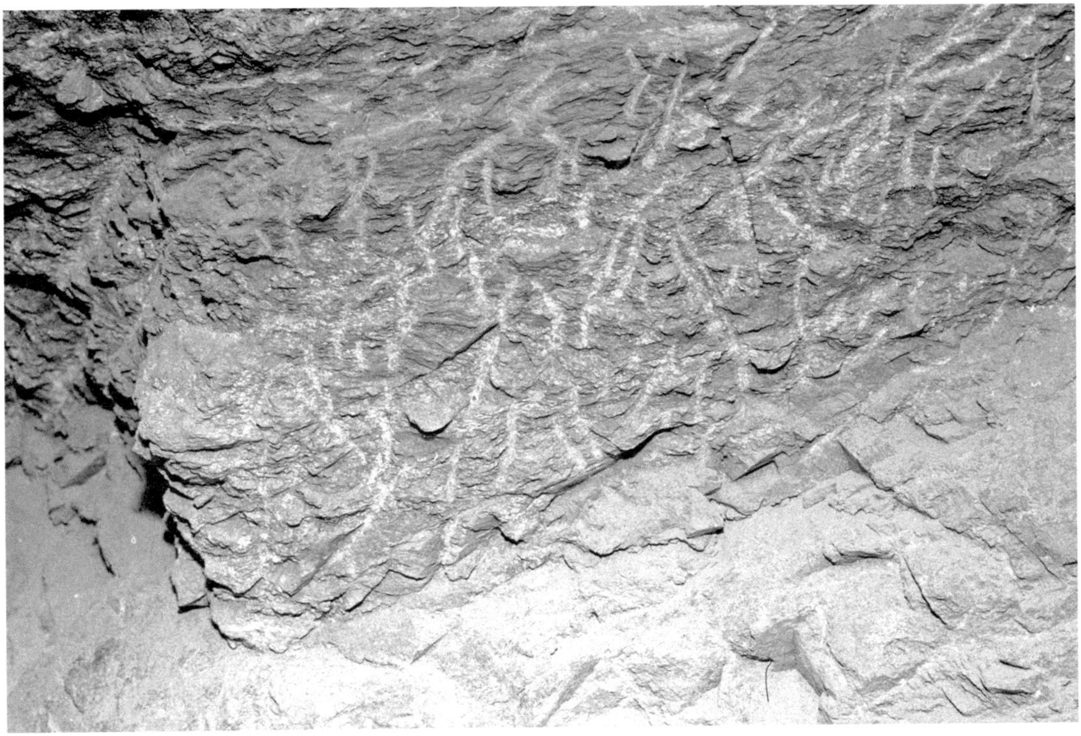

Fig. 5.43 Tool marks from sharp-edged chisels dating to the Early Arab Period in the deep mine west of Wadi Bahlog

straighten the narrow and sinuous course of the old shaft started in earlier times (Fig. 5.43). The first period of extraction apparently goes back to the New Kingdom which is confirmed by at least one elliptical, oval mill in the mine's vicinity as well as several later round mills from the Arab Period. The vein's course is followed over a length of about 100 m through the open terrain, although it was also partly mined below the surface.

A large tailing site is located about 350 m to the S of the settlement. Here again, a relatively large and horseshoe-shaped tailing heap of reddish quartz sands is associated with the remains of one of the two recorded, inclined washing tables. The second washing table is only some metres away to the N and surrounded by three, round, stone-lined basins. Three other delimited round areas are located at the eastern part of the site. At least five round mills, two broken rotor discs, and several, small anvil stones lie scattered here. Because of its close location, this tailing site probably belongs to the mentioned underground mine (Fig. 5.44).

In Wadi Bahlog too, one discerns two in essence diverging gold production methods. The first had been applied mainly during the New Kingdom with the mining of the wadi alluvium in Wadi Bahlog itself, as well as its tributary wadis. The second had probably also been used in the New Kingdom but is chiefly attested to the Early Arab Period. It consisted of trench workings and underground mining in a quartz vein system in the mountains to the W of the Arab settlement, as documented by the typical, irregular chisel mark patterns from the Early Arab Period (Fig. 5.43).

The auriferous quartz vein occurs in a transitional area defined near Wadi Bahlog by remnants

Fig. 5.44 Early Arab Period gold processing site with a large tailing and round work platforms in Wadi Bahlog

of metasediments and a prevailing hybrid granodiorite to tonalite. By assimilation of metasediments these had converted to hornfelsic rock inside the intrusion cover of granitic magmas and apparently altered their chemistry. The hybrid character of the magmatogenous surface rocks is still attested to by numerous xenoliths from former metasediments, which may be referred to as yet "undigested" or undissolved remnants of the intrusion cover.

Gold seems to have concentrated only to limited amounts in small but abundant, hydrothermal quartz veins, because during the gold mineralisation phase, the surrounding temperatures has apparently been too high for engendering precipitation. Only in cooler, marginal areas of hornfelsic metasediments, one finds more productive gold quartz veins. Through erosion, smaller, auriferous veins ended up as rubble in the wadi beds and thus too, became accessible to the ancient gold miners by selective gathering. Though this method is very demanding on personnel, it all the same led to substantially higher ore yields compared to those obtained from highly strenuous deep-mining methods, which after all had relied on rather primitive equipment.

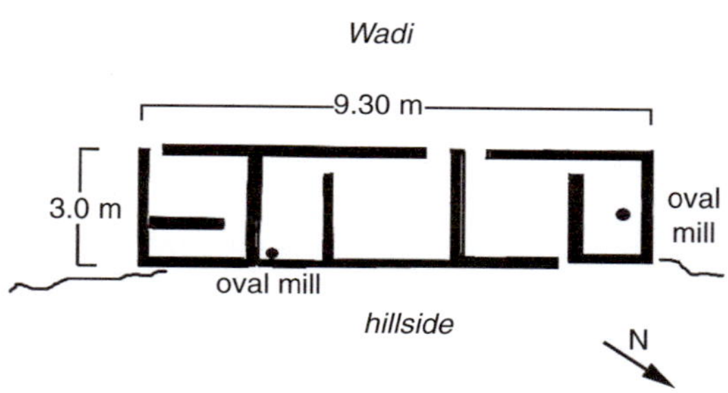

Fig. 5.45 Sketch of a typical New Kingdom house in Wadi Bahlog at Kab Amiri (R. Klemm)

5.2.12 Kab Amiri

Geographic position:	26°23′53″ N, 33°30′29″ E

Built between the granite rocks along the wadi edges at the confluents of small tributary wadis and mountain incisions of the western Wadi Bahlog, are the well-preserved house alignments from the New Kingdom whose walls still stand in heights of up to 1.50 m, which seemingly corresponds to their original height.

In all, we know of four, single, but closely neighbouring alignments. Each complex counts three to four relatively large rooms with extra subdivisions (Fig. 5.45).

Fig. 5.46 Location of the gold deposit at Gidami (modified Google-Earth image)

Between them there are rectangular, detached buildings measuring 6–10 m in length with no more than two rooms. Their entrances face the wadi and are often located very close to the next neighbouring house. In many instances, one observes that large, natural granite blocks had been integrated to the buildings' walls. The walls themselves tend to be of irregularly stacked, round granite boulders, which has a negative effect on their stability.

Numerous stone tools comprising oval stone mills, small grinding stones, pounding slabs, and surprisingly heavy cylindrical pestles are found inside and outside the buildings.

On both sides of the wadi in the nearby mountains there are several small, open but silted-up worked trenches.

5.2.13 Gidami (Fig. 5.46)

Geographic positions:	
Modern tailing:	26°23′13″ N, 33°26′59″ E
Ptolemaic settlement:	26°23′30″ N, 33°27′08″ E
North mine:	26°23′07″ N, 33°27′15″ E

As to the name of this site, one often refers to it as Aradiya (or Eradiya). On the topographic map of the Arab Republic of Egypt (scale 1:50,000), sheet NG36 K1d, the locality appears under the entry "Sirbakis". However, we opted for "Gidami" not only because the ancient settlement and the remains of the modern mine are located within Wadi Gidami, but also because it also is the name commonly used in published reports. A genuine deposit area is located at the modern uranium-prospecting site of Aradiya, right below the southern slopes of Gebel Aradiya. Therefore, the Early Arab gold mines to the E of Gebel Aradiya are referred to as Aradiya-East.

Gidami is among the larger gold deposits in the central part of the Egyptian Eastern Desert. It is located 17 km SE off the old route between Qena and Safaga (at kilometre 85) and close to a N-S connection leading through Wadi Atalla. This route has been rehabilitated only in recent years.

Mining had probably started here for the first time in the New Kingdom. Nevertheless, apart from the badly ruined settlement site on each side of Wadi Gidami in the SW of the district (Fig. 5.47) and a couple of oval stone mills near recent mining constructions, there is no further evidence to support this. Presumably, mining operations had concentrated on extracting ores from the wadi alluvium as indicated by the house alignments parallel to the wadi and characteristic for settlements from the New Kingdom. Because of intensive erosion, there are only few recognisable traces from this settlement in Wadi Gidami.

An approximately N-S oriented and up to 1.5 m wide gold quartz vein had been exploited over a length of about 1.8 km. Most of the vein had been mined in an open-cast method (Fig. 5.48), whereas deep mining techniques had been applied in other sections. With a yield of 15 g/t, the vein's gold grade is relatively high and was probably even higher within the already mined portions (Hume 1937).

Numerous, smaller quartz veins in the mountains to the N of Wadi Gidami and even much less important occurrences directly above the central settlement from the Ptolemaic Period had been exploited, as well.

As most of the rich archaeological material dates to the Ptolemaic Period, this was most probably the main mining period in the Gidami district. The deposits had continued exploitation in the Early Arab Period as testified to by later huts to the NW of the Ptolemaic settlement. Finally, mining took up again in the 1970s, as witnessed by isolated house ruins and a modern processing plant to the W of a modern shaft (Figs. 5.46 and 5.47).

The main archaeological site is located at a distinct bend of the wadi in the northern part of the district. Here are the ruins of the centralised Ptolemaic processing plant, together with a vast dwelling site (Fig. 5.47).

A large, today very much eroded tailing heap is seen in the site's SE corner. Rocks and tools litter its surface. In its centre is a basin measuring 4 × 4 m and 90 cm in depth. Its inner walls are lined with a smooth mortar (Fig. 5.49). Similar basins are known from other large Ptolemaic ore processing sites, such as Bir Semna and Barramiya. Near the middle of its western side and on its inside, the basin was equipped with a masoned, 1.60 m long socle whose surface slants towards the centre of the basin. The platform had presumably been associated to an inclined washing table next to the basin, of which though only very slight traces have been preserved.

Small tailings and especially concave Ptolemaic stone mills, including apron-shaped runner stones, are commonplace at the site. They were recorded between large groups of attached houses built on rectangular or round ground plans (Fig. 5.50). The mills are mostly of red granite porphyry, to a lesser extent of light granodiorite. In addition to the central washing table site, gold processing therefore also occurred at a series of other locations. The water supply had been assured from a well at the NW edge of the settlement at the location of a gaping, but dry hole surrounded by ancient spoils from its excavation (Fig. 5.51).

In a small wadi incision to the NW of the tailing site, there are a number of huts dating to the Early Arab Period. These separate, round architectural units had been built in double-face

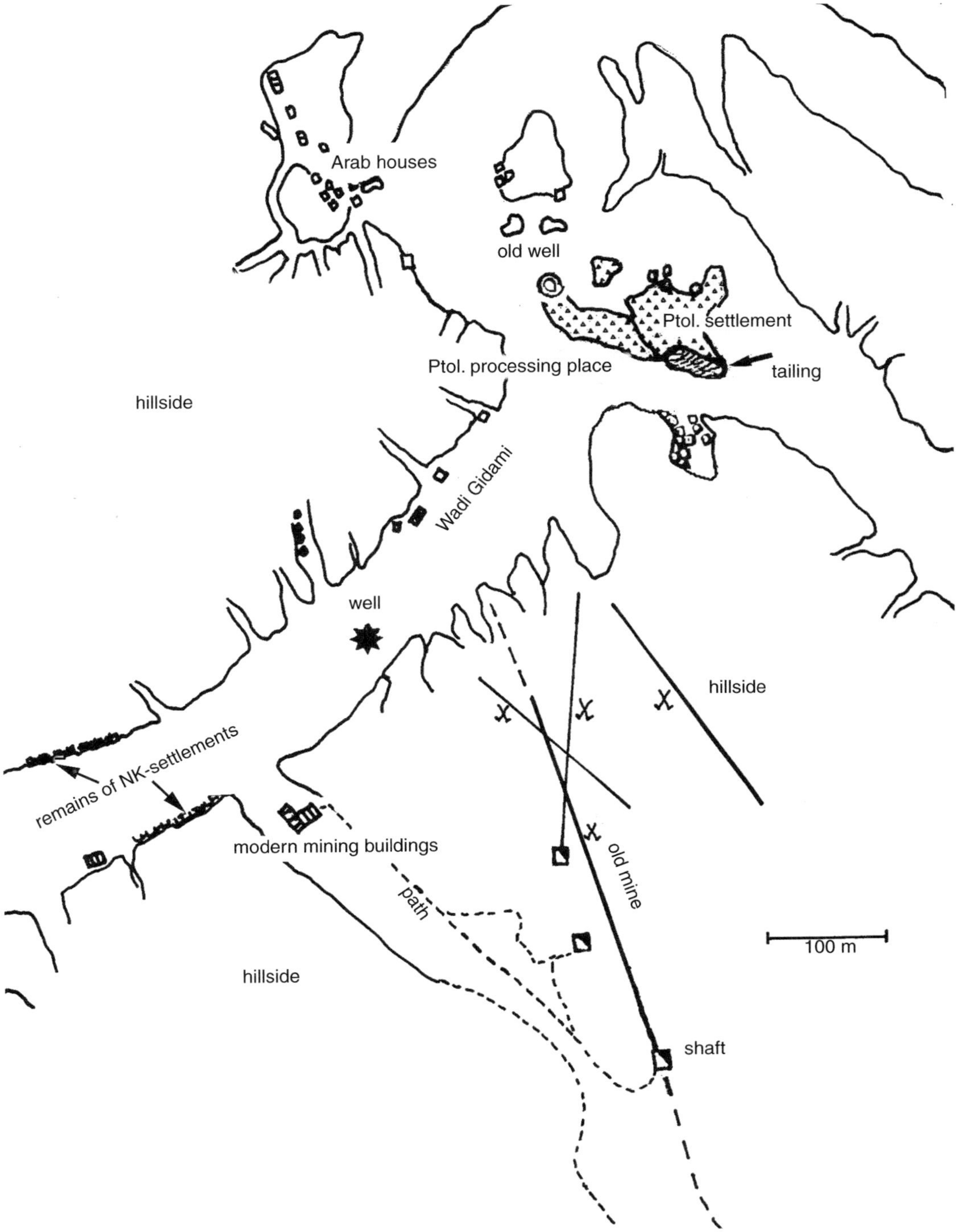

Fig. 5.47 Archaeological sketch map of Gidami (R. Klemm)

Fig. 5.48 View into the trench pit at Gidami, probably already operated in the New Kingdom. In the Ptolemaic and Early Arab Periods the mine was extended to a depth of ~35 m. The shafts and drifts date to the twentieth century

Fig. 5.49 Waterproof basin from the Ptolemaic Period, as part of the ancient gold processing site

Fig. 5.50 Typically concave gold mill with large, two-hand runner stone from the Ptolemaic processing site at Gidami

Fig. 5.51 Former waterhole with dry vegetation cover, probably the central well from the Ptolemaic Period

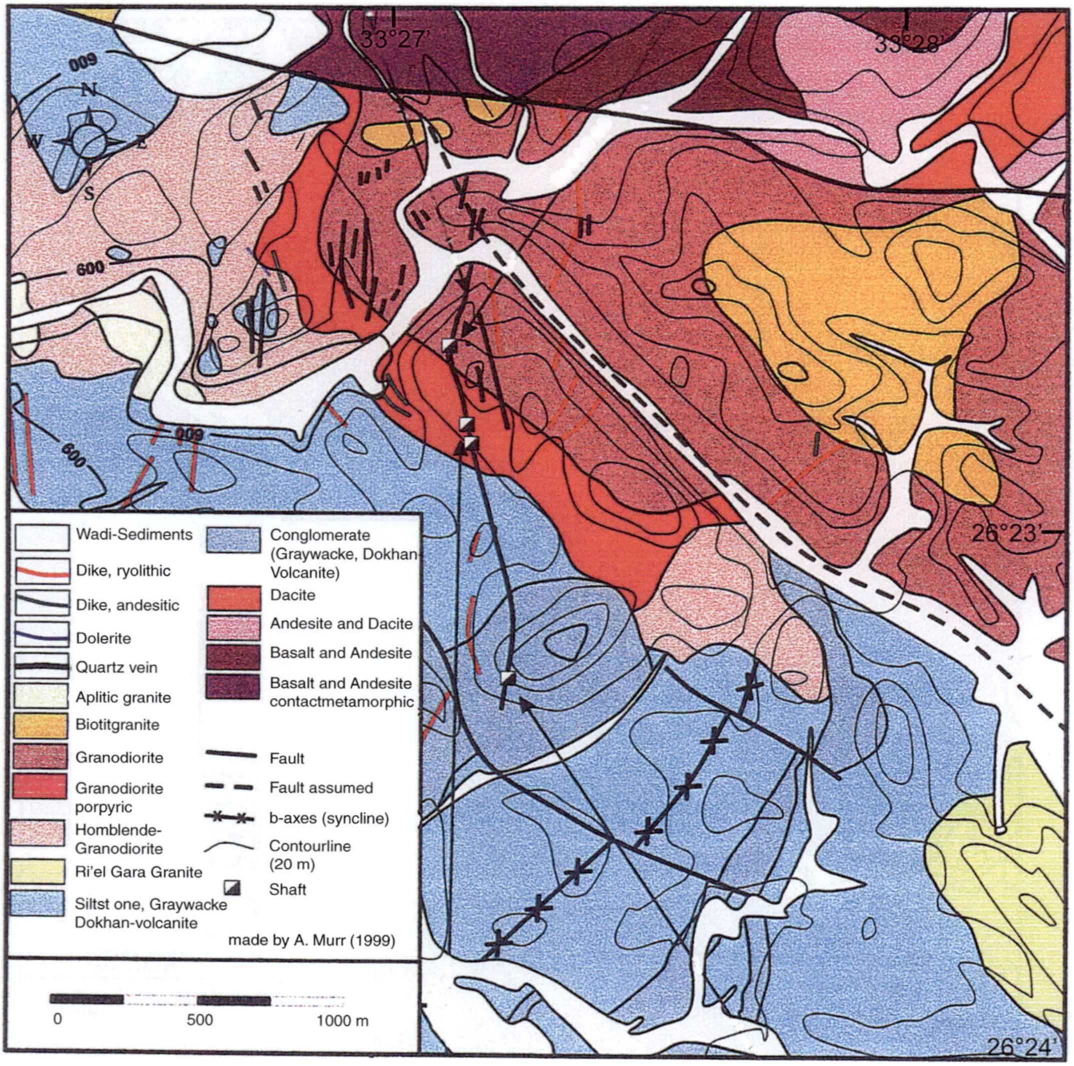

Fig. 5.52 Geological map of the surroundings of Gidami (Murr 1999)

masonry and are associated to round mills and small anvil stones.

A still operable well is located in the middle of the wadi, about 450 m SE of the Ptolemaic settlement. It is probably of recent date and may have some connection to the mining activities of the late twentieth century.

At the western margin of Wadi Gidami and close to the ruins of the recent miners' homes, whose walls contain Ptolemaic runner stones, there is an amassed heap of ore mills dating from various mining periods.

The pottery of the Ptolemaic settlement consists of light-red bowls and light-coloured amphorae with wide necks and remarkably thick handles. There are also some red-brown, obliquely brushed amphora shards dating to the Early Arab Period.

A. Murr (1999) carried out a detailed geologic description, whose report the following information draws upon:

Gidami's geologic situation (Figs. 5.52) is similar to that of Fatira. Molasse sediments of the Hammamat formation, which are intercalated with Dokhan volcanics, are intruded by granite

(Ri'el Gara Granit). The mineralisation is oriented along the hybrid, granodiorite margins of the granite and the thereby metamorphically affected Hammamat wallrock. The productive main vein strikes NNW-SSE and dips at 86° W. It is located in a left lateral shear zone. Because this granite intrudes molasse sediments of the Hammamat type, it also belongs to the group of post-tectonic granites. With decreasing distance to the quartz vein, the granodiorite becomes more deformed. The deformation is brittle which particularly diminishes the grain size of the feldspars. Quartz re-crystallises post-kinematically. It then displays no undular extinction and exhibits a pavement structure. In contact with the quartz vein, the granodiorite hydrothermally is altered to a sericite quartz rock. Relics of mafic minerals no longer occur. A more detailed description, especially of the geochemistry of the volcanics is given by Murr (1999).

The wallrock of the mined, 1–1.5 m thick gold quartz vein and the outer dike areas contain pyrite as the primary ore. It has partially or completely altered to lepidocrocite. Gold is contained in the lepidocrocite, as well as in fine clefts that pierce lepidocrocite crystals. This reveals a short-term mobility of gold during the oxidation of primary sulphides to lepidocrocite. In this case as well, hydrothermal gold transport occurred in sulphide complex fluids. Towards the centre of the vein, galena is increasingly associated to free gold, which is partly found in quartz, partly in pyrite. However, the centre of the vein itself consists of white quartz containing no ores.

More remains from Arab Period settlements within the upper Wadi Gidami, which apparently specialised on wadiworkings, are located at 26°24′31″ N, 33°24′10″ E (about 20 huts) and 26°24′10″ N, 33°24′26″ E (about 50 huts).

The Maghayir mine, located 4 km SE of Gidami is a recent site from the twentieth century and most probably contemporary with the operations at the latter.

5.2.14 Abu Gaharish

Geographic position:	26°22′27″ N, 33°38′48″ E (northern settlement)
	26°22′10″ N, 33°38′43″ E (southern settlement)

At the south-eastern edge of a plain within the granite body of Gaharish (Fig. 5.53) there are two settlements located at a distance of about 500 m to each other. The one to the N consists of around 30 house ruins and the remains of a rectangular water collecting basin. The one to the S consists of six to eight house ruins originally built from round granite nodes. Most houses have two rooms with inside doorways and occasionally, ill-preserved, masoned sleeping platforms. Intact oval stone mills and some mill fragments dated to the New Kingdom are discernible near the houses (Fig. 5.54). The sites are therefore seen as New Kingdom dwelling complexes.

The ancient mine trenches also occur in two different areas. The northern mine stretches over a distance of 350 m and follows a quartz vein displaced by shearing. The southern has two stopes close to the settlement (Fig. 5.55).

In recent years, several quartz vein mineralisations have been inspected at Abu Gaharish for tungsten ores (scheelite and wolframite). They all are located in the coarse-grained and pink-coloured granite stock near the contact zone with the Hammamat wallrock.

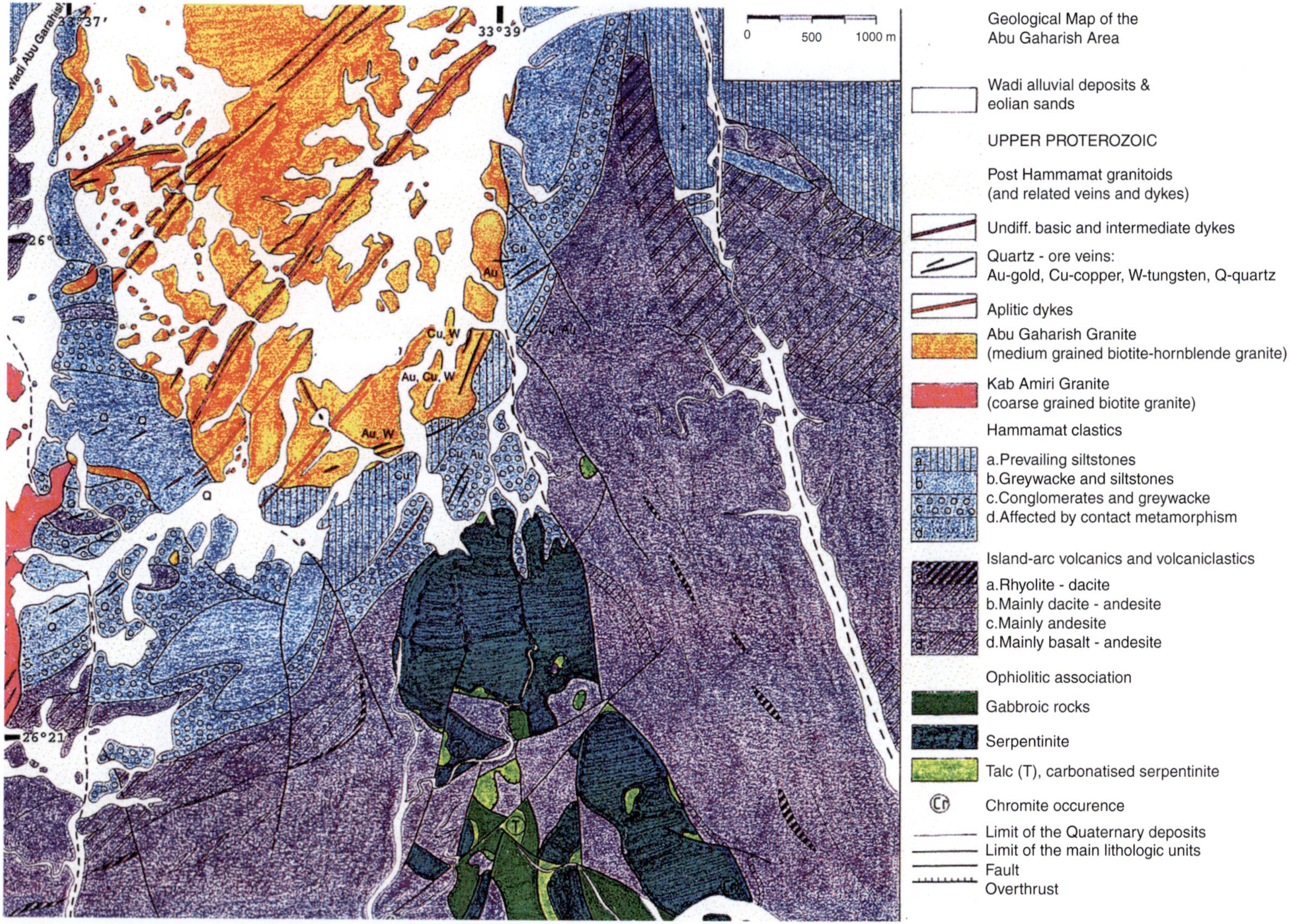

Fig. 5.53 Geological sketch map of the southern surroundings of the Abu Gaharish granite (H. Kräutner)

Fig. 5.54 New Kingdom houses bordering a plain in the granite complex of Abu Gaharish-South

Fig. 5.55 New Kingdom gold mine at Abu Gaharish-South. The pinch and swell shearing of the quartz vein is well-distinguishable in the photograph

5.2.15 Abu Gerida

Geographic position:	26°21′38″ N, 33°21′15″ E

Wadi Gerida has partly the appearance of a gorge as it leads through the fine-grained metasediments of the Hammamat sequences. Because of the neighbouring granite intrusion to the E, the Hammamat suite has converted to dense hornfelsic rocks. This hornfels is easily mistaken for andesitic metavolcanics. This happened to us, the Soviet-Egyptian Technoexport team, as well as the geologists of the MINEX Company. Traces from mining attempts as well as marginal, up to 30 cm thick pyrite impregnations are recognised along a NE-SW (53°/19°S) striking shear zone system through the metasediments (Fig. 5.56). The shear zone itself is mineralised with quartz and hematite. A rhyolitic porphyry dike intruded the shear zone parallel to this mineralisation. Later, the shear zone was again sheared causing a slight displacement of the quartz vein. Such rhyolite-porphyry dikes are quite common in the mountains. Another mining attempt had been made at the valley floor, about 500 m further W. This probably happened in the Early Arab Period, as supported by the sparse remains of the huts and some characteristic round mills.

Fig. 5.56 Trial mine of unknown age in Wadi Gerida (*arrow*). A pink rhyolite porphyry vein runs perpendicular to the mineralisation

5.2.16 Hamama I

Geographic position:	26°21′59″ N, 33°20′32″ E

A recent track established by the MINEX Company branches-off from Wadi Abu Gerida towards the SE and leads to the gold deposits at Hamama. Here, the company carried out prospecting work in the beginning of the 1990s. First, one arrives at the deposit Hamama I, located immediately next to the track. The mining area which is visible through a marked incision in the mountain appears at first sight to be rather disorderly (Fig. 5.57)

Nevertheless, two main extraction phases were distinguished:

1. Deep New Kingdom gold mines descending in narrow and sinuous galleries into the mountain. The sought after gold was contained in limonitic ores whose iron contents had no value in this period (Fig. 5.58).
2. Ptolemaic Period extraction phase, in which existing galleries had been extended and new lodes followed. Mining was still aimed at gold extraction. Here too, as in most cases of Ptolemaic take-overs of Egyptian mines, its practical location near a passageway wadi was to some extent important. The fact that the iron contained inside the ores still attracted little or no interest at all in this period can be recognised by the Ptolemaic huts just N of the extraction site. While on one hand limonitic ore blocks had been used as building material, as exemplified by a hut exclusively built from iron ore rocks, no traces from slags or any activity related to smelting can be detected on the other.

Like the settlement from the New Kingdom period, the Ptolemaic occupation is located on both sides of a new road. The road has been hastily laid out in a depression to the N of the

Fig. 5.57 Exploited New Kingdom and Ptolemaic trench pit at Hamama I

Fig. 5.58 New Kingdom and Ptolemaic mines at Hamama I

mines and cuts right through the ancient settlement with no regard taken its archaeological remains (Fig. 5.59).

Tools contained in the debris heaps from the roadworks date to the same occupation phases identified above. The oval New Kingdom mills

Fig. 5.59 New Kingdom settlement at Hamama I. The site was unknown at the time when a recent road was traced through its middle

as well as the runner stones and anvil stones which later were sometimes integrated to Ptolemaic wall masonry consist of a light, fine-grained diorite. The apron-shaped runner stones and fragments from associated, concave stone mills from the Ptolemaic Period are of reddish, coarse-grained granite. Round mills from a later extraction period are totally absent in these heaps. These finds were of course recorded independently from in their original archaeological contexts, thus rendering any further statement concerning the occupation phases impossible.

3. A third extraction phase dates to the Early Arab Period. According to the surface finds, mining seems however to have exclusively concentrated on iron production. Slag lumps are found in large amounts along the outer periphery of the mine. No round mills were found, which although otherwise may be expected within such gold ore processing contexts.

In the hills to the N of the main mines, several small vein shafts and ruins of adjacent, simple round huts were found.

The mining district at Hamama is located within the dense hornfels of former andesitic metavolcanics. This deeply mined extraction area of somewhat chaotic appearance is set within a highly hematite-rich quartz vein striking approximately N-S in a vertical dip. The richest ore parts had been extensively mined in lenticular sections between 1 and 3 m wide. At the surface, however, the widths may reach 20 m, whereby the single veins are separated by wide abutments of the heavily sheared, surrounding rock.

According to the left behind ores, there are considerable amounts of goethitic limonite next to the predominant hematite. This finding is evocative of the former presence of an iron-cap otherwise known as a gossan-deposit. This leads to suggest that next to hematite there were also large quantities of sulphide minerals such as

Fig. 5.60 Three Early Arab Period house complexes at Hamama II

possibly pyrite, considering the complete absence of malachite traces. MINEX was able to report up to 16 % Zn-contents in the form of smithonite ($ZnCO_3$) in a number of gossan samples.

5.2.17 Hamama II

Geographic position:	26°21′02″ N, 33°20′25″ E

On a wadi terrace about 1 km E of the mining district of Hamama I is an architectural complex consisting of ten interlinked rooms. The walls still attain heights of up to 1.20 m (Fig. 5.60).

An unusual finding about this building, which is located closely to an ancient quartz vein mine, is that many of its rooms still contain potsherds. Most belong to typical pot- and storage shapes known from mining areas of the Early Arab Period. A cooking place recently cleared outside its western wall, still exhibits a globular pot which had been lowered into the ground. About 300 m W of this site is another, presumably contemporary house with 12 rooms, although in a distinctly worse state of preservation. Additional house ruins are found about 130 m to the SW.

The mentioned, up to 4 m thick quartz vein to the E of the large Arab house complex comprises at least three quartz generations. The wallrock next to the vein is essentially made up of rhyolitic metaignimbrites with partly well-recognisable structures from pyroclastic welding, and to a lesser degree of more bulky metarhyolites, mafic metavolcanics, and metaarkoses.

The exploited vein is again associated with a shear zone system and follows a series of intensely sheared andesite dikes. These cut the surrounding, approximately NE-SW striking metavolcanics. At its NE side, the vein is exposed in a height of up to 5 m facing the valley, whereby only the zones along its margins had been mined. At the SW side it is up to 20 m deep and driven at a height of about 6 m into the mountain. In-between some mine shafts are distinguishable. Next to hematite-rich quartz, the minerals in the

Fig. 5.61 Ptolemaic processing site with a large tailing, about 5 km W of the Hamama mining district

untouched parts consist of carbonate whose decomposition by dissolution near the surface had shaped a cavernous and ferrous gossan matrix. The entire shear zone with its therein-contained quartz vein was slightly folded, which is observable down to the scale of a hand specimen.

The main vein was drill-probed by MINEX in 1989 to a depth of 75 m. Here, the ore minerals consist mainly of pyrite, subordinately chalcopyrite, galena and sphalerite in addition to a pinkish-red carbonate (presumably containing Mn). The gold grade varies around 7 g/t (J. Mc Hugh, personal communication). The drill cores revealed that the (?) Mn-carbonate content in the vein increases with depth, but above all that it becomes highly sulphidic. Consequently, it also showed that the formerly exploited, upper parts evidently belonged to the gossan cap of these sulphides. On the other hand, the intermittently richly available hematite points to a higher primary oxidation degree in these parts.

5.2.18 Processing Site Hamama-W (Fig. 5.61)

Geographic position:	26°21′26″ N, 33°17′20″ E

About 5.5 km downstream in Wadi Gerida a large processing plant is located in the middle of the valley. The area is situated already within the zone of the Nubian sandstone. The site was probably the best location for a water source in the Hamama mining district. The area is in fact geologically defined by an aquiclude at the interface between the impermeable bedrock and the overlying, permeable Nubian sandstone. A well at this location would thereby feed from the occasional runoff from two valleys and perennially from the water reservoirs within the sandstone.

Today the site measures about 100 × 100 m. Originally, it was considerably larger, as seen from the relatively steep edges at the wash sand heaps that have resulted from stream-cutting erosion.

Fig. 5.62 Ptolemaic runner stones in the tailing sands of the large processing site at Hamama-West

The tailing sands, which accumulated there, are much pervaded by rock fragments, probably due to deposition of wadi rubble after occasional rainstorms.

Numerous remains of Ptolemaic, concave mills and associated, apron-shaped runner stones certainly support a Ptolemaic occupation at the site (Fig. 5.62).

Along the edges of the tailings, and sometimes at their summits, one can barely make out single work platforms. They characterise by accumulations of mill fragments, often from dark-red, coarse-crystalline granite, apparently originating from Abu Had, whose rock is often observed in mills from the Ptolemaic Period.

5.2.19 Aradiya-East (Fig. 5.63)

Geographic position:	26°20′36″ N, 33°30′03″ E

Aradiya-East is exclusively dated to the Early Arab Period. It has at least seven tailings of a strikingly reddish-pinkish colour, which is indicative of very ferrous ores (Fig. 5.64). Three of the tailings are very well-preserved and respectively arranged in a horseshoe-shape around a washing table, whose inclined surface as usual faces the tailing heap (Fig. 5.65).

Scattered, small huts group along the edges of the mountains in narrow, relatively steep gorges leading to the valley bed. Their surface finds display the customary assemblage of round mills and small cubic anvil stones. As observed at other Early Arab sites, runoff from sporadic rainfall waters was collected at the foot of some of the gulches by aid of retention dams.

Thanks to its remote situation in a barely accessible valley bed, making it inaccessible even to cross-country vehicles, the site is remarkably well-preserved. The origin of the quartz ores processed here at an apparently large scale remains nevertheless obscure.

Despite intensive investigations with our guide Sheikh Salama who claimed he had seen a mine high on top of a nearby mountain to the S, we

Fig. 5.63 Early Arab Period gold mining site Aradiya-East with pinkish tailings and numerous huts (modified Google-Earth image)

were unable to find any traces related directly to mining inside the immediate perimeter around the site. Whether the ores had been extracted from the local wadi alluvium remains questionable. Since the wadis here are relatively narrow and are so extensively filled with rubble, any noteworthy wadiworking activity seems relatively unlikely here. However, the questions connected with this site ask for a more detailed answer.

The surrounding mountains consist of contact metamorphic hornfels rocks of former Hammamat sediments and possibly as well, island-arc volcanics whose identification is problematic due to modifications in the metamorphic structures. Generally, the hornfels structure has disintegrated by the intrusion of granites from below. Further, they are spread-through by granite apophyses. Therefore, the determination as to whether affected areas still consist of hornfels structures or whether they already are magmatic, is hard to decide. Such threshold areas between granitic magmas, hybridly affected through partial melting and former sediments exposed to contact metamorphism, are according to the hitherto observed conditions the ideal prerequisite for auriferous quartz vein mineralisations, provided that tectonics furnish the necessary space. Exploitable gold quartz veins are therefore quite conceivable in this case. However, where are they?

Fig. 5.64 Remains of Early Arab Period huts and tailings (*light pink*) at Aradiya-East

Fig. 5.65 Well-preserved Early Arab Period washing table at Aradiya-East, whose water collecting system is buried by debris

Fig. 5.66 Old Kingdom pottery shard at the trench pit on the N slope of Wadi Sagia

5.2.20 Wadi Sagia

Geographic position:	26°20′59″ N, 33°34′27″ E

Wadi Sagia is a small mining district consisting of several buried shafts. They are located way up the slope along the wadi. Due to weathering, the slate-like rock has disintegrated to small pieces, which together with the steep slopes transforms the mountain into difficult terrain. As in the case of Soleimat and Wadi Dara, because of the clearly visible, green stains along the wallrock, it again is debatable to which extent the demand for copper ores had motivated the mining activities here. In the mine's waste dumps, one finds large amounts of fist hammers and grooved stone axes, which together with a recovered fragment of a bowl with a red slip (Fig. 5.66) can be dated to the Old Kingdom and even earlier. At least 40 stone hammers were counted near the edge of a mine at an intermediary level (Fig. 5.67).

On a small outcrop near the foot of the lowest mine, which too was an underground operation, one barely distinguishes the ruins of several huts. Otherwise, no architectural remains are discernible in the surrounding wadi from the mountain top, which is the location of yet another small mine.

Geologically, the mined quartz veins are located in metasediments of the Hammamat formation, altered to hornfels by contact metamorphism. As in the area around Wadi Hammamat

Fig. 5.67 Fist hammers from the trench pit system in Wadi Sagia

itself, they are intensely imbricated with serpentinites and metavolcanics of the ophiolite series. This is well-documented by the occurrence of serpentinite lentils only few hundred metres away from the shaft mines and about 2 km to the N in a larger mountain massif.

Typical for the early gold mines is again the appearance of green malachite linings at the joints of the hornfels wallrock along the quartz veins. Together with internal, brown iron stains they reveal former pyrite- and chalcopyrite contents, which often are connected with productive gold levels.

5.2.21 Abu Had

Geographic position:	26°16′30″ N, 33°30′58″ E

A small Early Arab site is located in a small tributary valley at the southern margin of the granite plain to the S of the Aradiya granite stock. It is located within mafic metavolcanics and consists of only few hut remains and adjacent extraction trenches in quartz mineral shear zones. Carbonated shear zones and open, butte-like quartz had not been mined here.

However, the trenches alone hardly justify the presence of the settlement. The main production area probably concentrated in the wadi alluvium where numerous hollow depressions and particularly, conical rubble heaps measuring about 50 cm in height testify to former mining activities. As a whole, the site does not give the impression of a major gold producing site, though its existence underscores the general viability of this deposit area.

5.2.22 El-Rebschi

Although the general map established by EGSMA states a place called el-Rebschi, no such gold producing site was found during our investigations at neither this location nor its closer surroundings. At any rate, according the geologic framework data, no important gold mineralisation can be expected within this area.

5.2.23 Wadi Atalla el-Mur

Geographic position:	26°12′46″ N, 33°31′22″ E

To judge by the settlement remains in this district, the gold quartz vein located here had been mined in the Early Arab Period. It strikes 36° NE-SW and dips at 37° NW. It is up to 1.2 m wide and displays traces from exploitation over an approximate distance of 50 m. Because it is buried under heavy debris, the actual extraction depth cannot be determined.

The geology of the deposit's surroundings is somewhat problematic, principally because of the classification of the wallrock. At first sight, it seems to be composed of acid to intermediary metatuffs. Even so, in reality we are here dealing with siltstones and greywacke of the Hammamat formation altered to hornfels through contact metamorphism through the intrusion of granodiorite visible in the SW of the deposit. Numerous apophyses of this intrusion are also found in the hills NE to the exploited quartz vein. This vein is associated to a felsitic dike of which numerous more are observable in the vicinity. The entire setting with regard to the dikes and the wallrock is very much reminiscent of that encountered at Fatira, although with considerably less prolific mineralisations. To the NW impressive mafic metavolcanics with partially well-recognisable, pillow-shaped structures underlie these metasediments.

5.2.24 Atalla (Fig. 5.68)

Geographic position:	
Early Arab settlement:	26°09′17″ N, 33°30′45″ E
Main shaft:	26°09′14″ N, 33°30′52″ E

The mining district at Atalla had been under exploitation between the New Kingdom and the Early Arab Period, and later in the early twentieth century. A number of fist hammers from the Old/Middle Kingdom even suggest an earlier gold extraction, although these are so far the only evidence for this.

Modern mining has regrettably again caused much destruction at the ancient sites in the Atalla district. It must, however, be accorded that generally more care has been given to heritage protection here than elsewhere in development schemes of modern, industrial hardware.

The Atalla mining district lies in a tributary wadi of the wide Wadi Atalla, a traditional route connecting Wadi Hammamat to the Qena-Safaga highway. The road has recently been improved by the MINEX Company so that comfortable driving is now possible.

A number of ancient mines of the district are discernible on the archaeological sketch map (Fig. 5.69). They run almost parallel to each other and are oriented in a NNE-SSW direction. A tailing site with two, still recognisable washing tables is located in the middle of the district, between two, sizable Early Arab settlements on

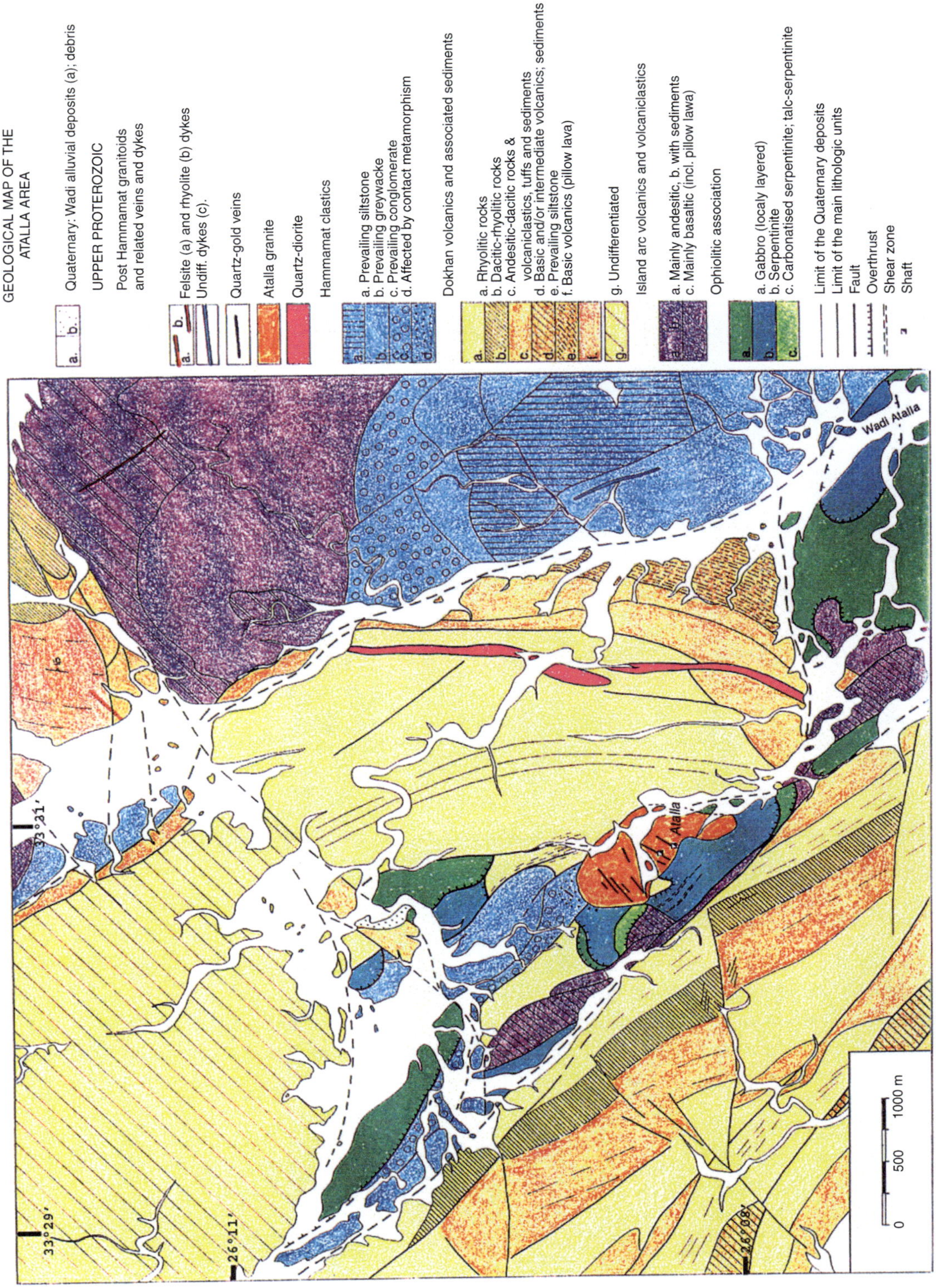

Fig. 5.68 Geological sketch map of the of the Atalla deposit area (H. Kräutner)

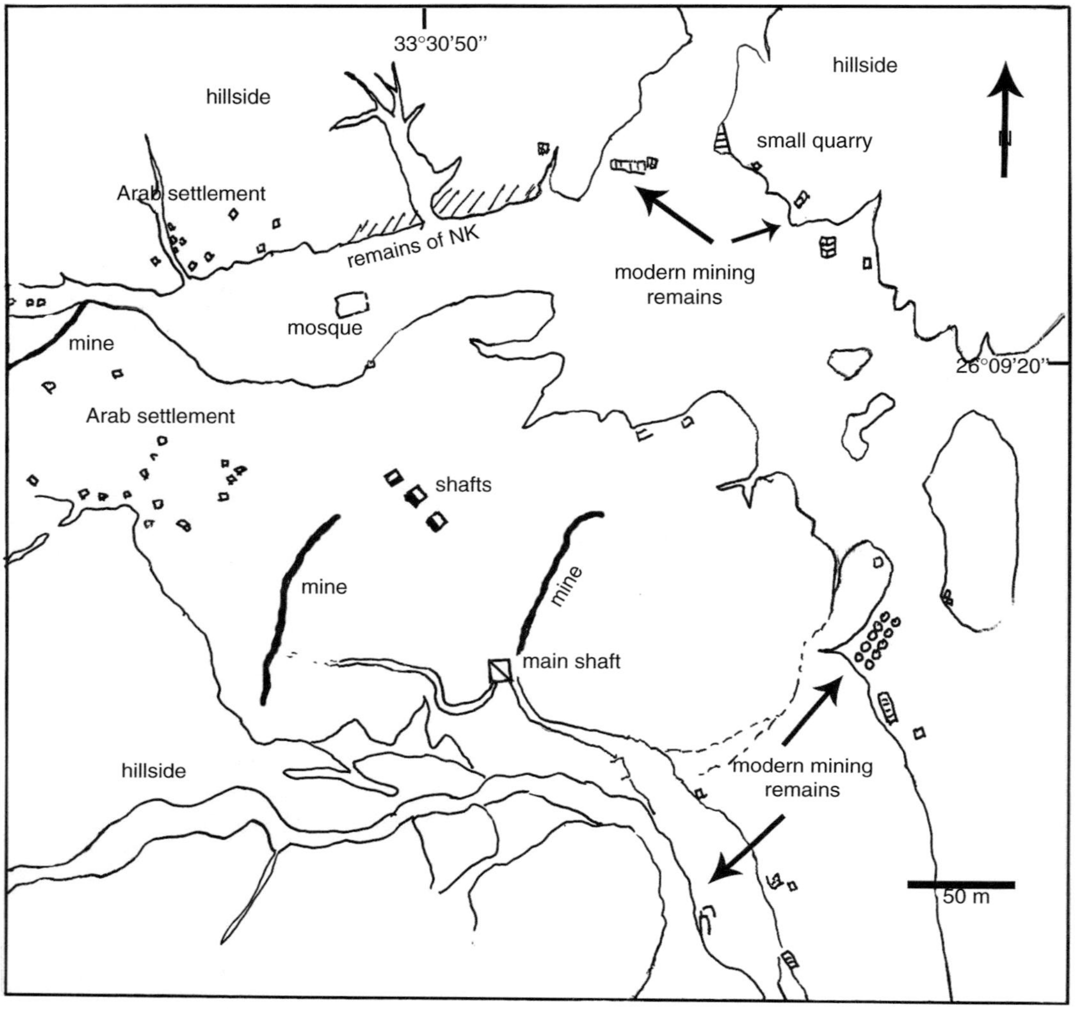

Fig. 5.69 Archaeological sketch map of the Atalla deposit area (R. Klemm)

both sides of a wadi. The washing tables are surrounded by round mills and the typical anvil stones. Shards from diagonally brushed, dark-red pottery, greenish faïence vessels with pale- or dark-green speckled glaze, as well as greenish glass shards compose the Arab Period assemblage.

Remains from single and clustered huts dating to the Early Arab Period occupy numerous sites beyond this area. Their ground plans are more or

Fig. 5.70 Remains of recently destroyed inscription by Ramesses III on a serpentinite rock wall in Wadi Atalla

less circular and, depending on the distance to the available building material, their walls are built from either granite nodes or andesite rock slabs. Generally, the ones from andesite rock are in a better state of preservation. Here and there, New Kingdom oval stone mills form part of the ruins' masonry.

To judge by the morphology of the mines as well as the oval mill fragments, mining at Atalla had started in the New Kingdom. Some evidence from fist hammers observed at the site surface may, however, even suggest an Old/Middle Kingdom occupation. Except for a typical house alignment associated with numerous New Kingdom mills in a tributary wadi to the S of Wadi Atalla, houses or huts from the New Kingdom were otherwise hardly identified in the main mining district. It is therefore likely that the modern ore processing complexes for the most have destroyed the extensive evidence from the New Kingdom.

A significant indication for a nearby New Kingdom mining industry is furnished by a hieroglyphic inscription containing the cartouche of Ramesses III (Green 1909) on a rock in the main wadi, shortly before the inlet to the mining district. The inscription was still there in 1989, but by 1995 it had for the most been chopped out (Fig. 5.70).

A survey in the bordering wadis delivered no hints suggestive of ancient wadiworkings or

Fig. 5.71 Exploited New Kingdom trench pit in the Atalla deposit

additional traces from New Kingdom settlements. Mining in this period was therefore probably carried out exclusively underground, which is consistent with the typically sinuous and narrow layout of the old mines at Atalla. (Fig. 5.71).

At the beginning of the twentieth century when mining at Atalla resumed, several, meanwhile dilapidated buildings were erected in the S of the area. The most conspicuous are two alignments of five round houses, each consisting of one room and whose walls are preserved to heights of up to 1.70 m (Fig. 5.72). A similar complex can also be seen at Semna. Within the perimeter of these houses, one encounters masses of glass shards from purple, English mineral water bottles, beer bottles, and English porcelain in addition to numerous ceramic crucibles used for gold assaying (Fig. 5.73). A large, modern tailing and connecting ways to deep shafts, lye

Fig. 5.72 Miner's houses from the early twentieth century at Atalla

Fig. 5.73 Bottle shards and gold assaying crucible from Atalla dating to the British mining period in the first half of the twentieth century

Fig. 5.74 Remains of a Ptolemaic processing site at the eastern slope of Wadi Atalla

basins, debris heaps etc. brand the ambience of this recent and derelict mining area.

Old Alamat towers scatter around the mountains of the entire district.

An interesting building, which from a distance may be taken for a watch-tower, is located at the eastern slopes of Wadi Atalla, near Bir Kubaniya, a little further S to the Atalla mine district. A closer view in fact shows that it had once probably been a processing site, in which the slope had been cleverly used for integrating the water reservoir system. It is composed of two basins in the upper part, in whose centre remaining deposits of tailing sands are clearly visible. In the lower part two pillar stubs frame another basin (Fig. 5.74). Finally, two rectangular huts are located in a symmetrical arrangement at the foot of the slope. Their wall masonry contains oval stone mills from the New Kingdom. A fragment of a Ptolemaic runner stone converted from a former New Kingdom mill lies amidst the slope rubble. The complex probably represents a Ptolemaic gold washing site that had functioned with the water from the adjacent well (Bir Kubaniye).

Murr (1999) thoroughly examined the Atalla deposit, on whose observations the following report essentially draws:

The geology at Atalla is determined by a regional NW-SE striking shear system marked by numerous, formerly exploited gold mineralisations that stretch from Hamama in the NW via Bir Umm El Fawakhir/El Sid to Hammuda in the SE (Fig. 5.75). It is a west-vergent (Loizenbauer and Neumayr 1996; Loizenbauer et al. 2002) nappe containing an ophiolite sequence. A later reactivation is reflected by a left-lateral strike-slip fault (El-Gaby et al. 1988; Loizenbauer and Neumayr 1996). Serpentinites and gabbros of the ophiolitic sequence are not the only rocks contained in the shear zone of the Atalla mining district, but also Hammamat sediment rocks, and interlocked Dokhan-volcanics. The age of the shear zone is thus younger than the Hammamat formation. Granite intrudes the Atalla area. It is not deformed and is therefore younger than the shear tectonics. Extensive contact metamorphism within the Dokhan-volcanics of the studied zone suggests that the granite continues below the terrain surface. Post-granite tectonics become apparent in NE-SW oriented fractures and are well-developed within the granite.

Loizenbauer and Neumayr (1996) view the parallel running partly mineralised joints in the Fawakhir area as an indication for extensional tectonics. The same was observed at Atalla. Mineralisation processes had taken place inside these open joints. The fissures running parallel to the strike-slip shear zone, which thereby are older than the granite, are not mineralised. A thorough analysis of the geometry and texture of the quartz veins in Wadi Atalla was recently carried out by Akawy (2007) who principally came to the same conclusions, especially with regard to the auriferous quartz veins.

The Atalla granite is coarsely and evenly grained, as well as holocrystalline. The primary stock consists of potash feldspar, plagioclase, quartz, muscovite, titanite, and zircon. Feldspar sericitising occurs with increasing proximity to the quartz veins. At contact with the auriferous quartz veins, the granite becomes subject to brittle deformation, as formation of secondary ores, consisting of arsenopyrite or pyrite occurs. Lack of deformation declares the granite as post-tectonic. The Fawakhir pluton is found further S in the same shear zone. It is of the same relative age as the Atalla granite. The Fawakhir pluton has been dated by Fullager and Greenberg (1978) through Rb/Sr isotope analysis to 586 ± 9 my. Thereby, the granite corresponds to the G-2 granites established by Hussein et al. (1982).

There are several, parallel running quartz veins in the Atalla district that have been mined. One of them had been mined in antiquity as well as recently. Its thickness varies between 0.20 and 1 m. Whereas the ancient mining zone follows the outcrop of the vein, the modern one is represented by a modern shaft lowered to a depth of about 25 m.

Two parageneses have been distinguished (Murr 1999). The first consists essentially of pyrite, often converted to lepidocrocite at the edges. Arsenopyrite and galena are represented to a lesser extent. Gold is found in the lepidocrocite but rarely as visible gold within the quartz. The second generation contains pyrite, galena, low Fe-sphalerite, and chalcopyrite. The latter forms independent crystals as drop-like ex-solution bodies in the sphalerite. Now and then galena has grown together with boulangerite. Gold appears as free gold in quartz together with sphalerite, galena, as well as in some zones within pyrite. Its association with sulphides is primarily given through transport in fluids as a sulphide complex.

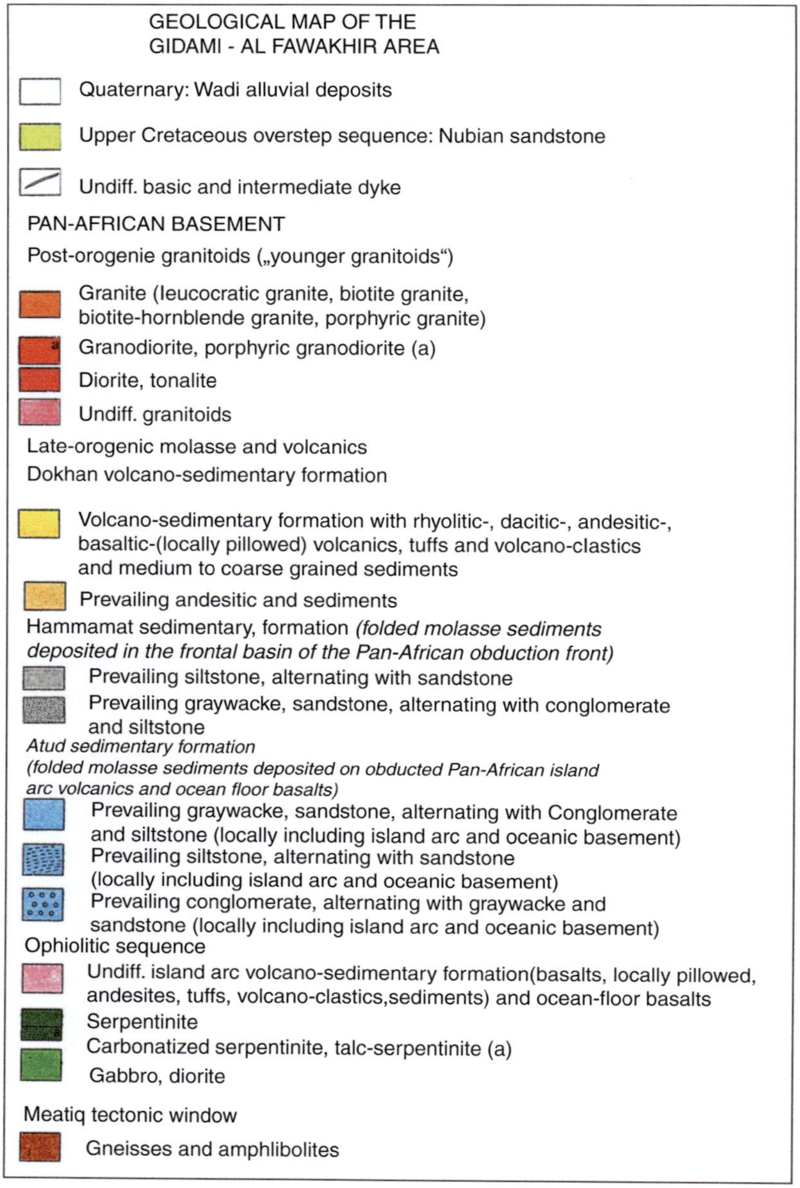

5.2.25 Umm Esh el-Zarqa I (Fig. 5.76)

Geographic position:	26°10′04″ N, 33°34′38″ E

The geologic surroundings of the mined quartz vein is principally marked by hornfelses of the Hammamat formation which merge into basaltic metavolcanics i.e. are overlapped by the latter in the NW.

The surface find assemblages at this site are limited exclusively to fist stone hammers and sporadic grooved stone axes, which are typical for mining activities in the Old/Middle Kingdom. They concentrated particularly along a NE-SW

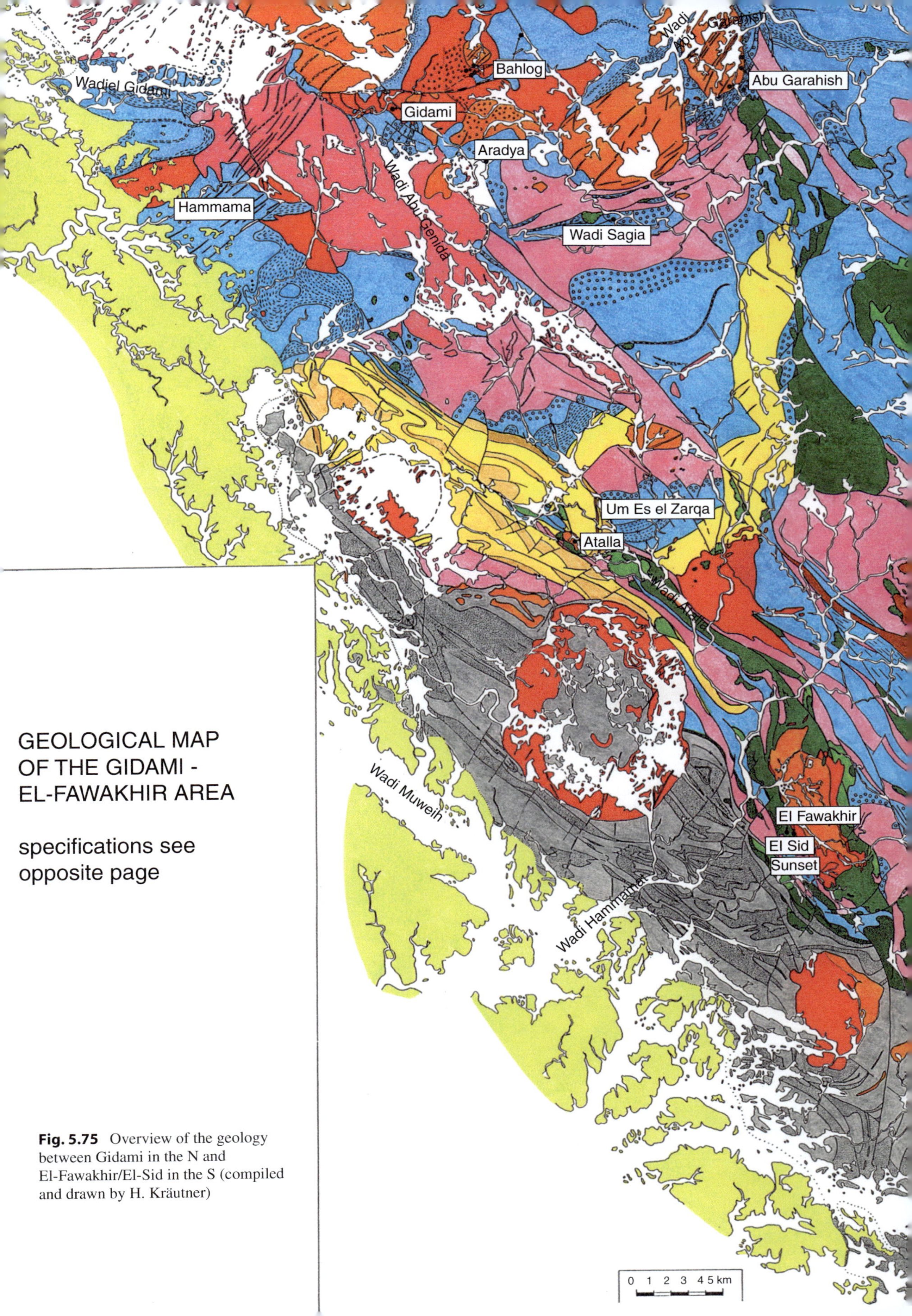

GEOLOGICAL MAP OF THE GIDAMI - EL-FAWAKHIR AREA

specifications see opposite page

Fig. 5.75 Overview of the geology between Gidami in the N and El-Fawakhir/El-Sid in the S (compiled and drawn by H. Kräutner)

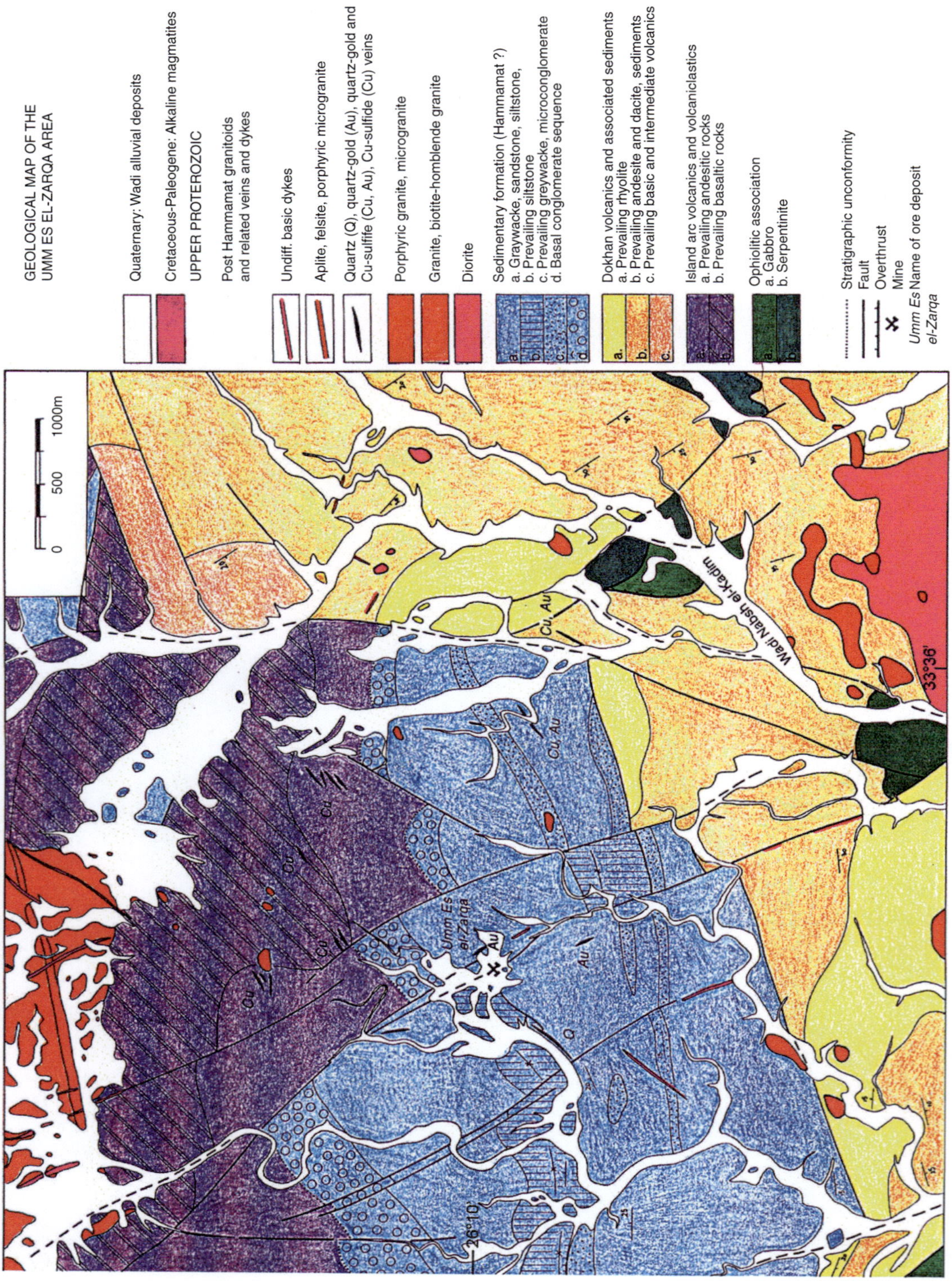

Fig. 5.76 Geological sketch map of the Umm Esh el-Zarqa area (H. Kräutner)

Fig. 5.77 Tools (grinding mill, grinding stone and cylindrical pestle) bordering wadiworkings in Wadi Umm Esh el-Zarqa II

(42°) striking dike, which dips at 43° SE. The dike was quarried over a length of more than 50 m. Though due to the heavily sheared wallrock, the quarried section ended up by collapsing, thus prohibiting any estimation as to the original depth of the extraction zone. Clearly distinguishable however, are the malachite linings in the joint structures with the wallrock so characteristic for such ancient mines. The second vein with a 106° strike and a dip of 54° S, about 30 m further to the W, was probably only prospected over a distance of 10 m without any resulting extraction.

5.2.26 Umm Esh el-Zarqa II

Geographic position:	
New Kingdom settlement:	26°09′38″ N, 33°34′38″ E

In the wider surroundings around the mentioned mine, virtually all wadi beds are littered with still 50–80 cm high debris heaps in as far as these have not been levelled by later wadi floods. According to the findings from the huts recorded in considerable numbers as well as the surface finds, gold production has been carried here on a relatively large scale during the New Kingdom by selective picking out of auriferous quartz lumps from the wadi beds (Fig. 5.77). An exploration of the closeby hills revealed abundant, although small, few-centimetre-wide quartz veins. They are believed to contain gold (Koshin and Bassyuni 1968) though not enough for warranting viable extraction, even for that period. However, a large workforce for processing the wadi alluvium might have produced sufficient amounts of ore, as suggested by widespread remains of domestic architecture.

5.2.27 Wadi Sodmein

Geographic position:	
New Kingdom-main settlement:	26°10′39″ N, 33°54′38″ E
Ptolemaic (?) complex:	26°10′55″ N, 34°54′02″ E

We were not able to visit this site. It was discovered through Google-Earth images.

To judge by the location and arrangement of the houses, the main settlement located in a side wadi seems to belong to the New Kingdom category based on wadiworkings. A smaller settlement (partially washed away) surrounded by a heavy, rectangular wall in the main wadi probably dates to the Ptolemaic Period.

5.2.28 Hammamat

During our fieldwork, which for the most part was carried out within a systematic geologic survey of the entire region around Wadi Hammamat and Wadi Atalla by Langwieder (1994), we were unable to trace the deposit at the location reported by Koshin and Bassyuni (1968).

They had described an approximately 350-m-long gold quartz vein near the Kuft-Quseir road, at the crossing Wadi Atalla/Wadi Hammamat. It was reported to strike NW-SE (150°E) with a 45° SW dip. They also claimed that it had produced traces from gold. According to our survey, it can be affirmed that no such vein had been exploited here in ancient times.

5.2.29 Umm Had-South

Geographic position:	26°01′ N, 33°31′ E (Koshin and Bassyuni 1968)

The SW rim of the open and weathered granite plain of Umm Had is reached only through crossing the plain. From there it has been reported that a quartz vein in the contact zone to the Hammamat-(hornfels) sediments had apparently been mined in the Old Kingdom. We were not able to find the vein during our field survey, but a Bedouin from the Ababda tribe showed us a two-hand hammer, allegedly originating from this site. It measured about 40 cm in length, and weighed about 10 kg. This hammer type was used exclusively during the Early Dynastic Period. Nevertheless, according to the information by the Bedouin, normal fist hammers typical for Old/Middle Kingdom period are also found at that site.

The strikingly dark-red, coarse-speckled Umm Had granite was the preferred raw material for mining tools in the Fawakhir region during the Ptolemaic Period.

5.2.30 Umm el-Fawakhir and El-Sid (Fig. 5.78)

Geographic positions:	
Fawakhir mine:	26°00′48″ N, 33°36′10″ E
Centre of the Roman/ Byzantine settlement:	26°00′36″ N, 33°36′27″E
El-Sid mine, administrative building:	25°59′15″ N, 33°36′23″ E
New Kingdom settlement in Wadi el-Sid:	25°58′47″ N, 33°37′04″ E

The area of Fawkhir-El-Sid is one of the most important gold producing districts in Ancient Egypt. Mining had begun here at the latest in the Old Kingdom, although it certainly shifted its main focal points of extraction along the advance of history.

The mine district around Wadi el-Sid is referred to in a unique epigraphic document from the twentieth dynasty (about 1150 BC), commonly known as the Turin Papyrus Map. Its central part renders a sketch plan of a goldmining area (Fig. 5.79). The papyrus, which among other also gives a road description with the positions of a stele by Sethos I and a well, is today widely acclaimed as the world's oldest geologic map. In fact, the map differentiates between various rock types by rendering them in different colours. However, neither the proportions of the road distances nor their geographic orientation are fully consistent with the features in the terrain. This is no doubt linked to the scroll's format whose limited width hardly allows for a precision, on-scale representation of the road system. An inclusion of a highest possible amount of

Fig. 5.78 Bir Umm el-Fawakhir and Wadi el-Sid area with the various ancient sites (modified Google-Earth image)

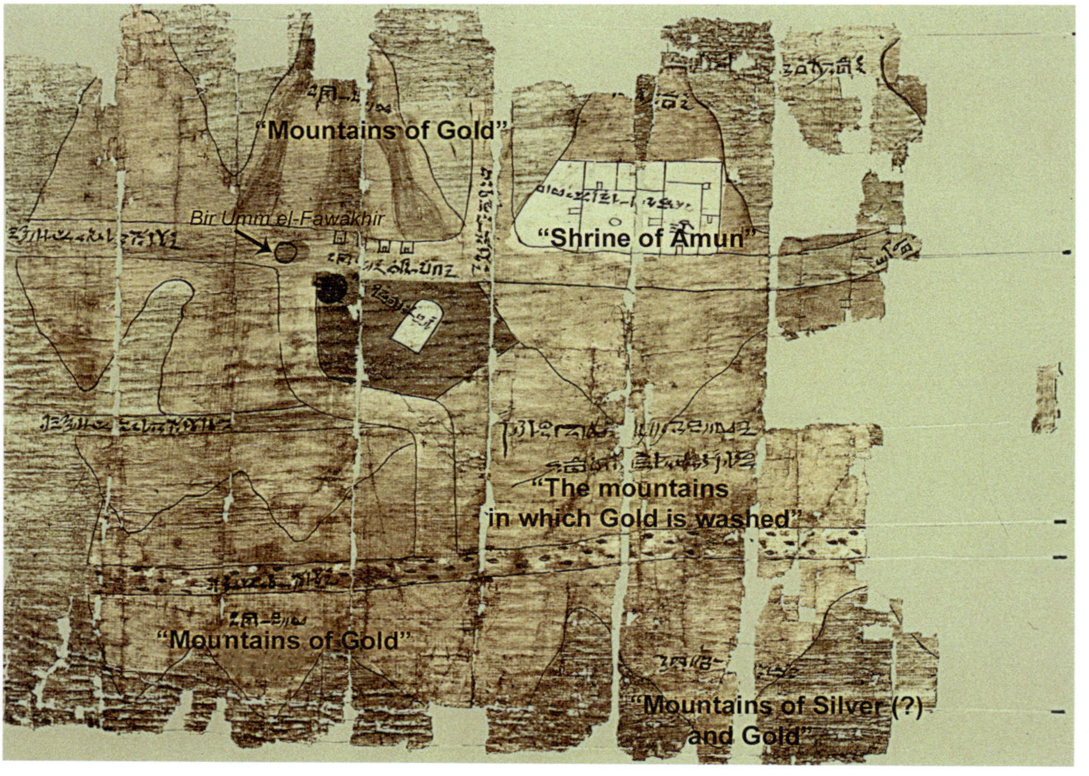

Fig. 5.79 Central part of the Turin Papyrus from the twelfth century BC with entries concerning gold deposits

information to the scroll map was therefore obtained through bundling, which inevitably entailed a distortion of distance proportions and orientation-related aspects of the Wadi Hammamat area. Familiarisation with this method of representation therefore solves many difficulties as to the map's readability. The papyrus is therefore better seen as a sketch map.

Because of the map's approximate nature, Egyptologists have come up with diverging identifications as to its quoted place names (Thomas 1913; Ferrar 1913; Goyon 1949). In one case, the area was considered to correspond to the Derahib area in Wadi Allaqi (Llewellyn 1903; Castiglioni and Vercoutter 1998). The first to identify the map correctly with Bir Umm el-Fawakhir was Gardiner (1914) whose proposition was supported by Murray (1942).

Finally, through direct assessment with modern topographic maps, the particular entries on the papyrus are now known and conclusively assigned to localities in the area around Bir Umm el-Fawakhir and Wadi el-Sid (Klemm and Klemm 1988; Harrell and Brown 1992). It appeared that entries on the papyrus, as for example the one stating "a road which leads to the sea" fully correspond to wadis leading to the Red Sea at the indicated location. The only difference with the

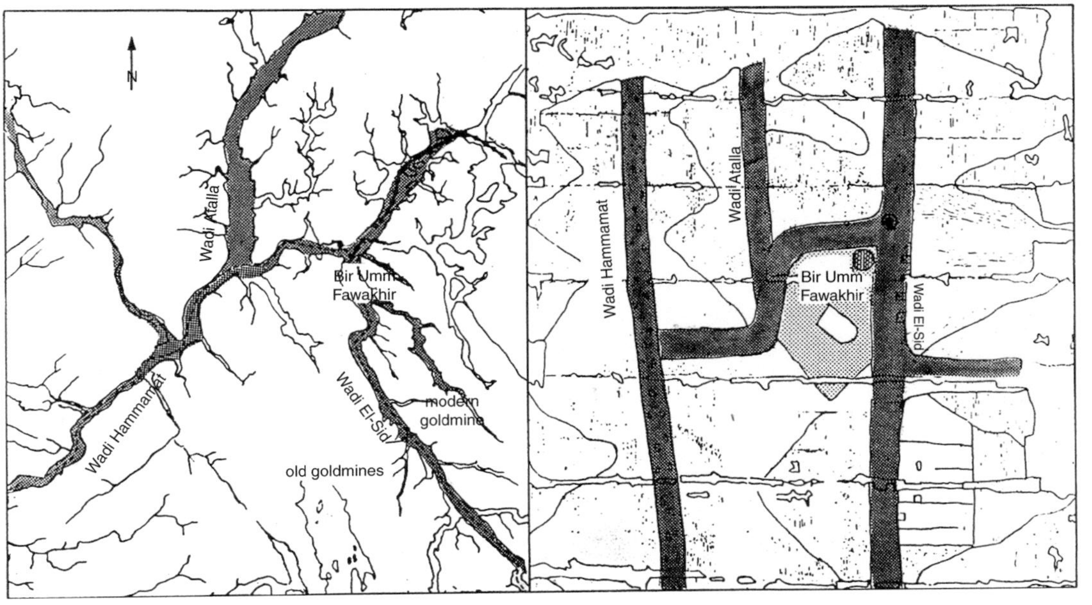

Fig. 5.80 The papyrus (*right*) compared to the actual topography of the area. Because the narrow shape of the scroll prohibits any representation of perpendicularly oriented wadis and roads, such features are simply "folded" into the margins of the papyrus

actual topography of the region is that, as otherwise observed on long and narrow papyrus scrolls, the required information was forced into the available space of the sheet through a folding technique. This was a commonplace method used in ancient itineraries, of which for instance the Roman Tabula Peutingeriana is one of the most famous representatives. If one "unfolds" the wadi courses shown on the map back to their original orientation, they would truncate on the papyrus scroll and much information would consequently go lost. On the other hand, their general locations would in essence match with those of any modern topographic map (Fig. 5.80). Depicted is a main road, which progresses as one scrolls through the papyrus. The branching-off sideways or wadis chiefly serve for orientation purposes as the traveller makes headway along the central highway.

The statement "mountains where gold is washed" is one of the map's key declarations. Evaluated with the aerial image, that location would match with the central part of green rock sequences 4 km to the S of today's Bir Umm el-Fawakhir, a site, however, where sufficient quantities of water for washing gold are quite unlikely. The indication would seem more appropriate in a closer location to the well on the map, which is near today's Bir Umm el-Fawakhir. However, own detailed area investigations have revealed that not only the pharaoh's gold mines had been located at the site in the mountains, but that it also had been the site where gold was washed, as unequivocally confirmed by large and well-preserved tailing

Fig. 5.81 Ancient tailing heap in the "mountains where gold is washed" (as per mining papyrus) west of Wadi el-Sid

heaps of reddish quartz sand (Fig. 5.81). Water therefore had needed to be brought to the site in order to separate the gold flakes from the ground quartz. Because there are more large tailings about 4 km away, near Bir Umm el-Fawakhir, which however, had been re-processed by cyanide leaching at the beginning of the twentieth century, it is plausible that water had been transported through Wadi el-Sid to the processing sites near the mines. From here quartz ore chunks had then been brought back (probably with donkeys) to the well to avoid idle convoys.

The area near the modern, though abandoned El-Sid mine corresponds to the settled area near the ancient extraction zone (cf. below).

The mines within the red granite area near the well Bir Umm el-Fawakhir date to the Late Ptolemaic and Roman-Byzantine Periods. However, neither the scale at which they were run nor their gold credits are comparable to those of the old mines in Wadi el-Sid.

The following observations list the most important archaeological mining sites at the locations figured in the Google-Earth satellite image (Figs. 5.78 and 5.82).

In recent years, the area has undergone thorough archaeological investigation. Ceramic analyses have served to a large extent for dating purposes and are available in a number of published reports (Meyer 1995; Meyer et al. 2000, 2003).

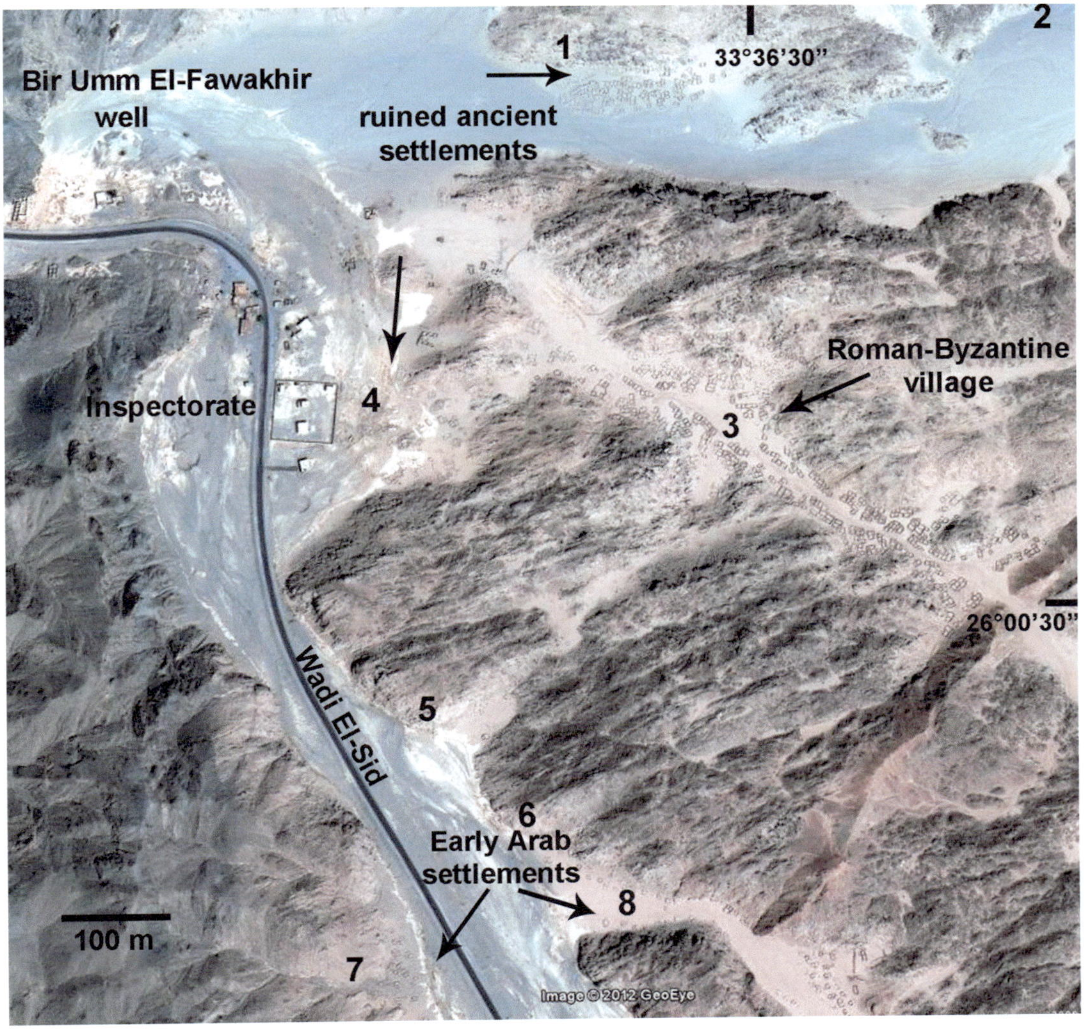

Fig. 5.82 Detail of the environment around Bir Umm el-Fawakhir (modified Google-Earth image)

5.2.30.1 Position 1

Location 1 is a large site with partly very well-preserved, mostly rectangular houses consisting of two rooms. Between them, are repeated occurrences of badly damaged single-roomed, round huts. As a rule, the houses are detached, in few cases they had been joined together to form alignments. Outside the generally well-preserved

Fig. 5.83 Elongated Roman/Byzantine settlement, being the largest and best preserved in the Bir Umm el-Fawakhir district (pos. 3)

walls, one recognises cooking places and masoned stoves (Fig. 5.83). The pottery consists of many dark, markedly grooved amphorae and simple, mostly unpainted shards dating to the Late Roman/Byzantine Period (Meyer et al. 2003). In spite of its nearness to the processing site at Bir Umm el-Fawakhir, the absence of mining tools may indicate that the settlement had links with the bordering granite stone quarry, which, according to the recorded tool marks and two unfinished columns, still was under operation during the Roman Empire (Klemm and Klemm 2008a, b). The rock from this quarry has actually been identified among the building material used in Rome (Gnoli 1989).

5.2.30.2 Position 2

Hidden in the mountains to the NE of location 1 is a vast settlement of roundish huts near the edges of a valley bed. About 12 houses are located immediately behind the access way leading to the wide valley basin, where another 60–70 buildings are found. The huts are built from heavily weathered, local granite, and in spite the rocks' round shapes, the walls are preserved in heights around 1.60 m, which is probably explained through the protected situation of the site. In this case too, the site seems to have served exclusively for dwelling purposes, since no mining tools could be found (Fig. 5.84). Finds consist next to small fragments of blue faïence, large amounts of partly intact pottery from amphorae and a dark-red slip ware with black lines dating chiefly to the Late Roman/Byzantine Period (Meyer et al. 2000, 2003).

5.2.30.3 Position 3

In a narrow wadi, about 250 m in a straight line S of locations 1 and 2, lies the largest settlement within the district. It consists of approximately 300, partly multi-chambered houses distributed on either side of the NW-SE oriented wadi. At its rear end, the wadi displays a clearly noticeable aplite dike.

Fig. 5.84 Well-preserved ruins of dry walled houses near the Roman granite quarries (pos. 2) where unfinished ashlars still scatter in the plain (*arrow*)

Inside the settlement, several large rubble and pottery heaps pile up between the house walls and the slope. A cursory inspection revealed that the main pottery assemblage consisted of a dark-red slip ware with fine grooves and small handles, partly decorated with black blotches, and very light-coloured amphorae and dishes with almost white slips. According to this assemblage, the settlement dates to the Late Roman Period, and according to the ostracon from Fawakhir (Guéraud 1942) probably to the first to second century AD, which has been confirmed by C. Meyer et al. (2000, 2003).

In contrast to the two locations discussed afore, location 3 shows some evidence of mining. Small heaps containing quartz ore lumps were sporadically observed, and in some huts there were the small, six-sided stone anvils from the Early Arab Period. In addition, there were some intact round mills as well as apron-shaped runner stones from the Ptolemaic Period.

5.2.30.4 Position 4

Here the remains of approximately 150 huts are located dispersed in a loose arrangement. Masoned clay stoves and recent refuse mark this site (Fig. 5.85).

According to the indications on the Turin map, the "shrine of Amun" must have been located somewhere within this area. Along the bordering mountain facade to the S, one can make out dome-like niches with wall foundations in the foreground.

5.2.30.5 Positions 5 and 6

To the E of Wadi el-Sid are extensive remains of roundish house ruins, which according to the round mills and cubic anvils, date to the Early Arab Period. Because of their immediate location next to the asphalt road in Wadi el-Sid, the site is severely damaged. As it seems, the settlement area originally spread over the entire wadi before the central area was destroyed by the construction of the modern road. This new road

Fig. 5.85 Circular hearth with clay mortar in the settlement at position 4

replaces the old one leading along locations 1 and 2 eastwards to the Red Sea from Bir Umm el-Fawakhir.

5.2.30.6 Position 7

The somewhat better preserved areas of this site are located to the W of Wadi el-Sid and the asphalt road. In this case too, the architectural features consist of round huts which erstwhile probably spread over the entire wadi (Fig. 5.86).

5.2.30.7 Position 8

The huts at location 8, in a more remote area in an eastern side wadi, are in a visibly better state of preservation.

5.2.30.8 Position 9 (Fig. 5.78)

Position 9 denotes several Early Arab huts clustering around two small adits (trial mines) near the mountains, as well as three larger house ruins in the wadi bed.

5.2.31 Bir Umm el-Fawakhir

The old tailing heap at the well of Umm el-Fawakhir was reprocessed around the turn of the last century and later, between the 1930s and 1950s by the Italian count Louison, whom commonly the local Bedouins refer to as "il Cont". During these campaigns considerable gold quantities were gained (Langwieder 1994). At the same occasion, a large Ptolemaic settlement was extensively destroyed, to judge by the flimsy remains left behind. Presumably, the site had once occupied the entire width of the wadi and covered an even older site from pharaonic times as inferred from similar findings from other mining districts (e.g. Sukkari, Dungash). The original tailing, which like the settlement remains was completely cleared away in the course of the chemical reprocessing (cyanide leaching), must once have been located near the well in the plain. The sifted-out debris from the settlement was

Fig. 5.86 Western part of a settlement that once probably spread over the entire wadi. Early Arab Period (pos. 7)

deposited for the most part on the southern side of the wadi, on top of settlement layers. The debris piles still contain the entire spectrum of tools from the site's mining history. Beside fist hammers from the oldest extraction phase in the Old/Middle Kingdom, one also finds numerous oval stone mills from the New Kingdom. Most common, however, are the Ptolemaic mills whose apron-shaped runner stones are mostly from the red Umm Had coarse-crystalline granite, naturally available just few kilometres away to the N. Some round stone mills from the Roman, or rather Early Arab Period, are also found here.

The walls of the modern well construction display the same tool panoply, presenting a cross-section of all rocks used.

The surroundings of Bir Umm el-Fawakhir never played a meaningful role as a gold mining district. Only two reasonably large mines cut into the red granite. They had apparently been established by Conte Louison as adits to ancient mines for exploration and development purposes. Additional gold extraction in the Bir Umm el-Fawakhir area is demonstrable only to a very limited extent. Contrary to the assumptions by Harrell and Brown (1992), more recent investigations have shown that the yields of auriferous quartz veins in the granite are very low (5 ppb) and thereby of no interest, especially for the ancient miners (Langwieder 1994). The main extraction area was in Wadi el-Sid where quartz veins running through the green rocks may occasionally attain extremely high gold grades (to 30 g/t).

5.2.32 The Pharaonic Mining District in Wadi el-Sid

The chief mining area of the New Kingdom is located in Wadi el-Sid, in the hinterland of the modern mine which itself foots on old mines. The terrain is completely riddled with holes and debris heaps. The mines themselves are for the most very narrow and thus accessible only

Fig. 5.87 Narrow, winding mine in the El-Sid mining district, probably from the New Kingdom

through crawling. They are 10–15 m deep and often follow the mineralised quartz veins at an angle, while secured by occasional abutments. Larger mines have irregularly laid out corridor systems in which small ore pockets were mined in recesses and side fissures (Fig. 5.87).

In the countless debris heaps, of which some overlie and fill older mine entrances, ancient tools including fist hammers, pounding globes, and small mortars are found. In the vicinities of the larger mines one furthermore distinguishes next to ceramics and fist hammers from the Old Kingdom, traces from metal chisels dating to the New Kingdom, as well as single, oval mills and fragments thereof.

The mining district stands out through a long prospection trench that runs over two hills and the valley between, thereby cutting all quartz veins perpendicularly and exposing the yielding quartz veins. Now and then along the edge of the ditch one notices small prospection mills and stone hammers, which had probably been utilised

Fig. 5.88 Remains of a large New Kingdom settlement on the W side of the asphalt road in Wadi el-Sid. Because of additional vestiges on the other side of the road it is probable that the site once occupied the entire width of the wadi. The pharaonic mining district is located just behind the hills to the W

for checking gold contents. This testifies to meticulously and systematically carried out gold prospection methods. At least two huts were observed next to the edge of the prospecting trench. Otherwise no architecture was recognised within this area.

Several tailing heaps of reddish quartz sand, though evidently eroded are surprisingly well-preserved, probably thanks to their protected situation by the mountains. They probably served as temporary stockpiles. Their compact texture reveal their high age (see Fig. 5.81). As indicated by the local geology and geo-morphology, water had to be transported to the site for the washing processes. From an archaeological standpoint, only the well at Bir Umm el-Fawakhir, 4 km away may be viewed as a potential water source. The main share of the gold may therefore have been processed at the well, as this was the site of the markedly larger tailing heap reprocessed and relocated in the early years of the twentieth century.

Remains of the New Kingdom settlement in the ancient mining district were identified on both sides of the Wadi el-Sid, not far from the old mines (Fig. 5.88). The site is barely visible at the surface and consists of attached architectural units with both round and rectangular plans in spacious arrangements. Originally, the site must have stretched over some distance further up the wadi, to a location where it widens considerably. The asphalt road built in the 1960s in this case is not responsible for the widespread destruction of the site. Long-ongoing water erosion reducing the original settlement down to two strips along the wadi edges is the more likely reason. The largest strip is on the W side, close to the "mountains in which gold is washed" according to the Turin Papyrus. The state of preservation is distinctly better on the E side. There the site reaches slightly into the northern edge of a side branch, called Wadi Azab. Somewhat to the N a burial site can be made out inside a wadi recess.

Numerous, mostly severely corroded, oval mills including smaller runner stones as well as very fragmented and sparse New Kingdom pottery were found in both parts of the settlement. Among the recorded pottery was a fragment of a red, polished Meidum bowl. Many

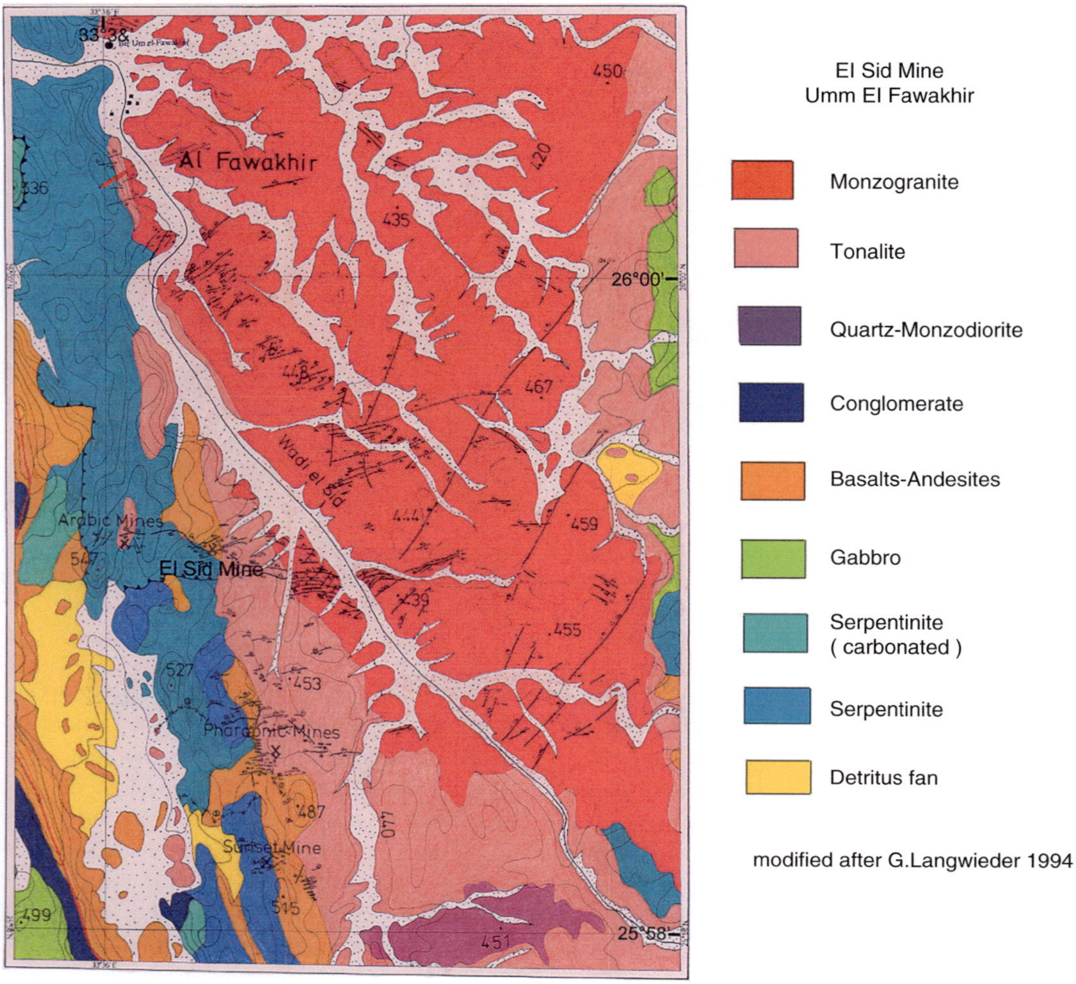

Fig. 5.89 Geological map of the area around the gold deposits at Bir Umm el-Fawakhir and El-Sid (Langwieder 1994)

mills are buried in the sand, and countless more are possibly concealed under the wadi surface. This high concentration of mills may therefore suggest that not only gold ores from the adjacent mines, had been processed here, but also ones collected in the wadi alluvium. However, direct evidence able of confirming this has yet to be furnished.

The discovery of this hitherto unknown, large mining settlement from the New Kingdom near the "mountains in which gold is washed", yet again confirms the reliability of the geographic indications in the Turin Papyrus.

The importance of the gold mining district at El-Sid lies primarily in the present and former high gold grades of up to 30 g/t within the mineralised quartz veins of the green rock sequences. In the main mine of the modern extraction area, one can actually see with the naked eye gold particles as reflections of a flashlight in the remaining parts of the mined quartz vein.

Langwieder (1994) thoroughly mapped the geologic surroundings of the deposit (Fig. 5.89). Later, Harraz (2000) as well as Zoheir and Moritz (2007) published interesting observations as to the formational conditions of the gold mineralisation processes, especially with regard to the El-Sid mine. Furthermore, Harraz (1995) examined the geochemical environments around the El Sid gold mineralisations.

All gold quartz mineralisations of El-Sid appear in intensely folded and sheared ophiolite sequences of serpentinite and metabasalt, into which Hammamat sediments are partly imbricated. These sequences are in direct contact with the tonalite margins of the Fawakhir granite. El-Bouseily et al. (1985) were the first to thoroughly examine the ore parageneses and establish that the gold mineralisation occurred in two stages: initially, together with pyrite and arsenopyrite and then with sphalerite and galena, whereby the gold itself appears mainly as free gold in both stages inside fissures of the sulphide minerals.

Langwieder (1994) distinguishes between three separate mining areas: the El-Sid mine itself, which predominantly had been exploited in the Ptolemaic Period as well as the first half of the twentieth century. Furthermore the "pharaonic mine", which corresponding to the mine district denoted by the Turin Papyrus, had probably been exploited in conjunction with the "sunshine mine" further S, already as early as the Old Kingdom but especially since the New Kingdom. This is confirmed by the pottery from the debris heaps, fist hammers and the typical New Kingdom oval mills.

The Fawakhir ophiolite extends over a distance of approximately 15 km between the Hammamat sequences in the W and the Meatiq-dome in the E. The ophiolite is intruded by Fawakhir pluton. The ophiolite contacts the Hammamat group in the W in an eastwards dipping overthrust zone, whereas the eastern contact with the Meatiq rocks is represented by a tectonic mélange.

During orogeny, molasse-sediments (Hammamat group) deposited steadily in an intramountainous basin. The ophiolite nappe at Fawakhir-El-Sid emplaced between 850 and 770 my (Hassan and Hashad 1990). The effect of this emplacement in the Fawakhir-El-Sid area was the development of an approximately 1.5 km wide overthrust zone dipping E. The consequences for this area are intensive deformations such as elongation of components within conglomerates, shearing, breccia formation, mylonitisation, mélange formations and intensive exfoliation of sediments, volcanics, and serpentinites. To Vail (1988) these ophiolites together with the overlying volcanics are structurally part of a fold- and overthrust belt, which formed during an E-W crustal contraction at the western margin of the Pre-Pan-African craton.

After emplacement of the ophiolite units, Fawakhir pluton intruded at 574+9 my (Rb/Sr age; Fullagar 1980). This caused contact metamorphic and metasomatic processes in the area around the Fawakhir pluton.

This entire complex was intruded by a coarse-grained, pink monzogranite that assimilated to tonalite at the margins through hybridisation (partial melting of the wallrock; Langwieder 1994). Shortly after this event followed a NNW-SSE oriented, intensive shearing activity, that tectonically deformed the entire area over more than 100 km by 15 km, and generated countless mylonitic micro shear zones. A clear geographic bond of the deposits of Hammama, Wadi Atalla el-Mur, Atalla, Umm Esh el-Zarga, Fawakhir, El-Sid, Umm Soleimat, and Hammuda to this shear zone system evidently reveals its genetic impact on the gold mineralisations.

Whereas the Fawakhir monzogranite and its periphery were affected by shearing, the Umm Had granite located to the NW clearly penetrated the Hammamat suite in a post-tectonic phase. Auriferous vein quartz occurrences appear therefore here in very complex geologic environments consisting of mafic to ultra-mafic ophiolite suites tectonically imbricated with Neoproterozoic, clastic sediment series (Hammamat sediments).

On the other hand, the gold mineralisations at Fawakhir/El-Sid are not the result from this shearing, but owe their existence directly to a post-tectonic extension, leading to an open fissure system virtually perpendicularly to the shear direction, i.e. in a more E-W direction (Langwieder 1994).

As for the interpretation of the genesis of the gold quartz mineralisations of the entire Egyptian Eastern Desert and parts of Nubian Sudan, the trace element analyses of gold based on the neutron activation method, which was carried out by the Canadian ACTMIN Company, proved extremely effective. The method in fact detects gold traces at extremely low ppb levels (ppb = parts per billion, i.e. one thousandth of a gram per ton of rock).

The results displayed highest primary gold grades in serpentinites (averaging at 40 ppb, occasionally reaching 180 ppb) and somewhat lower ones in mafic volcanites (with sporadic high values up to 140 ppb) from sediments of the Hammamat suite.

Gold contents in the granitic Fawakhir pluton averaged at 5 ppb and rarely exceeded that value. In the hybrid peripheries on the other hand, they concentrated around 90 ppb and in few cases, even reached markedly higher values.

Because the mean values of gold in normal basalt and granitoids are around and below 5 ppb, and the data show that the actual primary gold enrichment is found in ophiolites and sediments generated by their erosion as well as in "gold-inheriting" marginal areas through melting of plutonites. The gold contents in the different mine areas vary naturally within relatively wide brackets. At any rate, the Louison Company, which was mandated with the mining until 1956, established a mean gold content at ~30 g/t, which indeed would correspond to a very copious occurrence (Zoheir et al. 2007 gives ~28 g/t). In fact, we were able to occasionally record millimetre-large gold inclusions within in the milky-white quartz vein. As a rule, gold becomes visible to the naked eye at grades equal and above 1 oz/t (troy ounce: ~31.1 g) in freshly extracted quartz rocks.

A recurring geologic environment emerges from these observations with respect to the prospectors' identification of significant gold mineralisations. Within an immediate perimeter of up to 500 m of granitoid intrusive bodies, auriferous quartz vein mineralisation are found predominantly in ophiolites or their direct vicinity. Otherwise, they are also found in the molasse-like sediments of the Hammamat series, if these are intruded by granite magmas. In addition, generally less important gold mineralisations are found in the hybrid, marginal zones of the granitoids as seen near Bir Umm el-Fawakhir.

Beside gold, typical sulphide hydrothermal paragenesis as well, appear in association with pyrite-arsenopyrite and pyrite-sphalerite-galena within productive quartz veins. This indicates that the hydrothermal fluids transported gold as gold-sulphide complexes. In almost all cases, these quartz veins had emplaced in several generations. In this connection, the investigations by Harraz (2000) and Zoheir and Moritz (2007) presented useful additional data concerning the composition of ore-forming fluids and formational temperatures of the various sulphide parageneses. According to Harraz, auriferous parageneses with pyrite sphalerite galena had formed between 330 and 365 °C, and according to Zoheir and Moritz at 320 ±20 °C, and those with pyrite-arsenopyrite respectively between 265 and 295 °C, and 185 ± 15 °C.

Furthermore, tectonic conditions that generate the opening of dike fissures and shear zone systems and thereby permit the emplacement of auriferous quartz mineralisation, however, also play a decisive role. This is connected to a final tectonic phase marked by an approximate N-S extension after the intrusion of the Fawakhir pluton. The early prospectors had been much aware of this dependency, as quartz veins oriented in this direction only, display traces from prospecting, while those with differing orientations have remained untouched.

By the time of the New Kingdom, the Egyptian prospectors had acquired full and detailed knowledge of the fundamental geologic principles for this category of deposit formation.

Fig. 5.90 Three ergonomically diverging fist hammers from Umm Soleimat

5.3 Middle Central Group

5.3.1 Umm Soleimat (also known as Hamuda)

Geographic position:	25°54′54″ N, 33°44′46″ E

From a perspective of mining archaeology, farther reaching investigations at this site seem quite promising. This is particularly true regarding the question as to the simultaneous extraction of gold and copper ores, but also the location of possible smelting furnaces.

A vertical, up to 1 m wide quartz vein had been exploited in an open trench at this site. Its extraction depth can no longer be established as the shaft had evidently been filled-in intentionally, a relatively rare phenomenon for ancient Egyptian mines.

Numerous fist hammers measuring between 15 and 20 cm in length were observed on the slopes (Fig. 5.90). Because of the sharply edged percussion marks at their surfaces, these tools seem to have been used for extraction purposes rather than ore processing. There are no recognisable signs for ore processing with water at the surface near this mine.

In this case too, it strikes that the mine's walls are covered with the residues of a green malachite layer (Fig. 5.91) and that the surrounding wallrock rubble covering the slope is clearly green coloured. This for instance corroborates with the findings reported previously at the early mines in Wadi Dara and Abu Mureiwat. At Wadi Dara, copper smelting was evidenced by furnaces located in a propitious wind channel, about 1.5 km away from the mine (Castel et al. 1995). Such ovens remain nonetheless yet unaccounted for within the close surroundings of Umm Soleimat.

In the heavily sheared alternating, slaty sequences of greywacke and intermediate conglomerate layers of the Hammamat formation, the quartz veins appear as modified ac-clefts oriented perpendicularly to the shearing. Occasional granitoid apophyses indicate the geologic proximity of a granitoid intrusion somewhat deeper down whose assumed outcrops appear at the surface about 3 km to the NE.

Intensive malachite and other copper minerals within the cleft of the wallrock lead to assume that the mine was aimed exclusively at copper extraction. This is supported by the complete absence of grinding tools, which admittedly are

Fig. 5.91 Characteristic blue-green malachite linings in the cleft structures of the wall rock had probably been the most important indicators for copper and gold ores in the period of Old/Middle Kingdom. Umm Soleimat

generally absent in Old Kingdom mines. There are no visible remains from domestic architecture, neither in the wadi nor its surroundings to permit for extrapolations as to the date and type of the mine. To judge by the relative narrowness of the wadi, it is however conceivable that the houses have been washed away by the wadi floods.

In the nearby, contiguous valley course some recent prospection tests have been carried out in similar geologic surroundings at small quartz veins striking perpendicularly to the shearing, though without conclusive results.

5.3.2 Wadi Karim

Geographic position:	
Ptolemaic settlement remains and modern plant:	25°55′55″ N, 34°03′23″ E

The Wadi Karim widens up at the inlet of a tributary wadi coming in from the N. As a consequence of the construction of a modern processing plant at the site, a presumably Ptolemaic settlement was destroyed to such an extent that its original function is no longer clearly determinable. The modern plant presumably served for processing banded iron ores (BIF) located in the Hammamat series to the N (Fig. 5.94). Although extraction attempts are still clearly distinguishable there, the processing plant itself apparently never really came into an operating mode.

The site at Wadi Karim consists of several large, rectangular building complexes, arranged at right angles to each other with large spaces left open between them (Fig. 5.92). In the valley plain is a rectangular enclosure built from wadi rubble much reminiscent of similar Ptolemaic structures and processing sites discussed at Bir Semna (Fig. 5.93). Adjacent to that, especially towards

Fig. 5.92 House complexes in Wadi Karim (modified Goggle-Earth image)

the flanking mountain, are several, single complexes with less distinctive layouts. In all, about eight architectural units partially eroded away by floods have been counted in the wadi. Tools indicative of gold mining were found neither inside nor outside the buildings, whose rooms measure in some cases up to 8 × 8 m.

More, large housing complexes as well as a number of single buildings are discernible close to the mountain slope to the N. The only archaeological hint for possible gold processing found at this site was a Ptolemaic grinding stone.

Several debris heaps apparently resulting from recent excavation soundings contained pottery shards from the Ptolemaic to Roman Periods.

To judge by the entire site, including its buildings and the mentioned enclosure, a preliminary date to the Ptolemaic Period seems acceptable for the present because of the lack of any further archaeological investigations.

The site is quite unusual for a gold mining settlement. It therefore needs to be speculated whether it actually had functioned as a protection or supply station along a desert route during the Ptolemaic and Roman Periods.

Hume (1937, p. 733) remarked the following on Karim and Zeidun: "Ancient workings are said to exist at these places, but they have apparently not been considered as worth development. There is no geologic information concerning them".

Only few years ago, El Shimi and Soliman (2002) were able to identify gold contents averaging between 4 and 6 g/t in a larger number of BIF (banded iron formation) samples, which from an economic stand could prove quite crucial for the BIF occurrences at Wadi Karim (Fig. 5.94). However, it can't be excluded that the primary

Fig. 5.93 Rectangular wall of Ptolemaic water reservoir and processing place (?) in Wadi Karim

Fig. 5.94 "Banded Iron Formation" (*BIF*) in Wadi Karim. Light chert layers alternate with dark iron ore

gold contents of the BIF lead to productive gold mineralisation only by hydrothermal redistribution. Despite that, during our investigations we had not become aware of any ancient mining activity within the common banded iron formations of the Egyptian East Desert.

From a geological view, the settlement sites in Wadi Karim are located respectively at the margins of substantial metasediment sequences of the Hammamat formation, in a peripheral zone of the large molasse folds of the Karim basin, where they are in visible contact with granitoid intrusions. The metasediments have accordingly developed to hornfelses, which truly justifies prospects of finding gold within this zone.

The influence from contact metamorphism though, decreases quickly, which results to a better perceptibility of the siltstones, greywackes, and conglomerates of the intensively folded Hammamat series exposed in the mountain ravines. It is hence within this zone of banded iron ores where traces from extraction attempts are found.

5.3.3 Kab el-Abiad

Geographic position (New Kingdom settlement):	25°55′26″ N, 34°01′00″ E

Kab el-Abiad is located about 4 km SW to the afore discussed site in Wadi Karim. It consists of heavily eroded remains of a New Kingdom settlement along both valley flanks. There and in the mouth area of a small tributary valley coming from the NW, two recent mining attempts are distinguishable in outcropping quartz veins, which may possibly have been motivated by the presence of the archaelogical site. Inspite of our intensive search efforts, no ancient mining traces could be found, which may suggest that gold production dated to the New Kingdom probably concentrated exclusively on operations in the wadi alluvium (wadiworkings). About 9 km away from the site, the wadi catchment crosses a potentially auriferous zone marked by the roof zone of a granodioritic intrusion contacting the overlying Hammamat series.

5.3.4 Terfawi

Geographic position:	25°52′ N, 34°03′ E (Koshin and Bassyuni 1968)

The deposit at Terfawi is referred to as site no.174 in Koshin and Bassyuni (1968). It is described as located to the N of Bir Terfawi, in the upper portion of Wadi Terfawi, a tributary valley of Wadi Essel.

A relatively persistent quartz vein with an average thickness of 30 cm is followed at the surface over a distance of 150 m. It has a NE orientation and dips NW at 50°. Most of the vein had been left untouched, probably because of the low gold contents below 4.65 g/t. The vein quartz was identified to contain hematite (Neubauer 1958).

In Wadi Terfawi small tailings are noticed S of Bir Hamdalla. Architecture ruins with mills are seen in other parts of the wadi, which apparently hints to more gold mineralisation in the area.

Because of the low gold grades, the site was not short-listed for an immediate re-examination (Koshin and Bassyuni 1968).

For the lack of exploitable maps and the inadequate knowledge of the local terrain by our Bedouin guide, we saw ourselves unable to find Wadi Terfawi. The close vicinity between Hammamat series in Wadi Terfawi and granodiorite intrusions make the prospects for finding gold in the area nevertheless look quite good.

5.3.5 Wadi Zeidun I

Geographic position (Bir Zeidun):	25°42′41″ N, 33°44′42″ E

Close to the confluence with Wadi el-Esh (about 25°43′N, 33°43′E), in the central part of Wadi Zeidun, Koshin and Bassyuni (1968) refer to a site under their entry no. 176, which was already mentioned by Jenkins (1925) and Hume (1937) as consisting of a series of old extractions. In spite of our efforts during our stay in the area, we were unable to find any trace whatsoever of either ancient mines or other human occupation. On the other hand, not far from Bir Zeidun we did however, record several looted, circular Bedouin graves that to the untrained eye may appear like ruins from round huts. It is therefore conceivable that the above information is erroneous, especially considering that the local geologic environment is hardly favourable for the formation of gold minerals.

Hume (1937, p. 733) stated old "workings" close to the well at Zeidun. He added no information concerning the geology of this area, and it cannot be ruled out that the above mentioned architectural remains only, inspired him to make this statement.

5.3.6 Wadi Zeidun II

Geographic position:	25°36′ N, 33°57′ E (after Koshin and Bassyuni 1968)

We were not able to find this deposit. But since Koshin and Bassyuni (1968) described the site, their account is rendered here:

"The occurrence lies in the area of the Wadi Mukheit and Wadi Assima, the Wadi Zeidun's tributaries in Latitude 25°36′ N and Longitude 33°57′ E."

"Known here are four small quartz auriferous veins lying widely apart from each other. They were extensively worked out in ancient times. The largest of all is the 30 cm-thick quartz vein in Wadi Mukheit. It occurs in red granite being traced for 150 m in old workings being 30 cm thick (Neubauer 1958) on the whole the occurrence is of small scale."

5.3.7 Sharm el-Bahari

Geographic position:	25°47′ N, 34°16′ E. (after Koshin and Bassyuni, 1968)

This occurrence was not visited by us. Therefore, only the information by Koshin and Bassyuni (1968) is rendered here.

"This small occurrence lies in the upper courses of the Wadi Sherm el-Bahari, at the foot of the Gebel Abu Tiur…. The area is composed of sandy shales and conglomerates intruded by granites of the Abu Tiur massif. Several steeply dipping and vertical lens-shaped quartz veins of the thickness in places amounting to 2 m are encountered at the contact of sediments and granites. Most of the veins were mined in ancient times. (Neubauer 1958). Samples taken from the tailing heaps on the side of these workings showed, on analyses, 10.2 and 15,81 gr per ton of ore (Sabet 1961)."

"A number of the old workings exist in the area:

1. On in the northern side of Wadi Abu Markat in quartz-epidiorite rocks (lat. 25°51′ N, long. 34°16′ E);
2. On both sides of Wadi Abu Mireiwain metasedimentary rocks close to their contact with metavolcanics (lat. 25°47′ N, long. 34°19′ E).
3. In a schist bel, 1,7 km north-west of Gebel el-Nusla (lat. 25°43′ N, long. 34°18′ E).
4. In conglomerates of Wadi Abu Gheryan (lat. 25°40′ N, long. 34°29′ E)."

5.3.8 El-Nur

Geographic position (Bir el-Nur):	25°52′09″ N, 33°41′57″ E

Bir el-Nur is an approximately 25 m deep well with a masoned lining. Although presently dry, it was probably used until recently by Bedouins as may be inferred from the two hut ruins nearby. This recent occupation was probably also the reason for the severe damage inflicted to older hut ruins, some 60 m S of the well and probably dating to the Early Arab Period.

The geologic surroundings of el-Nur are marked by reddish silt sediments of the Igla formation, conglomerates, arkoses, and greywackes of the Hammamat formation. Shearing is well-discernible inside the conglomerates, which has lead to the formation of quartz minerals. Inspite of our intensive exploration, no traces from ancient mining were found here. From a standpoint of geology, however, the area may have potentially viable gold deposits. Yet, Neubauer (1958) reported traces from old workings and an almost horizontal quartz vein containing sulphide minerals and gold.

5.3.9 Umm Rus

Geographic positions:	
Modern and ancient main mine:	25°27′50″ N, 34°34′54″ E
Ptolemaic settlement Wadi Mubarak:	25°26′40″ N, 34°34′17″ E

At the outcropping surface of the Umm Rus intrusion, there are several mined, auriferous quartz vein mineralisations (Fig. 5.95) that particularly in northern and western parts dip at a remarkably gentle angle. The mined veins are quite easy to make out in the terrain thanks to the numerous access shafts and debris heaps. The general geologic map shows the explored and exploited mining districts from the New Kingdom (marked with a star) on dated evidence from processing mills. Although architectural remains are scanty in the immediate vicinity of the modern mine, some isolated remains seem nonetheless to date to the New Kingdom.

Substantial New Kingdom settlement remains are on the other hand known from a small tributary valley branching-off to the S from Wadi Dib, just N of the intrusion, and from Wadi Mubarak, although buildings from the Ptolemaic Period extensively cover the latter. Within the virtually round complex of the granodiorite stock on the map, evidence from the New Kingdom is preserved in the form of house ruins and typical oval grinding mills found near rudimentary tailing sites (Fig. 5.96).

Nevertheless, much of the evidence from ancient activities has been destroyed, particularly near the modern mine at Umm Rus, in the SE part of this complex. Chisel mark patterns at the shaft and mine walls yet here again, provide some exploitable information as to the date of these activities to the New Kingdom, even in this part of the mining district.

The main period of extraction in Umm Rus seems to have taken place during the Ptolemaic Period. The most conclusive indications are found about 1.8 km to the S of the granodiorite complex in Wadi Mubarak, where a large Ptolemaic settlement probably initially occupied the entire width of the valley in the form of a fortress-like complex. Today, the only distinguishable fortification ruins and settlement remains are located at the northern side of the valley. One clearly discerns a rubble fan coming in come in from the northern tributary valley into Wadi Mubarak which evidently has devastated the former settlement installations.

Numerous, characteristic concave stone mills and apron-shaped runner stones date the occupation of this site to the Ptolemaic Period (Figs. 5.97 and 5.98). A closer scrutiny of the site surface moreover reveals that the house ruins at least partly cover older structures from the New Kingdom, a finding supported by the presence of re-used stone mills from that period.

As it seems, mining resumed at Umm Rus in the Early Arab Period, as evidenced by the round mills and wall masonry containing Ptolemaic runner stones. Without further investigations, it no longer is possible to determine to which degree the relatively low number of recorded round mills actually reflects the intensity of this late mining phase. In the aftermath of the occupation, these mills in fact enjoyed some popularity as flourmills among local Bedouin tribes and over the years continued to be withdrawn from the original gold processing sites.

As to the way the mining activity was organised, the settlement stands out through its remote location, away from the actual mining sites. Availability of water had evidently an important role to play in this respect. Wadi Mubarak in fact has a highly yielding well, around which processing platforms and tailing heaps group. The fortifications on the slope to the S of this large settlement had manifestly served for the protection of this important processing site.

Many tailing residues and washing table remains recorded around the mines in the granodiorite stock seem to indicate that a shuttle transport similar to that reconstructed for Bir Umm el-Fawakhir and El-Sid was also in place at Umm Rus. In other words, ores may have been transported from the mines to the main processing site and vice versa, water to the processing sites near the mines.

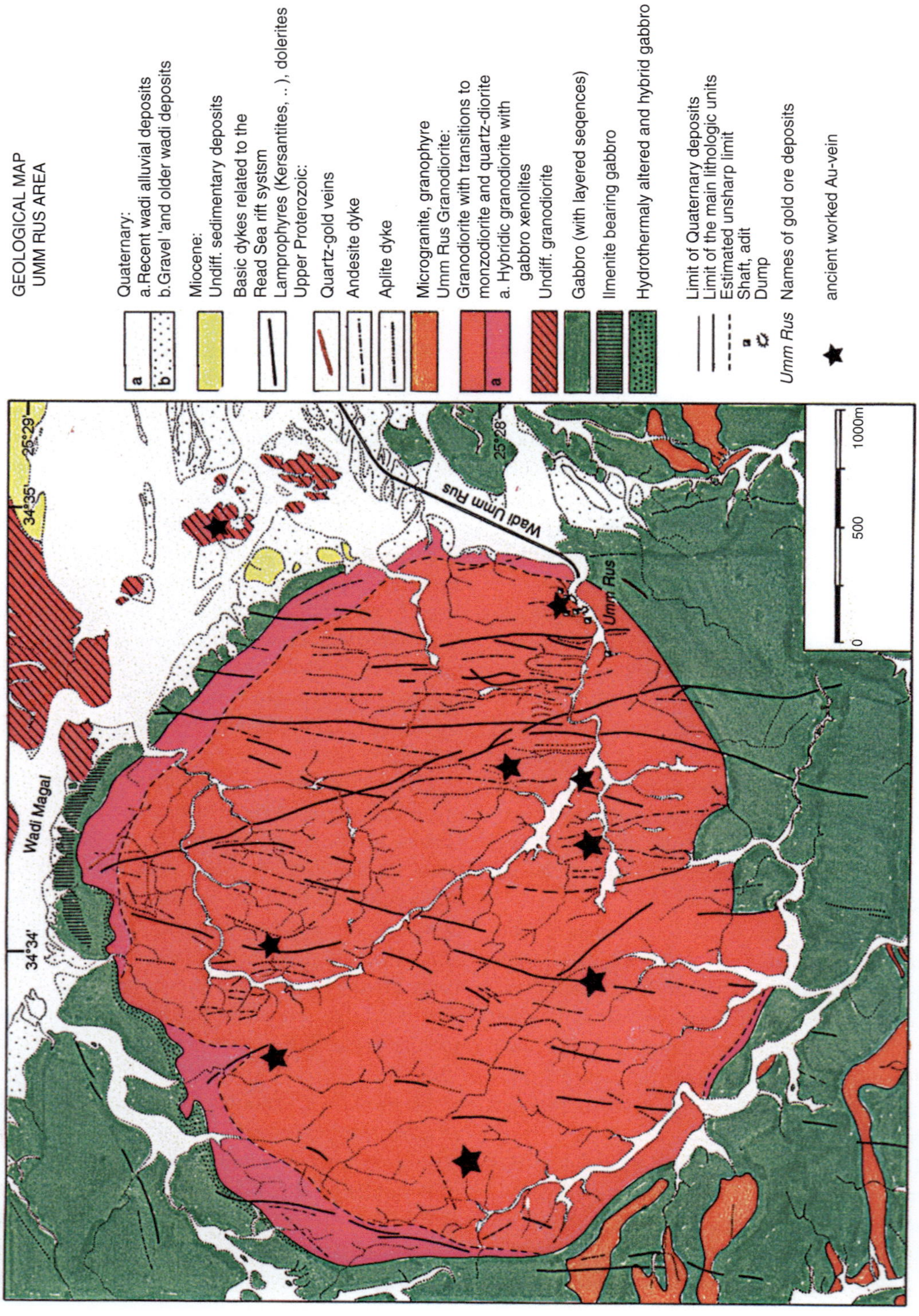

Fig. 5.95 Geological sketch map of Umm Rus with the most important gold quartz deposits (*black stars*) exploited in antiquity (H. Kräutner)

Fig. 5.96 Ptolemaic settlements and remains of a fort in Wadi Mubarak, south of the granodiorite stock (modified Google-Earth image)

Fig. 5.97 Concave ore mill with two-handed runner stone at the Ptolemaic settlement of Umm Rus

Fig. 5.98 Quartz ore ground down to a fine powder on a Ptolemaic mill

The occurrence of auriferous quartz veins in the Umm Rus deposit is chiefly associated to the marginal areas of the virtually circular granodiorite intrusion into a vast gabbro sequence (Fig. 5.95). It represents a distinct exception with respect to the lithologic environment of auriferous quartz veins in the Egyptian Eastern Desert.

The composition of the coarse-grained and light-coloured granodiorite varies occasionally without recognisable tendencies towards monzodiorites or quartzdiorites. At its margins near the gabbro it frequently shows a hybrid zone, which may often be mapped with a decreasing frequency of gabbro-xenoliths in the basic granodioritic mass. This leads to the assumption that the entire granodiorite with its varying mineralogical composition merely represents a hybridly modified intrusion dome of a former granitic magma.

The granodiorite is by no means homogeneous, but vary in its composition from ilmenite-rich to normal, medium-grained gabbro with undisturbed magma structures. In the marginal areas near the granodiorite however, there appear xenoliths of gabbro and hybridisation of the granodiorite to almost diorite.

The gabbro for its part, seems to tectonically overlie a sequence of metasediments of the Hammamat formation imbricated with serpentinite lentils to the SW of Umm Rus. The numerous small intrusions of micrograntes and granophyres inside the gabbros and metasediments probably ought to be viewed as apophyses of underlying, large granite bodies.

Approximately N-S striking, steep porphyric, andesitic, and aplitic dike swarms cross the entire sequence of granodiorites and gabbros. In their own turn, the dike swarms are frequently cut by mostly slightly inclined quartz veins. The same dike swarm directions are also observable in the neighbouring granite-pluton to the W (Kadaburah al-Hamra intrusion).

In the whole region of the Umm Rus granodioritic intrusion stock there are numerous quartz veins that too strike in a N-S direction, although

Fig. 5.99 Flat lying, northern vein system in the Umm Rus granodiorite stock with numerous openings (modified Google-Earth image)

in a remarkably gentle dip mostly to the W (Fig. 5.99). They also are limited to the granodiorite, i.e. in as far as these dikes continue into the gabbro, they thin-out and disappear after a short distance.

The auriferous quartz veins consist of milky-white to grey quartz. In addition, but in smaller amounts, there is ankerite carbonate and to a lesser extent pyrite, pyrrhotite, arsenopyrite, and chalcopyrite. Free metal gold is contained sparesly but also well-distributed at submicroscopic levels in sulphide minerals.

The quartz veins are supplemented by intensive alteration zones in the wallrock. They often contain wallrock xenoliths. Saad et al. (1996) have subdivided these alteration processes persuasively into four stadia, which we endorse, though with minor modifications:

Hydrothermal fluids came into motion apparently through intrusion of granitoid magma in the form of a volatile restphase as well as from the wall-rock. Along fault zones they transform available biotite to chlorite and feldspar to epidote, in this case chiefly plagioclase. The authors determined the formational temperatures of this stage at around 225 °C, according to the chlorite-thermometry.

As a result, the now K-enriched fluids reacted with plagioclase, forming sericite and K-feldspar. In both stages, since the fluids were apparently poor in CO_2, no carbonates developed.

Furthermore, the formation of ankerite carbonate in a later stage displays higher CO_2-values, which according to Saad et al. increased through a reduction of H_2O during formation of the hydrosilicates in the first two stages. In this stage, the sulphide and gold mineralisations precipitate, whereby the herefore necessary iron was probably obtained from the decomposition of ferromagnesium minerals.

The main quartz mineralisation process took place during the final stage.

Harraz and El-Dahbar (1993) carried out investigations concerning the nature and composition of auriferous fluids. They described them in contrast to Saad et al. (1996) as rich in CO_2-H_2O with a molar portion of 30–46 % CO_2, and a low NaCl salinity equivalent to around 7 %. The temperatures generating gold mineralisation were calculated at 250–300 °C at relatively low pressures around 0.35 kbars. El Tokhi and El Muslem (2002) came to similar formational temperatures within a 270–325 °C although basing their conclusions on triple-phase fluids (?).

The modern (in 20th century) gold extraction at Umm Rus concentrated around several vein systems striking NE-SW and dipping 30° W, chiefly within the marginal zones of the granodiorite intrusion. The main dike running approximately 500 m along the escarpment towards the Red Sea was mined in ancient times in depths reaching 20–25 m. Under modern exploitation three inclined shafts were driven 170 m into the vein. A vein located further up was mined in a 30 m deep shaft. At the surface, the vein can be followed along an outcrop over a distance of at least 220 m.

Other vein systems located to the N and NW of the intrusive stock had also been intensively exploitated in antiquity. As seen in Fig. 5.99, both systems dip at an unusually gentle angle. The extracted ores had probably been processed at the New Kingdom settlement of Wadi Dib further to the N of the complex. However, because of the severe wadi erosion, only little of this settlement has been preserved along the wadi edges and the somewhat higher lying terraces. Furthermore, we were unable to identify with confidence the location of a potential well at the site.

The ore reserves have repeatedly been checked over the past decades. Koshin and Bassyuni (1968) estimated them at 10,000 t with gold yields near 12 g/t.

5.3.10 Wadi Raheiya

Geographic position:	25°26′24″ N, 33°41′34″ E

Along the eastern and western edges of a wide wadi section, lie scattered the ruins of about 40 single houses and larger buildings within an area of approximately 500 m.

On the eastern side, they are built of weathered diorite boulders. On the western side, the prevailing building material consists of rather flat metavolcanics, thus assuring a better preservation and allowing for more rectangular ground plans than the round rock nodes.

The eastern houses exhibit the complete assemblage of New Kingdom tools. It comprises oval stone mills, runner stones and cylindrical pestles with associated, flat pounding stones with central depressions. On the W side, on the other hand, New Kingdom mills had occasionally been integrated to the relatively tall walls. The house inventories consist exclusively of round mills and typical, six-sided, small anvils, which probably earned the site its name (Raheiya = round mill). The findings are clearly indicative of two occupational phases. As suggested by the large number of mills, the first occurred during the New Kingdom at which the wadi alluvium had been exploited in addition to a nearby quartz vein. The second apparently only had a short duration in the Early Arab Period (Fig. 5.100).

The auriferous quartz vein mined during the New Kingdom as well as the Early Arab Period is located on a small ridge in a valley basin of much crusted granodiorite. It probably forms the roof of a large, granitoid intrusion, since the red-brown granodiorite appears as a small protrusion within

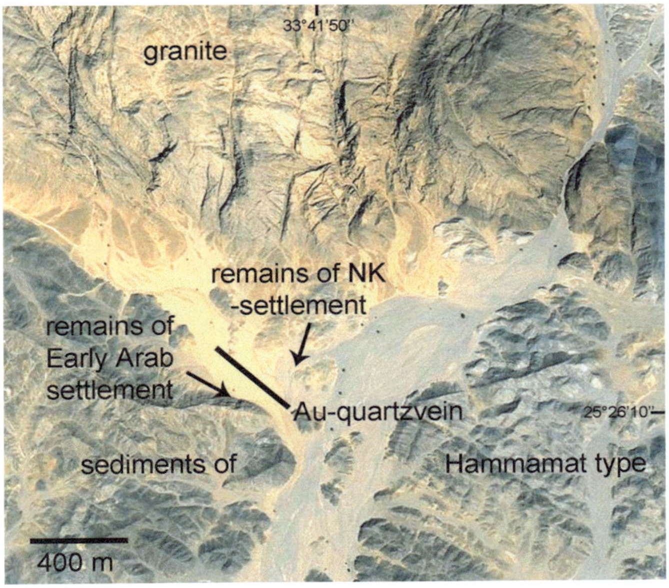

Fig. 5.100 Gold deposits and ancient settlement remains in Wadi Rahaiya in the transition area between the granodiorite intrusion and the Hammamat metasediments (modified Google-Earth image)

the surrounding hornfelsic, grey-green metasediments of former greywackes and conglomerates. In the terrain the volcano-sedimentary units form a dome-like arch which accentuates this roof structure.

The quartz vein mined in antiquity is located within the immediate contact of the granodiorite to the metavolcanics and metasediments in a NW-SE strike (120°/dip: 50–75° W). The 0.3–1 m wide vein had been mined over a length of about 80 m in a depth of up to 5 m. The quartz displays at least two mineralisation generations. As expected, no visible traces from gold were recorded, since a previous investigation by Neubauer (1958) could only document gold contents between 1.5 and 3 g/t. About 60 m to the N of this vein, one notices another swarm of eight more veins within the granodiorite. Because of their perpendicular strike to the granodiorite and their inadequate widths between 10 and 30 cm, they had apparently been ignored.

The small quantities of mined gold from the dikes alone would hardly account for a durable extraction period. Though, because of the size of the New Kingdom settlement, it is probable that the relatively wide valley basin was large enough to uphold wadiworkings on a worth-while scale. Potential traces in the form of small rubble heaps in the wadi bed are in any case no longer visible as they would have been washed away by water erosion.

5.3.11 Sigdit and Wadi Miyah

Geographic positions:	
Sigdit, settlement:	25°27′25″ N, 34°04′47″ E
Bir Sigdit:	25°27′35″ N, 34°04′45″ E
Wadi Miyah (fort):	25°24′42″ N, 34°05′06″ E

Immediately to the S of the silted-up well at Bir Sigdit one comes across a vast settlement site consisting of buildings of carefully arranged, rectangular layouts. They cover large parts of the wadi rims. The wadi itself is only about 50 m wide at this section. Their strong masonry, the remains of a paved street in the lower course of Wadi Miyah, as well as a handful of pottery

Fig. 5.101 Remains of a paved road from the Roman Period at Sigdit, Wadi Miyah

shards suffice to date the site confidently to the Roman Period. It is specially the paved street (Fig. 5.101) that reveals its function as a Roman road station. Signs suggestive of gold production are found nowhere here.

About 6 km farther to the S shortly after the confluence of Wadi Sigdit with Wadi Miyah is a large fortress-like building measuring 40 m by 15 m (Fig. 5.102). An alignment of additional buildings resembling a garrison is seen at the eastern side of Wadi Miyah. A small fortified complex equipped with a watchtower on a low hill marks the southern border of the area. Two more fortified watchtowers (Fig. 5.103) are visible on hillocks at the western wadi side. The problem as to the exact function of these buildings in this remote region needs to be solved by means of a systematic archaeological approach.

Fig. 5.102 Ruins of the fortress-like Ptolermaic/Roman complex at Sigdit, Wadi Miyah

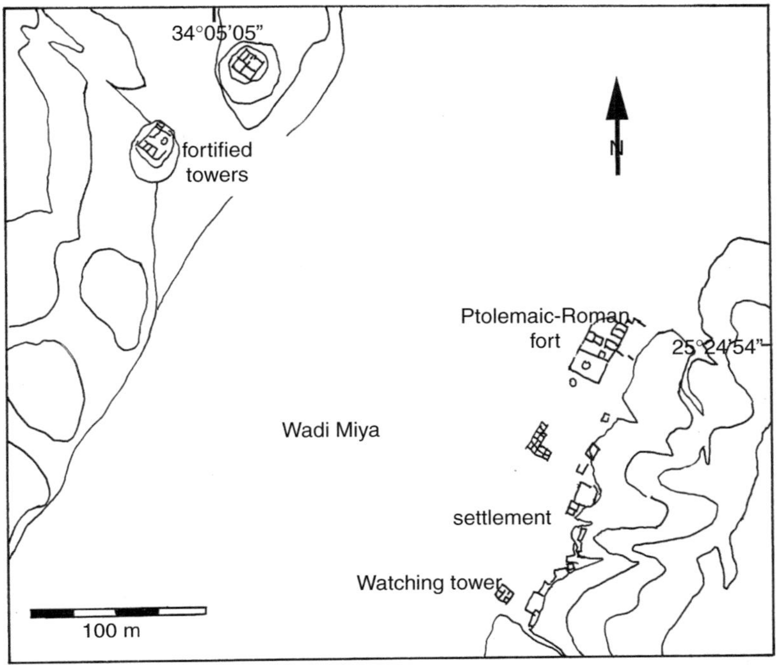

Fig. 5.103 Archaeological sketch map of Sigdit in Wadi Miyah (R. Klemm)

The later integration of Ptolemaic concave mills and apron-shaped runner stones into the masonry of these fortress-like Roman complexes strongly suggests the presence of a gold production industry already during the Ptolemaic Period. For the lack of any genuine mines in the area, it had apparently been based exclusively on exploiting the wadi sediments (wadiworkings).

The entire surroundings are dominated by contact metamorphic, mainly rhyolitic metavolcanics traversed by fault zones striking N-S. Numerous aplitic, vertical veins as well as a small number of andesite dikes are oriented in the same direction. The area is surrounded by three large granite intrusion stocks, which also seem to underlie it, as indicated by abounding granitoid apophyses.

We observed minor irregular quartz veins within the metavolcanics, though without traces from mining. We therefore assume that the original Ptolemaic settlement covered its gold production from processing the wadi alluvium, since the entire area with its metavolcanics altered by contact metamorphism may basically be viewed as potentially auriferous.

5.3.12 Daghbag

Geographic positions:	
Daghbag I	
Fort:	25°23′52″ N, 33°49′07″ E
New Kingdom settlement near the mine:	25°24′12″ N, 33°49′06″ E
New Kingdom settlement at the confluence of Wadi Daghbag and Wadi Hamdallah:	25°22′50″ N, 33°49′29″ E
Daghbag II, mine:	25°24′29″ N, 33°51′17″ E
Daghbag II (house ruins):	25°24′07″ N, 33°51′34″ E
Daghbag III, mine:	25°24′27″ N, 33°51′45″ E
Daghbag IV, mine:	25°22′39″ N, 33°48′30″ E

5.3.12.1 Daghbag I

The Roman fort Compasi (Meredith 1952) in Wadi Daghbag has typically been established around a well. Today only the eastern side of its ramparts have survived, the rest being washed away by wadi erosion. The well too, has almost entirely collapsed, and is dry at its base. Both, the well's masoned lining as well as the preserved

Fig. 5.104 New Kingdom settlements (*arrows*) and the mine at Dagbagh I (modified Google-Earth image)

part of the wall contain countless runner stones from the Ptolemaic Period. They are also found together with their concave socles in the terrain further down the wadi.

Typical for Roman forts in the Egyptian Eastern Desert is the use of burnt bricks, which here too, mingle with the stone rubble from the collapsed walls.

According to Koshin and Bassyuni (1968), 39 auriferous quartz veins are located within the deposit district of Daghbag. Unambiguous traces from ancient extraction are however made out in only four separate quartz vein systems. We had the opportunity of examining three of them more thoroughly. Although all occurrences appear in similar geologic contexts, they respectively reveal specific features, thus requiring separate descriptions.

Mining activity may already have reached an apex in the New Kingdom. This is inferred from the substantial house alignment at the western wadi edge opposite the mine at Daghbag I (Fig. 5.104). The lower parts of the settlement have been devastated by water erosion, though, numerous oval stone mills and fist grinders still lie scattered near the house ruins close to the slope.

The single houses are approximately 10 m long and in general subdivided into three rooms. The outer and even inner walls are preserved in heights of up to 1 m.

Fig. 5.105 Deep trench pit at Daghbag I. The barren central quartz generation was left behind

The mine at Daghbag I is located on the eastern slope the wadi. It is relatively narrow but very deep to judge by the large waste dump located just outside (Fig. 5.105). As mentioned above, the artefact assemblage from the immediately adjacent house ruins date its exploitation at the latest to the New Kingdom. As it seems, mining had been resumed during the Ptolemaic Period, which is supported by the abundant tools at the nearby processing site in Wadi Daghbag.

The exploited gold quartz vein strikes NW-SE (118°) and dips at 70 °S. It has a lenticular structure and its width varies between 10 cm and 1.5 m. The gold yields fluctuate between virtually nothing and 39 g/t (Koshin and Bassyuni 1968). This mine exposes an interesting feature. Obviously mining took place only within the quartz portions of the hanging wall and lying foot and most probably also within the alteration zones next to them. The barren central part of the vein remained untouched. This kind of selective mining could be observed frequently within the ancient oprerations.

The vein is located within a granodiorite - diorite stock (Fig. 5.106) intruded into metagreywackes and metasiltstones. Due to its current cross-bearing, it seems to be connected to them in an alternating layering. Particularly in the dioritic parts, numerous xenoliths are located within the contact zone to the metasediments. All metasediments within the contact area have altered to hornfels.

Numerous andesite dikes strike parallel to the vein and cut the granodiorite/diorite as well as the metasediments and smaller quartz veins. Moreover, aplitic and kersantitic veins pass through the area.

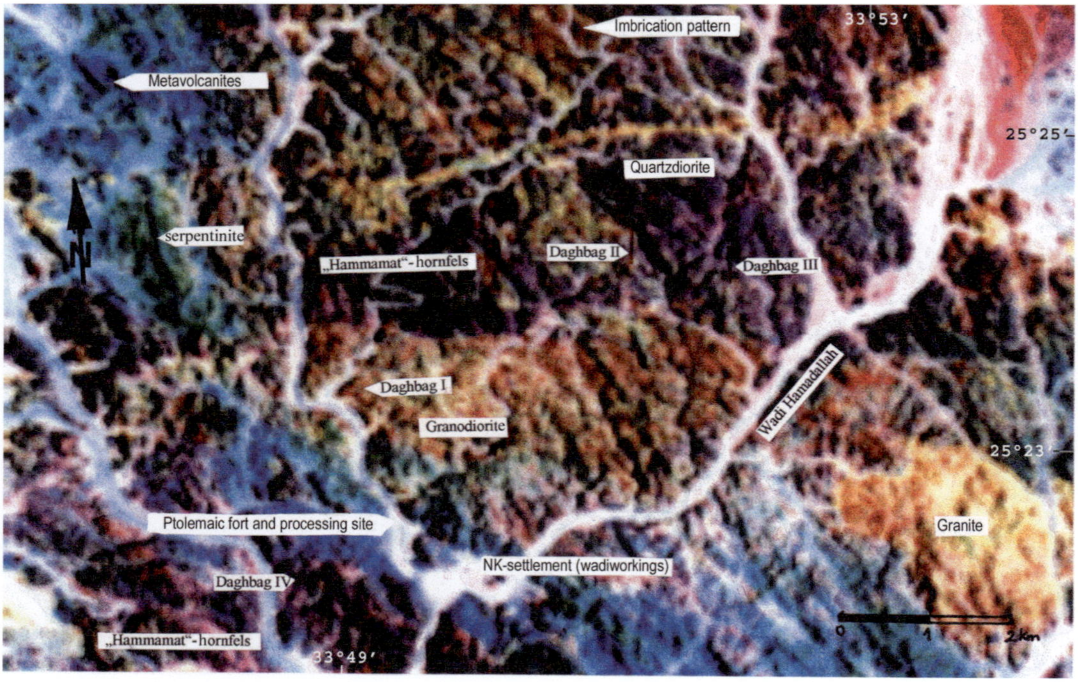

Fig. 5.106 Lithologically processed satellite image (TM 174/42, channels 7-4-1) of the Daghbag area displaying the geologic features

5.3.12.2 Daghbag II and III

The occurrences at Daghbag II and III are located in two tributary valleys to the N of Wadi Hamadallah. Both quartz vein systems emerge in quartzdiorites, which particularly near the vein mineralisations display severe shearing. In this case as well, the veins have lenticular shapes whose thicknesses vary between few centimetres and 1 m. In contrast to Daghbag I, they approximately strike N-S.

Daghbag II (strike: 4°, dip: 75° E) is followed over a distance of more than 300 m with larger numbers of partially collapsed shaft openings. The shafts are oriented in such way that sufficient daylight penetrates to the bottom. According to the archaeological findings, mining at Daghbag II presumably dates to the Old/Middle Kingdom. This early date is justified by finds from the mine's debris, which among other included some older fist hammers and a mortar used for crushing and grinding auriferous quartz (Fig. 5.107). Ore extraction had probably peaked in the New Kingdom, as attested to by distinctive mining methods, such as the numerous abutments in the shafts and a restriction of the extracted rock to the quartz vein itself. The New Kingdom mining operations had by and large deleted the traces from the older mine.

Daghbag III (strike: 15°, dip: 75° W) apparently consists of several small dikes which group in parallel orientation to the main vein. Here too, the adjacent (quartz-) diorites exhibit frequent and intensive shearing, especially within the surroundings of the vein. It is predominantly within these shear zones that the above mentioned alteration structures develop.

According to the archaeological context, Daghbag III was mined chiefly during the Ptolemaic Period. It provides some information as to the reconstruction of the characteristic mining methods in this period. At first, the sheared, hence brittle, overlying zones of the dike had been mined in a dome-like structure. In a second step, the quartz of the underlying, main dike had then been extracted simply by taking advantage of its clefts to avoid extra exertion because of the quartz's extreme hardness. The thereby resulting dome- and barrel-shaped vaults had furthermore

Fig. 5.107 Broken Old/Middle Kingdom mortar at the Daghbag II mine

Fig. 5.108 Ptolemaic Period mine at Daghbag III displaying dome-like structures

increased the stability of the cavity, which in addition significantly contributed to the reduction of the need to leave abutments. The technique also increased the amount of extractable ores (Fig. 5.108). However, in some cases apparently, especially in very yielding parts, even necessary abutments had been removed and replaced by dry-masoned wall pillars. This happened nevertheless at the expense of stability and sometimes even lead to an eventual collapse of the mine. At Daghbag II and III the small quartz veins that run perpendicular to the main strike are ostensibly free of gold, as none of them show traces from mining. In Daghbag III the shearing of the diorites displays a slight flexure into whose quartz veins an aplite dike penetrated at some later stage.

Fig. 5.109 Parts of a circular installation for gravity separation of heavy minerals at Daghbag IV, Ptolemaic Period

5.3.12.3 Daghbag IV

We did not visit this occurrence. According to Koshin and Bassyuni (1968) it is about 1.5 km to the W of Bir Daghbag. It thereby lies within a sequence of Hammamat metasediments altered by contact metamorphism, which alternates with amphibolites. All together, ten small quartz veins have been located, of which one was intensively exploited in ancient times. This vein had been pursued over more than 40 m, its thickness measuring between 15 and 25 cm. It strikes NE-SW and dips at 40° SE. The vein's blue-grey quartz contains clearly distinguishable impregnations from galena, which in paragenetic associations often appears in auriferous quartz veins in Egypt.

From a perspective of mining archaeology, this galena occurrence is particularly interesting. Immediately below a masoned well from the Ptolemaic Period are the clearly reconstructable ruins of a water-driven, heavy mineral processing plant, measuring about 8 m in diameter (Fig. 5.109). Such devices are well-known from Laurion/Attica (Fig. 5.110) and have been thoroughly described by Conophagos (1988).

It consists of a stone circle, whose original diameter had probably measured between 10 and 12 m. It consisted of about 60 cm wide ashlars that together had formed a waterproof compound. At its surface was an approximately 15 cm wide and 4 cm deep channel with perpendicular, coarse grooves. Numerous apron-shaped runner stones that stem from ore mills at this Ptolemaic processing plant are known from the wall masonry of the Roman fort at Compasi. A similar, circular heavy mineral separator is also known from Barramiya (cf. Fig. 5.137), of which one segment was integrated to the wall masonry at the Roman fort. We documented another such segment at Bokari. According to their respective archaeological findings, all date to the Ptolemaic Period. Additional processing devices of the same type are possibly also represented by similar structures at Samut and the much silted-up ones at Shehat (see Fig. 5.24).

The structure's channel has a relatively gentle gradient so as to induce the deposition of the heavier mineral particles, including gold, on the channel bed and thereby its separation from

Fig. 5.110 Circular installation for gravity separation of heavy minerals in Laurion/Attica, ~400 BC

the lighter quartz meal suspended in the water. Along the exterior margin of the structure, the ashlars display numerous plain depressions, which apparently originate from ore crushing.

At Laurion it is known that the so obtained galena was utilised to extract its silver contents by the so-called cupellation method. In this process, galena is isolated mechanically in the method just described. Lead is then extracted by roasting the galena in pans under high ventilation in order to oxidise its sulphur-content in the form of SO_2 gas. In a next step, the lead is melted in a well-ventilated area, thus obtaining lead oxide. The lead oxide is removed, leaving behind pure silver in a small smelt cake.

A similar extraction procedure is also conceivable for Wadi Daghbag. Nonetheless, during our short visit there, no evidence hinting to the presence of furnaces, lead oxides, or slag residues were found.

Finally, along the wadi edges at the confluence of Wadi Hamadallah and Wadi Daghbag, at the location of the dried up well, Bir Daghbag, there are 20–30, severely damaged New Kingdom houses, which nevertheless have survived the intensive wadi erosion. Tailings are still sufficiently discernible to propose a main processing phase in the New Kingdom. The relatively numerous buildings also doubtless reflect intensive gold ore processing connected to the nearby wadi deposits. However, because of the mentioned erosion in the wadi, the traces from these wadiworkings have disappeared.

At the western mountain flank near the gold washing station, there is a cavity whose rectangular opening was obstructed by a wall. The slope in front of it is marked by a waste dump, consisting of Ptolemaic mill fragments and large quantities of pottery shards. A thorough investigation will have to establish the original function of this structure.

Geologically (Fig. 5.106), the entire area is characterised by an intensive imbrication zone of Hammamat metasediments with an ophiolite sequence of sporadically carbonated serpentinites and predominantly mafic metavolcanics. Granite magmas penetrated into this zone, as visible in a wider perimeter by large intrusion complexes that altered their chemistry through assimilation with the respective wallrocks. In this case, granodiorite to occasionally (quartz-) diorite intrusive bodies were able to develop representing nothing else but roof areas of apophysis-like, partial intrusions. Typical for this area is the immediate contact with the metasediments and the ophiolites, all showing hornfelsic structures, and with the hybridised magmatites. It is especially striking to the N of Daghbag where magmas have intruded into a steeply oriented backdrop of metasediments and created a banded pattern of hornfelses, former metasediments, and granitoid magmatites.

According to the described genetical models, this geologic environment is potentially ideal for the quartz-vein-mineralisation containing gold. It therefore hardly surprises that currently 39 auriferous quartz veins are known from the area (Koshin and Bassyuni 1968).

Alteration phenomena in the wallrock are observable next to all mineralised quartz veins. They consist of chloritisation, sericitisation, and usually to a lesser degree, carbonation.

5.3.13 Talet (also: Tila) Gadalla (Fig. 5.111)

Geographic positions:	
Middle position for the mines:	25°23′16″ N, 33°42′37″ E
Middle position in Wadi Gadalla:	25°23′19″ N, 33°43′32″ E
Gadalla southern settlement:	25°22′26″ N, 33°43′32″ E

The well-preserved New Kingdom settlement at Gadalla S is located about 750 m to the S of the actual mining district. It consists of about 20 large houses (Fig. 5.112) built of flat metadiorite slabs. As already observed elsewhere, whenever flat building stones had been used, the houses are relatively well-preserved. Yet, the wadi erosion has partially washed away other buildings leaving only the slope-sided walls in upright positions.

The houses display average lengths between 10 and 12 m and widths between 3 and 4 m. They are subdivided into three to six rooms, and in many cases a walled terrace is distinguishable at their front sides. To judge by the large numbers of the therein-contained tools, the terraces apparently served as work platforms (Fig. 5.113).

Inside the rooms, one repeatedly finds several, oval New Kingdom mills and grinding stones, flat pounding stones, as well as pestles. In some cases, the mills had been transformed and used as flat pounding stones, which seems to hint to a prolonged occupation.

Next to these large houses, there are also a number of small, single-room huts. On the ground, they measure ca. 3 × 4 m. In many cases, a small extension had been added to the huts. In the settlement, we noticed that many of the often, well-worn, oval grinding mills had been recycled to pounding stones, used for crushing ore lumps. Because of the relatively long distance to the actually mined quartz veins to the N, it seems likely that the miners in this southern settlement had concentrated for the most on wadiworkings.

The very few pottery shards also date to the New Kingdom

Along the wadi edge near the mine itself, are several, relatively small house ruins, which too, had been built from flat rock slabs gathered from of the surrounding surface. In a gulch-like side wadi one recognises an artificial dam, which on one hand probably had served to protect the huts from floods. On the other, it had certainly also functioned as a water reservoir, because Wadi Gadalla displays next to underground mining, also distinct traces from wadiworkings, which due to the considerable workforce associated with this mining method, required adequately large amounts of water.

Gadalla North consists of about four groups of differently preserved houses, mostly along the northern rim of Wadi Talet Gadalla. Human occupation at this site seems to have been significantly more substantial than suggested by today's evidence, since again many huts seem to have been flushed away by occasionally occurring floods. The remaining, quite well-preserved houses are mostly located on somewhat higher ground, near the slope. The ones built in the wadi bed have almost all been destroyed beyond recognition. The find inventory dates to the New Kingdom and again comprises conspicuously large mills and

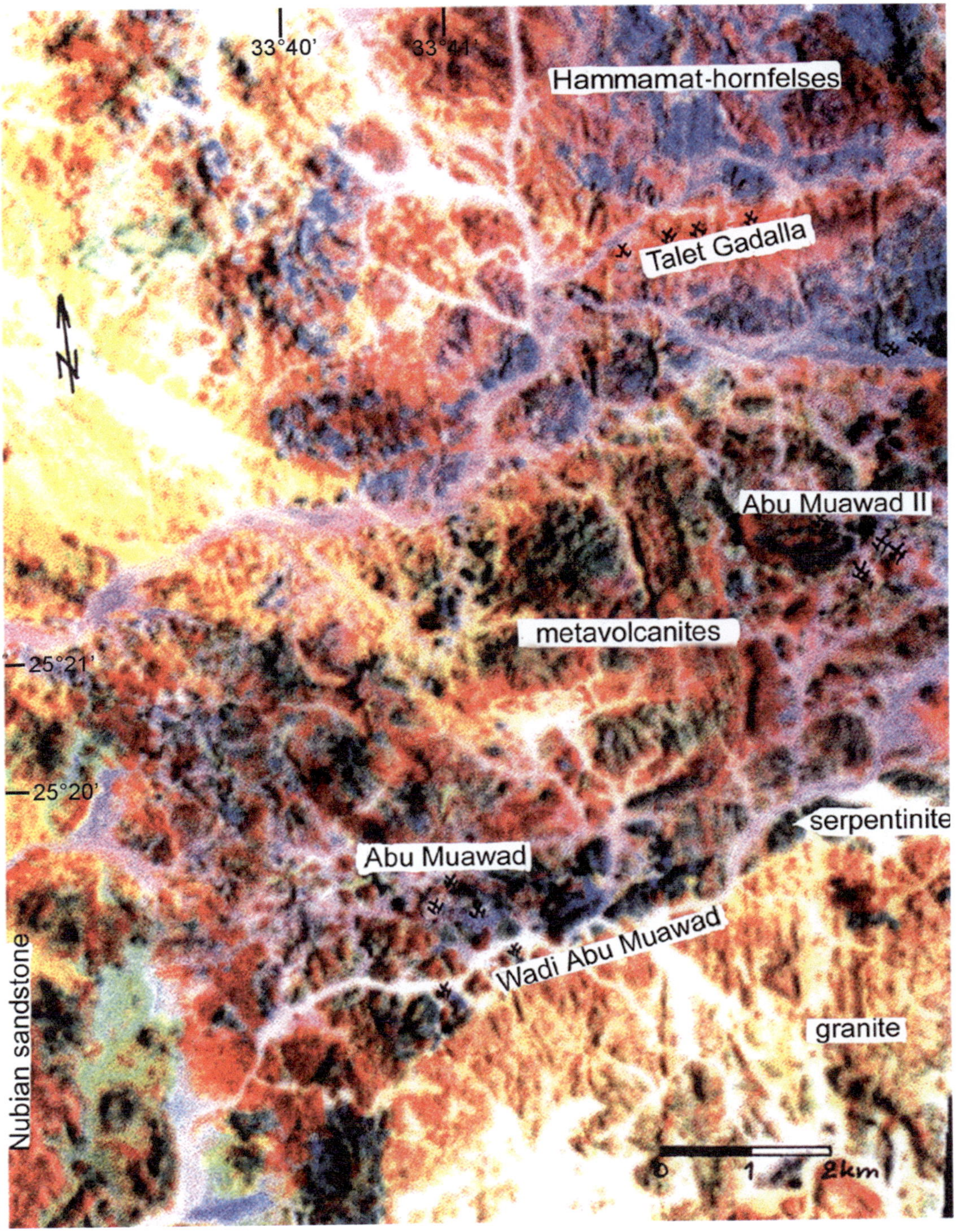

Fig. 5.111 Lithologically processed satellite image of the area around Talet Gadalla and Abu Muawad (TM 174/42, channels 7-4-1)

Fig. 5.112 Large houses and house complexes at the southern settlement of Talet Gadalla (modified Google-Earth image)

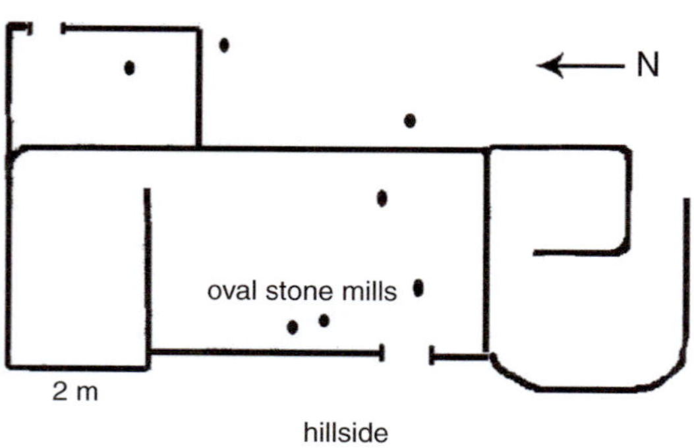

Fig. 5.113 Plan of a well-preserved New Kingdom house in Wadi Talet Gadalla

pounding stones, whereas some mills had later been re-used as flat pounding slabs.

The actual gold quartz occurrences in the area are located in the mountain barrier to the S of the wadi, in a narrow gorge close to which group 12 round as well as rectangular houses, containing find assemblages from the New Kingdom. Some of the mills found here stand out by their impressive sizes and an integrated seat at one of their ends. Other mills display several, parallel running grinding grooves.

About 1.3 km further to the SE, in a number of small tributary valleys one comes across loose groups of rectangular houses with two to three rooms, that

Fig. 5.114 Part of the main mine at Talet Gadalla. The exploited vein continues beyond the mountain ridge

seem to be associated with clearly visible debris heaps from intensively operated wadiworkings.

The architectural remains near the mines located in the mountains undoubtedly formed the dwelling site of its miners, as numerous mining tools had been discarded here. The activities of the occupants in the various settlements along the wadi edges on the other hand, had concentrated more on processing ores from wadiworkings.

The Talet Gadalla district stretches out between 0.5 and 2 km to the N of the wadi with the same name. It is located at the northern rim of a sporadically sheared diorite which had intruded into an environment marked by metasediments and metavolcanics. The metasediments are former greywackes and siltstones of the Hammamat formation. The other metamorphites appear as serpentinites, talcum-tremolite-carbonates, amphibolites, and rhyolitic slates. Numerous, chiefly aplitic and kersantitic, but also andesite dikes cut through the mountains in an approximate N-S direction.

The main, productive quartz vein consists actually of a swarm of parallel running veins with an approximate E-W strike within a shear zone (Fig. 5.114). In the western part of the area, the principal strike of the vein system is oriented at 85°, its dip between 60° and 70° S. In the eastern part the strike is at 75°, whereas its dip remains unchanged. For the most part the vein itself can be followed over relatively short distances only (in one case up to 250 m in the W). This is because it more accurately consists of an elongated, though intermittently stalling vein swarm, which then reappears in a slightly deviating strike. Towards the wallrock, these vein systems exhibit intensive alterations with chlorite and sericite formations. The veins' thicknesses vary between 15 and 50 cm, and the gold grades apparently average around 15 g/t with peak values exceeding 100 g/t (Koshin and Bassyuni 1968; Tawab 1991). The vein quartz is mostly milky-white and occasionally speckled with dark-grey stains, which is connected to the distribution of its ore contents. The up to 20 cm thick quartz veins oriented perpendicularly to this main direction contain gold. The mines themselves have unfortunately for the most caved in.

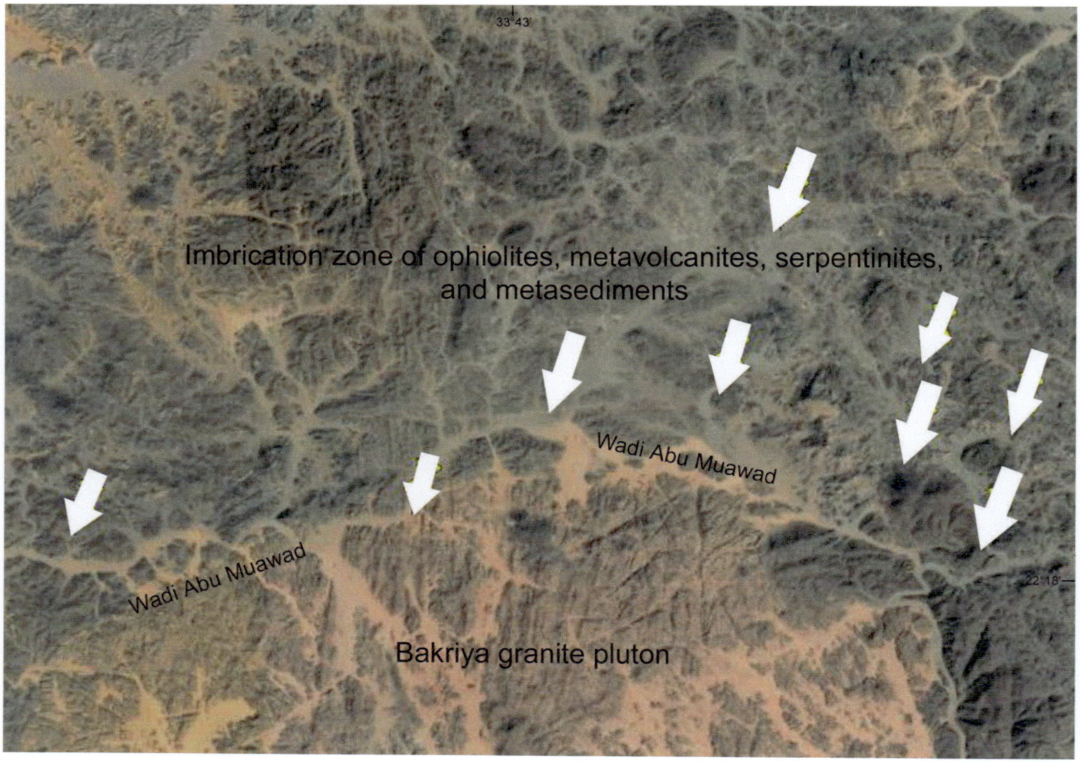

Fig. 5.115 Positions of the gold processing sites in Wadi Muawad (modified Google-Earth image)

5.3.14 Abu Muawad (Fig. 5.111)

Geographic position (New Kingdom settlement):	25°18′36″ N, 33°42′08″ E

The probably most extensive wadiworkings in this part of the Egyptian Eastern Desert are found in the region around Wadi Abu Muawad. Otherwise, there are no recognisable traces from underground mining inside this zone. Gold had been mined not only in the main wadi but also in virtually all large and small side wadis. The evidence is preseverved in small waste dumps in the wadi beds themselves and above all, in the numerous settlement traces along the wadi edges (Fig. 5.115).

The centre of activities is located in Abu Muawad-East. In small tributary wadis there are several large, New Kingdom house alignments, of which the most important one comprises about 40 house units on the S side of Wadi Abu Muawad (Fig. 5.116).

Essentially, two occupational phases can be distinguished; an early one dating to the New Kingdom and a late, less extensive one from the Early Arab Period. Human occupation is represented principally by a rather large, relatively well-preserved house cluster in a protected area behind a cemetery, in the northern part of the main wadi. Characteristic tools, house structures, and pottery clearly date to the respectively divergent occupation phases, although, the ceramic evidence from the New Kingdom Period is almost lacking in the site's surface.

As it rises from the Nubian sandstone in the W, Wadi Abu Muawad enters a geologically complex zone marked by volcano-sedimentary

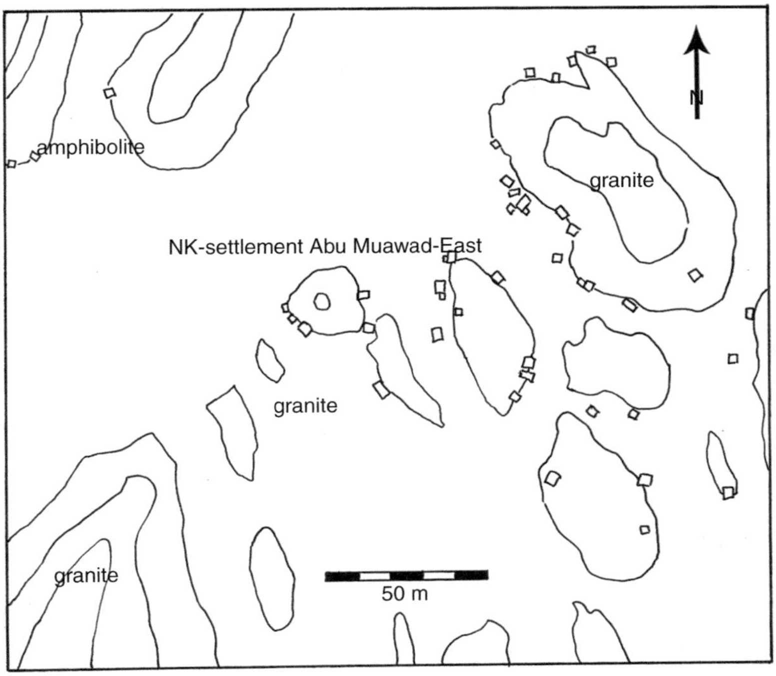

Fig. 5.116 Sketch plan of the New Kingdom settlement at Abu Muawad-East (R. Klemm)

series altered to hornfelses through contact metamorphism caused by the Gebel Bakriya granite. The wadi itself then ramifies within the hornfelses and eventually runs out after about 15 km.

Except for the approximately 3 km long course alongside the granite towards the N, architectural remains and traces from processing in the wadi alluvium are almost everywhere omnipresent.

Petrographically, the area is an intensively imbricated zone, composed of partly carbonated serpentinites, metavolcanics, talcum-, and graphite slates with small intercalations of marble lenses, diorites, and gabbros, which are covered in the W by the Nubian sandstone. As a result from contact-metamorphism through the large Bakriya-granite intrusion, which probably flatly underlies the entire zone, even examinations under the microscope are not always useful for distinguishing between former acid volcanites and metasediments. An almost omnipresent foliation is oriented in an approximate E-W direction. Numerous, mainly andesite dikes striking with little divergence from N to S and dipping almost vertically, cut through the entire relief.

The areas characterised by wadiworkings tend to concentrate in metasediment zones affected by contact metamorphism.

Small quartz veins are common, but in general only few centimetres thick as they adhere to the shear direction of the hostrocks. Wherever their strike diverges, their widths may be considerably larger, and their quartz is barren.

Ancient mining was ascertained to none of these quartz vein occurrences. It is therefore likely that gold production had been based exclusively on the processing of the area's wadi alluvium (wadiworkings).

Noteworthy is the presence of a thick, pale-green amazonite-pegmatite of about 1.5 m thickness, revealing evident traces from mining.

5.3.15 El-Hisinat (Fig. 5.117)

Geographic positions:	
N settlement:	25°17′42″ N, 33°48′28″ E
SE settlement:	25°16′52″ N, 33°48′59″ E
Mine area:	25°16′53″ N, 33°49′17″ E

5.3.15.1 El-Hisinat-North

This site is characterised by its large, relatively well-preserved houses from the New Kingdom (Fig. 5.118). It is located at the northern flank of the wide Wadi Miyah as well as the edges of a tributary side wadi.

The walls are preserved to heights around 1 m thanks to their dry masonry from flat rock slabs, which proves to be especially beneficial for their conservation. We counted around 40 houses measuring around 7 × 4 m and generally comprising two to three rooms. Many have a 2 × 2 m large annex added to their outside wall facing the mountain. Inside, doorways interconnect the rooms. All houses are accessed from the mountainside, and stand between 5 and 8 m away from each other. This arrangement corroborates with the pattern already observed at the wadi edges of New Kingdom settlement sites, specialised on wadiworkings.

A quartz vein too, was mined in this district. It strikes at 105° and dips at 86° S. Its thickness was established at 80 cm over a length of more than 30 m. However, it shows mining traces at only about 10 m and therefore had certainly been of minor importance.

In contrast to other New Kingdom sites along the wadi edges whose economies relied on wadiworkings, surprisingly few processing tools are found here. This may be connected to the site's exposed location at the edge of this much-frequented wadi, as in addition to the later round mills, New Kingdom millstones have until today remained particularly popular for grinding cereals among Bedouins.

5.3.15.2 El-Hisinat-South and Mine

Opposite Hisinat-North, in the mouth area of a side wadi leading to the SE, one comes across nine New Kingdom houses with the corresponding millstone inventory. Farther into the wadi, the terrain reveals evident traces from former wadiworkings.

Near a mined quartz vein, there is a group of at least ten houses on the eastern side of the wadi. As already seen elsewhere, Ptolemaic Period miners had taken up the activities of their predecessors of the New Kingdom and continued to exploit the deposits. Mining had also begun at a small quartz vein nearby, although evidently without any significant consequences. A number of Ptolemaic Period runner stones are found within the rubble of this part of the wadi. Moreover, stone hammers and pounding stones from the New Kingdom period were identified here too. Mansour et al. (1956) recorded up to 5.5 g Au/t from the small quartz veins. However, the findings by Zalata et al. (1972), whose figures amount to only 0.25 g/t Au, were not able to substantiate this.

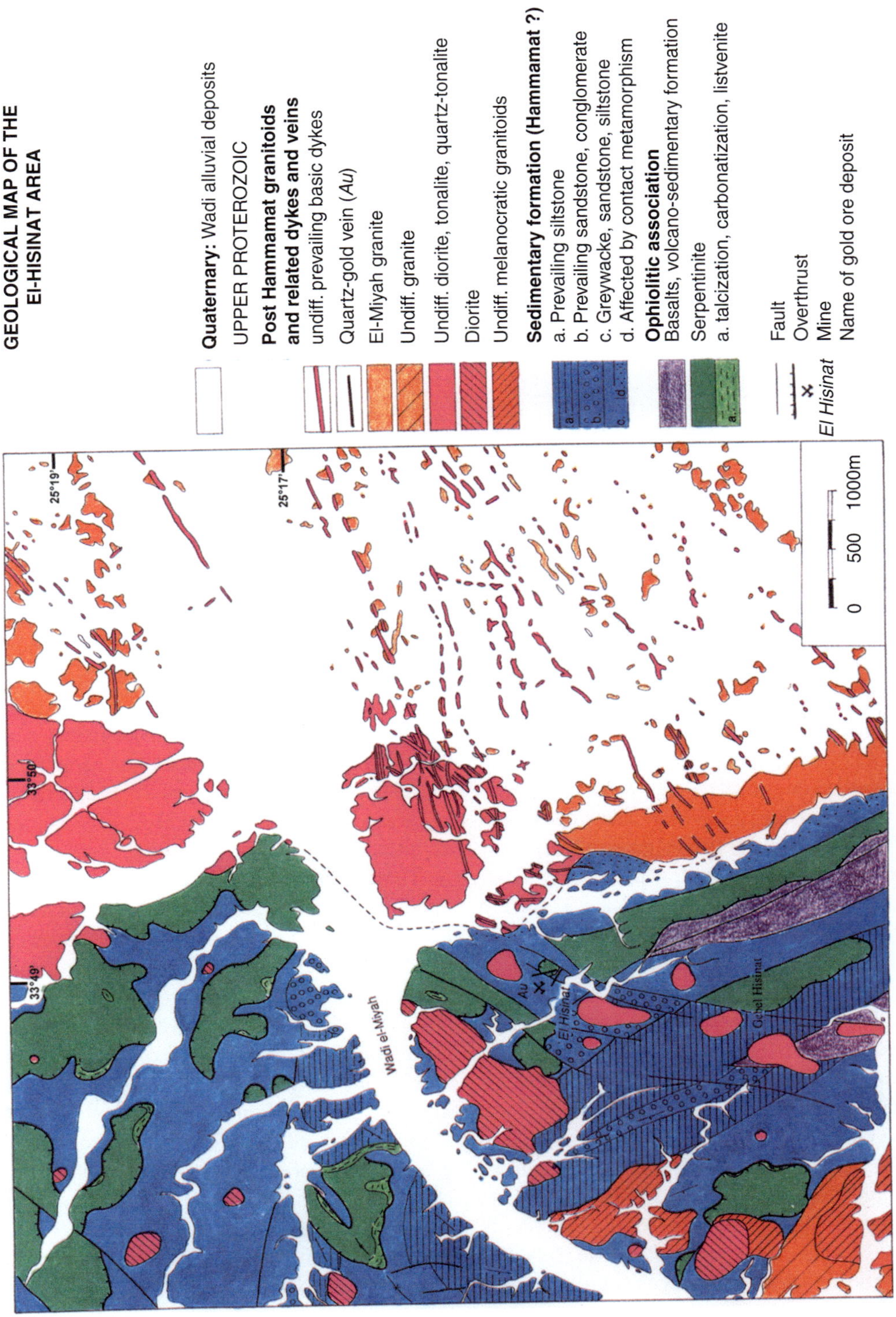

Fig. 5.117 Geological sketch map of the El-Hisinat area, Wadi Miyah (H. Kräutner)

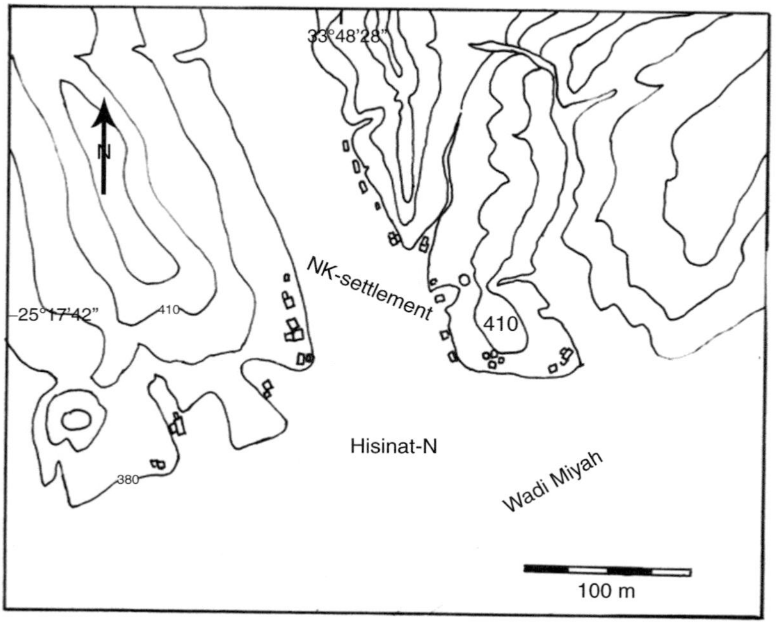

Fig. 5.118 New Kingdom settlement at El-Hisinat-North, Wadi Miyah (R. Klemm)

5.3.15.3 El-Hisinat-Southeast

At the western rim of a wide wadi leading southwards, some 1,200 m away from the southern edge of Wadi Miyah, is a large and well-preserved New Kingdom settlement site comprising about 25 houses (Fig. 5.119). They group together along the wadi edge as well as on a small hill. The houses are rectangular, relatively large, and in general have two rooms. The find inventory consists of unusually regularly shaped, oval stone mills, each with an easily distinguishable, raised seat at one of its ends. This has a comfort-enhancing effect compared to squatting or kneeling position (Fig. 5.120). Because we found no signs for underground mining in the surrounding mountains, gold had probably been obtained from wadiworkings only. However, in contrast to the aforementioned, western wadi, the traces from these workings have by and large been eroded away.

The Hisinat deposit is in geologic terms at the western rim of the massive Miyah granite-pluton, just N of Gebel Hisinat. The ancient gold quartz mine appears in Hammamat metasediments clearly marked by contact metamorphism. The metasediments consist of former greywackes, siltstones, and conglomerates, which in spite of significant metamorphism are relatively well differentiable in the lithologically processed satellite image (see Fig. 5.117). The series are steeply folded forming a long fault structure in the Hammamat series which furnishes excellent building material for dry masonry and thereby explains the excellent state of preservation of the mainly New Kingdom architecture.

These sediment units were folded with the older ophiolite nappe sequences in a late phase of the Pan-African orogenesis. Among them are residues overlying the Hammamat sequences, of mainly basalt and serpentinite, that occasionally are carbonated and sometimes transformed to talcum. Both units, but above all the serpentinites and the amphibolite occasionally are intensely sheared and therefore extremely carbonated. To the SW of the Hisinat ridge, the Hammamat series contact the Bokari-granodiorite. Here the granitoids are much hybridised by assimilation of Hammamat rocks. The marginal assimilation of mafic roof sections by the Miyah granite, through which the latter altered to diorites and tonalities, is also noteworthy.

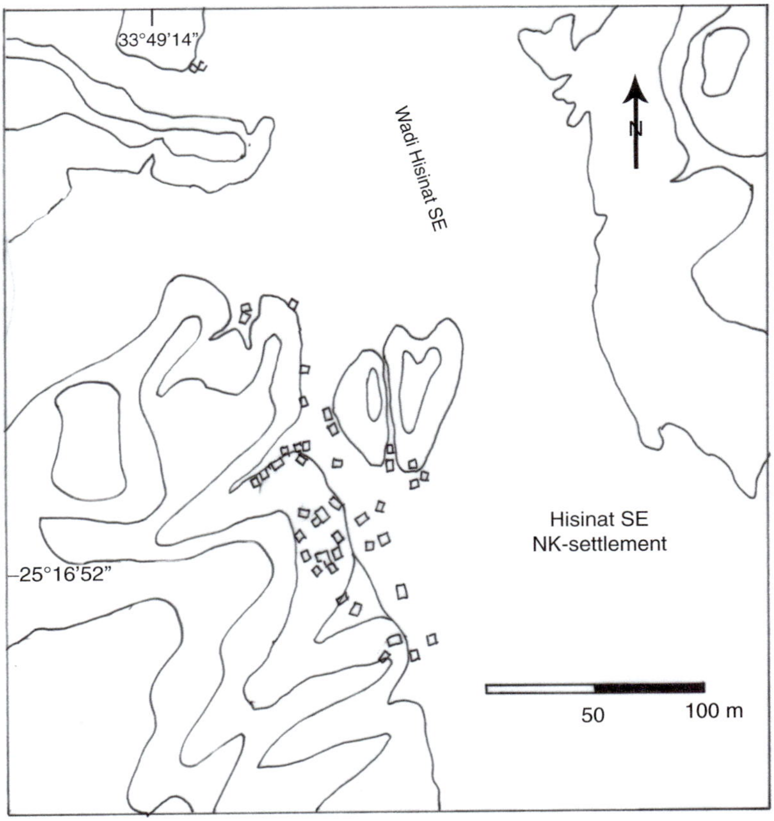

Fig. 5.119 New Kingdom settlement in the SE of Wadi Hisinat (R. Klemm)

Fig. 5.120 New Kingdom grinding mill with an integrated seat. The white quartz powder was produced with a small runner stone at El-Hisinat-Southeast

The mined quartz vein strikes at 112° and dips vertically. It is tracked over a distance of about 150 m through the mountains. Its average thickness is about 80 cm (max. 1.5 m) and its explotation goes back to the Old/Middle Kingdom. The main mining periods, however, had taken place during the New Kingdom and Ptolemaic Period with deep mining reaching depths around 25 m but for the most fluctuating between 5 and 10 m below ground surface. To the W of this mine, there are a number of quartz veins that strike at 175° and dip at 45° E, though none of them show traces from mining.

5.3.16 Abu Dabab

Geographic position:	25°19′04″ N, 34°33′33″ E

We were unable to visit the deposit at Abu Dabab. The following information on this site was therefore taken from Koshin and Bassyuni (1968):

"The area of the deposit is composed of greenstone rock represented by epidiorite-dioritic complex and amphibolite. The rocks are intruded by a pink granite and a series of dykes of volcanics. The N of the deposit is an auriferous quartz vein trending north-west (326°) and dipping north-east at 25–30° is traced in the greenstone rock. It is traced for 950 m along the strike. Another vein, probably the continuation of the former, is encountered in the southern part of the area."

"The veins consist of milky-white dense quartz; in places the latter contains numerous cavities. Galenite impregnation is also noted. Near the veins enclosing rocks are highly sericitised and chloritised."

Modern high resolving Google-Earth satellite image clearly shows the vein, but no remains of ancient huts and working traces. However, intensive prospection activities started recently about 3.5 km further N.

5.3.17 Umm Samra

Geographic position:	
Mine:	25°17′38″ N, ′34°07′07″ E
New Kingdom settlement:	25°17′08″ N, 34°07′23″ E

A number of minor quartz veins appear within the NW part of a very coarse-grained, pinkish-red biotite-granite stock at Gebel Umm Samra and some 100 m towards the contact zone of an intruded metarhyolite. Although mining did occur here under the New Kingdom, it apparently didn't grow to full-scale operations. Nearby, one recognises houses associated in context with millstone assemblages.

Eventhough significant gold contents are only very rarely contained in quartz veins of such younger granites, the EGSMA-Technoexport group was able to affirm grades of up 24 g/t Au in two of them.

The discovery of this atypical occurrence by the New Kingdom prospectors again testifies to their assiduousness, which eventually contributed to a significant increase in the gold production in that period. This case, however, seems to reflect an unsuccessful, or aborted prospecting campaign.

As the mills barely display abrasions from use, they probably had not served for a very long time. Moreover, the absence of secondarily used mills and the relatively low number of houses indicates that Umm Samra probably never developed beyond the rank of a prospecting zone to an established mining district.

Contrasting to this apparently unsuccessful mining attempt, a New Kingdom settlement with over 50, partially well-preserved house ruins and displaying the usual finds (Fig. 5.121), is located about 1 km further S, in a trough-like extension on the granite floor. Because no traces from mining activities were found in the surroundings' quartz veins, the settlement probably focused exclusively on wadiworkings.

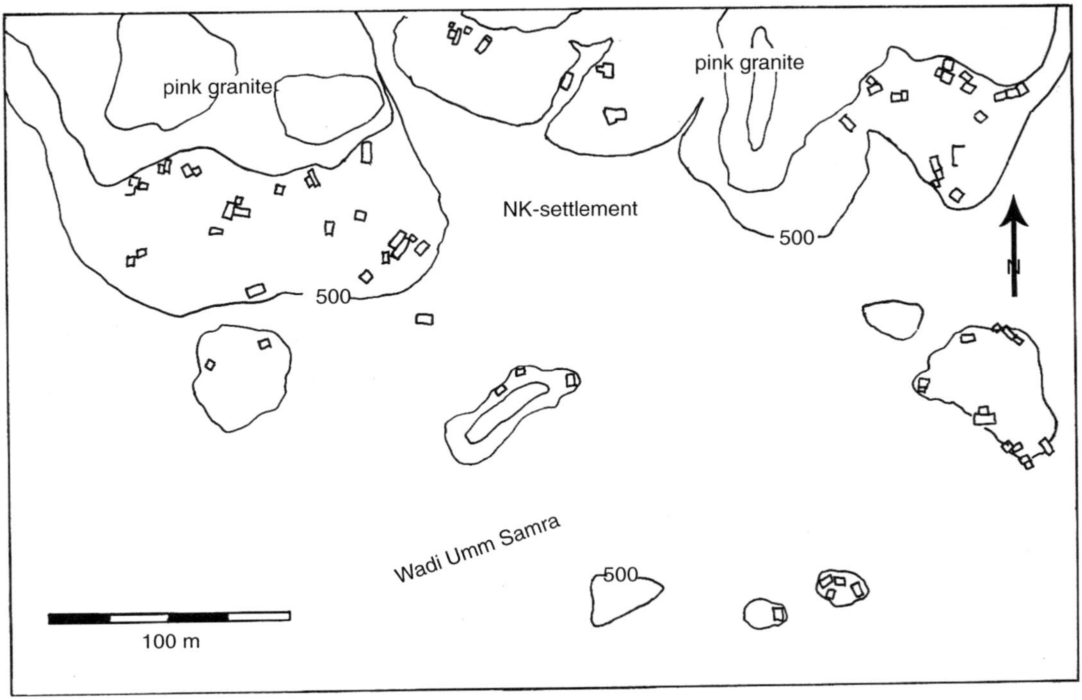

Fig. 5.121 Sketch map of the New Kingdom settlement remains at Umm Samra next to wadiworkings (R. Klemm)

5.3.18 **Bokari** (Fig. 5.122)

Geographic positions Bokari I:	
Old Kingdom mines:	25°15′54″ N, 33°45′05″ E
New Kingdom settlement in Wadi Bokari:	25°15′44″ N, 33°45′07″ E
Ptolemaic processing site:	25°15′08″ N, 33°45′22″ E
New Kingdom settlement in Wadi Miyah:	25°15′07″ N, 33°46′10″ E
Geographic position (Bokari II)	25°13′15″ N, 33°43′57″ E

The Bokari deposit area (also Bakriya, Bakari) which had been under exploitation from the Early Dynastic Period to the Early Arab Period, is located to the N of Wadi Miyah in its middle reaches.

5.3.18.1 Bokari I

The extensive mine district at Bokari stretches out along the northwestern rim of Wadi Miyah. It comprises several separate areas, of which all are marked to a different degree by mining traces dating from the Predynastic to Early Arab Periods, with a distinct low from the Roman-Byzantine era.

A central tailing site had been established around a now no longer discernible well to the W of Wadi Bokari, a tributary valley branching-off NW from Wadi Miyah. It displays three distinct processing sites consisting of wash-sand heaps that, according to the surface finds, had formed during the Ptolemaic Period. At the perimeter of this tailing site, particularly in the E and SW, one

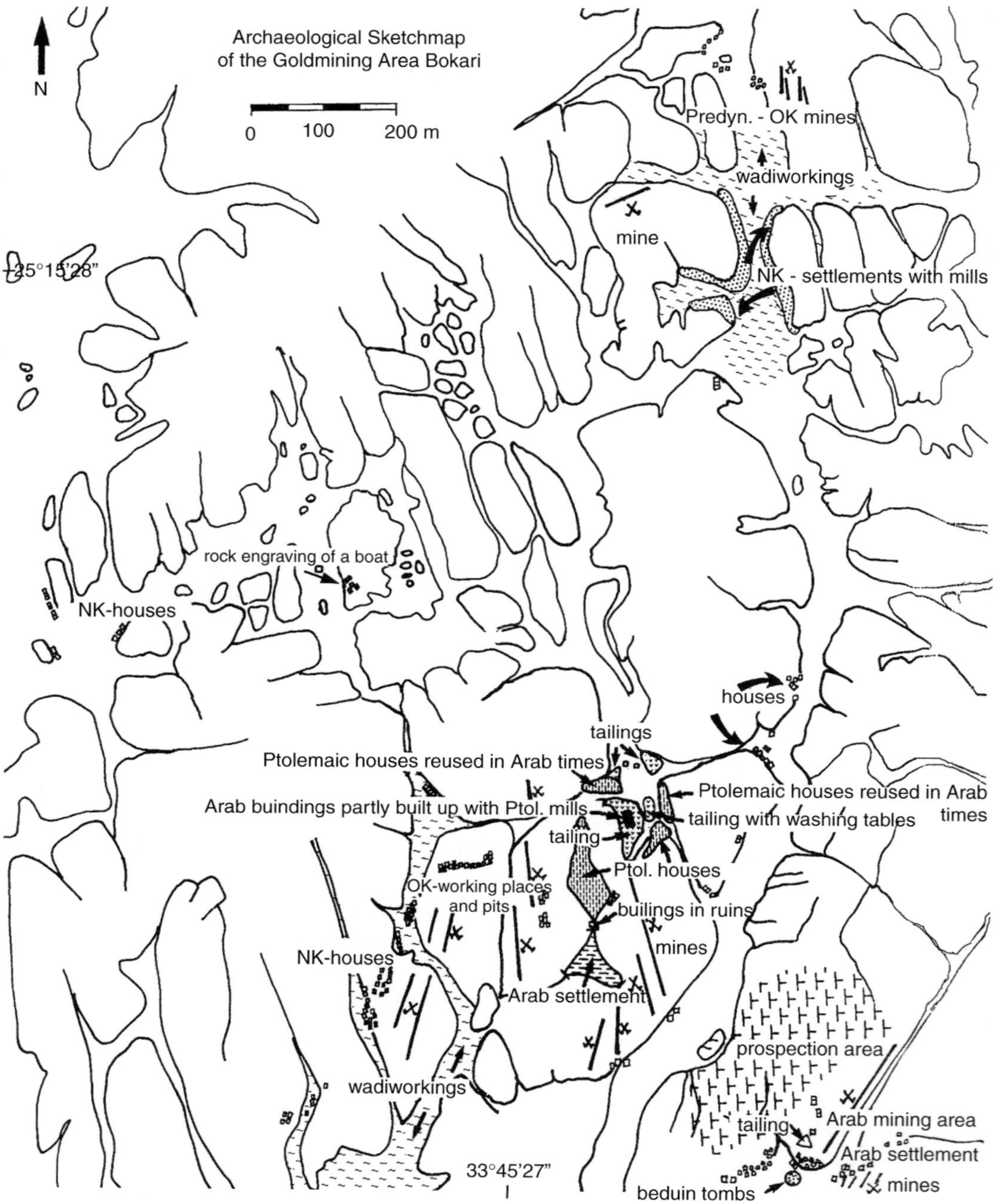

Fig. 5.122 Archaeological map of the ancient Bokari 1 mining district (R. Klemm)

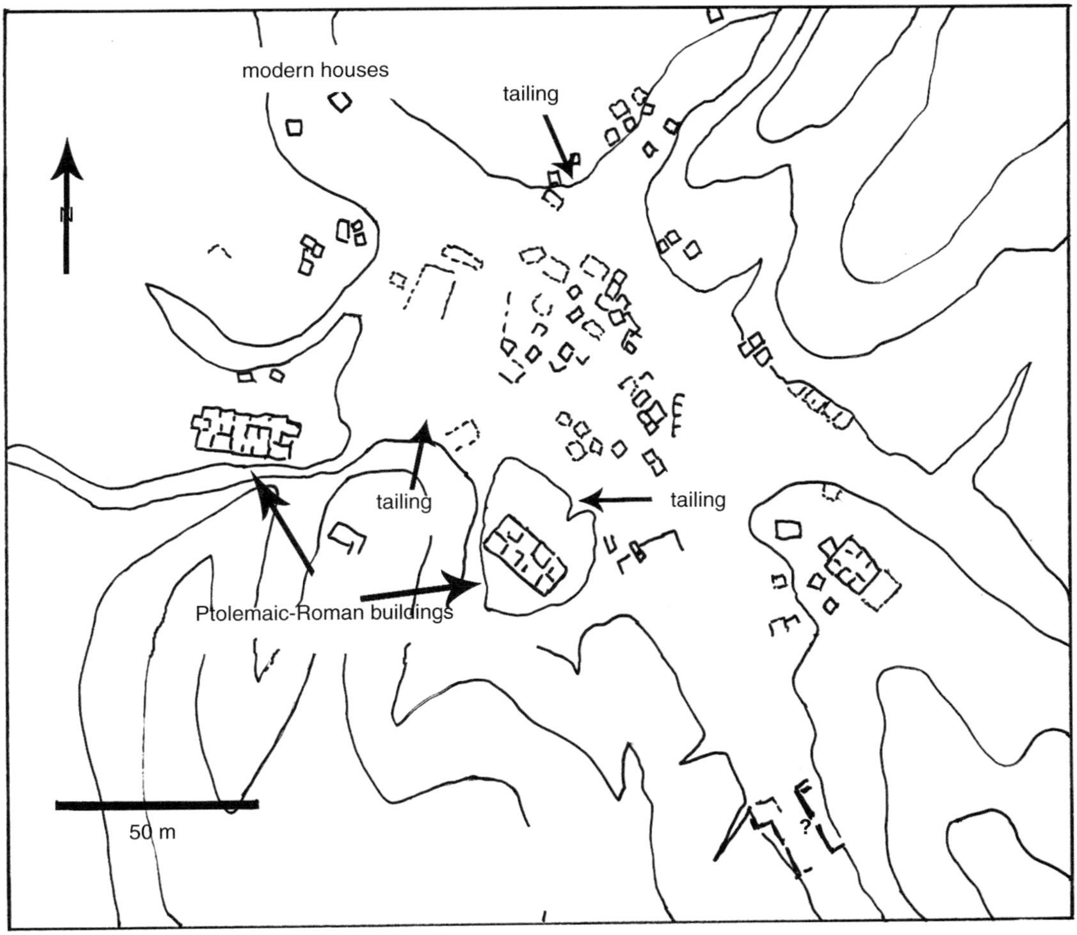

Fig. 5.123 Detail of the central processing site at Bokari 1 (R. Klemm)

notices several, large Ptolemaic house complexes of which two are remarkably well-preserved (Fig. 5.123). Both owe their good condition to perceptible restoration work. Lower masonry layers displaying a higher degree of weathering represent an early building phase, presumably in the Ptolemaic Period. The walls had then probably been replaced in the Roman Period by a noticeably better preserved masonry, which included a large number of Ptolemaic apron-shaped runner stones and concave stone mills. At the eastern edge of the tailings, there are at least two washing tables, of which one is a double version with two inclined surfaces oriented in opposite directions.

The ruins of two recent houses in the NW of the area containing Ptolemaic apron-shaped runner stones reveal that prospecting work had also been carried out in the twentieth century. Though, other than inflicting destruction to the site, it did not lead to the success of the ancient miners.

About 100 m to the W of the tailing site one comes across a mined quartz vein striking almost NS. Coming from the S, it is followed over a distance of about 1.5 km, of which various sections had been mined in evidently different phases.

In a mountain depression to the S of the central processing area there is a settlement yielding mainly Roman (or Early Arab?) evidence. Its three flanks facing the valley are protected by

Fig. 5.124 New Kingdom settlement remains at Bokari-North. Many ruins have been washed away by water erosion

walls. At the foot its shortest access path leading from the tailing site in the N are the remains of a large building with an integrated passageway, whose function is unknown. According to a personal communication by S. Sidebotham, its pottery evidence points to an occupation in the Roman Period. However, Early Arab Period ceramics are found here too, therefore also suggesting later occupational phases.

The settlement itself consists of about 100 huts, each comprising between one and three rooms. Inside their double-shell walls, the finds include at least one but more often several round mills and anvils. In addition, there are fragments from amphorae and domestic pottery. Another tailing site with a relatively well-preserved, masoned washing table is found at the eastern edge of the settlement.

Several old mines in the mountains bordering to the W had been established along quartz veins oriented N-S and NNW-SSE. Zalata et al. (1972) report maximum gold grades at 59 g/t. However, according to them, most quartz veins in the Bokari district turned out to be barren.

About 1 km to the N of this area, one comes across extensive, though severely disturbed New Kingdom settlements and wadiworkings, spreading from the margins of Wadi Bokari over the entire valley bed and into some of the side valleys (Fig. 5.124). As usual, the archaeological context deteriorates from valley edge towards its centre. Noteworthy is a small, slightly elevated kiosk from dry masonry, somewhat off the main wadi, as well as a conspicuous New Kingdom building set on top of a projecting rock at the southern side of this area. In conjunction with other, comparably isolated architectural units recorded elsewhere in mining districts of the Eastern Desert, this house may possibly represent that of an overseeing official, whose responsibility it was to watch over the entire wadiworking district.

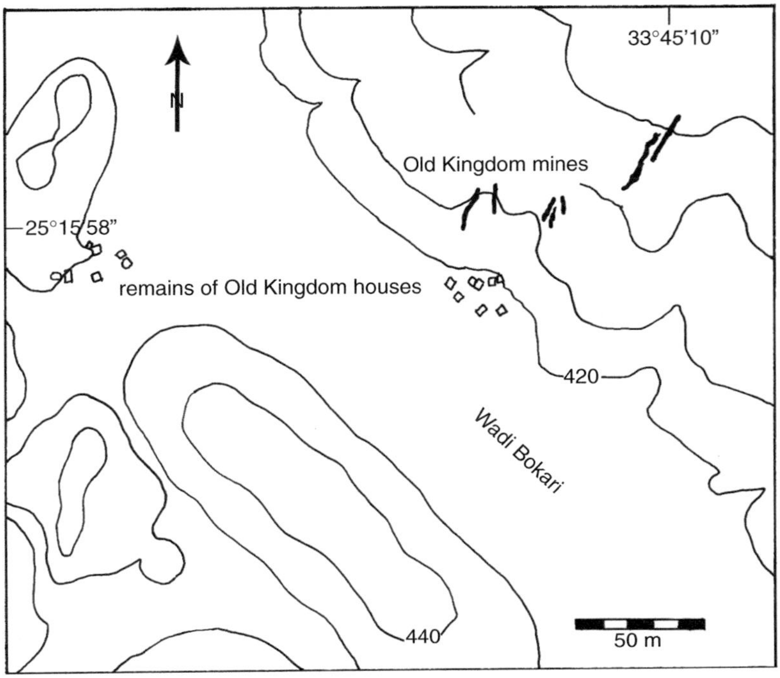

Fig. 5.125 Remains of an Old Kingdom settlement and mines in upper Wadi Bokari (R. Klemm)

Additional large settlements from the New Kingdom are found in the adjacent tributary valley that runs parallel to Wadi Bokari in the W. They too, had probably been established with the purpose of exploiting the gold ores in the wadi sediments. Because of their immediate location at the wadi edge as well as in a small side wadi, these houses are in particularly good condition. Evidently, though, they once partly spread-out over the entire wadi bed, as seen from frequently appearing remains from large structures washed away by repeated flooding.

To the N of the granodiorite area of Wadi Bokari itself, and about 200 m N of the New Kingdom settlement, and shortly before the pink Bakriya granite, a row of quartz veins with clearly recognisable malachite impregnations cross the granodiorite wallrock. According to the tool finds, mining within these veins had taken place from the Early Dynastic Period to the Old Kingdom (Fig. 5.125). The diagnostic material consists of numerous hand-held hammers, grooved stone axes and pottery fragments. The outstandingly good preservation of the exploited areas lends itself to take a closer look at the ancient mining methods applied here. Evidently, the only tool used had been the rock hammer of which several representatives are displayed in Fig. 5.126. They had served for pounding the merely 5–10 cm wide, and vertical quartz vein directly into a fine grain inside a 50–90 cm wide shaft lowered into the granodiorite wallrock, down to a depth of 10 m. The thereby resulting smoothness of the sidewalls has been preserved to the present day (Fig. 5.127). In these depths, the early miners attained the primary sulphide ores whose gold contents they were unable to extract. This is why mining had stopped in this area and was not resumed at later periods. As for the question relating to the ancient prospecting methods, the mentioned malachite layers along the joints with the wallrock provide useful information. As a rule, such malachite layers always occur in context with Early Dynastic to Old/

Fig. 5.126 Typical Predynastic to Old Kingdom stone hammers from the mine in upper Wadi Bokari. Due to sharp-edged chipping, some had become useless for pounding the vein and had therefore been discarded

Middle Kingdom Period gold and copper mines. Due to the absence of any significant vegetation cover in this mountainous terrain, the ancient prospectors had undoubtedly been able to identify such discolorations as key benchmarks in their gold (and copper) ore quests.

The second Old Kingdom mining complex is located about 600 m SE of the central tailing site in a relatively small, tributary valley, which too runs parallel to Wadi Bokari. Here as well, two vertically dipping gold quartz veins had been mined in a similar method, using stone hammers in direct blows to the veins, which again had resulted to conspicuously smooth walls in deeply excavated mines (Fig. 5.128). In front of the mines are worn out tools, one of which being exceptionally large and comparable to a specimen from Higalig dated to the Early Dynastic Period

An Early Arab settlement with more than 100 round huts and rectangular houses with several rooms lies about 1 km SE of the central Ptolemaic site and N of Wadi Bokari's confluence into Wadi Miyah. It is hidden away, immediately behind the promontories of a tributary valley leading NE and associated to a Bedouin cemetery that also includes a large circular grave. This concealed setting is quite typical for Early Arab settlements in the Egyptian East Desert.

Another vast settlement of this Early Arab type is located directly at the NW edge of Wadi Miyah, on plateau-like hills below an enormous, andesite dike. There are no detectable mining traces associated with the mineralised quartz veins near this settlement. This seems to suggest that the ores had exclusively been extracted from the erosion rubble of the wadi, because in most dwellings and especially around the numerous tailings, typically Early Arab processing tools such as round mills and compact stone anvils.

Immediately to the N of this site, in the mouth area of the next, relatively large, parallel wadi to Wadi Bokari, numerous remains of New Kingdom houses have barely survived the wadi floods on

Fig. 5.127 Old Kingdom mine in upper Wadi Bokari with smoothed walls from the crushing effect through the use of heavy and blunt hammers

Fig. 5.128 Old Kingdom mine in green-stained wallrock with characteristically smooth walls within a small western side wadi of Wadi Bokari

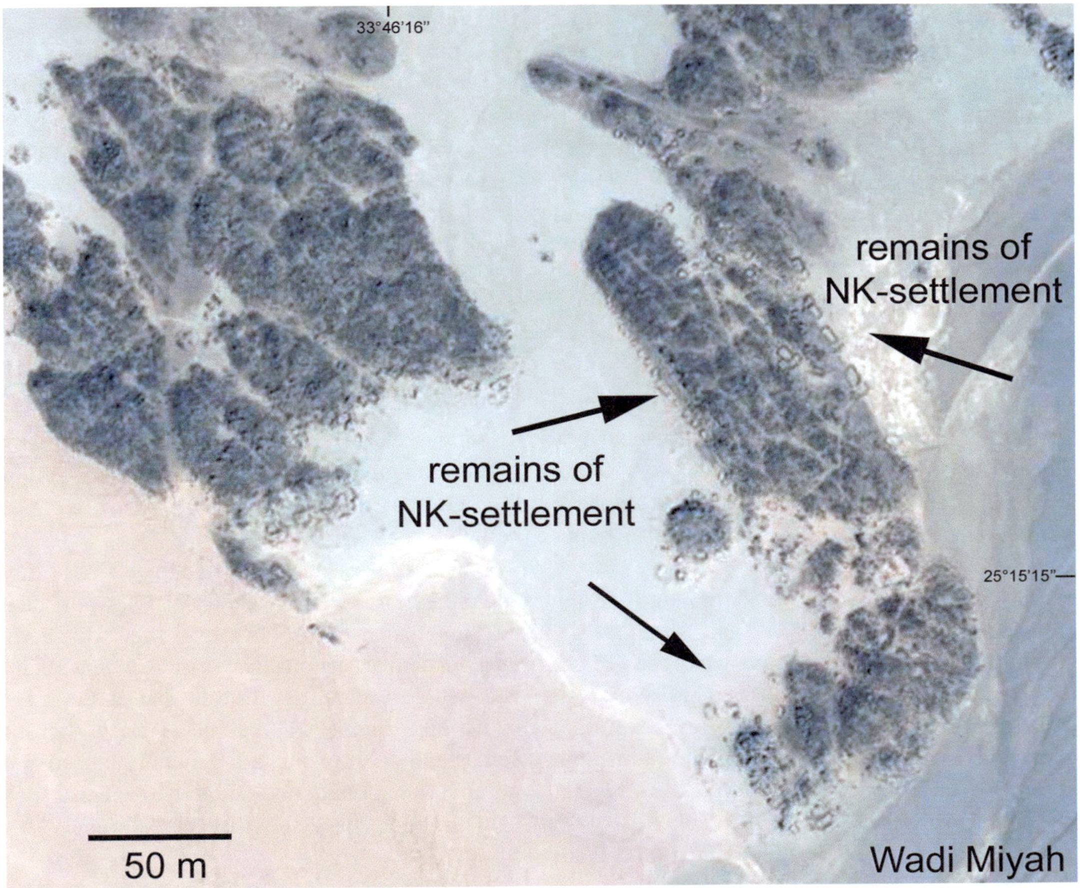

Fig. 5.129 New Kingdom settlement partly covered by sand drifts to the NE of Wadi Bokari, near Wadi Miyah (modified Google-Earth image)

the projecting lower foothills (Fig. 5.129). Yet again, as no mined quartz veins could be detected in the closer surroundings, it may be surmised that gold production was exclusively based on the wadi sediments.

Apparently, there had been much settling activity under the New Kingdom, especially along the NW flank of Wadi Miyah, though as a result of the wadi erosion in this water-rich valley (arab.: miyah = water), most of it is virtually beyond recognition. Still well-preserved house plans are nonetheless found here and there along the wadi rims. This for instance is the case further S, about 5 km away in Bokari II where an additional 50, rectangular New Kingdom houses are found with for the most two rooms containing oval stone mills and unusually large pestles of 30 cm length. Protected by the house corners, residues from tailings have yet resisted time, thus indicating that here just as at other New Kingdom, wadiworking settlements, processing had not been carried out at a central place but rather within the dwelling houses.

With some reluctance for the Roman Period, we may reiterate that all major historical periods of Ancient Egyptian right up to the Early Arab Period are represented in the Bokari district. Next to genuine mining of mineralised quartz veins, there was an extensive wadiworking activity, which undoubtedly had peaked in the New Kingdom and the Early Arab Period. Today in fact, Bokari counts as one of the chief gold deposits of Ancient Egypt.

The mining district is therefore ideally suited for further archaeological investigations and systematic excavations focussing on gold production. The fact that only few of its sites have suffered from modern mining may raise hopes for interesting results.

The area is geologically dominated by the intrusion of an apparently very massive granite-pluton of which the roof is only exposed in the Bakriya granite (Fig. 5.130). The entire deposit area exploited in ancient times forms a broad, marginal zone around the pluton which through intensive melting of the mainly Hammamat sediment wallrocks, altered its chemistry to granodiorite. To the SE of the deposit and Wadi Miyah this granitic magma has absorbed Hammamat sediments as roof material to such a degree that it crystallised to an extremely xenolith-rich, granitoid rock of heterogeneous composition. The position of this intrusion roof is indicated by numerous granitoid apophyses, which appear through roof rock assimilation as tonalite and diorite rocks in the Hammamat sequences in the NE of the mapped area. However, the Bakriya-intrusion too, is exposed only by its roof, which becomes apparent by the overlying hornfels remains of former Hammamat rocks. These partly mix to such an extent with the magma material that a dark, diorite-like rock originates from that process.

The anciently mined quartz veins are in a medium-grained, grey granodiorite, which is traversed by lengthy dark streaks of "undigested" metasediments. Due to a somewhat higher resistance to weathering than the light-grey variety, these streaks protrude at the surface. Numerous steep, mafic and acid dikes striking NE-SW, penetrate the entire, granitoid area. In one of the dikes, there are masses of light-coloured, medium-grained granite inclusions that should be viewed as witnesses of the original granite below. From this, it may be derived that the surface granodiorites too, with their dark streaks merely represent the hybrid, marginal facies of an originally granitic magma, and that the dark streaks are to be interpreted as not fully absorbed roof xenoliths.

All auriferous quartz veins strike NNE-SSW and dip almost vertically at 80° W. They post-date the magmatic veins, because they distinctly intrude them at several sites. A more recent quartz vein with an identical strike to that of the magmatic dike unmistakably crosses an auriferous vein. Such quartz veins are generally barren and we are aware of only one example in which a such vein had been checked by the ancient miners.

5.3.18.2 Bokari II

The number and sizes of the settlements alone reveal that this district is much less an important gold mining district than Bokari I. Bokari II is located SW to the latter and to the N of Wadi Miyah. As to its geology, the area stands out by its more dioritic and even gabbroid rocks, through which numerous, small quartz veins strike in an approximate N-S direction. None of them displays traces from mining. It is therefore likely that this moderately extensive New Kingdom settlement in a small tributary valley to the N of Wadi Miyah too, obtained its gold ores from the surrounding wadi alluvium (Fig. 5.130)

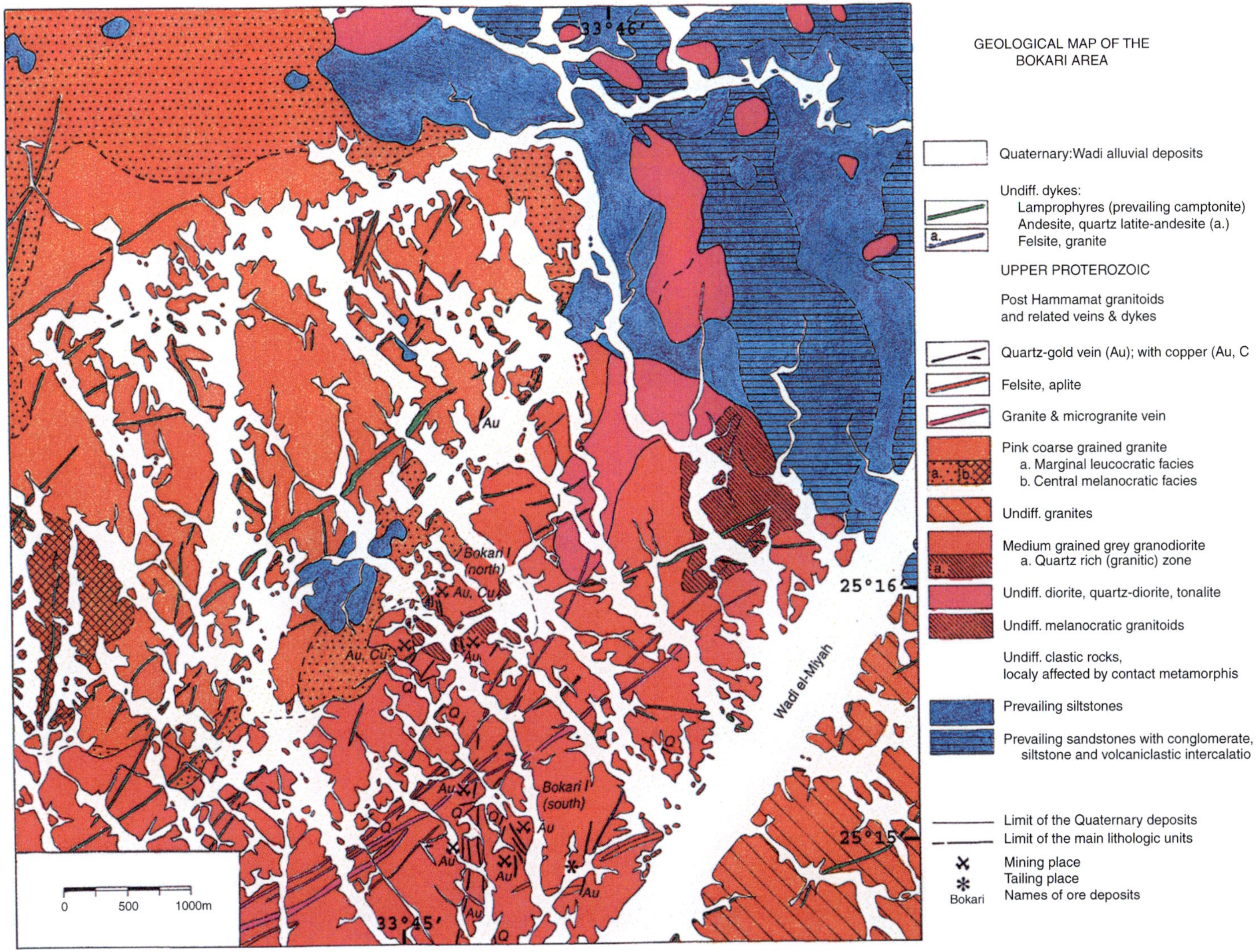

Fig. 5.130 Geological sketch map of the Bokari area (H. Kräutner)

Fig. 5.131 New Kingdom settlement Bokari II in hidden location in southeastern Wadi Miyah (modified Google-Earth image)

In Wadi Miyah and its tributary valleys gold production was evidently based on ores from the wadi alluvium, which benefited from better water supply and high ground-water levels (Fig. 5.131).

5.3.19 Abu Qareiya

Geographic position:	25°13′ N, 34° 02′ E (Koshin and Bassyuni 1968)

Geologically, the area is characterised by a zone of serpentinites and mafic metavolcanics, delimited in the NE and the SW by granite intrusions. Noteworthy are distinct alteration areas in the metavolcanics with predominant sericitisation and to a lesser grade chloritisation.

Quarried quartz veins were not found here. Ancient gold production had therefore probably been restricted exclusively to the processing of the wadi alluvium.

5.4 Southern Central Group

5.4.1 Umm Salatit

Geographic position:	25°10′45″ N, 33°54′49″ E

The architecture and artefacts at this site date to the New Kingdom and the Ptolemaic Period. In the immediate vicinity to the oval, New Kingdom mills one still finds small heaps of quartz ores that had been selected for processing. Remarkably, the tailings are light-grey and not pale-pink as usually observed at other gold processing sites. Next to the New Kingdom grinding mills and numerous Ptolemaic runner stones, there are countless amphora cones and light-coloured amphora handles, also from the Ptolemaic Period. No evidence from any later occupation is known from Salatit.

The site is located at the S edge of the large Umm Aid granite-pluton, within its contact zone to the metasedimentary and metavolcanic hostrocks, appearing here in the form of horn-

Fig. 5.132 Large, multi-chambered houses at the New Kingdom settlement of Umm Salim, W slope of Wadi Umm Salim

felses. The findings unambiguously reveal the site's function as a gold producing site, even though traces from deep mining were found nowhere in the surrounding area. There are numerous, small, but unexploited quartz veins striking 130° and dipping 70° NE. To judge by the size of the tailing heaps, substantial gold production had exclusively concentrated on the processing of the wadi alluvia, which in this case are very coarse.

About 500 m to the N of the site, the area is characterised by an outstanding criss-cross pattern made-up of magmatic dikes striking NE-SW and NNW-SSE through the granite. Among them is also a remarkable pegmatite dike within the immediate neighbourhood of the settlement site.

5.4.2 Beza

Geographic position: (Fort Beza):	25°10′22″ N, 34°04′40″ E

Koshin and Bassyuni (1968) and Jenkins (1925) reported ancient ruins (Fort Beza) and remains from ancient gold mines within the upper reaches of Wadi Beza, near a well (Bir Beza). Unfortunately, we didn't succeed in finding this site and suspect the information to be erroneous.

5.4.3 Umm Salim

Geographic position:	25°09′ 38″ N, 33°58′00″ E

Around 15, partly large and multi-chambered New Kingdom houses were counted on the slopes at both sides of Wadi Umm Salim (also referred to as Salem or Selem, signifying as much as "thorned tree"). Two had been built on a wadi terrace on the W side of the valley and distinctly stand out from the rest of the buildings (Fig. 5.132). Oval grinding mills lie scattered inside and between the houses. Their hillside situation probably results from the relatively wealthy hydrology in the wadi, which causes frequent floods as can also be told from the comparatively lush vegetation cover.

By the E slope one notices a small tailing site.

We found no underground mines, but in the NW reaches of the wadi yet clear residues from wadiworkings group near a series of quartz veins, which can be traced through the entire wadi.

The few mills, the mentioned tailing, the absence of surface pottery, and the houses' good

Fig. 5.133 Well preserved rectangular houses probably from New Kingdom at Abu Ashayir

state of preservation indicate that this district had been occupied only for a short phase in the New Kingdom. We detected no signs of any later occupation.

Zalata et al. (1972) carried out examinations on 15 quartz samples, of which only one proved to contain very low amounts of gold, at around 0.2 g/t. This finding may explain why the small quartz veins had been neglected.

To the NE of Gebel Umm Salim and the S of the wide plain of Wadi Himri, in the uppermost course of Wadi Umm Salim, one comes across the ruins of a small New Kingdom settlement, set in an ophiolite surrounding of partly intensely carbonated serpentinites (listvenite) and metavolcanics. It had probably only produced gold from the wadi alluvium, as no vein mining was found within its wider surroundings.

5.4.4 Abu Ashayir and Wadi Suwayqat

Geographic position (Abu Ashayir):	25°06′42″ N, 33°49′41″ E
Geographic position (Wadi Suwayqat):	25°08′08″ N, 33°50′47″ E

At km 2.8 on the road from Wadi Barramiya to Bokari, one arrives at a very well-preserved and relatively large settlement of more than 45 house units we decided to name after the nearby Wadi Abu Ashayir (Fig. 5.133). Though, because of the absence of diagnostic finds, we were unable to estimate its date. Since no traces from either mines or wadiworkings were found nearby, its function still needs to be determined. The possibility that its inhabitants had once been engaged in wadiworkings seems however, unlikely. However, according to the architecture of the houses an New Kingdom occupation may not be ruled out.

About 3.5 km further to the NE are the remains of another New Kingdom settlement located in the border area of the Wadi Suwayqat granite-pluton with severely altered metasediments from contact metamorphism. The site evidently subsisted on gold production from wadiworkings. As a result from floods, only the house ruins near the immediate wadi edge have survived.

5.4.5 **Barramiya** (Fig. 5.134)

Geographic positions:	
Ptolemaic processing plant:	25°04′10″ N, 33°47′35″ E
New Kingdom extractions:	25°04′26″ N, 33°47′00″ E

The deposit area of Barramiya is one of the most important ones in the Egyptian Eastern Desert. Apart from some lengthy interruptions it stayed under exploitation from Egypt's earliest history until the 1950s.

Mining under the British occupation in the early twentieth century has severely damaged and even completely wiped out ancient archaeological traces in the entire district, particularly in the valley plain leading from Wadi Barramiya to the mines.

The archaeological site at Barramiya has been identified with a gold mining district indirectly referred to in an inscription on a stele by Sethos I in the rock temple at Kanais (Schott 1961). According to the inscription, a well had been dug close to the temple in Wadi Miyah for the miners on their way to the gold mines. The well has survived until today, and according to our own measurements water is reached there in a depth of 49 m.

First, in Barramiya there is an impressive water basin surrounded by a large, rectangular wall measuring approximately 66 × 37 m (Fig. 5.135). The basin had probably enclosed a former well but even more likely a Ptolemaic gold washing device for ore processing. The fill debris at the periphery of the rectangular structure essentially consists of tailing sands in which numerous Ptolemaic mills and apron-shaped runner stones are found in addition to countless oval New Kingdom mills. Apparently, a former well and a processing site from the New Kingdom had been re-activated at this location in the Ptolemaic Period and transformed to an ore washing installation. This multi-phase occupation is exemplified by an apron-shaped runner stone of a Ptolemaic mill which had been made from a former, New Kingdom oval grinding mill (Fig. 5.136).

Similar rectangular gold-washing installations are known from the ancient mining district of Laurion/Attica, to the S of Athens (Conophagos 1988) and thus date the one at Barramiya to the Ptolemaic Period.

Outside the northern, narrow side of the wall rectangle is a large tailing site behind a double washing table with a reservoir in waterproof masonry. The finding clearly evidences that inclined washing tables had also been used in the Ptolemaic Period.

There are no signs of any later, Roman or Early Arab occupation at Barramiya.

To the W of the walled well there is another large tailing which once probably formed one unit together with the one to the N of the wall. It has been cut in two by a road and the recent well, Bir Barramiya.

NW to the tailings, close to the former well is a third tailing heap ontop of an elevation. Next to it are at least two fragments from a circular, Ptolemaic processing device (for heavy mineral separation) of the Laurion type (Fig. 5.137). Similar segments of such installations forming perfect circles are also known from Daghbag, Bokari and Samut. The site is surrounded by the severely damaged ruins of a large Ptolemaic settlement which spreads all the way to the tailing heaps in the W. Here, stones from earlier architecture appear in the masonry of modern buildings built by the 20th century mining company. As a result, they also include a profusion of stone tools dating to the New

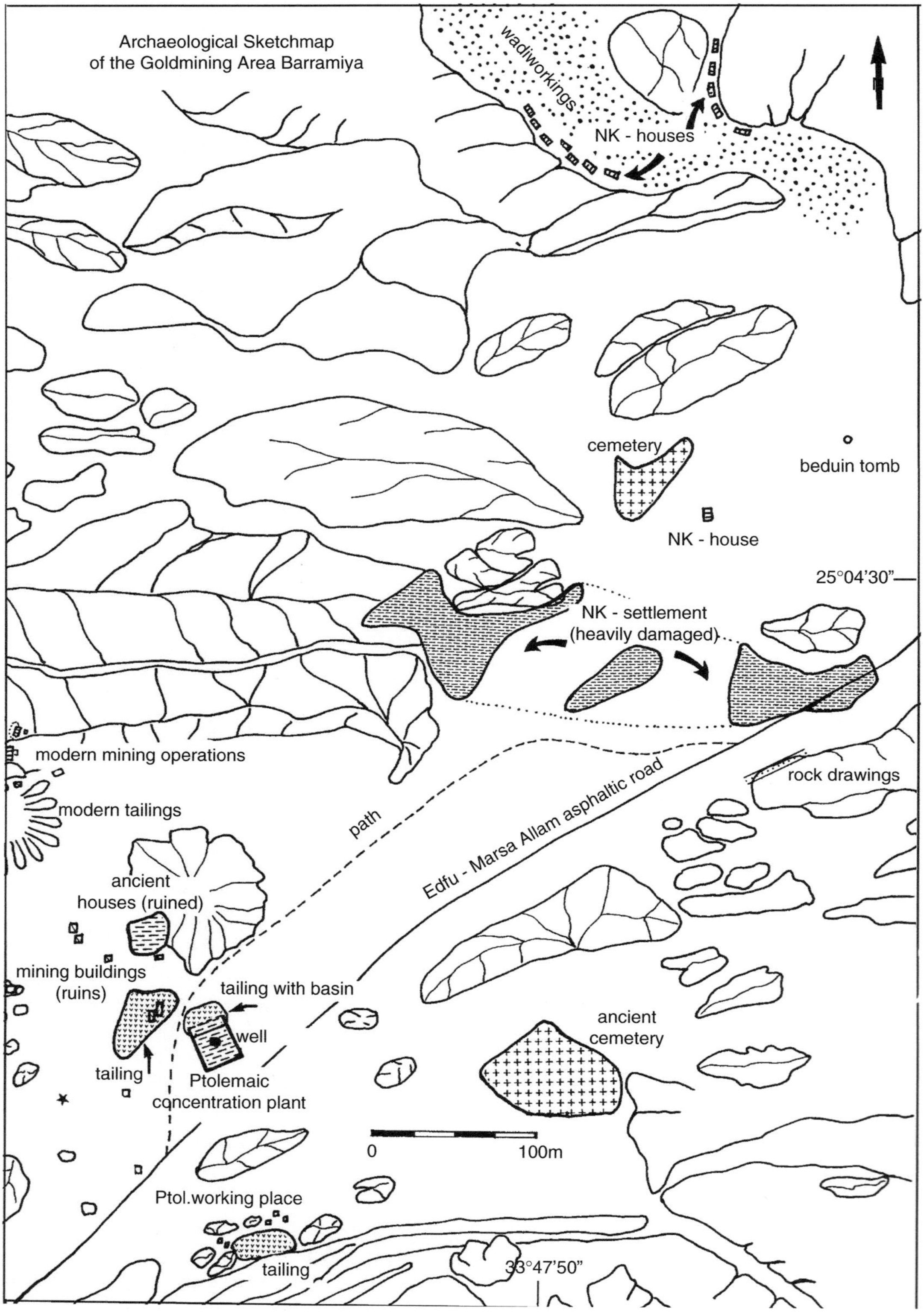

Fig. 5.134 Archaeological sketch map of the Barramiya mining district (R. Klemm)

Fig. 5.135 Ptolemaic gold processing site at Barramiya

Fig. 5.136 New Kingdom grinding mill transformed into a two handed runner stone in the Ptolemaic Period to match a concave ore mill

Fig. 5.137 Segment of a circular installation for gravity separation of heavy minerals in the debris from Ptolemaic Period gold processing at Barramiya

Kingdom as well as some from the Ptolemaic Period (Fig. 5.138).

About 80 m W to the tailings and close to a recent construction belonging to the agriculture authorities, one recognises the foundation socle of a building, which as it seems, had once also served as a surrounding terrasse of the building. The socle still contains residues from a former plaster, and the terrasse is well identifiable. Towards the middle one barely distinguishes the layout of wall structures under the debris. Future investigations will need to concentrate around the question as to the building's identification with a former, small temple complex.

Regardless of its wide-spread destruction and in spite of the predominance of Ptolemaic Period find contexts, the New Kingdom settlement seems with regard to size to have been by far the more important occupation at Barramiya. As seen in the archaeological map in Fig. 5.134, we were able to make out two New Kingdom settlements in the area of which the three sectors forming the southern one covers a distance of about 400 m from E to W.

The enormous settlement area probably results from wadiworking activities, whose traces are still well-discernible in more sheltered locations to the N. There, along the former wadiworkings, several singular New Kingdom houses occur in two alignments. In the main wadi, however, through which an asphalt road leads, recent mining activity has by and large wiped out all traces from former ore processing (Fig. 5.139).

Typical finds from the New Kingdom settlements at Barramiya include pounding stones with circular depressions in their centres that had been partly recycled from discarded grinding mills (see Fig. 5.138). They are frequently found at the central tailing heap near the well. Moreover, typically oval grinding mills in often, well-advanced states of weathering are found everywhere, as well as matching fist grinders (Fig. 5.140).

Fig. 5.138 House-wall in Barramiya from the first half of the twentieth century containing ancient ore processing tools (*arrows*) from the New Kingdom and the Ptolemaic Period

Fig. 5.139 Large New Kingdom settlement in Wadi Barramiya displaying a rich mill inventory. The asphalt road in the background partly runs through the site

Fig. 5.140 One hand grinding stones from the large New Kingdom settlement at Barramiya

Next to wadiworkings, Barramiya also reveals an intensive underground mining activity whose locations are rendered as black lines in Fig. 5.141.

The main deposit area is situated at the N flank of the valley plain, in the upper part of a tributary wadi branching-off N of Wadi Barramiya, only few kilometres E of the border between the Nubian sandstone and the Precambrian basement rocks. The quartz veins mined in the New Kingdom and Ptolemaic Period generally orient in an E-W direction to the N of the old well. They are identified by trenches of various depths and lengths over a distance of about 900 m. Furthermore, an especially impressive vein striking NNE-SSW and running through the middle section of the area was mined over a distance of about 100 m down to the level of the wadi bed. This resulted to an accessible gap in the mountain slope with up to 15 m high side walls (Fig. 5.142).

As observed in gold mining districts elsewhere (e.g., Bokari, Wadi Dara, Higalig), the typically smooth vein walls had probably fashioned as far back as the Old Kingdom, as a result from the crushing effect through the use of heavy stone hammers and axes to separate the quartz ores from the vein.

Due to its complex overthrust-, fold-, and fault tectonics, the geology in the area around Barramiya is somewhat difficult to understand. A slightly simplified situation is rendered in the geologic map Fig. 5.143. The area is basically partly built up from a dense, alternating sequence of metasediments at times reminiscent of the Hammamat formation due to occurring metagreywackes, siltstones, and conglomerates. The metasediments however, also include several hundred metres of thick marble layers (comprising black marbles) and partly graphite containing quartz-chlorite-sericite schists

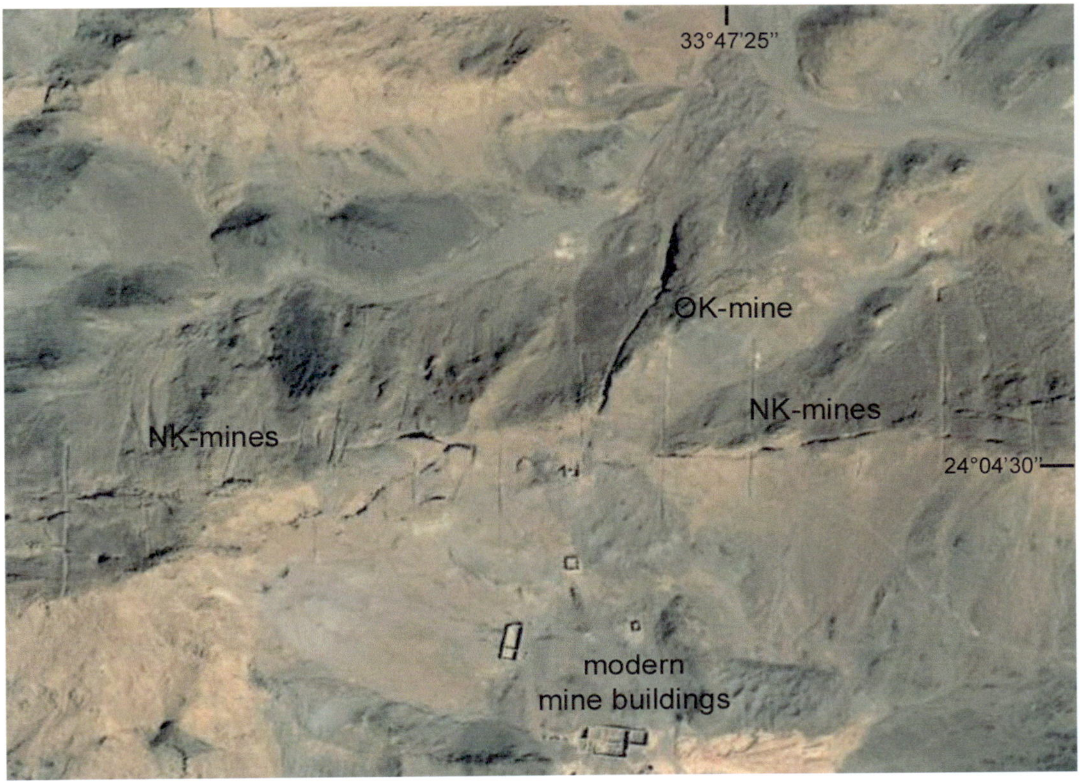

Fig. 5.141 Main mining district at Barramiya. The trench pits appear as black lines (modified Google-Earth image)

as well as actinolite-graphite schists. Both may formerly have been tuff-rich anaerobic marine sediments. The degree of metamorphism is markedly higher here than discussed in the more northern districts of the basement rocks to the S of the line Hamama - Gidami - Abu Mureiwat - Safaga. This line corresponds approximately to the border dividing the northern, mainly Meso-Proterozoic and the southern Neo-Proterozoic basement rocks of the Eastern Desert of Egypt, the latter of which is marked by the Hammamat sequences, ophiolites and island-arc volcanism, and not least, by intensive intrusions of granitoid magmatites.

This Hammamat-like sequence is intermittently covered by a massive ophiolite nappe, clearly dominated by serpentinites with partly well-preserved pyroxenites, but also containing to a lesser degree metabasalts. This nappe is imbricated to such an extent with the underlying units that mapping of certain areas within either of both groups proved extremely problematic. As a result, the black marbles which for instance appear in the northern zone of the geologic map of Barramiya (Fig. 5.143) thus alternate with an imbricated wedge of amphibole pyroxenites, which in its own turn borders with metagreywackes.

Presumably during the nappe overthrust, the entire mountainous backdrop became subject to isoclinal folding in shallow undulations with ENE-WSW striking axes. The folding affected in particular the gold mineral zone. This zone is heavily sheared and dips steeply N into a syncline structure of graphite-sericite and graphite-actinolite-tremolite schists. Both schists display complex internal layering and moreover may occasionally be talcumised and carbonated. Following their foliation, there are layers of well-developed pyrite crystal cubes. Because these are

Fig. 5.142 Exploited quartz vein at Barramiya. Its smooth walls seem to indicate an exploitation period to the Old/Middle Kingdom

well exposed only within the auriferous zone, the question concerning possible primary sedimentary gold contents seems no longer answerable. By any rate though, because of the striking bond of the partly, extremely rich gold mineralisation to these sediment sequences, the question imposes itself as to whether or not primary pre-enrichment had taken place.

Within an environment of intensive, CO_2-rich hydrotherms, the ENE-WSW striking serpentinites and ultramafic rocks had partly together with the imbricated metasediments completely or partially carbonated. During this process enormous amounts of SiO_2 were released. Silicon dioxide for its part, lead to silification of the schistose metasediments, thus shrouding their original identity in as far as they are very easily taken for former acid tuffs or rhyolites.

The thermal energy and the considerable CO_2 amounts were most probably supplied by

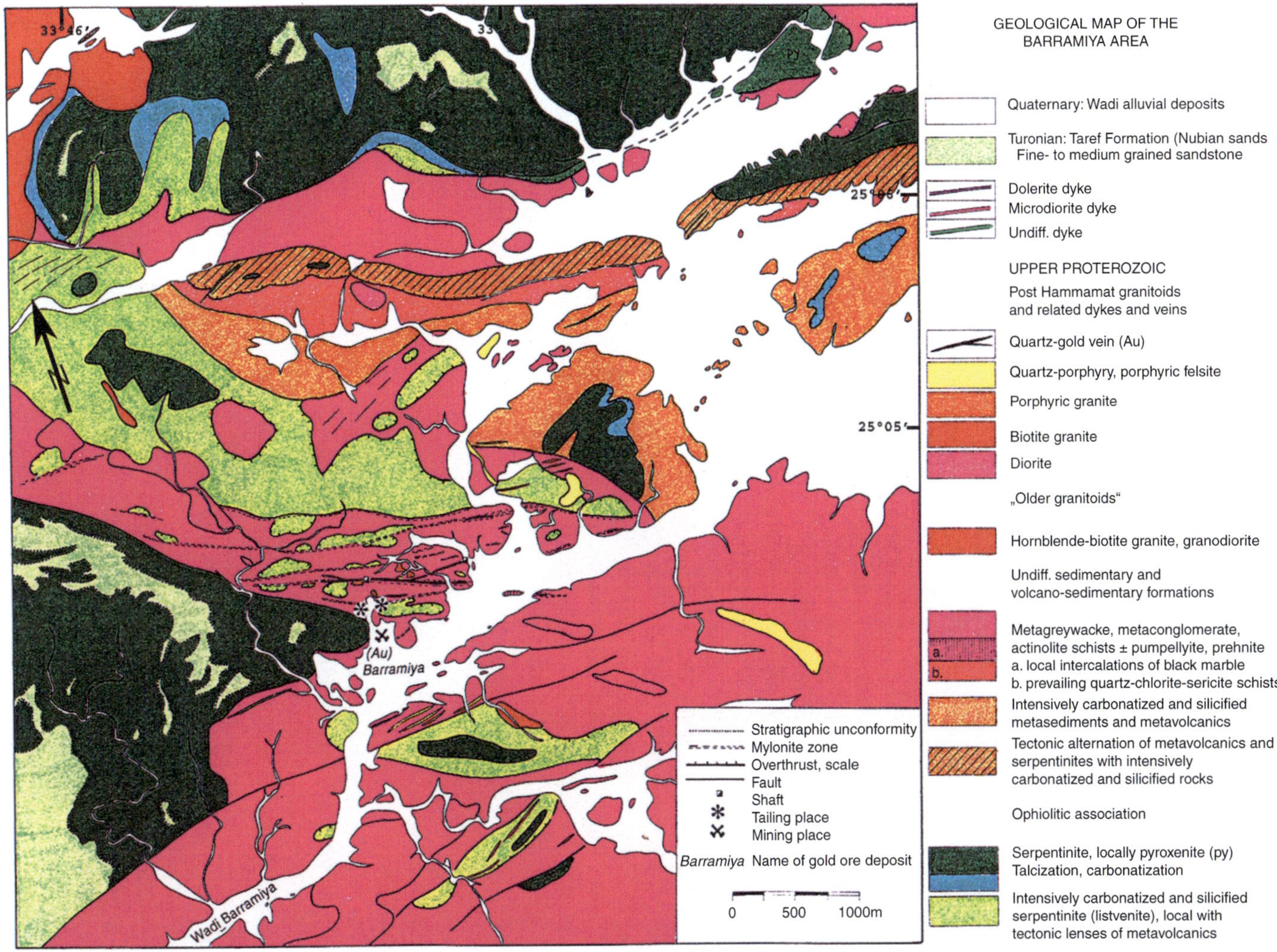

Fig. 5.143 Geological sketch map of the Barramiya surroundings (H. Kräutner)

granitoid intrusions, which occur in the deposit area as irregular and apophysial offshoots. Depending on the assimilated original roof material, the composition of such intrusive stocks may vary from granodiorite to diorite. The plagiogranite porphyries described by the EGSMA-Technoexport team (Lyssenko et al. 1975) in the deposit area are as well to be put in connection with this variability of young intrusions, and are therefore not connected to the ophiolite processes.

In addition to our own geologic record, the EGSMA-Technoexport group around Lyssenko et al. (1975) carried out thorough investigations in the deposit area at Barramiya and produced very informative geologic maps. The study by Gabra (1986) essentially foots on this work. Recently, Gad and Kusky (2006) revisited the area's lithology by means of Landsat Thematic Mapper images.

The main gold mineralisation is basically bound to an approximately E-W striking quartz vein system which dips steeply at an inclination of 80° N, i.e. it intersects the above mentioned syncline of graphite- and tuff-yielding metasediments. This vein system splits up at both its E and W ends. The gold mineralisation is however not exclusively bound to this quartz vein system itself, but also to a considerable degree to the tectonically sheared and fissured sectors, and above all to its graphite-yielding, recumbent horizons.

Next to this main vein system, there are two other subsidiary, auriferous veins in a NNE-SSW strike, so-called caunter lodes which are viewed as large Riedel shears (Bridges et al. 1986). They had been almost completely mined already in antiquity. Although both main and N lodes had too been mined to very deep levels in ancient times, sufficient ore reserves survived to account for a very profitable exploitation between 1907 and 1919.

The ore mineralisation processes of the auriferous zone seem to have been relatively simple. Next to the dominating pyrite, there were moreover arsenopyrite, to a lesser extent galena, some sphalerite, and chalcopyrite. It was probably the latter, which by its secondary malachite layers in the wallrock led to the first discovery of the deposit in the Old Kingdom Period.

According to the above discussed geologic-tectonic criteria we consider the following stadia in the genesis for the gold mineralisation phases as probable:

Possible primary gold contents in clay (?) and tuff metasediments, which had formed under anaerobic conditions, may have represented the basis for later, hydrothermal gold accumulation processes. The ophiolite sequence as well, may be viewed as an additional primary gold source.

The second precondition for the formation of the deposit may originate in the isoclinal folding of the predominantly, metasedimentary series in association with the ophiolite nappe overthrust phase and the therewith following, or even resulting intensive shearing.

The magmatic intrusion activity is followed by a CO_2 rich aftermath. It leads to carbonation of chiefly sheared areas within the serpentinites and metasediments. It also engenders subsequent silification resulting from the release of silicon dioxide with the carbonation of the serpentinites and eventually, leaching-out of possible gold pre-enrichments within these rocks.

Extension tectonics in the dome-like culminations of entire sequences from granite intrusions provokes discharge fractures with Riedel cross-faults. The zones weakened by shearing are particularly marked by a distinctive joint pattern.

The thermal potential from the intrusion bodies launches and accelerates hydrothermal convection systems containing mainly silicic acid, but also hydrogen sulphide, that leach-out available gold contents from the metasediments and ophiolites primarily in ppb amounts and then transport them as gold-sulphide complexes. Open veins and fissured shear zones offer at this stage the best hydrothermal pathways and space for precipitation.

As the gold-sulphide complexes get into contact with ferrous particles dissolved by hydrothermal reactions from the wallrock, they disintegrate under formation of pyrite, or in presence of additional arsenic, of arsenopyrite. The gold is then evenly dispersed into the pyrite or arsenopyrite lattices. On the other hand, through pressure reduction in the open clefts, the

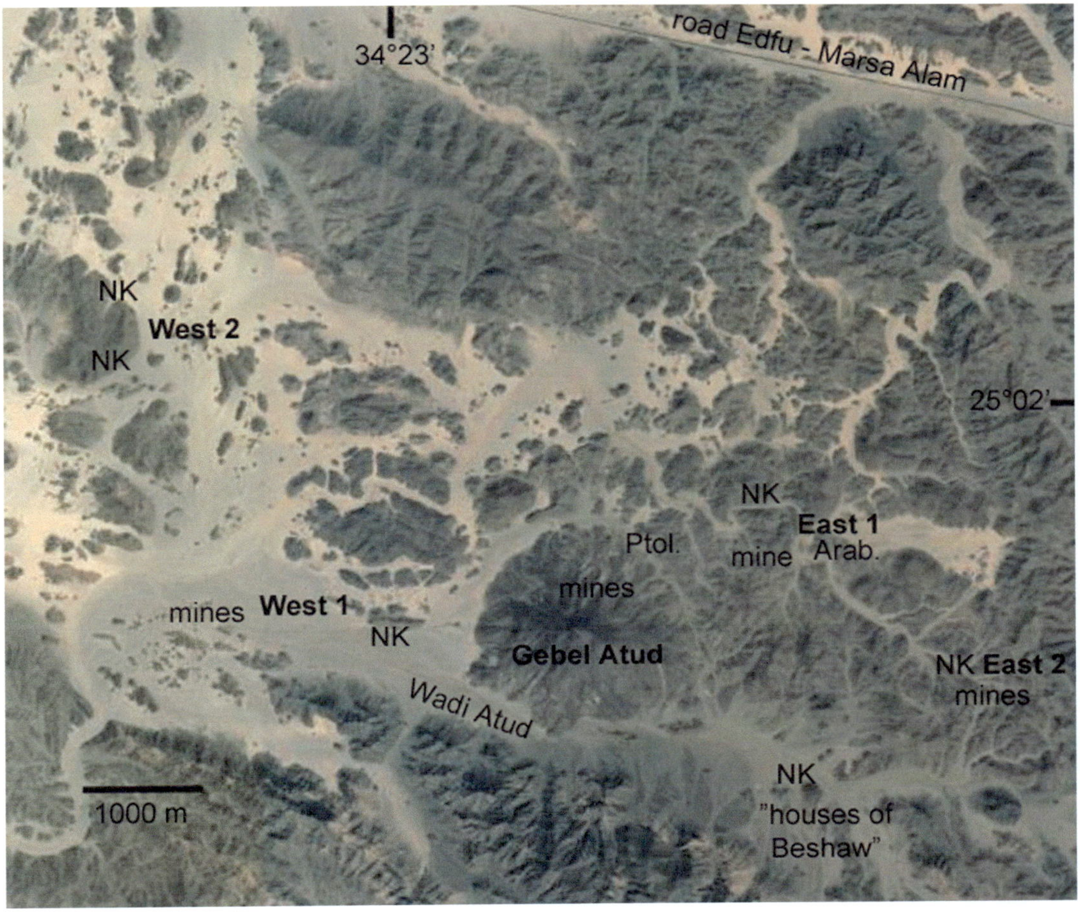

Fig. 5.144 The sites and gold mines in the Atud area (modified Google-Earth image)

hydrothermal fluids boil and subsequently cool down quickly, which then leads to the precipitation of silicic acid as well as the decomposition of the gold-sulphide complexes, and thereby also to the precipitation of gold.

In a last phase of gold emission, surface oxidation occurs, whereby the sulphide minerals pyrite or arsenopyrite decompose and re-deposit to iron hydroxides. In this process the dispersed gold contents are released and accumulate to fine spangles. Ancient technology only permitted the extraction of this released gold and of the precipitated native gold within the quartz veins, whereas the gold contained in sulphide minerals is only extractible by modern processes.

Bridges et al. (1986) propose a similar, though substantially simplified genetic model.

5.4.6 **Atud** (Fig. 5.144)

Geographic positions (mine):	25°01′18″ N, 34°25′03″ E
(Ptolemaic settlement):	25°01′18″ N, 34°24′13″ E
Atud-East 1 (large New Kingdom settlement):	25°01′31″ N, 34°24′49″ E
Arab settlement:	25°01′13″ N, 34°25′02″ E
Atud-East 2 (New Kingdom settlement + mine):	25°00′37″ N, 34°25′56″ E
Atud-West 1 (New Kingdom settlement):	25°00′57″ N, 34°23′02″ E
Atud-West 2 (New Kingdom settlements):	25°02′11″ N, 34°21′36″ E
Head of Wadi Atud	24°58′49″ N, 34°28′56″ E

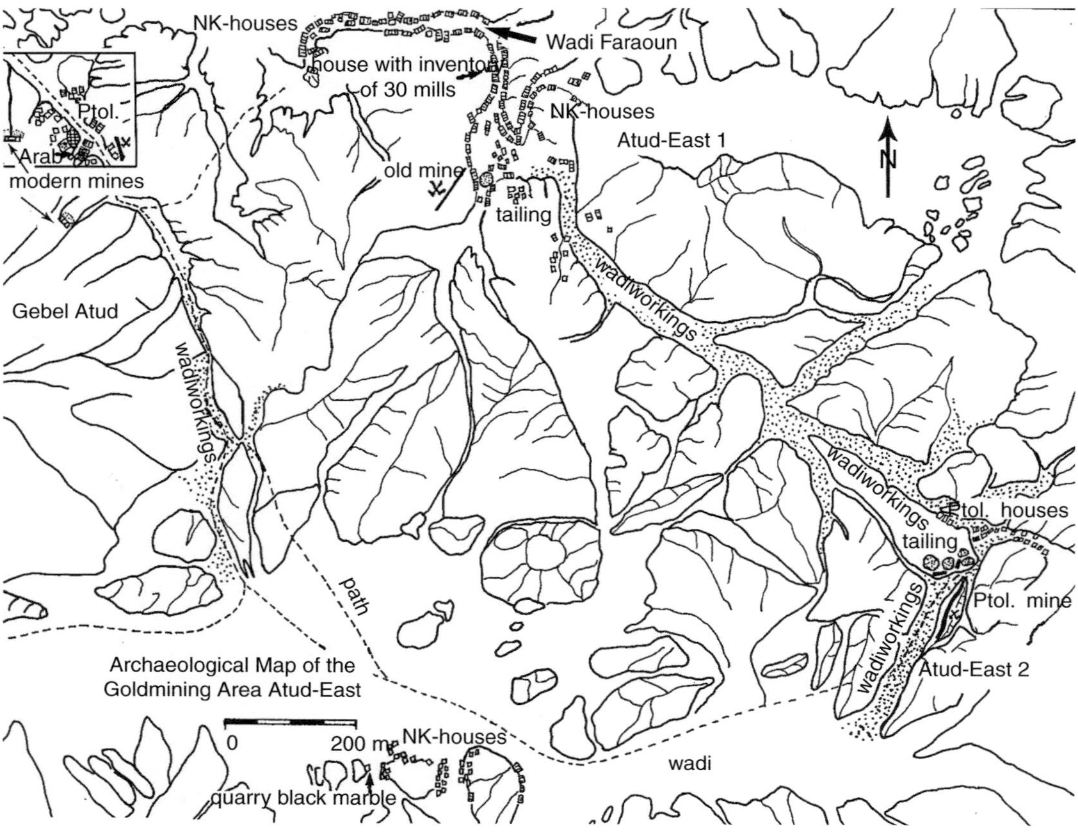

Fig. 5.145 Archaeological sketch map of the mining district at Atud East (R. Klemm)

In the district around the mountain cone of Gebel Atud, three clearly distinguishable, ancient extraction phases can each be attributed to the New Kingdom, the Ptolemaic, and Early Arab Period. Mining took up once again during the first half of the twentieth century (Fig. 5.145).

A processing site with considerable amounts of tailings (Fig. 5.146) is situated at the foot of the ancient and modern gold mine, on the NE edge of Gebel Atud. The site itself and its closer and more distant surroundings are littered with Ptolemaic stone mills. Some specimens have intriguingly survived as unfinished half-products and come in cubic shapes lacking all traces from use wear. This undeniable, Ptolemaic tool assemblage matches with that from another settlement at the S slope of Gebel Atud. The latter site had originally been fortified and reveals and its architecture is composed of adjoining, small rooms with high walls. The site had then been re-occupied during the Early Arab Period, as revealed by its countless round mills, anvils and small tailings found inside the rooms. Unfortunately, the site has suffered badly from recent mining activities (Fig. 5.147).

More Ptolemaic settlement remains are located further to the N, on the opposite side of the narrow wadi. As it seems, these Ptolemaic sites had once been part of an integrated, larger settlement system.

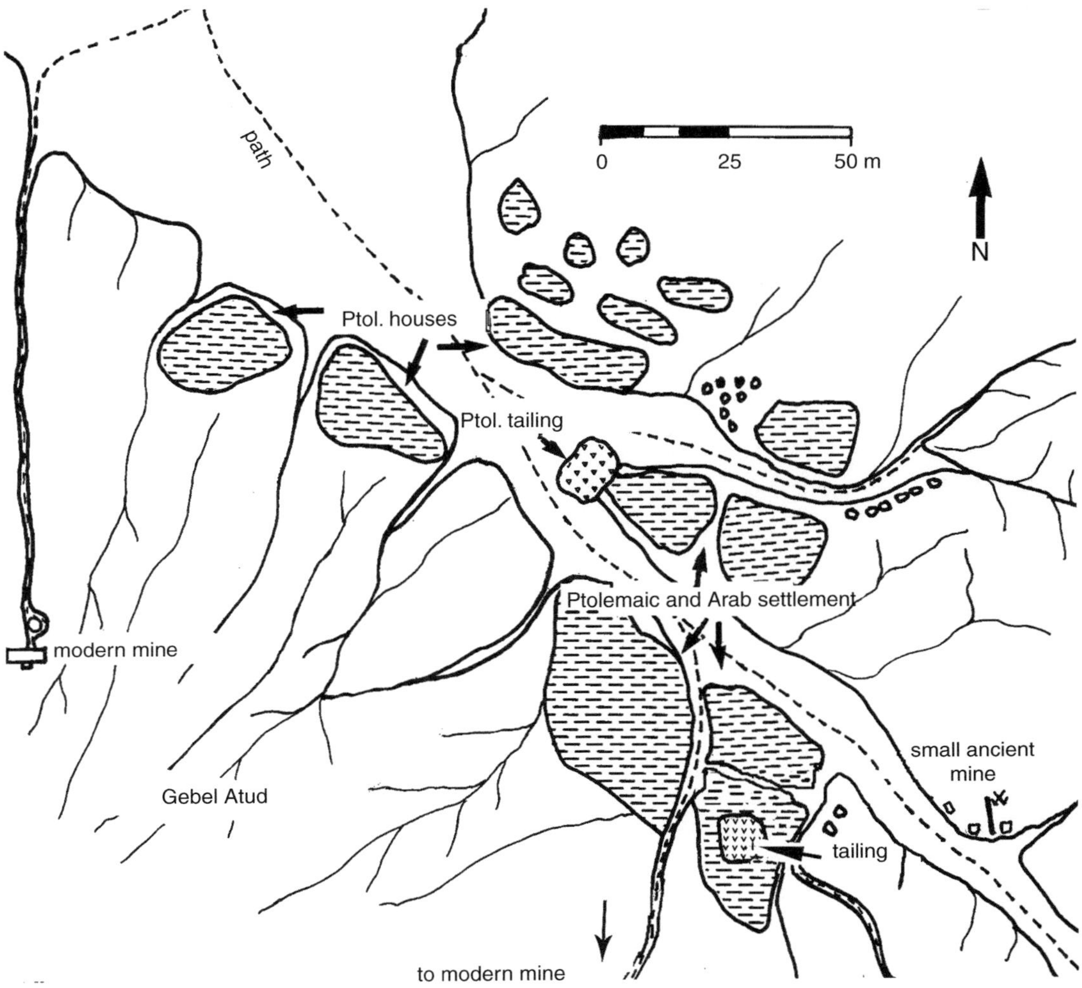

Fig. 5.146 Sketch map of the central processing site north-east of Gebel Atud (R. Klemm)

In the mountains to the SW of this settlement are a number of ancient and recently re-exploited mines than are reached via a road leading straight through the ancient settlement (Fig. 5.148).

Approximately 800 m NE of the Ptolemaic settlement at Gebel Atud is a large, New Kingdom settlement (Atud East 1) consisting of 80–90 houses lined-up on either side of a narrow wadi (Fig. 5.149) called Wadi Faraon. The settlement consists of characteristic, New Kingdom multi-chambered, long houses with terrace-like annexes facing the wadi. Most of the numerous oval stone mills are found on these terraces, which suggest that these had been the preferred locations for milling the collected quartz ores. A large number of these mills are also found in the rubble of the wadi bed. Although the wadi is relatively narrow with an average width of 20 m, the gold quartz-yield must nevertheless have been quite substantial, to judge by the size of the

Fig. 5.147 Ptolemaic processing site and settlement with Early Arab Period reoccupation east of Gebel Atud. The road to the modern mine leads straight through the middle

settlement as well as the countless numbers of distinguishable mills lying at the surface. The exploited area reaches evidently into the wadi's ramifications to the SE, about 1 km further up where it ends. This area is manifestly marked by traces from wadiworkings (Fig. 5.145).

The houses of this elongated settlement had been built from different rocks gathered up from their respective surroundings. One house was even exclusively built from white (barren) quartz rocks from its surrounding surface. The average size of a single house varies around 5 × 2.5 m, though it may reach 16 m in length, depending on the number of its rooms.

At the eastern end of this New Kingdom house alignment the wadi bends-off to the S. About 400 m from there, one recognises the residues of a tailing next to some round mills and a couple of hut ruins from the Early Arab Period. Two long, parallel trenches are located in the mountains bordering the W and display distinct traces from ancient mining. They were mined in depths between 6 and about 25 m. The vein system strikes approximately NNE-SSW (35°) and dips NW. The still verifiable gold grades amount according to Koshin and Bassyuni (1968) to an average of 2.7 g/t, in very few cases even to more than 11 g/t.

After about 1.5 km beyond the Early Arab settlement, one arrives at another, New Kingdom settlement, which is located in the SE part of Wadi Faraon (Atud East 2) close to several productive gold quartz veins (Fig. 5.150). In all, five pyrite-containing gold quartz veins were counted with widths between 15 and 50 cm and lengths between 120 and 300 m. They strike approximately NNE-SSW and dip at 40–55° NW (Koshin and Bassyuni 1968). According to Koshin and Bassyuni, the average gold grade is about 5.9 g/t, whereas its peak is at 22.5 g/t.

Fig. 5.148 Ptolemaic processing and settlement site at Gebel Atud with a road leading to the modern mine (modified Google-Earth image)

Thirteen more, large New Kingdom houses are lined-up along the southern slope of a long, narrow hill to the W of Gebel Atud. Though the houses contain the characteristic tool inventory, they strikingly lack surface pottery.

About 4 km NW of this New Kingdom settlement there is another alignment of New Kingdom-houses on a large hill just before the large plain of the Humr Wajjat granite-pluton. About 600 m further N another settlement of single houses lie hidden away, together with three larger house complexes in a small tributary wadi. All may be ascribed to wadiworking activities (Fig. 5.144).

Approximately 3 km SE from Bir Atud, in the SE part of Wadi Atud is the location of the so-called "houses of Beshaw" along the edges of two noticeably small hills. Some of the huts had been built with reddish, felsitic rocks. According to the surface inventory, the settlement dates chiefly to the New Kingdom (Fig. 5.151), and had presumably been specialised on processing ores from the wadi alluvium.

Fig. 5.149 New Kingdom house ruins along the edges of Wadi Faraon in Atud-East

Fig. 5.150 New Kingdom settlement in Atud-East 2. The darker patches in the background are remnants from former wadiworkings in Wadi Faraon

Fig. 5.151 The so-called "Houses of Beshaw", dated by their mills and architecture to the New Kingdom

Other huts consist of flat, andesite rocks. They are mostly rectangular and in a better state of preservation than the round ones from felsite. They are evidently later and no doubt connected to the black marble quarry in the close neighbourhood.

The deposit areas and gold producing sites spread over a comparatively large region within the Atud area. They are chiefly associated to diorites, which at Gebel Atud itself clearly turn to gabbro. Nevertheless, gabbro surrounded by diorite occurs at numerous sites in domes and hence, particularly in elevated positions (Fig. 5.152). Associated with the gabbro one observes repeated occurrences of relatively well-preserved, xenolith-like intercalations of pyroxenite.

These magmatites obviously penetrated into an ophiolite nappe which has disassembled into single, small NW-SE oriented nappes. Next to the dominating serpentinites, which are intensely carbonated and occasionally mapped as listvenites, these sheet segments contain intercalations of bulky, relatively well-preserved metabasalts and metasediments with partly graphite-yielding metasiltstones, as well as especially marbles in Wadi Atud, where the grey-black graphite marbles are quarried.

Leucocratic, porphyric microgranites intruded this entire sequence, in whose surroundings the auriferous quartz veins were mined. This is the only such occurrence in the greater part of the area around Atud-East 2, but is nevertheless distinguishable in other deposit areas through the appearance of granite apophyses.

In this district too, gold minerals are exclusively bound to hydrothermal quartz veins, which at least in the main occurrence (Main Atud), are linked to a shear zone system. The auriferous vein mineralisation at Atud-East is by contrast connected to local extension tectonics, causing a NE-SW striking fault younger than the intrusion of the young, porphyric granite stocks.

The characteristic ore paragenesis here is pyrite and to marginal degrees, chalcopyrite and arsenopyrite. Gold is mainly locked inside the sulphide minerals and is only rarely visible under a magnifying glass.

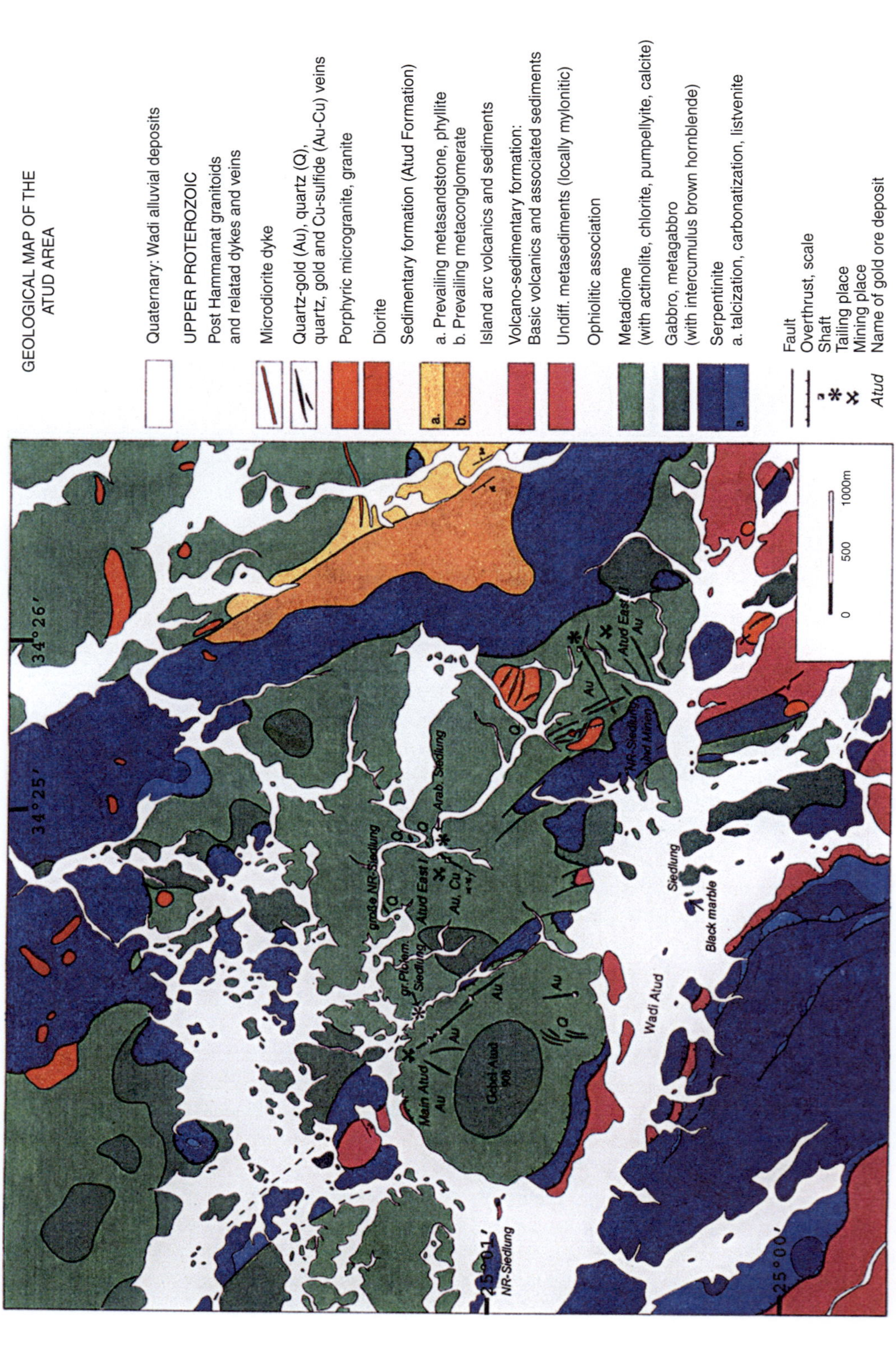

Fig. 5.152 Geological sketch map of Atud (H. Kräutner)

Modern (20th century) exploration work obviously concentrated exclusively on occurrences in the main vein (Main Lode). This vein actually composes of various narrow ones striking NW-SE with thicknesses in a range of only few centimetres. It dips steeply E while splitting up in a 10–60 m wide zone. The two smaller quartz vein occurrences further to the W (No. 1 and No. 2 W Lode) strike respectively NNE-SSW, while dipping vertically (No.1 W Lode), and virtually N-S with a dip to the W (No. 2 W Lode). Apparently only the first had been mined in antiquity. Today, however, the vein has caved-in to such an extent that any further investigation would have gone beyond the scope of our project.

The quartz vein occurrence Atud-East 1strikes NNE-SSW and dips gently at 25° NW. It is pursued over an approximate distance of 150 m at the surface in widths between 10 and 20 cm. Koshin and Bassyuni (1968) state that at its outcrop it yields an average of 3 and 9–12 g/t at more than 20 m depth.

5.4.7 Head of Wadi Atud

Small Arab Period hut groups loosely scatter between 24°58′50″ N, 34°28′56″ E and 24°58′29″N, 34°28′27″ E . They had apparently subsisted on intensive wadiworkings in the surrounding valleys and are located 6 km in a straight line SE from Atud-East. Mines were not found in this area.

5.4.8 Marsa Allam

Geographic position:	25°02′54″ N, 34°49′14″ E

About 8 km to the W of Marsa Allam, and 600 m after the asphalt road has entered the mountain range, a mined trench opens on top of a much eroded, Early Arab settlement in an easily overlooked, small tributary valley to the S. The thickness of the quartz vein fluctuates between 10 cm and 2 m, which according to the recorded mills and anvil stones from the settlement site had been mined for gold.

The vein strikes NNE-SSW and dips 60° W. About 100 m to its W appear two additional, small veins with respectively N-S and NNW-SSE strikes. Both, display traces from prospecting only, and no genuine mining had been carried out here. Unsuccessful mining attempts had been made here in the 1940s, according to the information of our accompanying geologist from EGSMA, Dr. M. Abu Bakr el-Hawari. Several large mounds consisting of unprocessed vein quartz still bear evidence for this.

Here as well, the vein is oriented approximately in N-S direction within a metavolcanic sequence of mostly former andesites and agglomerates. Noteworthy hydrothermal alteration features in the wallrock were not observed in its vicinity.

5.4.9 Wadi Umm Khariga

Geographic position:	25°01′30″ N, 34°41′46″ E

About 4 km to the E of the inlet of Wadi Sukkari to Wadi Umm Khariga, along the asphalt road from Edfu to Marsa Allam, are the remains of seven to ten New Kingdom houses in a flood-protected position near a rock promontory facing W. Its find inventory consists of the typical gold processing tools from this period.

The geologic surroundings are again relatively monotonous here and consist of heavily sheared, rhyolitic- dacitic metavolcanics, which toward the E, are imbricated with partially carbonated serpentinites and intermediary layers of serpentinised gabbro.

No mined quartz vein was found in the mountains, which would hint to a mining activity in the quartz-rich wadi alluvium.

5.4.10 Umm Quli

Geographic position (wadi centre):	25°04′26″ N, 34°39′12″ E

Only few hundred metres E from the site discussed above, a tributary wadi called Umm Quli flows into Wadi Umm Khariga. About 4 km to the S of this inlet, a quartz vein with some traces from mining was found near a small, Early Arab settlement site. The quartz vein is located at the immediate overthrust border between the serpentinites and the recumbent acid metavolcanics.

Further S in Wadi Umm Quli, towards Sukkari, several additional Early Arab settlement sites scatter over a distance of 8 km within the same geologic environment. Similar to the situation in Wadi Faraon, these sites had been specialised on wadiworkings only.

5.4.11 Wadi Faraon

Geographic position (wadi centre):	24°57′48″ N, 34°43′47″ E

Wadi Faraon stretches out to the E of the deposit area of Sukkari. It has many small tributary valleys in the NE and is set within a similar geologic environment to that W of Gebel Sukkari. The area is speckled with small granitoid and apophysis-like intrusions. Genuine underground mining is not reported from this district. Only some small quartz veins located directly to the E of the main Sukkari deposit had been scrutinised thoroughly in the Early Arab Period, though without any success.

Countless Early Arab working sites scatter in small groups over a wide area in Wadi Faraon and especially in its upper reaches. Mining had apparently been based on wadiworkings only, particularly considering that the wadi rubble is very rich in quartz.

The situation in the small tributary valley leading SW from Wadi Faraon directly to the E flank of the deposit area is somewhat different. This little valley is littered with debris heaps of exploited gold ores that, even today, are preserved in heights between 50 cm and 1 m, and are well distinguishable in aerial images. The flimsy evidence observed at the settlement sites seems to indicate a date in the New Kingdom.

5.4.12 Sukkari (Fig. 5.153)

Geographic position (Ptolemaic settlement):	24°57′02″ N, 34°42′44″ E
Mine	24°56′46″ N, 34°42′49″ E

About 700 m to the N of the both, modern (20th century) and ancient mine at Gebel Sukkari, in the broad valley of Wadi Sukkari, and at an incision of a smaller twin crest hill are located the settlement remains of an ancient gold mining site from the New Kingdom, Ptolemaic/Roman-Byzantine, and Early Arab Periods (Fig. 5.154).

The oldest New Kingdom occupation spreads-out in a semicircle immediately at the foot of these two hills. The much ruined architecture consists of irregular simple huts built in contiguous compounds. A small number of characteristic, oval stone mills were recorded (Fig. 5.155).

The settlement is partly overlapped by a significantly larger Ptolemaic occupation layer, of which most spreads-out in front of it. Although the Ptolemaic ruins as well are heavily disturbed, their general state of preservation is nonetheless better than that of the forerunner site. Today one gets the impression that the Ptolemaic settlement had been divided up into several quarters, but this is merely due to a modern road crossing the site in its western part. Originally, the site had formed a coherent unit. To the W and the E of this large settlement one clearly discerns large, fortified building complexes (Fig. 5.156). Without further possible differentiation at this stage, they in addition mostly reveal traces from Ptolemaic/Roman-Byzantine occupational phases as well as architectural enlargements dating to the Early Arab Period. Such relatively large complexes have also been observed at Bokari as originally being Ptolemaic, but later reoccupied in the Roman-Byzantine Period.

We only recorded mills in small numbers in all parts of the settlement. But near the outer building complexes, where they were observed located next to several tailing sites (Fig. 5.157), their numbers tended to increase

In spite of severe damages inflicted by modern (20th century) mining, at low sunlight one distinctly recognises flat gravel heaps in several sectors to the N and S of a promontory in the wadi. They indicate that gold mining at Wadi Sukkari was not only restricted to quartz veins, but that surrounding wadi alluvium was exploited as well. For reasons already stated above, this would also explain the relatively large settlement size (see Fig. 5.154).

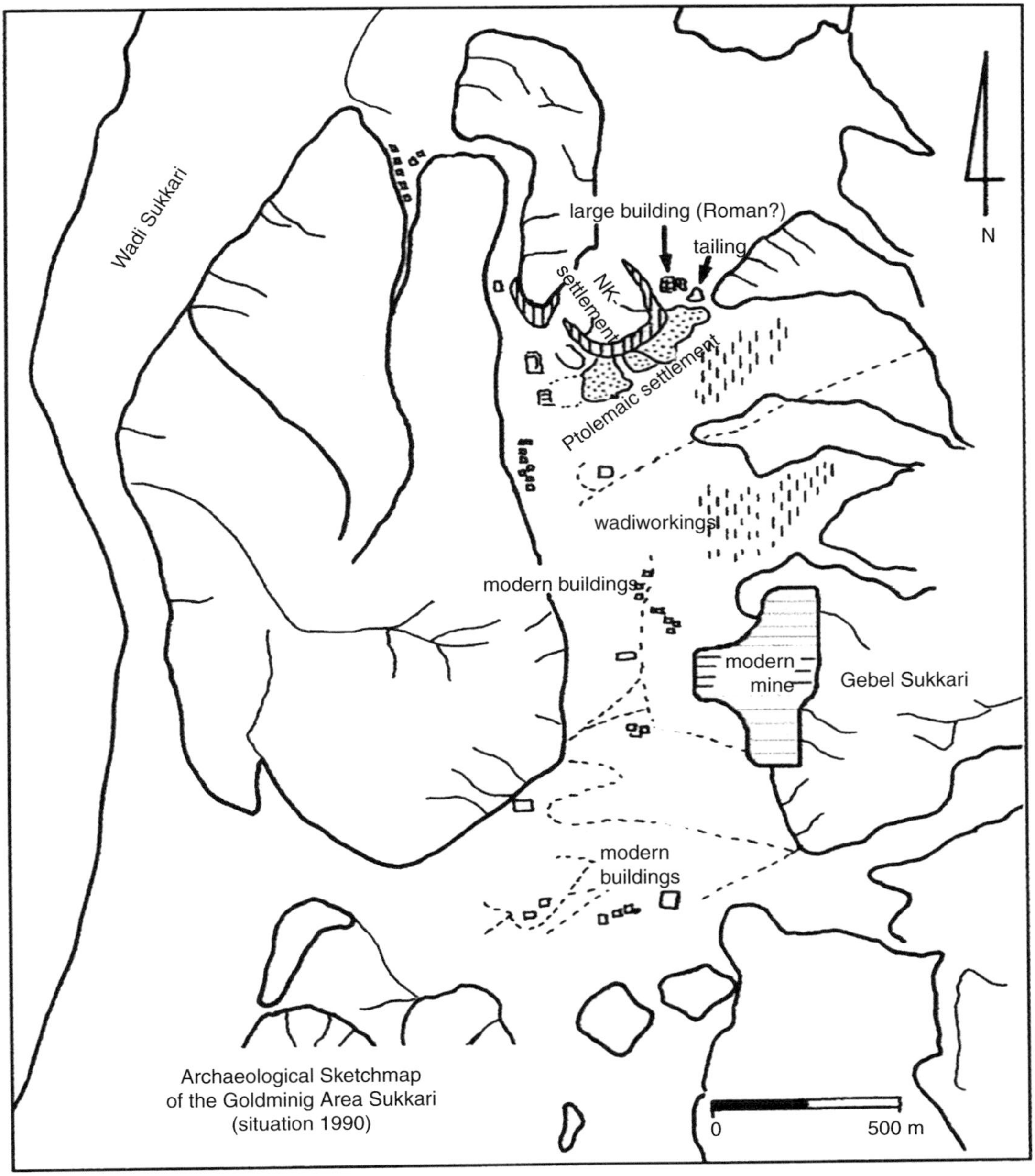

Fig. 5.153 Archaeological map of the Sukkari gold mining district (R. Klemm)

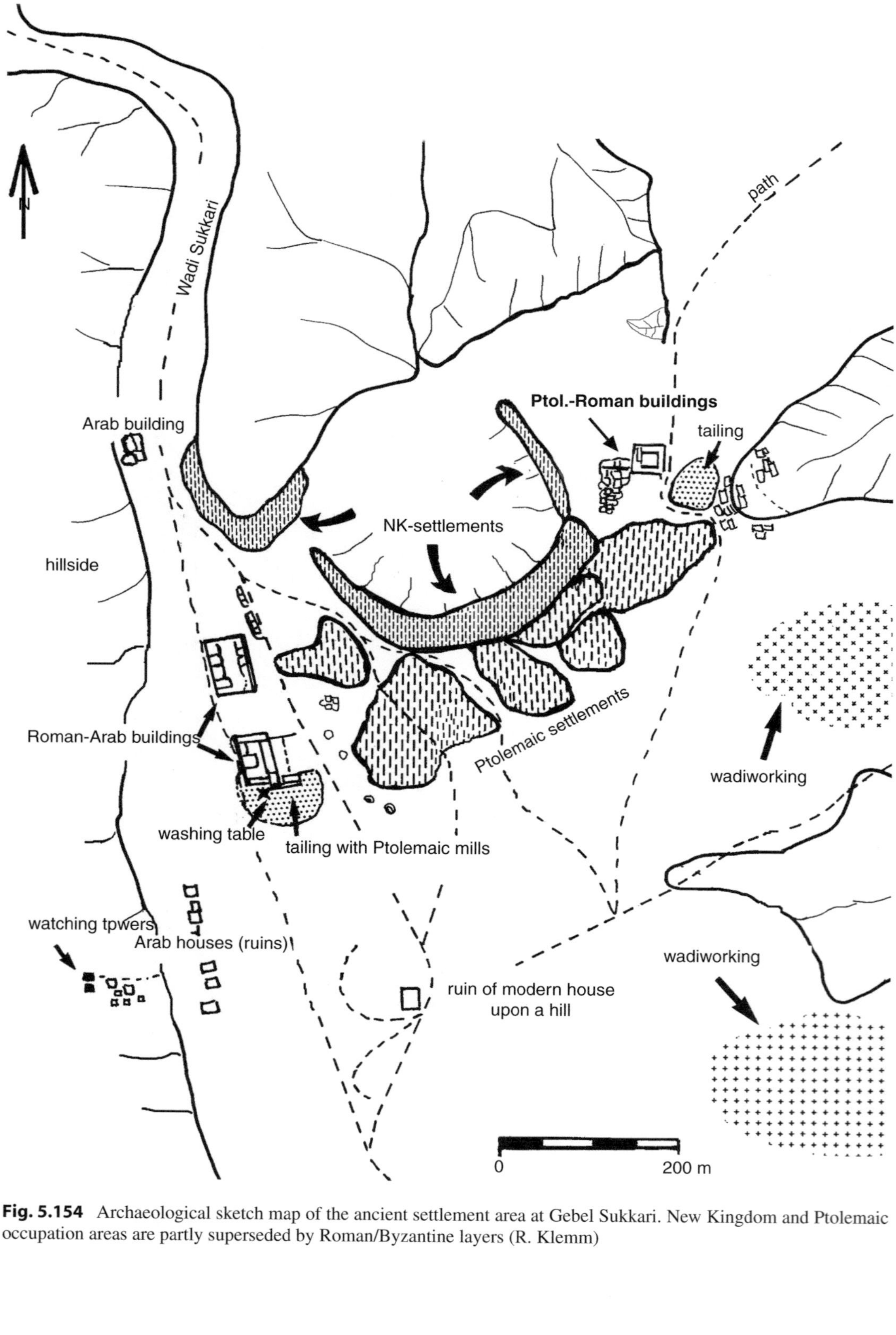

Fig. 5.154 Archaeological sketch map of the ancient settlement area at Gebel Sukkari. New Kingdom and Ptolemaic occupation areas are partly superseded by Roman/Byzantine layers (R. Klemm)

Fig. 5.155 Large New Kingdom and Ptolemaic settlement at the foot of Gebel Sukkari

Fig. 5.156 Large building complexes to the W of the central settlement at Sukkari

Fig. 5.157 Tailing containing tools from the New Kingdom and Ptolemaic Periods at Sukkari

The former mine is situated in the southern part of the area of the modern installations and had once occupied probably only a small segment of the local workforce.

According to our observations in the terrain, underground mining at Gebel Sukkari had apparently begun in the New Kingdom. It had continued through the Ptolemaic Period, as traces from this period even testify to a more intensive exploitation.

By the Roman-Byzantine Period, gold mining had declined, which is reflected by that period's extremely scanty tool evidence. Next to the mentioned, fortified house groups, a small number of single houses as well as two watchtowers in the mountains to the W probably also date to this period, as Sukkari lies at a passageway along the route to the emerald mines of Gebel Zabara (Smaragdus Mons; Murray 1925).

A remote, multi-chambered building to the NW of the area is dated by the pottery to the Early Arab Period.

Mining in the twentieth century has significantly diminished the evidence from earlier mines, which have been either destroyed or else been filled-in, thence prohibiting more thorough archaeological exploration without intensive excavations.

According to Abu Zeit (1987), the mining concessions for this district had been granted to a certain John Wells in 1908, who subsequently sold them in 1915 to the "Sukkari Gold Mine Company". Four inclined shafts were then lowered under this company's responsibilities into the main vein system, from where the ancient shafts could supposedly be followed.

In the period between 1912 and 1916, a total of 567 t of ore were extracted from the mine with a gold-yield exceeding barely 8 kg. In 1920 the mine was shut down, apparently due to water shortages. In 1929 the grants were acquired by a Darramelly Bey, though because of bad financing no significant mining followed. In 1932 estimations carried out under the control of the Egyptian Mining Department came up with ore reserves between 1,320 and 60,000 t, at an average grade of 18.2 g/t Au and 3.2 g/t Ag. A short production phase thus began in 1937 but came to an end in 1942 due to insufficient gold-yields.

Between 1967 and 1977 the occurrence was once again prospected thoroughly by a joint venture organised by the EGSMA and the Soviet Technoexport Company, and briefly again, in 1988 by the British led MINEX Minerals Egypt. Mining activities however, did not resume.

In 1996 the Australian-Egyptian firm Centamin Egypt Ltd. were granted prospecting rights first at Sukkari, that since have been extended to the districts of Hangaliya, Barramiya, Atut, Umm Ud, Semna, Abu Mureiwat and Hamama.

In a highly publicised campaign, Centamin came out in particular on the Internet with enormous reserves data. In a vast investigation involving an extensive diamond drill holes programme, estimations for the so-called Amun Zone around Sukkari only, foresaw 1.67 million ounces (1 ounce = 31.1 g) at a cut-off grade at 0.5 g/t, and 1.16 million ounces at a cut-off grade at 1.0 g/t.

In the meanwhile gold production has started there since 2009. As one can see from Google-Earth images (Fig. 5.158), the archaeological mining sites are widely destroyed but the area of the ancient settlements with its archaeological inventory is at least partly preserved. Thus, the chances for a systematic archaeological investigation of this important witness of ancient mining industry in Egypt are somewhat kept.

The gold mineralisation is noticeably linked to a border zone between a sequence similar to the Hammamat series of greywacke, siltstones, and occasional fine conglomerates with metatuffs and a therein intruded granodiorite stock (Fig. 5.159). Partially lens-shaped scales of intermittent ophiolites alternate with this metasediment sequence. Notably the occurrence of graphite-containing metatuff and graphite schists should be emphasised within the immediate deposit area, which there however, are marked by tectonic shears and hydrothermal alterations.

The intrusive body along the contact zone is characterised by intense silifications and a profusion of small hydrothermal quartz veins. It also displays pyrite and kaolin (Koshin and Bassyuni 1968). The auriferous "Main Lode" itself follows a large fault zone, which at its outcrop strikes about 40 m across the valley bed NNE-SSW and dips relatively gently at 40–45° SE, but with increasing depth becomes distinctly steeper (up to 60°).

In the main vein system three symmetrically structured quartz generations were distinguished. Firstly, a 5–10 cm wide, extremely fragile zone directly at the border zone to the hostrock (selvage), a second of equal width, and a third main zone more or less in the centre of the vein. The latter has an average thickness of 1.5 m, peaking at 3 m and minimum values near 50 cm. The two younger, marginal generations seem to be the most productive ones, as they are normally mined out. Parts of the central quartz band for their part, were mainly neglected by the miners and left standing as abutments. The main vein occasionally splits up into several small subsidiary streaks, which renders the three described generations barely distinguishable from each other.

The main ore mineral is pyrite, which though predominantly occurs in the altered wallrock and to a lesser degree in the vein quartz itself. It appears mostly as idiomorphic cubes up to about 1 cm large, singular crystals. On a lower scale occurs chalcopyrite, which was generally neglected by the ancient miners. In addition, some galena (mostly very fine-grained within the quartz matrix) and in the modern, deeper mine parts sporadically massive occurrences of idiomorphic-rhombohedral arsenopyrite, which because less auriferous, had been piled as debris on waste dumps.

The geologic map shows that the deposit area at Sukkari is set in a tectonically imbricated zone, characterised by a volcano-sedimentary (here undifferentiated) zone in the W and a presumably volcano-sedimentary island arc sequence in the SE. The latter is intensely penetrated by mafic to acid intrusive rocks, sheared in a NE-SW strike, and crossed approximately in the same direction by a younger series of mainly intermediary dike swarms (Fig. 5.159).

Helmy et al. (2004) submitted a detailed, economic study on the geology and petrography on the Sukkari district.

Fig. 5.158 Comparative Google-Earth images of the Sukkari mining site. Left, image from 2003 with the well preserved mining hill and some remains of the 20th century activities around it .The various ancient settlement remains are affected merely by modern roads. Right, image from 2009 with the complete ablated old mining hill and the at least partly preserved ancient sites

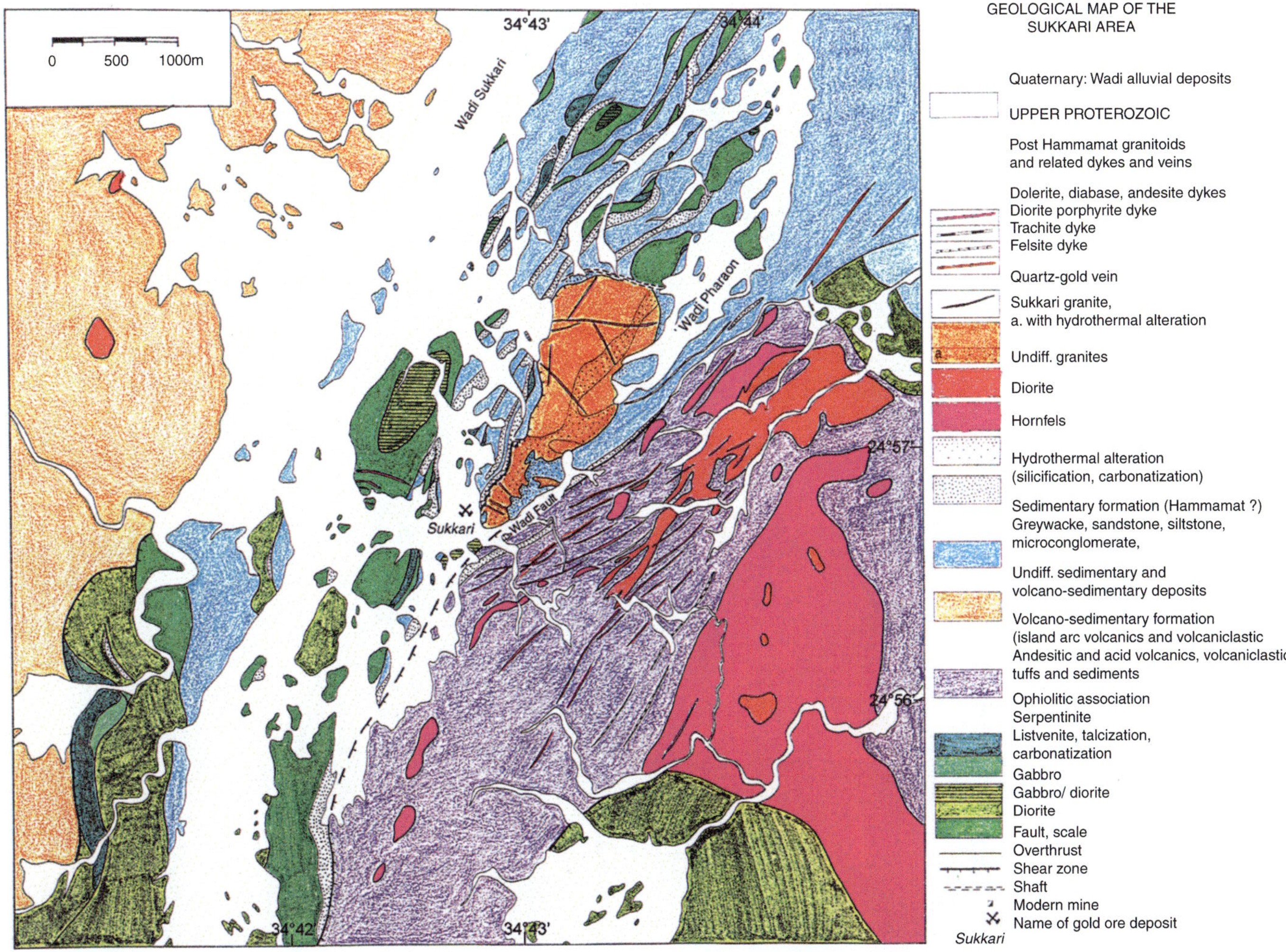

Fig. 5.159 Geological sketch map of Sukkari. The main vein system starts at the limits of the sedimentary series to the intrusive granite and plunges into the latter (H. Kräutner)

Fig. 5.160 Agglutinating huts built in the shell facing technique of Early Arab Period at the foot of the Tundub mine

5.4.13 Umm Tundub

Geographic position (mine):	24°54′55″ N, 34°45′58″ E

Below the mine, on the slope of a very narrow wadi is a confined Early Arab settlement consisting of 8–10 interconnected room units (Fig. 5.160), in which 20–25 round mills made from light-coloured, foreign granite, were observed.

Umm Tundub II (24°56′08″ N, 34°46′34″ E) and III (24°57′15″ N, 34°46′27″ E) are represented by relatively small Early Arab working and settlement sites, formerly focused exclusively on gold production from wadi rubble.

Geologically, the area of Wadi Tunduba is determined by mostly heavily sheared, steep, and spaciously folded metaandesites and metadacites. In the area of the Early Arab gold producing sites the series strike NE-SW, but 4 km N of Tundub II a large fault oriented in the same direction cuts-off a NNE-SSW striking unit of a matching lithological sequence.

The mined quartz vein approximately strikes in the same direction as its hostrock (20–30°) and dips 35–45° NW. It is marked by intense shearing, while the shear zone itself dips almost vertically jointly with the hostrock. Within the shear zone coarse-crystalline calcites occurs, though which follow the shear zone and not the vein and therefore have no linkage to the gold mineralisation.

5.4.14 Bir Ambaud

Geographic position:	24°55′44″ N, 34°50′01″ E

The well at Bir Ambaud is located about 3 km (downstream) to the NE of the Ambaud mine and close to the inlet of a relatively wide tributary wadi coming in from the NW. The confluence area is situated amidst a wide catchment area whose drainage once unquestionably warranted for a good water supply. Today the well has collapsed.

Formerly the site of the well was apparently perceptible from a distance by important volumes of surrounding tailings, which in the meantime have been sacrificed to a modern mining scheme as they were removed for cyanide processing by the Hawari Company.

Anyhow, we were able to record large amounts of Ptolemaic concave stone mills and runner stones in addition to Early Arab round mills, but none from the New Kingdom. Therefore more Ptolemaic extraction sites unknown to us must therefore exist in the area (cf. Ambaud). Bir Ambaud by any account had indipuatbly represented a central gold processing site for a greater catchment district.

5.4.15 Ambaud

Geographic position: (remaining settlement):	24°54′41″ N, 34°48′38″ E
Mine	24°54′50″ N, 34°48′42″ E

About 3 km S of Bir Ambaud is a modern mining district, which according to our Ababda guide and former EGSMA employee Taufik, had been thoroughly prospected by the Hawari Company in the 1940s in a test tunnel. The mine, however, never went into production. Hawari was until the nationalisations programme of the mining industry in1956 a small Egyptian firm active at many sites in the Eastern Desert since the early 1930s.

Judging by the widespread disturbances from the Early Arab Period yielding rich assemblages from gold processing, Ambaud on a whole was undoubtedly a location where mining had played an important role in antiquity. Regrettably though, nearly all relevant legacies from ancient mining have been erased through the company's utter lack of awareness for the archaeological heritage.

The auriferous quartz vein remains nonetheless easily accessible through the mentioned prospecting tunnel, which at first even follows the old mine. The ancient mine is still pursued over a distance of more than 50 m in vertical, opencast trenches.

The auriferous quartz vein that yields up to 10 g/t runs through a 30–50 cm wide carbonated shear zone with intense, red-brown stains from limonite. The vein strikes 120° E and dips 85° NE. It occurs in acid metavolcanics (chiefly former rhyolites), which in the deposit area is penetrated by a medium-grained and secondarily, much chloritised diorite. To judge from the ancient mining traces, the shear zone as well, is from time to time auriferous. The limonite staining is hence inferred from the oxidative decomposition of former pyrites to which preserved negative marks still testify inside the modern tunnel.

The geologic environment of the deposit area is dominated by acid and mafic metavolcanics, into which large granite and granodiorite intrusions are intruded. This is exemplified by the upcoming granite at Gebel el-Ambaud, 3 km further N. The chloritised diorite within the deposit area is probably once again merely a hybridised marginal facies of this intrusion, for which however, no clear-cut field data are available. Interesting are coarse-crystalline larvikite intercalations inside the diorite of the deposit surroundings, which may conclude to high fluid phases during crystallisation, as would be expected from an assimilation of H_2O-rich wallrock.

According to Jakubiak (1987), there are more ancient work platforms at the western flanks of the mountain range of Wadi Ambaud.

5.4.16 Kurduman

Geographic position:	24°54′15″ N, 34°40′58″ E

The Kurduman deposit consists essentially of a long quartz vein, which at times shows slight displacements. It can be followed through a depression between two hill chains (Fig. 5.161).

The quartz vein is located in a small, tributary valley immediately to the E of Wadi Ghadir, about 12 km S of Sukkari, from where it is easily accessible via a desert road.

About 20–25 small, round huts cluster to the S of an extensively collapsed mine. Together with their assemblage consisting of round mills and anvil stones, they date the mine to the Early Arab Period. However older mining phases cannot be ruled out.

The geologic surroundings are made up of predominantly rhyolitic metavolcanics, which to the S of the deposit go over to layers of andesitic

Fig. 5.161 The trench pit of Kurduman mine crosses several hills

metavolcanics. These base series are covered to the N of the deposit by heavily carbonated serpentinites and in the S by metasediments. The sequence is intruded by granodiorite to quartz-diorite bodies, which is indicated in the immediate deposit area by few, minor apophyses.

The mined, chief vein measures between 30 and 70 cm in width and consists of white and blue-grey quartz, as it strikes 127° E with a 75–80° dip SW. At the surface it is pursued over more than 100 m. It is easily accessed by two modern, vertical shafts. The mineralisation is bound to a shear zone in metarhyolites, along which the wallrock is intensely carbonated. Two more parallel running vein strains had been mined up to a maximum depth of 20 m. Within the shear zone the veins often swell-up in lenses. After that, the vein course continues in only barely distinguishable, brownish wallrock alterations.

At the intersection of a NE-SW striking side vein, which also was exploited in antiquity, is a small test shaft. Considered as barren, the latter had obviously been given up.

In addition to the principal mine, the vein had been pursued in several small shaft holes beyond the mountain ridge, although probably only for purposes of inspection.

About 250 m to the S of this main extraction, two more, yet N-S striking veins had been mined in shallow trenches under identical geologic conditions in the continuing shear zone system. Evidently though, the gold contents did not meet the expectations for deeper mining.

5.4.17 Zabahiya

Geographic position:	24°52′25″ N, 34°41′39″ E

A small, recent mine foots on a vein probably already exploited under the New Kingdom. The deeper-going, old mine, which strikes at an angle to the new tunnel, had probably produced considerable amounts of gold.

Even though there are no further visible hints in the surroundings as to its exact date, there is some chance that the evidence lies buried under the debris heaps from the modern shaft.

The geologic environment of this occurrence is similar to that in Kurduman and is marked by partly, severely sheared metavolcanics. Next to the hitherto dominating, former rhyolitic tuffs and lavas, the metavolcanics here also display numerous, intermediary layers. Furthermore, they contain concordant layers of gneiss and mica schist with well-discernible fuchsite whose Cr-contents are probably connected to nearby serpentinite occurrences, which are imbricated with these metavolcanics and even contain some subeconomic chromite outcrops.

Today the quartz vein is accessed by a recent tunnel. Its thickness varies between 0.5 and 1.2 m. Though it generally strikes in an approximate N-S direction as it progresses within the shearing of the layer planes, it also follows local flexures, resulting to short strike alterations to the NE-SW. At the ridge of the somewhat isolated hill in the deposit area one notices ancient prospecting trenches. These evidently did not lead to further mining as demonstrated by the fact that no settlement or processing sites are found in the immediate neighbourhood, and that samples collected by EGSMA by any standard delivered no contents for gold and only very low ones for silver.

The modern mine at Zabahiya seems to illustrate a case, in which deep mining was launched somewhat prematurely and whose hopes for profit rested more on the ancient extraction traces than on rational criteria required by the standards of modern prospecting. The identity of the responsible company is unknown to us.

5.4.18 Urf el-Fahid

Geographic position:	25°00′ N, 34°11′ E (Koshin and Bassyuni 1968)

Gebel Urf el-Fahid ("cockscomb") is discernible from a distance and marked by a prominent quartz crest which passes through its centre (Fig. 5.162).

But in spite of intensive surveying, neither the mountain nor the adjacent homonymous wadi delivered evidence from ancient mining. Koshin and Bassyuni (1968) state however under this site name "some old workings are known from Wadi Hadia, not far from its junction with Wadi Mueilha …."

Fig. 5.162 Hill Urf el-Fahid crossed by a protruding quartz vein. It has survived as it contains no ores

5.4.19 Wadi Umm Rashid

Geographic position:	
Settlement beginning:	24°58′15″ N, 34°05′23″ E
Settlement end:	24°59′06″ N, 34°04′58″ E

Wadi Umm Rashid is best accessed from Wadi Mueilha by bending-off directly to the N at position 24°56′53″ N, 34°07′07″ E.

Several multi-chambered, New Kingdom houses had been built over a distance of more than 1.8 km in the middle course of Wadi Rashid, especially along its edges and apparently progressively as advance from wadiworkings was being made. Many of the houses have two to three rooms and a round forecourt in which stone mills and other characteristic processing tools are found.

A small steatite occurrence, which presumably had been quarried in antiquity, is located nearby.

The geology of Wadi Umm Rashid, where the wadi alluvium was chiefly exploited during the New Kingdom, corresponds by and large to that in Wadi Faraon, near Sukkari. It is represented by a densely folded sequence of intensely imbricated ophiolite and metasediment layers. In shallow depths below this sequence are granitoid intrusions, apparent at the surface mostly in the form of sporadic apophyses.

Mined quartz mineralisations were not found, as opposed to the masses of quartz fragments in the wadi bed itself.

5.4.20 Marwat (Abu Mureiwa)

Geographic position:	24°59′22″ N, 33°56′34″ E
Mureiwa 2:	24°59′03″ N, 33°57′00″ E

Approx. 8.5 km in a straight line to the S of the asphalt road between Edfu and Marsa Allam, and S of a light, shimmering quartz vein, two settlements are located in a wadi that stands out by its large waste dumps left behind after wadiworkings. One distinguishes about 20, mostly two-roomed New Kingdom houses in an irregular line-up. Only few, adequately dated, oval mills were found. Few fragments of red-brown, densely grooved pottery seem to suggest that the site had been reoccupied for a short time in the Early Arab Period.

The geologic catchment area around this gold processing site is very much reminiscent of that at Barramiya. As in the case of the latter, the area is marked by a steeply folded and alternating sequence of metasediments with insertions of black marble and graphite schist and some metavolcanics. This sequence is covered by extremely carbonated graphite schists with which it is partially imbricated. A surprising occurrence of acid gneisses in these series is noteworthy. In the catchment area of the exploited wadi alluvium, the layers dip steeply at 70° S and strike together with the dominating fold axes at 80° in a virtual E-W direction.

5.4.21 Umm Hugab

Geographic position:	24°56′51″ N, 34°03′35″ E

Important New Kingdom house alignments along both flanks of Wadi Umm Hugab, particularly along the western one, associate with a yet, barely apparent landscape of debris heaps from wadiworkings in the valley bed (Figs. 5.163 and 5.164). Altogether, about 100 houses were counted, which were equipped with the characteristic artefacts such as oval stone mills, runner stones, or even a carefully manufactured stone mortar. The houses are divided up in two to three rooms, and the walls are in a comparatively good state of preservation (Fig. 5.165).

In addition to the New Kingdom houses are a number of round huts built in a shell-facing technique, which together with characteristic round mills date the buildings to the Early Arab Period. Nearby to the S is a prayer site.

The deposit is located in a small, tributary wadi turning W in the upper reaches of Wadi Umm Hugab (Fig. 5.166).

A tailing site with a group of small quartz ore heaps is seen at the foot of an approximately 10 m deep mine trench.

The geology around the deposit is determined by a recognisable fold structure which clearly tapers towards the SE. At the surface it is com-

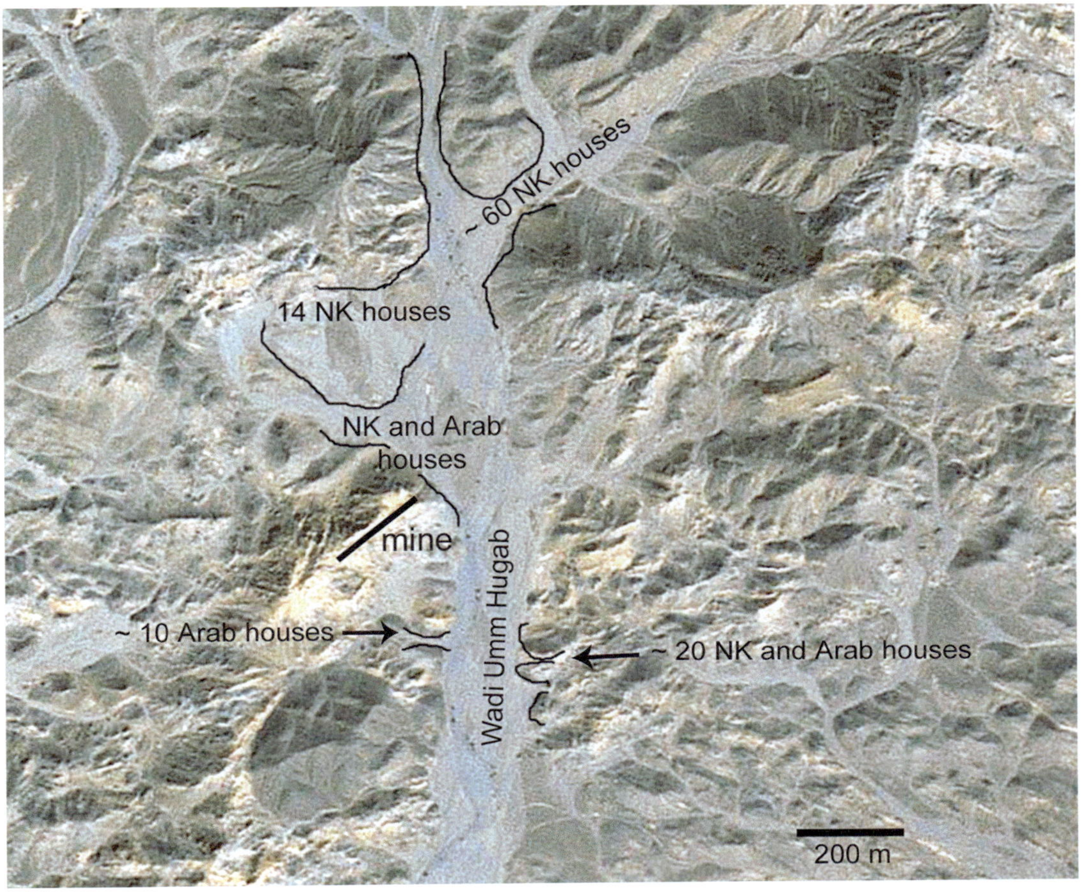

Fig. 5.163 The ancient settlement area and mine at Umm Hugab. The black lines along the wadi beds indicate ruins of ancient house alignments (modified Google-Earth image)

Fig. 5.164 View over Wadi Umm Hugab and the location of New Kingdom settlements at the wadi's edges

Fig. 5.165 Well-preserved New Kingdom house and a mortar on its outside at Umm Hugab

Fig. 5.166 Mine at Umm Hugab above large debris fans and a number of houses as well as a tailing site

posed of metamorphically overprinted volcano-sedimentary rock sequences which are intensely imbricated with serpentinites (Fig. 5.167). The series consist for the most of partially carbonated talcum-tremolite schists alternating with graphite schists in a syncline of metagreywackes and metaconglomerates extending to the SW. Yet again, these series are very evocative of the Hammamat sequence and probably even match with its stratigraphy.

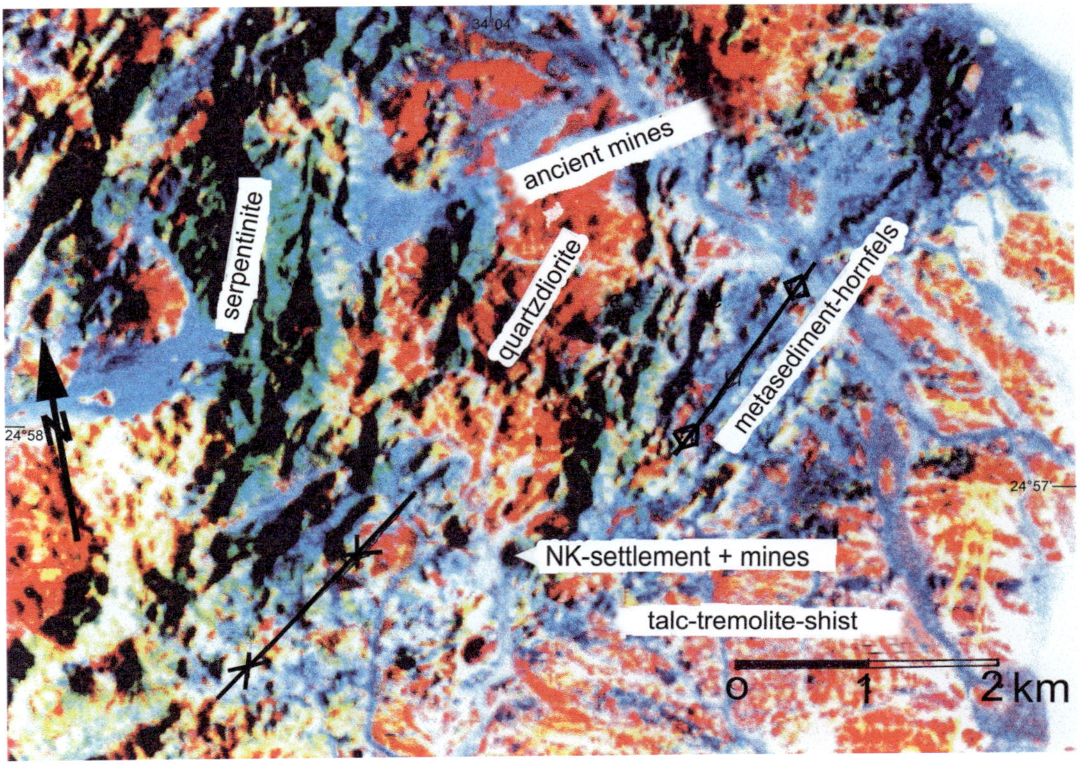

Fig. 5.167 Lithologically processed satellite image of the area around Umm Hugab (TM 174/43, channels 7-4-1)

The fold axes of the syncline and the adjacent anticline to the NE, strike approximately NE-SW and dip in gentle undulations.

A granitoid magma intruded into the lower parts of a today partly eroded anticline. Within the contact zone to the mentioned volcano-sedimentary series and serpentinites it is represented by granodioritic and quartzdioritic rocks.

Near the extraction appears a fine-grained andesite dike which breaks-up into blocks of about 1 m and displays smooth joint surfaces. These blocks had unquestionably furnished the raw material for the abundant large oval stone mills of the New Kingdom houses and for the round stone mills in a small Early Arab settlement somewhat further to the SW.

Auriferous, grey quartz occurs in close vicinity to small vein trenches. They strike 60° E and dip 45–50° NW. They therefore by and large follow the planes of the sheared talcum tremolite and graphite schists. Within the grey quartz vein, small particles of visible gold appear. A milky-white, apparently barren streak of a more recent generation runs through the vein. The vein's median width is close to 0.5 m. Wallrock alterations are blurred by the red-brown shear zone, which hints to a primary pyrite enrichment inside the alteration zone.

A geologic sketch map by Shablovsky (1973) shows yet another vein mineralisation about 4 km to the N, in the quartzdiorites along the southern flanks of a meandering side valley. This vein too, had apparently been mined in antiquity. Here as well, residues of talcum tremolite schists still overlie the intruded, granitoid magma. This case again demonstrates that the immediate roof area of a small intrusive body may specify the area for auriferous quartz mineralisations. We did not visit this occurrence during our stay in Umm Hugab, as we unluckily had been unaware of it. According to Shablovsky's sketch map, this area hosts a considerable number of smaller quartz veins, which never had been genuinely mined.

5.4.22 Dungash (Fig. 5.168)

Geographic positions:	
Ancient well	24°56′19″ N, 32°52′14″ E
NK - Ptol. settlement	24°56′20″ N, 33°51′40″ E
Main mine	from 24°56′00″ N, 33°51′50″ E to 24°56′08″ N, 33°52′08″ E

In travelling to Wadi Dungash from the Nile Valley, one drives up Wadi Shait, whose mouth fans enter the plain at Kom Ombo. One then turns-off to the NE, into Wadi Beza and after about 100 km from the Nile, Wadi Dungash branches-off to the E.

Next to the mining of the gold quartz veins in the districts southern mountains, the generally E-W orienting wadi as well as its tributaries been systematically delved through for gold quartz rocks ever since the beginning of New Kingdom (Fig. 5.168, dotted areas). The ores had been processed virtually on location, in the neighbouring, vast settlement complexes. This is for example illustrated by pounders found inside the house ruins that had been used for breaking up the ore lumps. The crushed chunks had then been ground down in a profusion of still present, oval grinding mills, thus liberating tiny gold flitters from a fine quartz meal.

During the Ptolemaic Period the veins along the higher slopes on the S side of the wadi were then mined in predominantly underground trenches. Mining technologies had undergone significant developments by this period, as witnessed by seven deep shafts that distinguish by their typical, vault-like roof-structures and abutments at regular intervals. Some of the features in the Dungash mines may indeed be viewed as representative for these innovations. The Early Arab Period saw a revival of gold mining at Dungash, and then once more much later, at the beginning of the twentieth century.

Approximately half-way up the 1 km long section much marked by ancient mining is a large New Kingdom settlement in a pronounced wadi bend. The site surface is speckled with oval mills, runner, and pounding stones. The few potshards date to the New Kingdom. Some small, recent houses with New Kingdom mills and building stones integrated to their walls (Fig. 5.169) are associated with a nearby Islamic prayer site. Unfortunately, due to "cleaning-up work" by the Hawari Company (1956) the original state of the site was so badly damaged that detailed archaeological investigations are no longer possible there. All tailings have in addition been washed away by erosion.

At the northern slopes in the background some huts with New Kingdom inventories are still to be made out.

The site's well was located further to the SE rather than on the settlement's terrace. It is preserved as a large, though completely silted up hole. It probably functioned as the settlement's central water supply. Because of the lack of pottery at this location, water was probably chiefly collected and transported in animal hides in the manner of Bedouins today.

A cistern (Fig. 5.170), which still fills up when it rains, is discerned near the mountain foot, to the S of the New Kingdom settlement. An adjacent rock face holding inscriptions from the Old- and Middle Kingdom (site C in Fig. 5.168) testifies that the cistern was used as a water supply station for expeditions, long before mining had started here in the New Kingdom. Accordingly, no allusion to gold mining is made in them (Klemm and Eichler 1998).

The second, large New Kingdom settlement lies to the N of these sites in a side wadi turning off to the E (Fig. 5.171). It is predominantly composed of single, rectangular houses built in parallel alignments to the wadi. Their lengths vary between 6 and 10 m, their widths between 4 and 6 m. Generally, they comprise four rooms, in few cases more, although their layouts and sizes are not easily determined at the surface. Some single huts group at an elevated position on the wadi terrace to the E of the northern house alignment. They presumably date to the Early Arab Period, between the ninth and eleventh centuries on grounds of their good state of preservation, their location apart, and above all, their typical round, shell-facing walls that had been filled with fine gravel.

An area composed of wadiworking heaps reveals itself under low sunlight over a distance of about 100 m into the wadi (Fig. 5.172). In the rear part of the wadi, an ancient small mine is visible in the mountains.

A cemetery presumably serving the New Kingdom settlements is located further to the

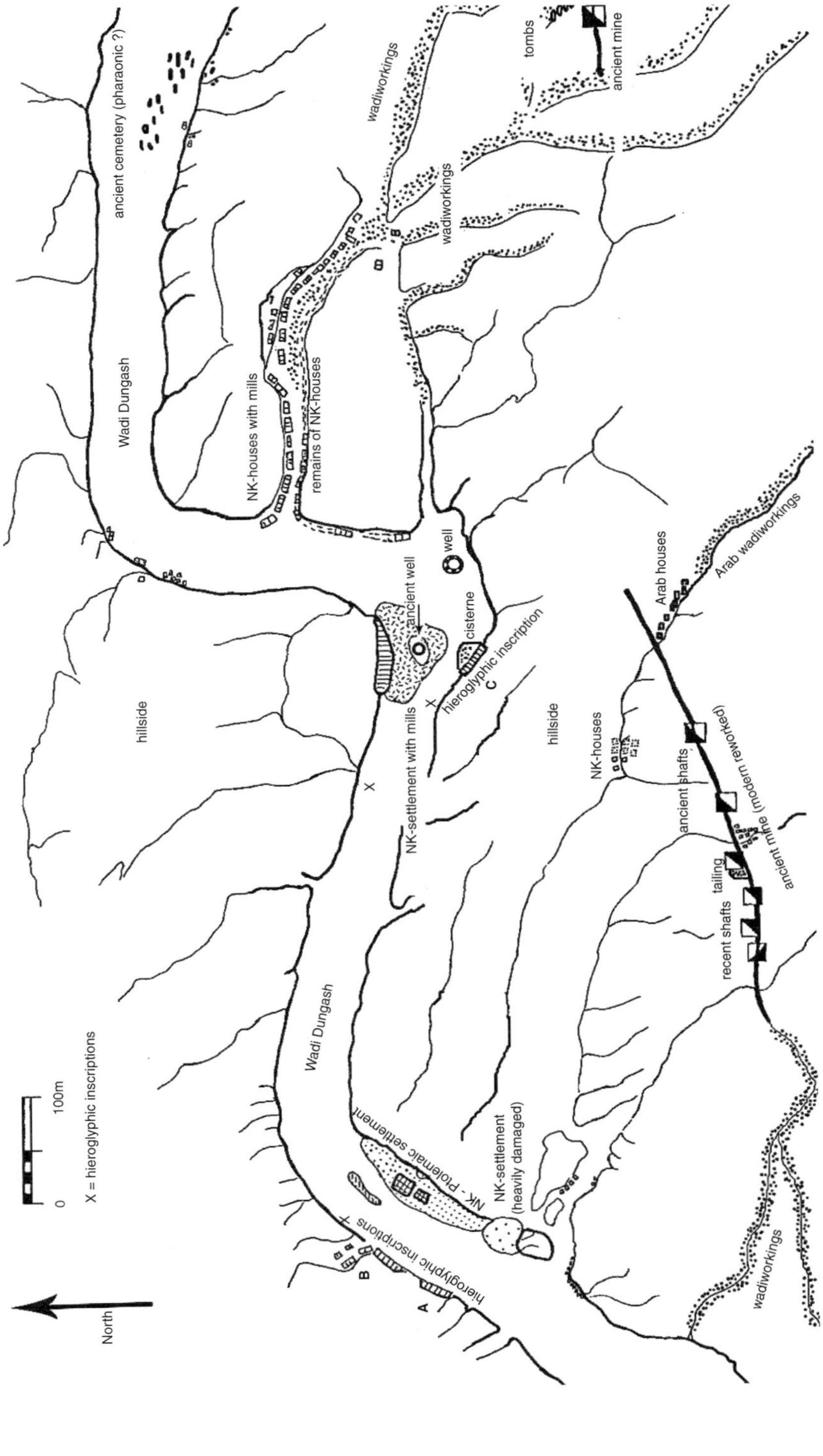

Fig. 5.168 Archaeological sketch map of Dungash (R. Klemm)

Fig. 5.169 Wadi Dungash, New Kingdom settlement. The two well-preserved huts are recent but built from New Kingdom mills and ancient building stones. More New Kingdom ruins are discernible along the wadi edge

Fig. 5.170 Cistern to the S of the main settlement. The desiccation cracks have formed as a result from a recent flood deposit. The rock faces from greywacke in the background bear inscriptions (marked as "C" in Fig. 5.168).

Fig. 5.171 New Kingdom settlement and wadiworkings in a eastern side valley of Wadi Dungash

Fig. 5.172 Well-distinguishable traces from wadiworkings further upstream in the side valley shown in Fig. 5.171

Fig. 5.173 New Kingdom hieroglyphic rock inscription at site B mentioning a "mayor Chau, goldsmith of Amun, Neshor" (Klemm and Eichler 1998, p.254, no. 11)

East in the main wadi. Some of the tombs mapped in Fig. 5.168 have recently been completely looted by Bedouins. Potshards from mostly red, burnished bowls lie scattered around the empty graves among countless mollusc shells. The cemetery consists of low grave mounds heaped up from unsorted wadi rubble. They spread in an uneven distribution in an E-W direction along the wadi and are best distinguished in leaning twilight.

Further upstream, there is another graveyard. It consists of widely spread, single and multiple burials. The information that they have been looted by local Bedouins suggests a relatively early date.

The inscription sites A and B in Fig. 5.168 are located in the western part of the area and to the NE of the New Kingdom mining zone. They had been executed on suitable rock surfaces of the Hammamat series (greywacke and conglomerates) and are probably linked to the gold mining activities. The inscriptions date almost exclusively to the New Kingdom (Fig. 5.173). Three isolated inscriptions to the E of site B date to the Old Kingdom and had apparently been executed by expedition members travelling through the area in similar contexts to those mentioned for the nearby cistern at site C (Klemm and Eichler 1998).

Relatively large settlement areas spread-out along the edge of Wadi Dungash, at western foot of the mined mountain, just opposite the inscription sites A and B.

We were able to identify a first, but barely discernible occupation phase from the New Kingdom in the southern area. It was covered in the N by better preserved structures from the Ptolemaic Period. The sector is also characterised by two fortified house complexes. The north-eastern one had been reoccupied in a later phase, since it displayed well-preserved corner bastions whose masonry included Ptolemaic, apron-shaped runner stones. The SW complex on the other hand, was dated exclusively to the Ptolemaic Period, based on the pottery evidence that had been washed out at its surface. The severely damaged ruins on a wadi terrace in front of this area also date to the Ptolemaic Period.

A noteworthy fragment from a small, round mill from vesicular lava was found here at

Fig. 5.174 New Kingdom mine at Dungash enlarged with a dome-like ceiling structure in the Ptolemaic Period, which significantly increased its stability

Dungash. Similar specimens are known from the Ptolemaic mining sites at Umm Rus and Sukkari as well as in the city of Berenice.

Compared to the magnitude of the wadiworkings in the eastern part of the district, the mine at Dungash seems in spite of its relatively important expanse, nevertheless quite small. We therefore think that its contribution to the total gold yield had been correspondingly insignificant, especially during the New Kingdom.

The first shaft in the W displays the typical mining features known from the Ptolemaic Period. Its tunnels had been executed in vault-like ceilings with numerous, though partially removed abutments (Fig. 5.174). The shaft descends in an inclined angle (approximately 45°) down to considerable depths. A New Kingdom grinding mill situated close to the next shaft opening indicates that the mine had probably already been operating in the New Kingdom.

But through its later reactivation, widespread damage was inflicted to relevant archaeological traces.

Additional remains from wadiworkings can be made out along the pathway leading to the mine (Fig. 5.168, dotted stretches to the W of the mine).

The geology in the Dungash area (Fig. 5.175) was described by Murr (1994) as a basement sequence affected by metamorphism consisting of dacites, trachyandesites, and andesites with island arc features, transgressively overlain by metasedimentary greywackes, siltstones and conglomerates of the Hammamat formation. Both sequences had been subject to intensive folding and were overthrust by an ophiolite nappe of serpentinites and gabbros.

The nappe was then intruded by (presently undeformed) diorites, which most probably had developed from assimilation of the roof parts of a large granitoid, magma body. In the thermal aftermath of this magmatism auriferous quartz veins had formed next to other quartz mineralisations in tectonic clefts of the wallrock.

The exploited chief quartz vein strikes WSW-ENE (78°) and dips between 30° and 60° SE. At the surface it is pursued over a distance of about 600 m. Its thickness varies between 1.5 and 2 m.

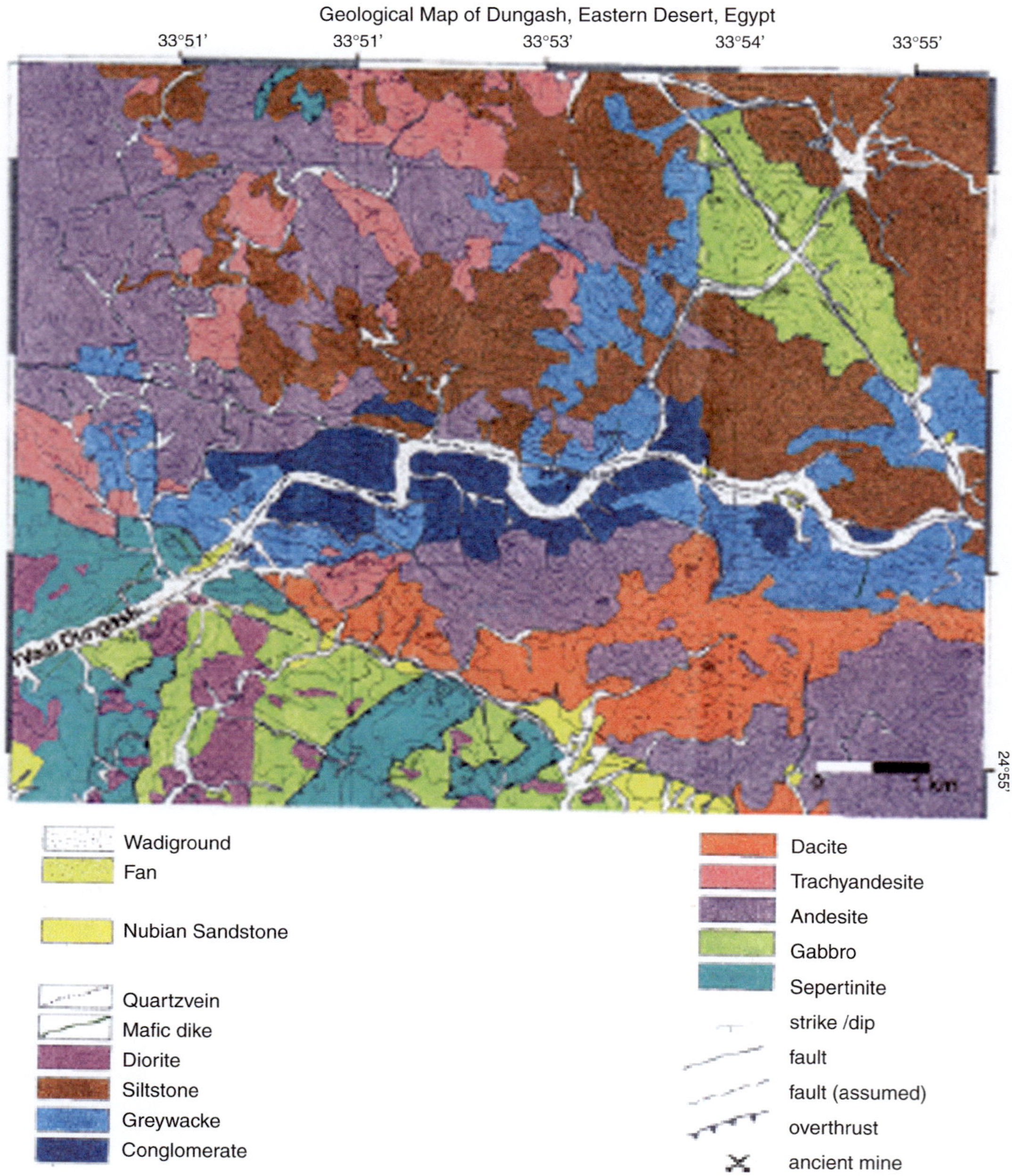

Fig. 5.175 Geological map of the Dungash area (A. Murr, 1994)

It consists of at least two quartz generations of which however, only one had been mined. At the surface the vein follows the border area between the metasediments and the trachyandesite and continues further E into the andesite where it plunges into the depths.

The older quartz generation consists of a milky-white quality, which after having brecciated was filled-in by bluish-grey quartz. The latter is characterised by scanty sulphide ores containing mainly pyrite and to a much lower degree chalcopyrite, arsenopyrite, galena, sphalerite, and very rarely pyrrhotite. The gold mineralisation is associated with these sulphide phases. According to the investigations on the fluids carried out by Murr (1994), the formational temperatures are located between 250° and 150 °C.

According to Zoheir et al. (2008a, b, c), however, wallrock alterations and ore formation occur between 400° and 250 °C according to As-arsenopyrite and Al-chlorite thermometry. Indeed, their method, just as much as that of Helba et al. (2008) is based on investigations of wallrock alterations and less on ones concerning the fluids regulating the quartz mineralisation of the ore minerals. This may explain the inconsistencies as to the formational temperatures.

At Dungash in the New Kingdom, in comparison to all following mining periods, gold production had concentrated next to underground mining probably particularly on the processing the wadi alluvium. This is notably supported by the findings at the large settlement sites at Bir Dungash, as well as in the tributary wadi turning-off to the E, little more than 100 m to the N. It was apparently only during the Ptolemaic and the later gold production periods that underground mining had intensified. Unfortunately though, many ancient mining traces have gone lost through the recent lowering of inclination shafts. In less affected sectors, ancient extraction depths exceeding 10 m were nonetheless recorded. Because some of these mines have caved-in, they had presumably been even deeper.

5.4.23 Wadi Dalalil

Geographic position (New Kingdom settlement):	24°54′45″ N, 34°00′48″ E
Mine:	24°54′30″ N, 34°01′28″ E

Wadi Dalalil ("valley of many shades") is a northern tributary valley to Wadi Mueilha, whose inlet is 10 km N of the Ptolemaic fort Mueilha. About 4.5 km NE from here one comes across New Kingdom settlement sites.

The New Kingdom site at Umm Dalalil is located NW to the old (and modern) mine and composed of approximately 30 houses. Twenty of them lie in a side wadi leading W, and ten within the main Wadi Dalalil in a loose alignment along the wadi edges. They thus correspond to the usually encountered pattern of rectangular buildings divided into two to three rooms, facing the wadi and equipped with the typical inventory of this period. The walls are preserved in several masonry layers, which might be explained with low water erosion and the generally remote location of the site. Because the houses occur in comparatively large numbers and furthermore are

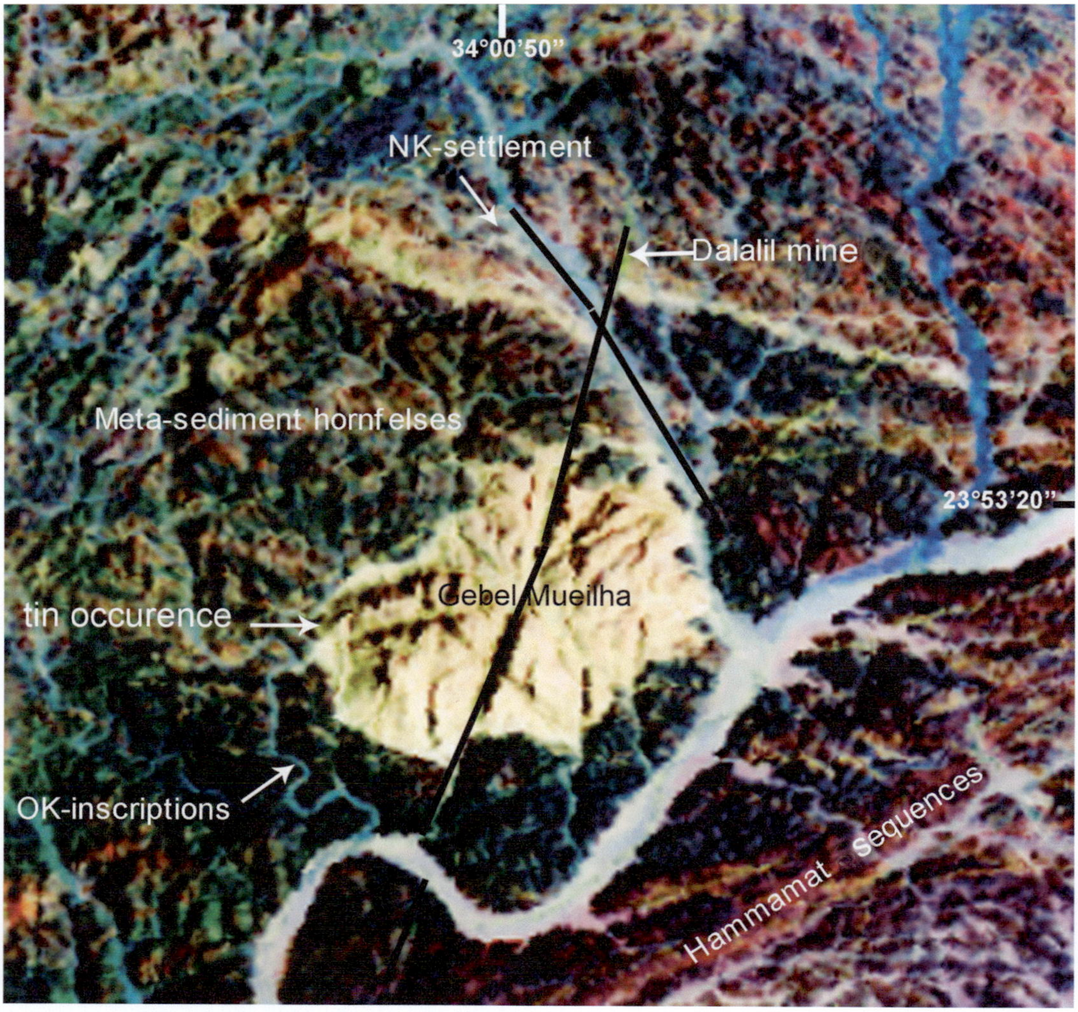

Fig. 5.176 Lithologically processed satellite image of the area around Gebel Mueilha showing the locations of the tin deposits, the Old Kingdom rock inscriptions, and the mining district of Wadi Dalalil (TM 174/43, channels 7-4-1)

located at some distance from the mine, one is yet again led to think that gold was gained mainly in wadiworkings.

The geologic framework of this district is marked by the immediate roof situation of granite intruded into the metasediments. In the lithologically processed satellite image one distinguishes quite well steeply folded, generally NE-SW striking metasediments in the vicinity of Wadi Mueilha. To the W of Wadi Dalalil they are intruded by a round granite stock, whereby a distinctly contrasted contact aureole appears around the intrusion. The metasediments assimilated in the roof of the intrusion formed in conjunction with the granitic magma a hybrid quartzdiorite rock. According to our observations, these are favourable preconditions for the formation of auriferous quartz veins (Fig. 5.176).

Fig. 5.177 Ancient pits in the Dalalil mining district. All show traces from modern exploitation

A mine had been opened with a wide trench in a quartz carbonate vein (strike: NNW-SSE, dip: 75°E) 800 m to the SE of the settlement and in the low lying hills, at the NE wadi edge (Fig. 5.177). In addition to calcite, which here and there occurs in well-developed crystals, the vein quartz also contains, green octahedral fluorite with perfect cleavage. Former copper sulphide contents are attested to by malachite linings along the wallrock and former pyrite through cubic limonite pseudomorphosis. Andesite and granodiorite dikes run through the entire area parallel to the quartz vein whereby the former often display up to 4 mm long hornblende crystal needles. Clear extraction traces were recorded in the eastern parts of the mine, and several ancient, small mines were observed in the surroundings.

Modern prospecting and mining attempts have not lead to any gold production, but undoubtedly have destroyed traces from ancient mining.

Fig. 5.178 The Ptolemaic sites at Samut showing the western and eastern forts and the Position of the circular separation complexes ("lavéries hélicoïdales") (modified Google-Earth image)

5.4.24 Samut

Geographic position (mine):	24°50′57″ N, 33°54′54″ E

From a distance one already makes out the outstanding ruins of Samut, which spread-out to the E and W of the N-S oriented old mine (Fig. 5.178). The architecture and the wealthy pottery finds allocate this site to the Ptolemaic Period. The pottery chiefly consists of light-coloured, and burnished fragments from amphorae with elongated, straight handles.

In the Early Arab Period the site had been reoccupied, as disclosed by dark-red amphorae with oblique grooves applied to a round base, and by a painted, light-coloured ware.

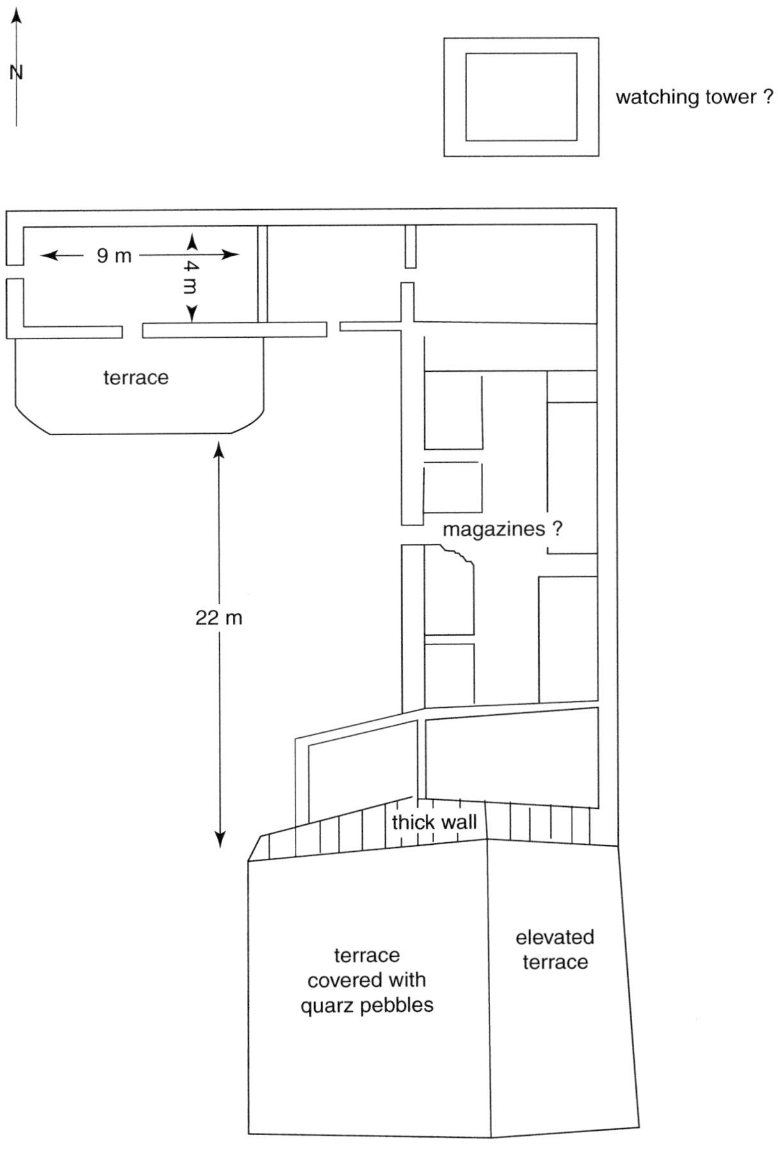

Fig. 5.179 Sketch plan of the western complex at Samut

Features in the western sector are severely damaged (Fig. 5.179). In the E, however, along the western side of a partly 2 m tall, outside wall, magazine-like and elongated room units are relatively well-preserved. Smaller, square rooms line-up along the E and S of the wall. All rooms group around a central courtyard, in which there once had been a well. Because of its elevated situation it had probably functioned as a cistern for runoff water. Parts of its masonry are still preserved (Fig. 5.180).

In a depression about 100 m N of the eastern building are two circular features (Fig. 5.181).

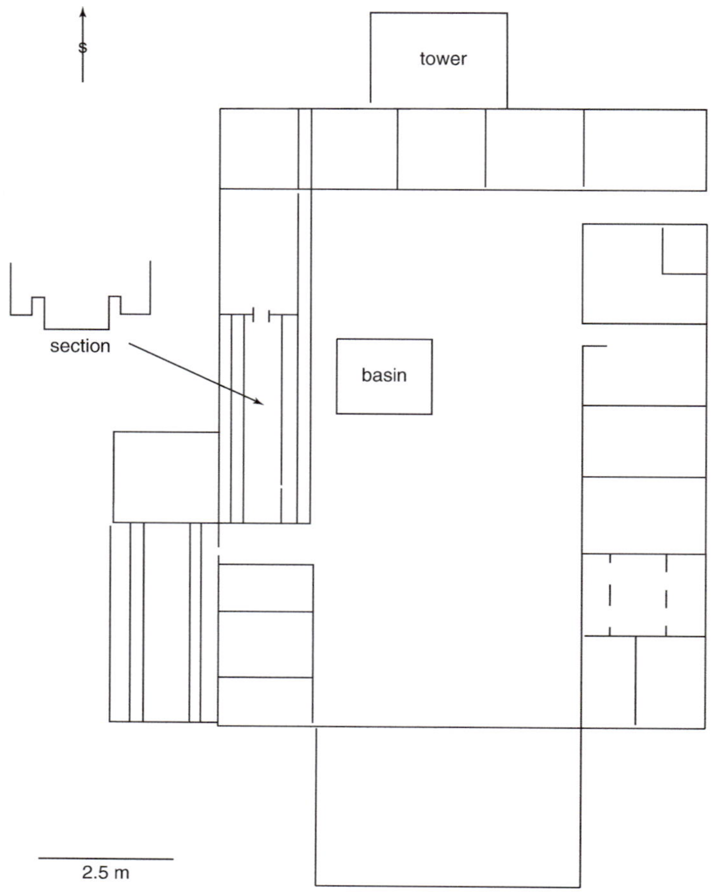

Fig. 5.180 Sketch plan of the eastern complex at Samut

Fig. 5.181 View of the depression with both circular separation devices (*arrows*) at Samut

Fig. 5.182 The best preserved "lavérie hélicoïdale" at Samut. Only its base has been preserved

Nearby, there is a large, inclined and flattened ramp which leads up to a house.

The circular structures resemble the Ptolemaic ore processing devices referred to as "lavéries hélicoïdales" at Laurion/Attica as well as the ones documented at Daghbag, Barramiya and Bokari in the Eastern Desert of Egypt. The ones at Samut had been masoned in the shell-facing technique. Today they have silted up to such an extent that they have become virtually level with the surrounding ground. Its former dressed stones with the washing grooves have disappeared (Fig. 5.182).

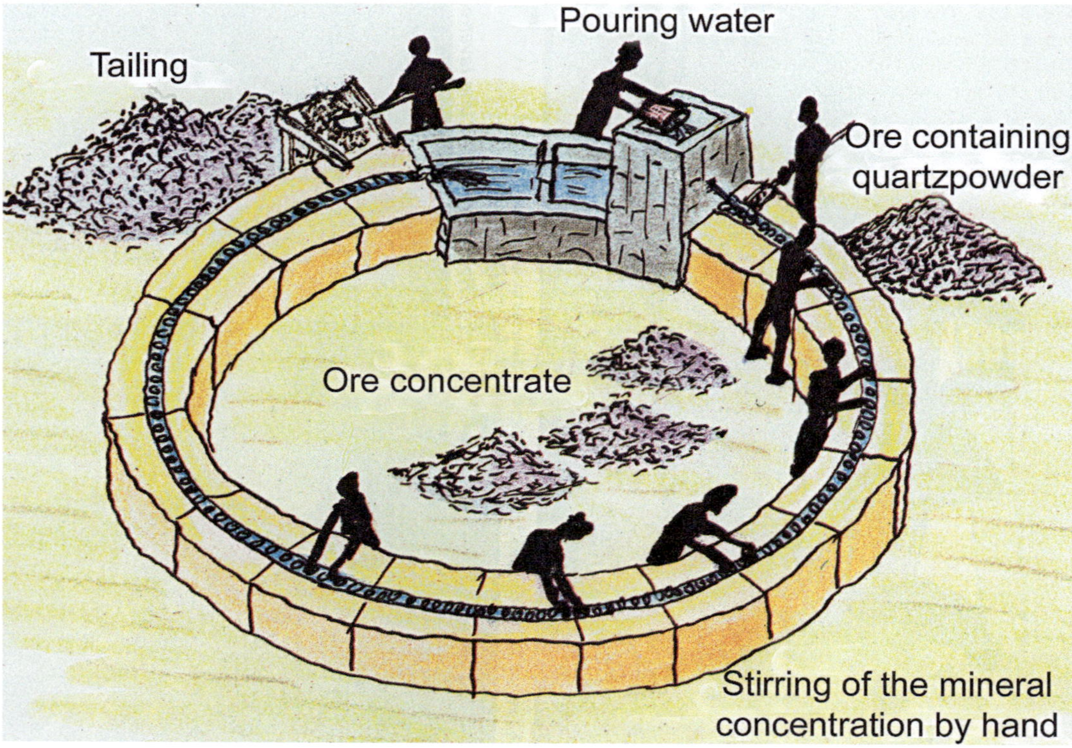

Fig. 5.183 Reconstruction of a "laverie hélicoïdale" at Laurion/Attica by Conophagos (1988)

The devices have a respective diameter of 12 m. The function of the structure at Laurion was described by Conophagos (1988) (Figs. 5.183 and 5.184).

Occasional fist hammers, grooved hammers and fragments from mortars found in the slope rubble of the mine and of smaller ones nearby, in the area of the circular washing devices, suggest a date to the Old/Middle Kingdoms. The mining history at Samut can thus be interpreted as follows:

Mining had probably begun sometime in the Old/Middle Kingdom and continued into the New Kingdom period, although with a clear emphasis on wadiworkings. Older traces from underground mining had been veiled by later activities, and may be revealed by detailed archaeological investigations only. Underground mining reached nonetheless its apex during the Ptolemaic Period which probably coincided with the construction of the two, fort-like buildings. Finally, a short exploitation phase had apparently taken place during the Early Arab Period, as evidenced in the surroundings of the mine only and by the reoccupation of the forts. Mining resumed at the beginning of the twentieth century under an Egyptian-Sudanese syndicate (Hume

Fig. 5.184 Detail of a "lavérie hélicoïdale" from Laurion/Attica with depressions in the continuous notch for deposition of heavy mineral concentrates (Photo credit: H. Lausch)

1937), which again led to widespread destruction of the archaeological matter. Several houses from this time have been left standing, especially in the southern and western sectors.

SE to the Samut mine one comes across another house group from the New Kingdom, including six to eight, typically rectangular buildings, of which some display two rooms and a small side niche. Close this relatively small settlement extending over a distance of approximately 150 m, there is a very light quartz vein with traces from mining (Fig. 5.185).

In the easternmost house a pounding stone associated to fine quartz sand was found, suggesting that quartz processing had actually taken place inside the houses. This finding again, would support that mining here had been carried only out on a test or prospecting level. Unfortunately, no diagnostic pottery was found. In and beyond the houses numerous New Kingdom, oval stone mills were found. We decided to refer to this site as "Samut-SE".

About 1.2 km NW from the main mine, there is another unusually large New Kingdom settle-

Fig. 5.185 Loosely scattered, large, New Kingdom houses in Samut SE. A small mine is located in the background (*arrow*)

ment of which the most northern part is displayed in Fig. 5.186. Here, more than 200 houses distribute over a distance of about 1 km (from 24°51′48″ N, 33°54′21″ E to 24°51′20″ N, 33°54′20″ E) on both sides of a N-S oriented, wide wadi called Wadi Samut el-Beda. Large rectangular, detached houses with walls from severely weather-beaten stones gathered in the granite surroundings are lined-up on flat elevations (Fig. 5.187). Almost all houses contain numerous New Kingdom oval mills, small grinding stones, anvil stones, and pestles from different rock types, among which coarse-grained reddish, mostly very weathered granite and dark porphyry rocks (Fig. 5.188).

On the gentle slopes and wadi terraces one barely distinguishes a number of graves. However, to judge by the size of the settlement, the cemetery had probably been much larger at the time it was in use.

The average dimensions of the houses measure 10 × 5 m. They generally have two to three, interconnected rooms. Their walls are preserved in heights to 1.20 m. They are by and large consistent with those usually encountered at comparable New Kingdom settlements engaged in wadiworkings. In this case it were ore fragments mainly from numerous, only few centimetre thick, and approximately N-S striking quartz veins in the surrounding mountains that assured the gold-yield from the wadi.

The settlement counts among the largest house alignment sites of its kind that we were able to visit in the Eastern Desert mining districts. The gold yield must therefore have been correspondingly important here.

In the satellite image the deposit district around Samut seems to be structured in an almost regular rhombic wadi pattern. This is due to par-

Fig. 5.186 Northern end of a large New Kingdom settlement in Wadi Samut-El Beda. Much of the site has been flushed away by floods (modified Google-Earth image)

ticularly distinctive vertical joints arranged in approximate N-S and NW-SE strikes. The pattern is enhanced by additional, large rhyolite to dacite dikes, for the most at identical strikes and a vertical dip.

The surface rocks are uniformly composed of quartzdiorite and diorite. Granite was not identified within the closer surroundings of the deposit.

All ancient underground operations, however, had concentrated on a single quartz vein with a bifurcation in the N, beyond which no extraction had occurred. The mined part of the vein strikes 10–18° and dips at 80° E. The vein itself consists

Fig. 5.187 New Kingdom houses aligned along the western edge of Wadi Samut el-Beda with abundant mill inventories

of milky-white quartz and to some extent of calcite. Near the surface it has marked, brownish-red stains from iron hydroxides which is probably due to a relatively high ratio of primary pyrite, which oxidised in a supergene process and whose negative shapes are well recognisable. The vein thickness varies between 0.5 and 1 m. The ancient extraction zone partly affects the wallrock to a depth of 2 m. This seems to suggest that the latter as well, had been identified as auriferous and had consequently been extracted. This hardly surprises, considering the well-developed alteration zone, which with its marked stains hardly reveals any sericite or chlorite within the wallrock. As known from several other gold deposits in the Egyptian Eastern Desert, not unimportant amounts of gold are bonded inside the former pyrite of the alteration zone before it was freed through oxidation of the surface of its host pyrite.

Fig. 5.188 Mills and pounding stone fragments inside a New Kingdom house in Wadi Samut el-Beda

5.4.25 Bir Samut

Geographic position:	24°48′25″ N, 33°54′12″ E

Numerous fragments from Ptolemaic concave grinding mills and therewith associated runner and pounding stones were observed within the stonework of the fortified enclosure around Bir Samut (Fig. 5.189). This is also the case for a number of houses and a relatively large, rectangular wall on a small hill, some hundred metres to the NW. Here too, countless fragments from Ptolemaic mills and pounding stones are integrated to the walls of the buildings (Fig. 5.190). Bir Samut had therefore once undoubtedly been a Ptolemaic gold processing plant. In the surrounding mountains however, there had been no mine to provide the site with ores, although the geologic surroundings yet match with the homogenous, granodioritic-dioritic rocks recorded at the Samut deposit, only few kilometres away to the N. Moreover, the rhombic pattern formed by joints and wadis and followed by numerous felsitic dikes is also observed in this area. Also, large numbers of thin, approximately N-S striking quartz veins cross the mountains here. Therefore, here as too, after having been eroded away from the quartz veins in the mountains the gold ores ended up in the wadi floors, where they were selectively picked out in wadi workings and processed near the central well at Bir Samut.

The well at Bir Samut may even have assured most of the water supply for the processing plant at the large Samut mine. This however, would require verification from systematic archaeological investigations.

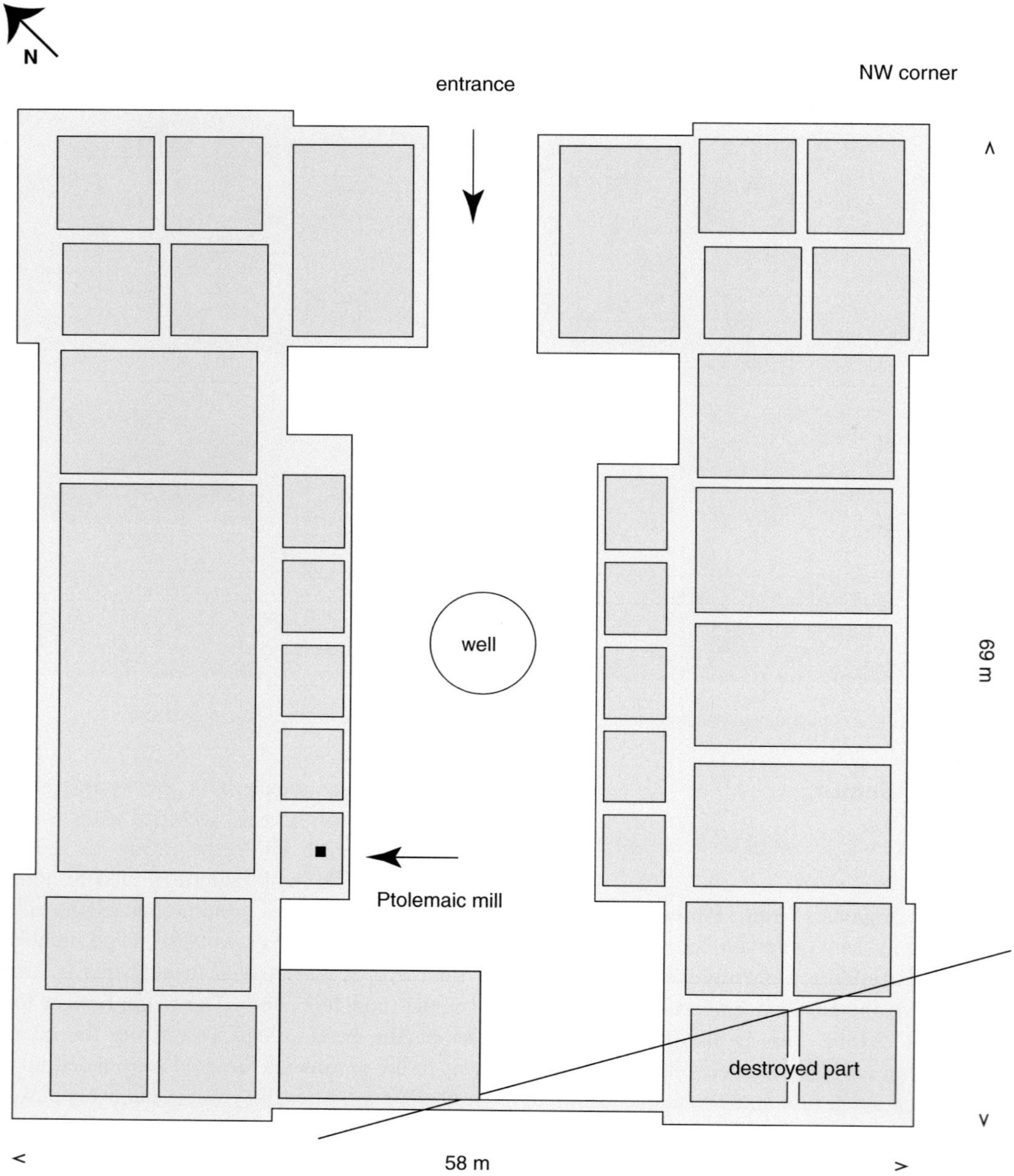

Fig. 5.189 Sketch plan of the Ptolemaic/Roman fort at Bir Samut, calibrated by aid of a satellite image (R. Klemm)

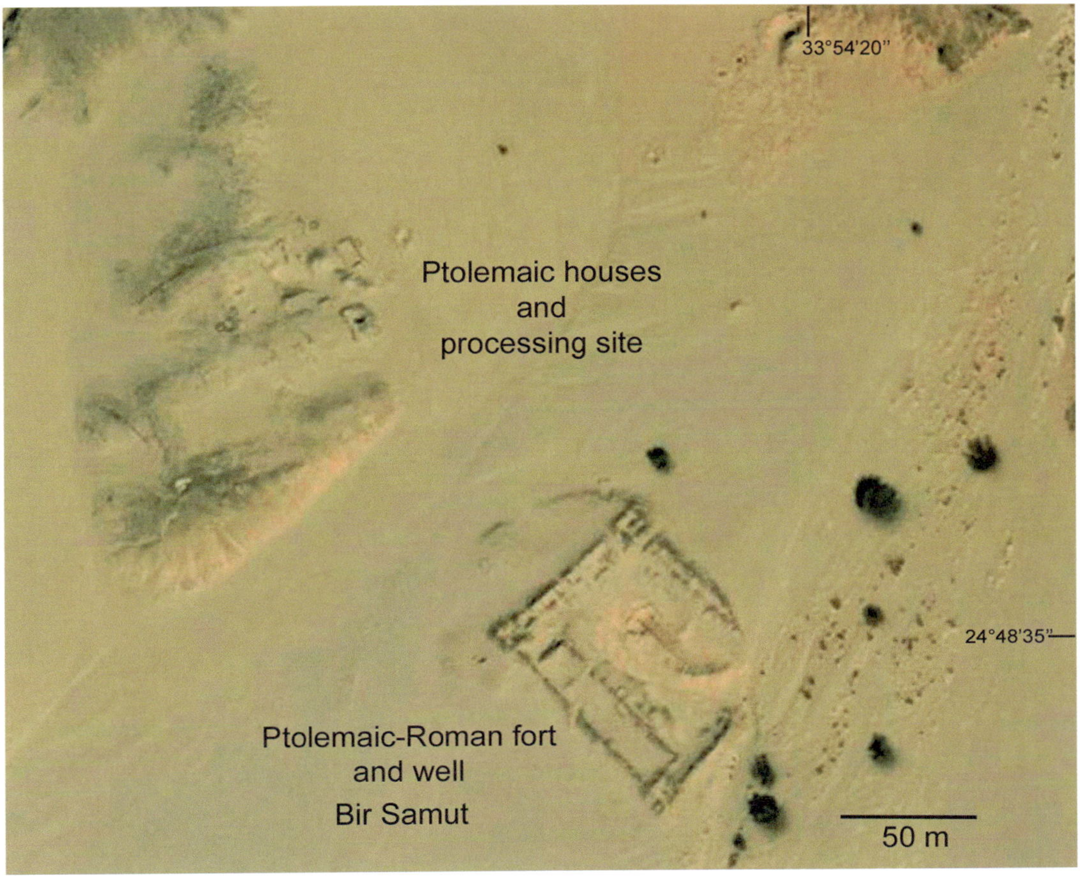

Fig. 5.190 Ptolemaic/Roman fort Bir Samut and sand covered houses as well as a potential gold processing device (modified Google-Earth image)

5.4.26 Wadi Mueilha

Geographic positions:	
New Kingdom settlement:	24°48′56″ N, 33°58′48″ E
Fort Jaharish:	24°49′14″ N, 33°59′21″ E

This large, New Kingdom settlement at the SW edge of Wadi Mueilha consists of 25–30 characteristic houses mostly composed of two to three rooms and arranged in alignments parallel to the wadi. Everywhere in its vicinity oval grinding mills are found with the connected, small runner stones. In contrast to the rectangular houses along the wadi front, ones located in the more remote valley incisions, are better preserved and diverge clearly from the former by their approximately square ground plans (Fig. 5.191).

The auriferous veins associated to the settlement occur in dark- to medium-grey, quartzdiorite hornfelses and are located about 500 m away from the large granite area of Gebel Urf Abu Haman, to the SW. However, the surface rocks here had in fact formed by assimilation of overlying metasedimentary sequences with the intrusive granitic magma, as discerned in the terrain only few hundred metres away to the NW. Thereby, the granitic magma altered its chemistry to quartzdiorite and even to diorite. Both, the intrusive granite-pluton as well as its metasedimentary roof rocks much metamorphosed through contact are crossed by up to 20 m wide dolerite dikes.

The exploited, extremely carbonate-rich quartz veins cross the quartzdiorite wallrock in a

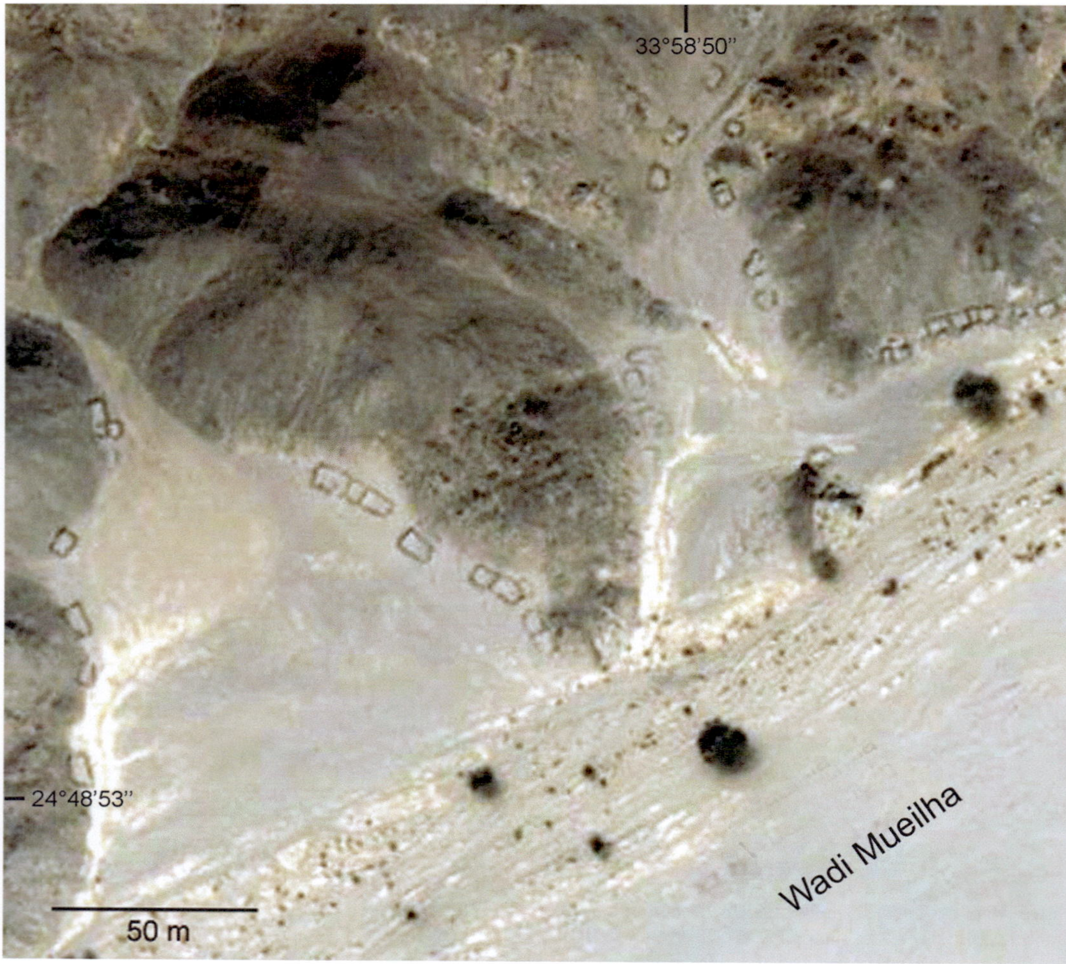

Fig. 5.191 Typical New Kingdom settlement in Wadi Mueilha consisting of attached house alignments along the wadi edge (*right*). In the smaller ravines this arrangement is less strictly followed (modified Google-Earth image)

NNW-SSE strike (330°) and a median dip at 85° SW. In the middle of the quartz veins frequently occurring, elongated cavities filled with crystal lawns point to a fast cleft opening during the hydrothermal quartz-ore precipitation. These cavities are also often filled-in with coarse spar ankerite/siderite. Next to that occur small, unexploited veins of almost pure carbonate, containing only little quartz. Through the bundling of vein swarms the debris had been dumped into the previously cleared ore trenches, with the effect that today the veins may only be pursued indirectly.

The discussed vein mineralisation at this site suggests that the extracted parts merely represent the uppermost zones of a vein system presumably yielding substantially richer ore yields in deeper levels. But without more specific data, a metric depth estimation of the rich-ore-zones proves problematic. However, the minimum depth of this zone would locate between 40 and 80 m and would therefore have

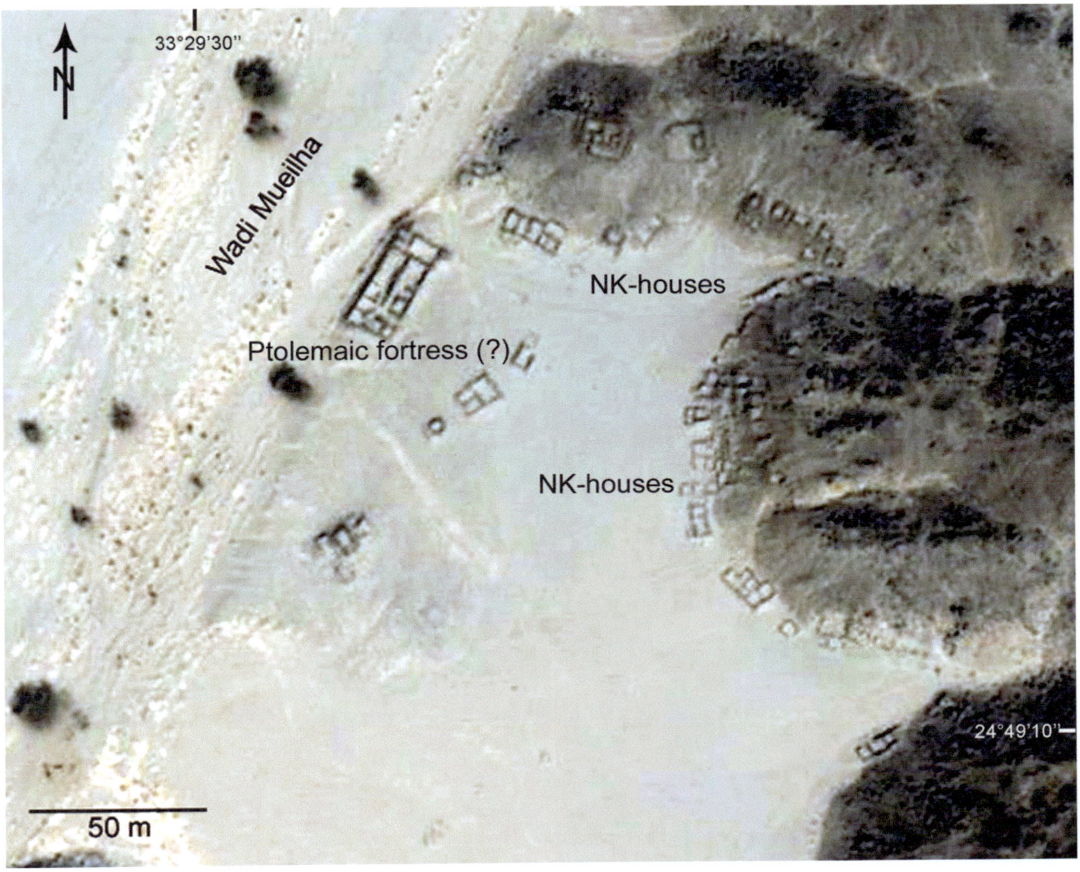

Fig. 5.192 Ptolemaic settlement with a fortified, large building, the so-called "Fort Mueilha". The predominantly multi-chambered houses are partly sanded in (modified Google-Earth image)

been already far beyond the scope of the New Kingdom miners.

After about 1 km NE of the New Kingdom settlement one arrives at "Fort Mueilha" (Jarahish) on the eastern side of the wide wadi. This Ptolemaic Period fortification had probably been built for the protection of a well while its function as a road station had survived until the Roman Period (Fig. 5.192).

Mined vein minerals were not found in this area. However, we did find the concave mills from the Ptolemaic Period that may to point to the exploitation of wadiworkings on a reduced scale.

5.4.27 Hangaliya-West

Geographic position:	24°50′23″ N, 34°33′05″ E

Bending-off to the S at Sheikh Salim from the Marsa Allam-Edfu road, one reaches the pilgrimage site at Sheikh Shasly after a drive of about 100 km on an asphalt road. Only 30 km down that road to Sheikh Shasly, a track branches-off to the E that leads to the Hangaliya mine.

Hangaliya-W is about 3 km W away from the main Hangaliya district, in a tributary valley leading N. The site is an Early Arab gold mining and processing place. Hidden away in a

Fig. 5.193 Early Arab Period work platforms with light-coloured tailings and remains of a washing table at Hangaliya-West

small, tributary valley oriented N-S, numerous huts cluster in a small group of hills near the southern end of Wadi Hangaliya. The huts are built in the shell-facing technique, in a loose distribution at the foot of the hills. Many of them are associated to small tailing heaps and at least five additional tailings were counted in the plain. One of them even displays a juxtaposed, barely distinguishable washing table and a masoned water reservoir (Fig. 5.193). Some of the round huts had been equipped with terraces. Otherwise numerous round mills scatter over the surface. The settlement stretches out over the saddle of an elevation and continues into the next valley incision, where an unusual-looking round mill is made out (Fig. 5.194).

An old cemetery, whose simple grave mounds probably date to a mining phase in the Early Arab Period is located next a Bedouin burial ground in the wadi centre. The small shallow mine in the neighbouring mountains had certainly not represented the only livelihood for this fairly large site. The settlers had therefore no doubt also been occupied in wadiworkings.

Fig. 5.194 Lower part of a round mill with carved grooves for enhancing its efficiency. As the grooves wear down with time, the grinding process gradually become less effective. Hangaliya-West

5.4.28 Hangaliya

Geographic positions	
Main mine:	24°50′12 ″N, 34°34′12″ E
New Kingdom settlement:	24°50′27 ″N, 34°34′04″ E
Large Ptolemaic settlement:	24°50′29 ″N, 34°35′19″ E

In addition to its important archaeological ruins, the Hangaliya district (Fig. 5.195) stands out by three, until recently exploited mines, which evidently date back to antiquity (Fig. 5.196). From W to E they are Shamaliya, Rabeiya and Shagara. They are easily discerned in the terrain by modern shafts, adits, and debris heaps.

The largest ancient settlement is located near Bir Hangaliya, in the eastern part of the district (Fig. 5.197). A Ptolemaic Period occupation clearly dominates the site (Fig. 5.198). It covers a settlement from the New Kingdom traceable only at two locations at the mountain foot to the N of the plain. It is possible that the New Kingdom settlement had once spread over the entire plain, as is indicated by the typical, oval millstones in the rubble near the well itself. Moreover, a New Kingdom settlement, again with numerous mills, is found further W, near the Shamaliya mine.

Scanty remains from the Early Arab Period concentrate near the edge of the mountain to the E. However, the main gold extraction site during this period seems to have been at Hangaliya -West.

Mills in all stages of wear dating to the New Kingdom and especially the Ptolemaic Period are found in all parts of this extensive settlement (Fig. 5.199). In the Ptolemaic Period the central tailing site was probably located near the well.

The capacity of the well at Bir Hangaliya has been able to satisfy most of the areas' water demands since antiquity. This is among others supported by the fact that Bedouins are very present in the area (Fig. 5.200) and that their large burials have been accumulating here in the style of round tumuli tombs (Fig. 5.201).

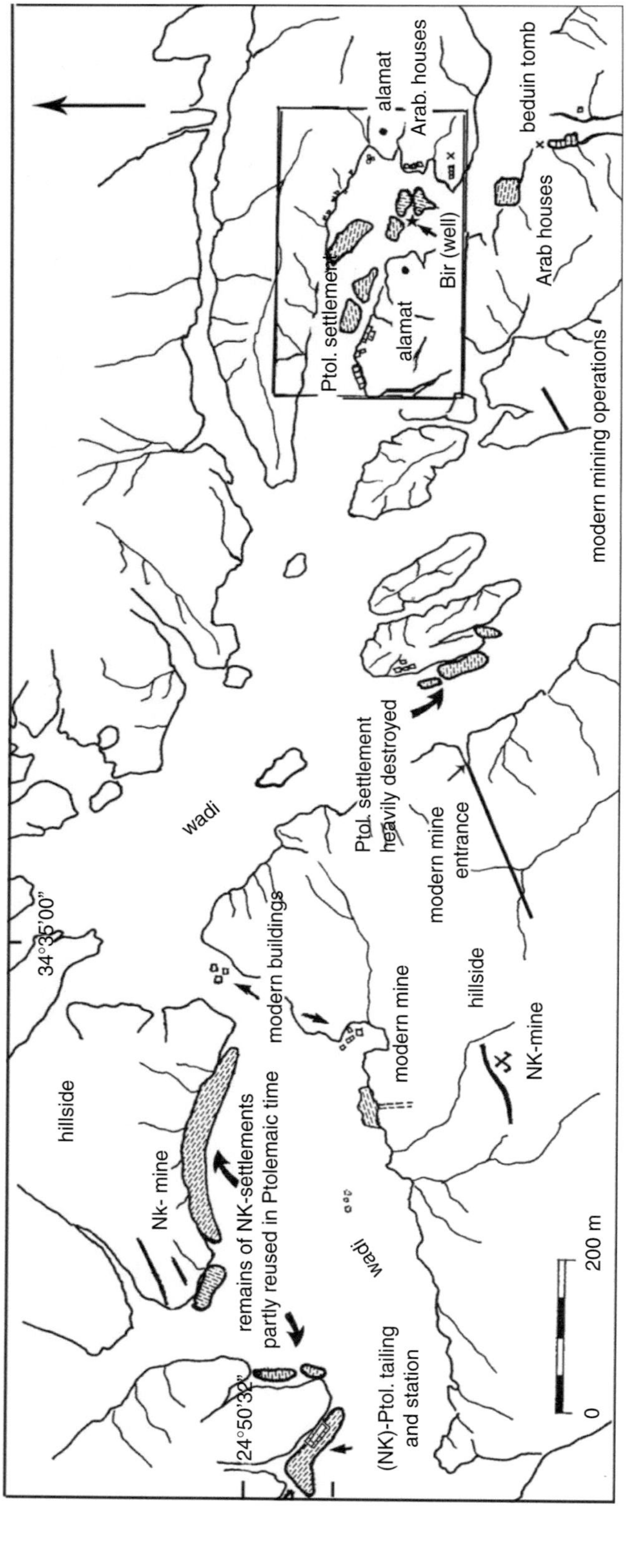

Fig. 5.195 Archaeological sketch map of the Hangaliya district (R. Klemm)

Fig. 5.196 Access to the ancient mine at Hangaliya. The modern adit is located at the hill's foot

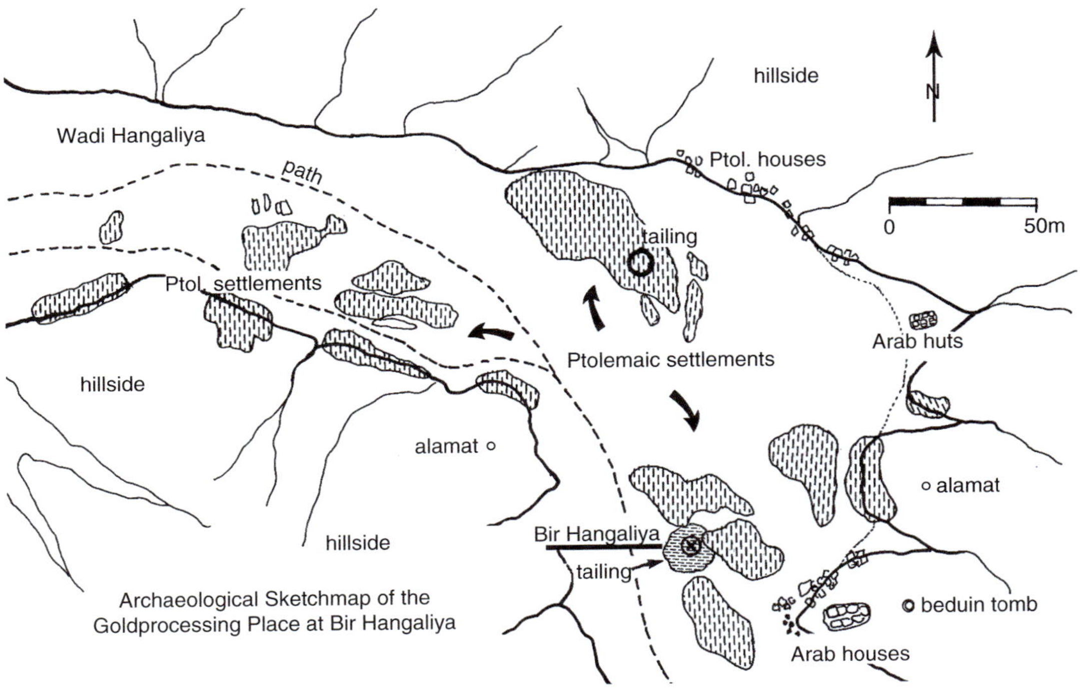

Fig. 5.197 Archaeological sketch detailing the area around Bir Hangaliya (R. Klemm)

Fig. 5.198 Ptolemaic dwelling structures at Bir Hangaliya

Fig. 5.199 Ptolemaic concave ore mill with apron shaped runner stone in the Bir Hangaliya settlement

Fig. 5.200 Deposited Bedouin property at Bir Hangaliya

Fig. 5.201 Looted circular cairn, near Bir Hangaliya. Such tombs may date up to Predynastic times and are quite scarce in the southern parts of the Egyptian Eastern Desert but more common in Nubia/Sudan

According to the archaeological context, underground mining at Hangaliya had continued to grow after the New Kingdom and had become the main extraction activity in the Ptolemaic Period. In view of the little evidence from the New Kingdom settlement, wadiworking seems if at all to have been carried out to a more limited extent in that period.

Otherwise, the site has yet again suffered severely from damages perpetrated by modern mining.

The geology of the Hangaliya deposit is determined by its situation between the Gebel Hafafit massif and an important ophiolite nappe system intruded by granite to the NE (Fig. 5.202). The following discussion is based on a thorough study by Murr (1999).

The metamorphic units in amphibolite facies of Gebel Hafafit are separated by the so-called Nugrus overthrust from an intensely folded ophiolite nappe of mainly serpentinite and basalt. Wherever serpentinite is much affected by tectonics it has altered to quartz-carbonate. This sequence is intruded by granitoid magmas. Two main tectonic lineations have been identified in the granite and the wallrock of the intrusion. The first runs more or less parallel to the Nugrus overthrust (NW-SE). The second runs ENE-WSW to almost E-W and is younger than the first one. The resulting extension is observable as faults in the granite and by the intrusion of dikes. The auriferous quartz vein system at Hangaliya belongs to this latter system within the biotite-granite.

The mined quartz vein at Hangaliya follows a shear zone within the granite. It strikes almost E-W and dips 60° N, and its thickness varies between 30 cm and 1 m. Ancient extractions follow the quartz vein along its outcrop. A modern extraction shaft follows the quartz vein over a distance of about 200 m.

A secondary arsenopyrite mineralisation after deformation is observed with increasing proximity to the quartz vein mineralisation. The conduct of iron is essential for the mineralisation process. Gold contents are mainly bonded to arsenopyrite. This increases inside the granite-mylonite with increasing proximity to the quartz vein. Thus, a positive correlation of the total iron content in the rock with arsenopyrite may be expected. In fact, however, it turns out that iron contents rather tend to decrease than rise with increasing proximity to the vein. Therefore, the Fe required for the arsenopyrite did not attain the wallrock via the mineralised fluid.

The results from the geochemical and petrographic analyses on the granites from Hangaliya lead according to Murr (1999) to following formational model for the genesis of auriferous arsenopyrite in the wallrock of the quartz mineralisation: The deformation and hydrothermal alteration of granite engenders the formation of a quartz-sericite rock. Because the total Fe content remains constant in the respective samples, in spite of clear secondary ore formation, it may for instance be assumed that the Fe released in the biotite alteration is bonded to the arsenopyrite. At the contact between fluid containing sulphide complexes and the wallrock, arsenopyrite forms, in the case of the absence of As, only pyrite. Fe would therefore come from the wallrock, As and S from the fluid. Because gold is transported in the fluid in a sulphide complex, it accumulates in arsenopyrite during its formation.

Two ore parageneses were distinguished in the quartz vein. The first contains mainly arsenopyrite. This is often converted to lepidocrocite at the vein margins. Arsenopyrite ore formation is most intensive at the contact with the wallrock. Toward the centre of the vein, the arsenopyrite content in the quartz decreases, but pyrite occurs occasionally. Macroscopic and microscopic gold is found in lepidocrocite only. The second paragenesis contains chalcopyrite, galena, sphalerite, gold, some pyrite, and in few instances arsenopyrite. This second gold generation occurs inside the quartz as free gold, as well as in compounds with galena and sphalerite. Chalcopyrite appears as both a single mineral inside quartz, as well as drop-shaped exsolution bodies in sphalerite. This paragenesis too, is subject to alteration. Thereby, chalcopyrite is transformed at the margins to digenite and covellite, and galena to cerrusite.

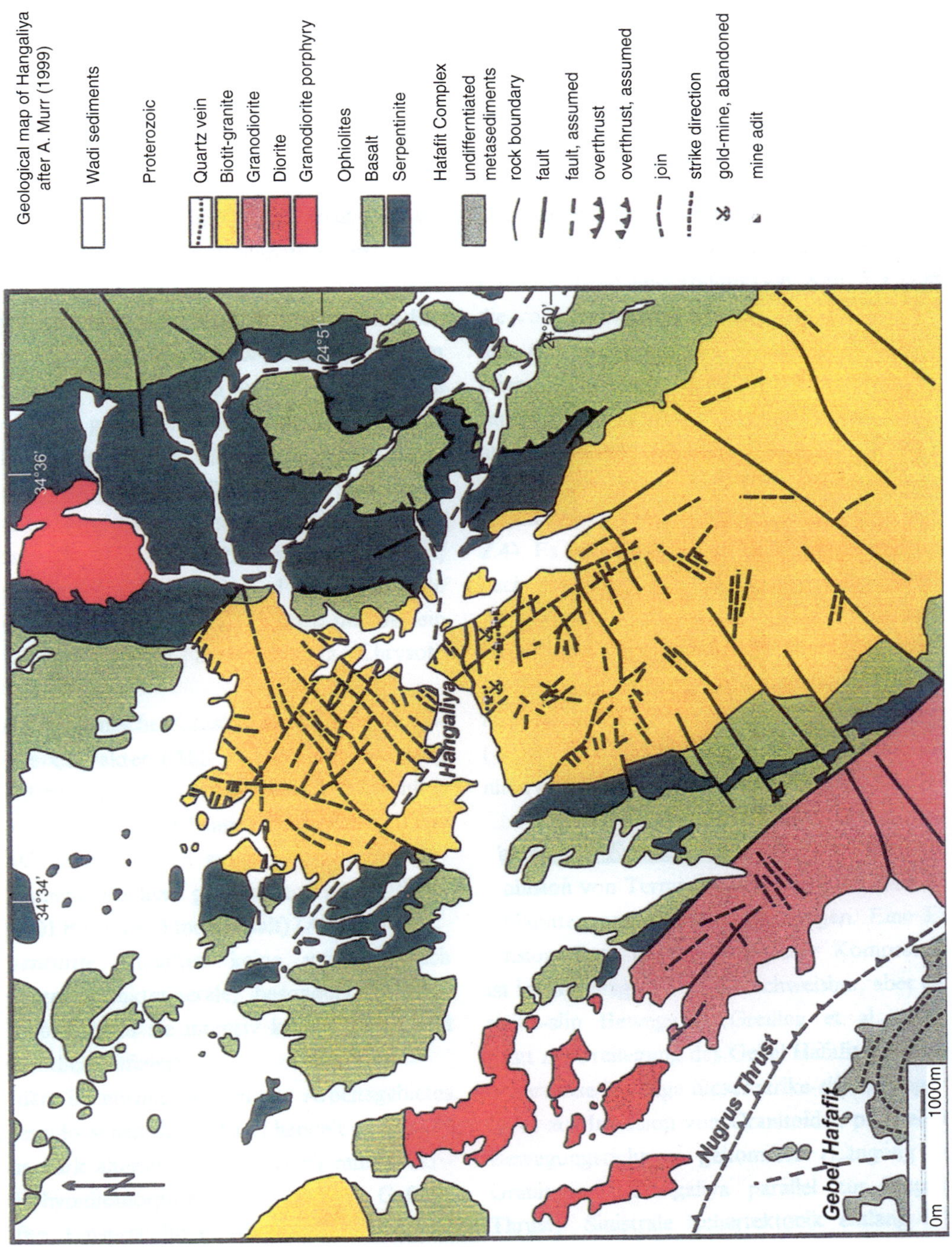

Fig. 5.202 Geological map of Hangaliya and its surroundings (A. Murr)

Fig. 5.203 Relics from the exploited mine at Umm Ud dating to the first half of the twentieth century

5.4.29 Umm Ud

Geographic position:	24°48′39 ″N, 34°41′42″ E

The deposit is located in the upper reaches of Wadi Umm Ud, in a tributary valley (Umm Regheiga) branching-off to the S and about 27 km S of Sukkari from where it is reached via a desert road.

On account of apparently ancient 4 m deep mines, the discovery is ascribed to the Hawari Company in 1941. The occurrence was subsequently mined until 1953 (Jakubiak 1987). In spite of our efforts, we found no extraction traces other than those from modern mining (Fig. 5.203).

The deposit area's geology again consists of prevailing acid metavolcanics, whereas closer to

Fig. 5.204 Adit at Umm Ud. Two inclined quartz veins are clearly distinguished. As both proved unproductive at this level, they were mined at deeper depths

the deposit itself of amphibolites, biotite-chlorite schists, and talcum schists intruded by quartz-diorites-diorites. Numerous basalt and andesite dikes strike NNW-SSE across the mountains.

The deposit consists of two auriferous veins in mutual angular orientations. The main quartz vein's thickness reaches 50–70 cm and splits up repeatedly within a 7 m wide, much disturbed zone into several small swarms (Fig. 5.204). It strikes 150° E and dips steeply at 75–80° SW. It consists of dark, blue-grey quartz. At its surface it may be tracked over more than 400 m. According to Jakubiak (1987), the vein has been exploited in five shafts and several intersecting tunnels over a distance of more than 230 m down to a depth of 85 m. The evaluation of the ore reserves vary according to Jakubiak (1987) between 15,000 and 21,000 t, at grades between

Fig. 5.205 Looted tumulus graves amassed from Ptolemaic settlement debris in Wadi Ghadir

23 g/t Au and 25 g/t Au. Francis et al. (1972) estimate them to range even between 19 g/t and 107 g/t Au.

The second vein ("Upper Lode") strikes 170° E and dips 25–60° W. It is pursued over more than 210 m at an average thickness of 24 cm. Koshin and Bassyuni (1968) believe its gold content to reach 30 g/t. A third dike in Wadi Hattabi ("Wadi Lode") is visible over 240 m at the surface and strikes NNE-SSW with a vertical dip. Its gold content is estimated at about 8 g/t Au. It had not been exploited.

In spite of the relatively high gold grades, it is quite surprising that the highly experienced New Kingdom prospectors, just as much as the modern ones, hadn't noticed this yet well-visible deposit.

5.4.30 Wadi Ghadir

Geographic positions:	
Beginning of the settlements:	24°37′41″ N, 34°49′13″ E
End of the settlements:	24°48′34″ N, 34°50′39″ E

At least seven, partly large and for the most part Ptolemaic settlements spread-out along the edges and terraces on both sides of Wadi Ghadir within a stretch of approximately 3.5 km. In spite of intensive surveying, no ancient mines in vein occurrence were discovered in the surrounding mountains.

The highest settlement density at the NW slope separates into two clear clusters, between which opens a gap of approximately 400 m length. In the SW cluster Ptolemaic Period concave mills and apron-shaped runner stones had been integrated to the structures of round tumuli Bedouin graves (Fig. 5.205). They add up to a number between 20 and 30 and diffuse over the entire surface of the former Ptolemaic settlement. All show signs of looting, resulting to the masoned mills ending up again in the surface debris (Fig. 5.20). This finding reveals following situation. In the midst of a Ptolemaic settlement which had developed from wadiworkings (as no other underground or trench mine was found in the surroundings), the stones from former Ptolemaic constructions as well as tools had been gathered to build so-called "Trogdolytai" graves. The striking appearance of these graves has probably so far prevented the detection of the older mining settlement on which they had been built. To our knowledge, the site is recorded nowhere and is hereby listed for the first time among Egypt's ancient gold producing sites.

In the NE sector, on a natural terrace on the same side of the wadi is another, almost untouched Ptolemaic settlement. Nevertheless, it too displays traces from later occupation through pottery matching with that left behind by the people associated to the graves at the above mentioned site to the W.

Almost directly opposite, in two small tributary valleys along the SE slope, there are more, however less noteworthy sites.

It may therefore be concluded from the so far discussed that gold production in the Ptolemaic Period was above all concentrated to the southern parts of the central Eastern Desert. In a wider implication, this is presumably connected to the development of routes leading through southern regions, down to the Red Sea port at Berenice (Sidebotham 1991, 2008). Wadi Ghadir too, which eventually flows into the Red Sea about 30 km S of Marsa Allam, belongs to the catchment area of the trade route between Lakeita and Berenice. Somewhat further S of Wadi Ghadir is Gebel Zabara with its emerald mines, which too had been exploited principally during the Ptolemaic Period.

5.4.31 Allawi (also Lawi)

Geographic position:	24°46′40″ N, 34°49′40″ E

We found no traces from either underground or trench at this position, but only extensive remains from wadiworkings. In spite of that, Koshin and Bassyuni (1968) report two parallel NW-SE striking veins from which "the upper or eastern lode was actively worked by the ancients."

Geologically is the area characterised by metasediments with occasional graphite schists which are intruded by diorites and intersected by a series of dikes.

5.4.32 Dweig and Abu Rahayah

Geographic position:	24°44′ N, 34°26′ E (Koshin and Bassyuni 1968)

We did not succeed in finding this site during our survey. The short descriptions by Koshin and Bassyuni (1968) are therefore rendered in the following:

"The ruins of, perhaps, an ancient mine was reported by Jenkins (1925) from one of the upper tributaries of Wadi Shait, on the southern slopes of the Gebel Dweig Mountain."

On the other hand, they mention the important Ptolemaic/Roman fort of Phalacro (also referred to as Dwieg 24°44′07″ N, 34°25′39″ E), in the mouth area of Wadi Talat Umm Rikhat into Wadi ad-Duwayq, a fortified road station at the road junction from respectively Koptos and Edfu (Apollinopolis Magna) to Berenice. In the satellite image, no hints as to ancient gold mining were however found in the surroundings of the fort.

It was only after the completion of our survey activity in the area that we became aware of a small gold occurrence near Bir Abu Rahayah (24°43′22″ N, 34°44′23″ E). Because Abu Rahayah translates as "father of mills", and "Rahayah" according to our knowledge, locally refers to rotor mills only, we ventured to conjecture the presence of such mills and consequently, an Early Arab gold processing site near this well. In any case, the EGSMA-Technoexport map by Francis et al. (1972) cites Bir Abu Rahaya as being a gold deposit.

5.4.33 Lewewi (also Umm Gamil)

Geographic position:	24°42′32″ N, 34°49′30″ E

Lewewi is about 5 km SE of Gebel Lewewi. The site is accessed with some difficulty only via Wadi Ghadir, Wadi al-Lawi, and Umm Gamil al-Bayda. It is located within the upper stretches of Wadi Umm Gamil al-Bayda.

According to the satellite image, there is a site here consisting of small, round huts, which probably date to the Early Arab Period. Another small settlement about 1.5 km further down the valley lies hidden away in a side wadi.

Fig. 5.206 Large house complex (20 × 20 m) at Hamash, probably from the Ptolemaic Period in spite of nearby remnants from the Early Arab Period

Since we were unable of visiting the area, the description by Jakubiak (1987) is given here: "Ancient workings follow two sub-parallel quartz veins intruded along the contact of 30 m wide granite dike intruded into shists."

5.4.34 Hamash

Geographic position:	24°40′55″ N, 34°05′41″ E

On the W side of Wadi Hamash, S of the inlet of an eastern side wadi, two separable installations regarding layout and age can be discerned. The southern one consists of a multi-chambered building with a roughly square plan measuring about 20 × 20 m (Fig. 5.206). In spite of the severe damages inside it, we managed to record about 20 room-units. It presumably dates to the Ptolemaic Period, although one notices two open spaces next to it with the remains from slag heaps as well as a prayer site, probably from the Early Arab Period. This may also be the case for a relatively small house cluster with a prayer site somewhat further to the N, although a more recent date associated with a mining phase in 1904–1905 cannot be ruled out.

Approximately 600 m farther to the N, settlement ruins spread-out extensively on the same side of the wadi. They consist of a long alignment of large rooms. The site is cut by erosion in the mouth area of a slope gully. Further upstream, there are more, probably contemporary houses separated by a wide road. A number of apron-shaped, double- handled runner stones ascribe them to the Ptolemaic Period.

Beyond this point, halfway up the slope of a promontory hill more, severely ruined houses

Fig. 5.207 Ptolemaic complex with ramp at Hamash of unknown function

can be distinguished. Two parallel running stone lines mark some sort of ramp leading up to a large stone heap, which indicates the location of a relatively large building (Fig. 5.207). A similar structure was also recorded at Wadi Atalla (see Fig. 5.74) and interpreted as a former Ptolemaic processing complex. This may also have been the case for the present complex at Hamash, although further investigations are admittedly necessary to clarify the exact function of both complexes.

Under the entry "Hamash", Meredith (1939) described a site consisting of "ancient gold mines, perhaps Roman". We are unable to confirm this observation during our visit to the area.

At Hamash a copper-rich, auriferous quartz vein had been exploited with a main NNW-SSW strike, which due to slight flexures, varies between 5° and 30°. Its dip was recorded between 30° and 60° W. The vein therefore, follows the main joint orientation of the granitoid wallrock and is visible at the surface for over more than 350 m. Its width varies only little between 30 and 60 cm.

The occurrence appears at the NE edge of a small granite intrusive body in contact with metavolcanics in an advanced stage of decomposition (Fig. 5.208). The intrusion itself is connected for its part, to the large granitepluton of Gebel Ras Sibrit. It probably only represents the intrusion roof to the overlying metavolcanics. These consist mainly of andesite lava flows whose former surface areas in spite of the contact metamorphism, still display relatively well-discernible amygdaloidal structures.

The occurrence is accessed today by four inclined shafts that follow the dip to a depth of 60 m. At this level the vein width is reported to increase to 0.5–1.1 m (Koshin and Bassyuni 1968). The mentioned flexures allude to the presence of several veins. Although this is not the case, the vein does however split up into in several small branches. The quartz itself is predominantly greenish to white. The green colour is

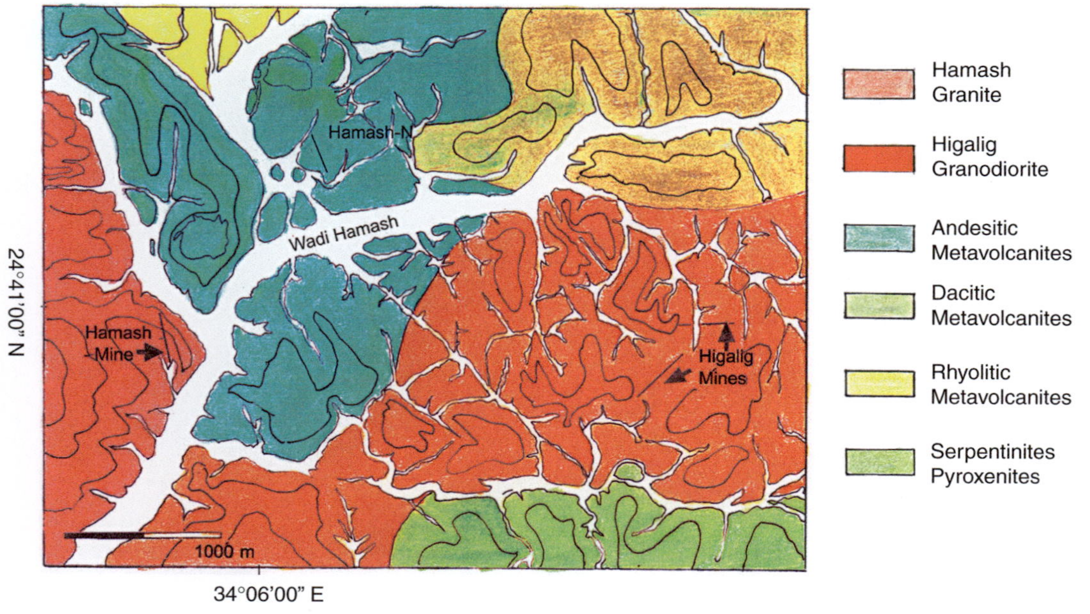

Fig. 5.208 Geological sketch map of the deposit area of Hamash and Higalig

caused by epidote, which too, tends to stain the wallrock and particularly its joints in a depth of up to several metres.

With pyrite, chalcopyrite, bornite, tetrahedrite, chalcocite (Hilmy and Osman 1989), and arsenopyrite, in addition to very well-dispersed and hence virtually invisible gold, the ore mineralisation is relatively multi-faceted, as compared to the usually monotonous pyrite/chalcopyrite parageneses. Sulphide mineralisations together with gold seem to be particularly bound to pocket-like enrichment zones, and to judge by some remaining handpieces, the sulphide ore ratio increases with depth. Generally, the gold contents do not seem to be particularly high and hardly exceed 8–10 g/t. Malachite and iron hydroxides are secondary decomposition ores. At the surface in the joints of the granite wallrock they appear respectively in the form of green malachite impregnations (Fig. 5.209) and of brownish-red stains, because of their iron oxide content. Hilmy and Osman (1989) describe an early phase of auriferous pyrite which was substantially enriched by remobilisation in a later phase. This however, was of no value for the ancient miners, as gold contained in pyrite was not exploitable in antiquity.

Traces from ancient underground mining are still visible in depths to 20–25 m, where abutments had been mindfully left standing. The actual extraction depths however, cannot be verified, on one hand because of cautiously executed dry wall offsets from these depths onwards, and on the other, the lack of traces at the side walls of a wide shaft.

Originally, mining itself seems primarily to have concentrated on gold. But to judge by the slags associated to a small house complex directly below the mine and probably dating to the first half of the nineteenth century, copper production at least at that time was certainly being pursued. Modern prospecting work in the twentieth century comprising the lowering of access shafts, has by far produced most mining traces. Considering the limited sizes of the quartz debris heaps from this period, the prospecting work had on no account led to a large scale production.

Fig. 5.209 Old mine at Hamash with modern adit (*top*) and malachite linings in the clefts of the granite wallrock

Fig. 5.210 Spoil heap from recent prospecting at Higalig with malachite staining in granodiorite cleft structures

5.4.35 Hamash-North

Geographic position:	24°41′50″ N, 34°05′53″ E

About 1.7 km N of Hamash a quartz vein mined at least since the Old/Middle Kingdom is embedded in andesitic hornfelses of an intrusion roof. However, it strikes in the same direction as the mines at Hamash. The wallrock alteration zone is intensely stained by brown iron oxide parallel to a lateral andesite dike. This has a tarnishing effect on the green malachite impregnations in the joints of the immediate wallrock. Though as already mentioned with regard to all extraction sites predating and contemporary with the Middle Kingdom, it were these green malachite impregnations that had served as main markers for the early prospectors in their decision whether or not to launch gold production. Richer malachite deposits also led to copper mining (see Wadi Dara).

The vein itself is pursued for more than 120 m and leads over a small ridge while being studded with 4–5 m deep, but caved in trenches.

In recent years (after 2000) extensive prospecting work in the district and in particular in Hamash-N by the Centamin Company, have led to extensive destruction within that area, but it seems as if the ancient mining traces remained generally conserved.

5.4.36 Higalig (also Hagalik)

Geographic position:	24°41′14″ N, 34°07′35″ E

About 3 km NE of Hamash, a small side wadi bends-off to the E and leads over a distance of 1.4 km to the deposit area of Higalig, in the mountains just S of the wadi.

The gold mines at Higalig are among the oldest known in the Egyptian Eastern Desert. All mined occurrences are associated to green malachite mineralisations (Fig. 5.210), being the most important prospecting criterion in this early phase of mining toward beginning of the third millennium BC. The recurring question as to whether copper had been mined in this district before

Fig. 5.211 Early Dynastic stone hammers from the Higalig mining district. Some are chipped near the impact surface which made them less usable

gold, remains yet again unanswered. The excavated evidence from the furnaces in Wadi Dara (Castel et al. 1992, 1995) seems for that matter to support this latter sequence.

The most remarkable mining tools from the Higalig district are represented by up to 30 cm long stone hammers weighing between 8 and 10 kg with blow surfaces rounded from wear (Fig. 5.211). Because of their unusually large size and weight, the tools required manipulation with two hands. They seem to have been used in a crushing action rather than for breaking-off ore or rock chunks. In the event that the round blow surface was damaged through flaking, the stone hammer lost its effectiveness and was subsequently disposed of. Most finds therefore consist of such discarded hammers. Several such large hammers have been found close to one of the strikingly smooth and rounded walls of a mine shaft at Higalig (Fig. 5.212). It may therefore be assumed that the gold ores had been considered to be crushed in a fine enough meal at their withdrawal from the mines, and that any further grinding had thereby become superfluous. For the lack of archaeological evidence, whatever the method by which the gold had been separated from the quartz meal, it had undoubtedly involved the use of water.

In comparing the mining methods at Higalig and Bokari, one first of all notes the different sizes of the tools. The reason may be linked to the available space inside the mine. At Bokari the verticality and the narrowness of the mine limited excessive, swinging movements. Therefore preference was given to smaller fist hammers and stone axes. At Higalig enough space was available to allow for a swinging motion while holding the hammer with both hands as it struck the mine's wall. At both mines the ore had been crushed already during the extraction process. At Bokari too, many hammers rendered obsolete from inadequate use wear were found in refuse contexts just outside the mine.

Fig. 5.212 Early Dynastic gold mine at Higalig. The smooth wall surfaces result from the crushing effect of the tools used in the extraction of the vein quartz

Furthermore, at Higalig, disc-shaped stone hammers with worn edges were found that had served for breaking down the countless, very fine quartz veins, which cross the area in plentiful numbers.

Higalig may be treated as a type-site for gold mining in the Early Dynastic Period. Except for traces from modern prospecting, our investigations revealed no signs of later extraction activities here. Because of the unusually large stone hammers, we decided to label this extraction period as the "two-hand-hammer" period.

In petrographic terms, the hornblende-rich hostrock to the various mined quartz veins is generally granodioritic. At its margins near the contact zone with the wallrock, it transforms to a diorite variety. In terms of geology, these granitoid rocks belong to the hybrid margins of a vast granite intrusion, which is only sporadically exposed by erosion in this area. The intrusion is covered by vestige streaks of mostly andesitic metavolcanics of the roof rock and preserved as hornfels. Varying degrees of assimilation of the granite magma with this roof material engendered the barely differentiable magmatite variants mentioned above. Moreover, this part of the Eastern Desert is speckled with numerous, much younger (oligocene?) basalt intrusions of which the pipes have been left standing by the erosion as round, virtually black hills (not on the geol. sketch map).

Almost all mined quartz veins display easily discernible malachite layers in the joints of the wallrock. This probably represented the most consequential feature for their ensuing mining development. Apart from the one in the W, which strikes NE-SW, they all strike almost E-W and dip 45° N. In the untouched sectors, the outcrops reveal conspicuous, brown iron hydroxide stains, which derive from decomposition of pyrite and chalcopyrite, the two prevailing ore minerals.

5.5 Southeastern Group

5.5.1 Wadi Geili (also Geli)

Geographic position:	24°31′30″ N, 34°43′30″ E (Koshin and Bassyuni 1968)

Without giving more details, Koshin and Bassyuni (1968) refer to remains from ancient work platforms in the upper reaches of Wadi Gamal. We did not visit this site.

At the indicated position the Google-Earth satellite images actually reveal a relatively large

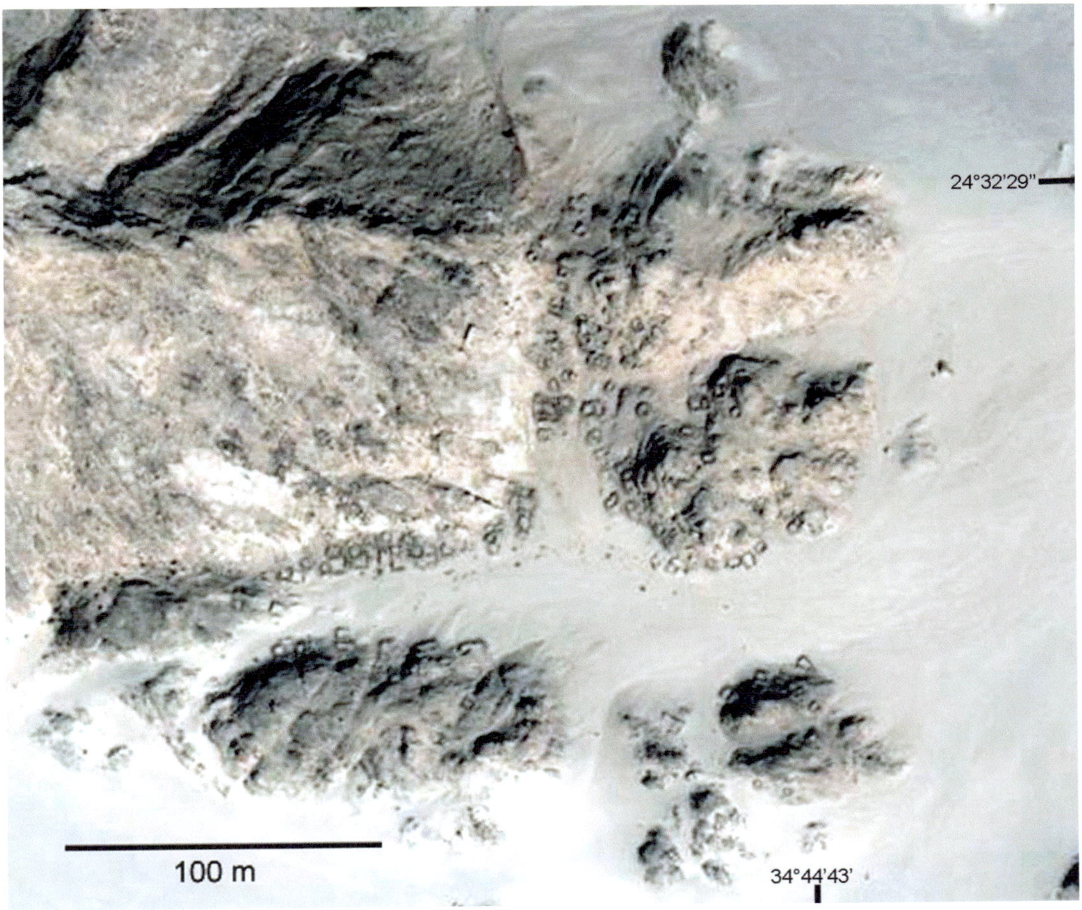

Fig. 5.213 The large, probably Early Arab Period settlement at Geili to the N of Wadi Gamal is hidden behind granite hills (modified Google-Earth image)

burial ground, consisting mainly of round graves. However, 2.5 km to the NE an extensive, presumably Early Arab settlement (24°32′26″ N, 34°44′40″ E) of more than 80 huts (Fig. 5.213) spreads-out over three hills and only 700 m NW another supposable more recent Arab-Beduin village. No mined veins were found in the surroundings. Indeed, as its geologic location is set within extensive granites, the area cannot be considered to be particularly auriferous. Even gold production from wadiworkings seems therefore hardly likely at this location.

About 9 km further down in Wadi Gamal another settlement on the N side of the wadi spreading into a tributary wadi reveals unusual house plans (Fig. 5.214). Although its geologic location within metasediments in proximity to granite significantly increases the chances of finding gold, we found no traces that might stem from ancient mining.

Fig. 5.214 Early Arab Period settlement with an unusual feature for a mining site. Its function was possibly connected to camel rearing in Wadi Gamal (modified Google-Earth image)

5.5.2 Wadi Dendekan

Geographic position:	24°16′56″ N, 35°10′32″ E

The upper part of Wadi Rian leads to Wadi Dendekan and into its tribuatry valley, Wadi Rusus. Here too, the geologic context is characterised by vast sequences of rhyolitic and mainly andesite lava with well recognisable layer structures and amygdaloidal textures. Numerous pink-coloured, felsitic dikes strike through the area. Small quartz veins had formed parallel next to them and probably furnished the ores for the nearby, New Kingdom and Early Arab gold processing plants.

Due to the high dependability of the New Kingdom prospectors, we checked the surroundings very carefully for ancient extractions. The results were somewhat disappointing, as the few trench mines we were able to find were in an advanced stage of collapse, thus rendering the identification of the extracted material extremely difficult.

5.5.3 Qualan (also Qulan)

Geographic position:	24°18′16″ N, 35°11′54″ E

The Early Arab extraction area is located on the S slope of the watershed towards the S end of Wadi Qualan and Wadi Dendekan.

A series of auriferous quartz veins occur in the surroundings of hornfelsic, andesitic and acid metavolcanics in close contact with intrusive magmatites.

The main vein is pursued for more than 160 m at the surface. It strikes above 70° ENE-WSW and falls 35° N. Two samples revealed grades at 18 and 22 g/t Au respectively. A second, 10–30 cm thick vein to the S of it is followed over more than 200 m and contains an average content of 8 g/t Au (Koshin and Bassyuni 1968).

5.5.4 Abu Rahaiya

Geographic position:	
Mine:	24°08′59″ N, 35°13′42″ E

Thirty to forty single, round huts, with diameters between 2 and 2.5 m group together on a flat hill. The huts are often associated to small prayer sites with vertically placed stones serving as mihrabs. Round mills and pounding stones scatter profusely over the site's surface. The walls consist of roundish nodes of a light- coloured granodiorite and are built in the shell-facing technique. The name "father of mills" presumably refers to a heap-like accumulation of round mills within the settlement area.

The deposit area of Abu Rahaiya locates in the central sector of Wadi Kashir, in the plain of el-Bayatat. It occurs in a medium- to coarse-grained granodiorite, structured by an unusually high number of wide andesite dikes in a rhombic pattern. The oldest ones strike NNE-SSW and are intersected by more andesite dikes with NE-SW and NW-SE strikes. All have a vertical dip.

The mined quartz vein swarms too, strike NNE-SSW and are generally only few centimetres wide. At the surface they are often discernible only by brownish-red stains in the alteration zones of the wallrock whose cleft structure frequently reveals malachite layers. There are seemingly two quartz generations, one with a marked tendency to chalcopyrite, the other to pyrite. The latter tends to occur in the iron hydroxide stains within the surface outcrops. Unfortunately, the surface is covered by drift sands, which have thus filled-up most trenches and prohibit all observation prior to any excavation. According to Koshin and Bassyuni (1968), visible gold can found at these vein swarms.

5.5.5 Bitan

Geographic position:	23°37′53″ N, 35°03′39″ E

The geologic context of Bitan is virtually identical to that of Eleiga. Its description is therefore rendered in the discussion on Eleiga hereafter.

About 700 m NW to Bir Bitan, at the eastern mountain face, as well as in the hills behind it, is an Early Arab settlement. There, right at the valley escarpment is a large but today collapsed corridor trench. New Kingdom mills within the masonry of the Early Arab house ruins indicate that the site had probably been exploited during this earlier period. The quartz vein mineralisation striking from almost N to S in a vertical dip occurs in a quartzdiorite wallrock with faintly parallel textures surrounding in a wide circumference the gabbroid core of the Eleiga intrusion. At the surface nearby, is a second vein revealing no traces from ancient mining but instead ones from recent, unsuccessful prospecting.

The mountains are crossed by numerous andesite dikes in NE-SW and NW-SE strikes, creating the above described, rhombic pattern. These dikes therewith follow a different tectonic regime to that of the quartz veins.

5.5.6 Umm Eleiga

Geographic position:	
Large building complex:	24°37′06 ″N, 34°59′16″ E
Mine area:	24°36′44 ″N, 35°03′19″ E

The site at Umm Eleiga (Fig. 5.215) in some sense forms an exception among the gold deposit sites in the Egyptian Eastern Desert. The geologic environment at this location seems to indicate that genuine gold nuggets had been gathered here over lengthy periods. In this high plain a web of fine, auriferous quartz veins were eroded, partially reprocessed and forming nuggets by intensive flowing river systems during a younger geological past, probably Pleistocene. Today the landscape is dominated by meter-thick layers of river rubble among which gold nuggets accumulated to be found by roaming nomads.

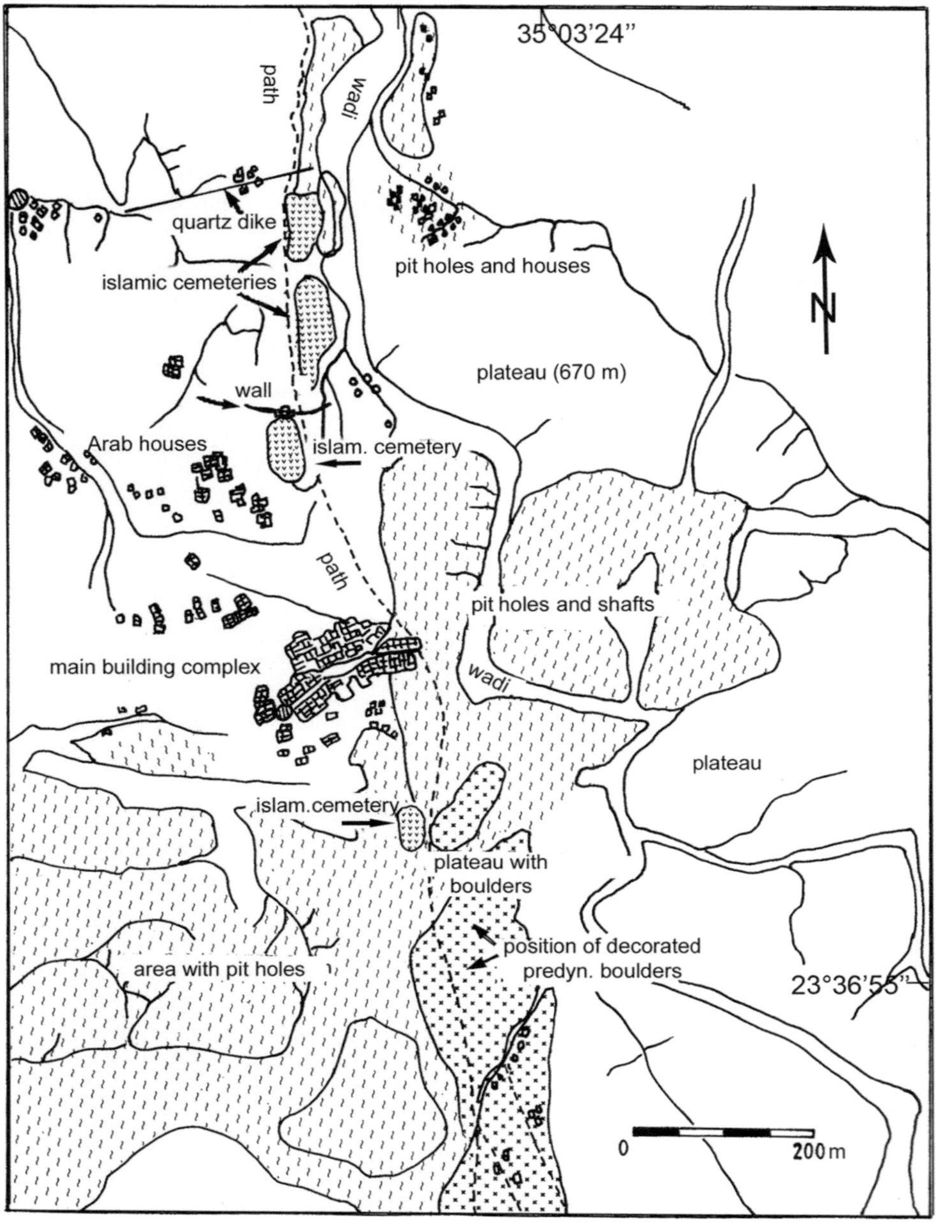

Fig. 5.215 Archaeological sketch map of the gold mining site Umm Eleiga (R. Klemm)

Boulders with engraved depictions of figurative and abstract motifs in fact testify to an early presence of such possibly mobile groups on this plateau. Winkler (1938) refers to this Predynastic population as the so-called "Earliest Hunters" of the Amratian towards the middle of the fourth millennium BC. The up to 80 cm long stones display among others, undulating, linear patterns frequently ending in spirals (Fig. 5.216). The motifs, which also depict wild animals like ostriches and elephants in loosely arranged compositions, may be younger than the linear ones (Fig. 5.217). So far, no representations of herds have been found. The fact that about 80–100 such

Fig. 5.216 Decorated boulder showing a carved snake motif at Umm Eleiga

Fig. 5.217 The geologist A. Kraus visualises with chalk a number of animal motifs carved onto a boulder at Umm Eleiga

Fig. 5.218 The large Early Arab Period building complex at Umm Eleiga. Relatively small room units are crossed by a central road

decorated stones are still found in an area covering about 50,000 square metres suggests that a majority has already disappeared, chiefly through erosion. This high density of rock art representations therefore seems to imply that they had been executed over lengthy periods by gold seeking inhabitants, and that the area had therefore once been more or less rich in gold (Klemm 1995).

Meagre occurrences of gold artefacts in the Predynastic and Early Dynastic Periods may generally be explained by the hypothesis that gold mining had yet to develop to a fully-fledged industry in the Eastern Desert, as is reflected by the low number of documented mines from this period. So rather than being acquired through methodical prospecting and mining, gold would initially have attained the Nile Valley in the form of occasionally collected placers through trade with the nomadic population. This assumption in fact corroborates with the evidence that gold bead jewellery from Predynastic tombs had actually been forged from nuggets, and not cast (Petrie 1901). Direct archaeological evidence for this type of gold production remains therefore exceptional.

In the Early Arab Period systematic mining in the plain had begun in the form of open pit hole mines. Numerous huts and large complexes were built, including two large cemeteries, a large prayer site, and an enormous supply centre, being the district's most striking complex (Fig. 5.218). To the W of it, is a tailing site with numerous round mills.

The large, central complex consists of a suite of relatively small rooms averaging 2 × 3 m in size, in a regular layout on both sides of an approximately 5 m wide, central alley. The 80 cm wide, shell-facing walls are remarkably solid, which is why they are preserved in relatively tall heights (occasionally up to 1.80 m). Particularly the northern parts of the complex are characterised by a concentration of round mills, anvil stones, and quartz lumps scattered in many rooms.

A British company carried out some short-term mining at Umm Eleiga in the twentieth century, as is attested to by some recent house ruins.

The geology of Eleiga and Bir Bitan is distinguished by a 6 km wide intrusive body with a gabbroid core (Fig. 5.219). It has a concentric structure with increasingly acid zones towards

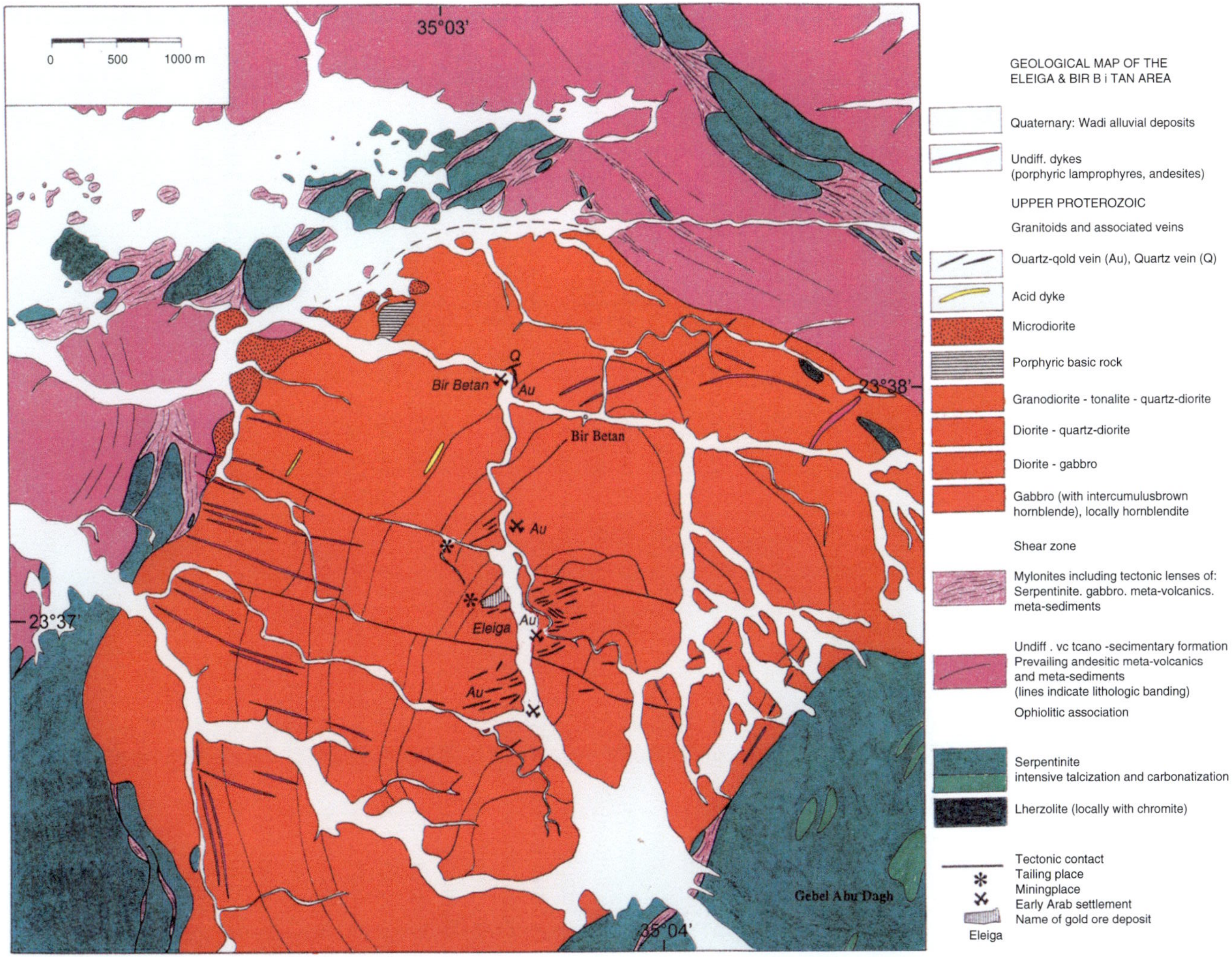

Fig. 5.219 Geological sketch map of the Eleiga surroundings (H. Kräutner)

the outside, ranging from dioritegabbros, diorites, granodiorites to tonalites. A microdiorite marginal facies sharply bordered towards the adjacent magmatites occurs moreover as a final individual intrusive phase at its NW edge.

This magma complex had penetrated into the bedrock, which consists of locally much diverging volcano-sedimentary series. They are partly deeply imbricated with overthrust ophiolite covers and sheared mainly in a NW-SE direction.

Unfortunately, no deeper parts of this intrusive body are exposed, which might contribute to the cognition of its original composition. In accordance with the hitherto experience on plutonic, Neoproterozoic intrusions in the Egyptian bedrock, all are close to granitic. In this case we therefore yet assume that a granitoid magma had penetrated into the nappe mentioned above, and by assimilation of larger portions of the ophiolite was locally modified inside the intrusion roof to form the gabbro-diorite-tonalite-suite at Umm Eleiga. In spite of the diverging melting temperatures of granitoid and gabbroid magmas, this was all the same possible, provided the presence of volatile constituents (H_2O) that decrease the melting temperatures. This had indeed been the case, because the unusually coarse-grained gabbros contain abnormally high amounts of brown and green hornblende in cumulate phases, which point distinctly to raised H_2O contents during crystallisation.

If this gabbro genesis applies, then the gold quartz mineralisation at Bir Bitan-Eleiga would by no means represent an exception, but merely a special case within our general concept of gold deposit formations in Egypt.

Zoheir et al. (2008a, b, c) who presented similar petrographic results, calculated formational pressures between 4 and 5 kbar for the gabbros and diorites, and between 3 and 4 kbar for the more acid rocks at intrusion temperatures between 720 and 800 °C.

The geologic map (Fig. 5.219) furthermore discloses an intrusion of numerous dike swarms after the crystallisation of the magma. It penetrated the three dimensional rhombic gap structures, which had formed after the cooling of the magma. This dike swarm intrusion continued into the wallrock as is typically observable on all the other comparable magma intrusion bodies in the Egyptian bedrock.

Along the courses of the mafic dikes are furthermore found pegmatite with hornblende peripheries as well as micrographic intergrowths inside the cores, which furnishes an additional argument for highly volatile processes.

The quartz vein mineralisations too, follow a three-dimensional rhombic pattern, which in contrast differs to the mafic dikes by a ENE-WSW and NNW-SSE strike and a generally vertical dip.

Widespread trench mines from the Early Arab Period follow both strike directions plus a third one, running as a flat lode. Because this vein spreads-out much like a sill sheet at depths between 1.5 and 3 m below the plain surface, it was accessed and exploited laterally through numerous, shallow shafts over short distances. Seen from above, the shafts distribute regularly over the plain surface, thereby forming a honeycomb pattern found nowhere else in the Egyptian Eastern Desert.

During the New Kingdom, no mining seems to have taken place in such shallow gold quartz veins. Settlement ruins from this period can however be attributed to a site at the NE edge of Wadi Abu Dagh and in a large plain into which leads the area with the auriferous veins. The New Kingdom activities concentrated presumably mainly on wadiworkings.

5.5.7 Hutit I and II (also Huzama or Rahaba) (Fig. 5.220)

Geographic positions:	
Hutit I, mine entrance:	23°27′26″ N, 35°11′33″ E
Hutit II, shaft:	23°27′16″ N, 35°11′41″ E

The deposit is located at the valley ends of Wadi Huzama and Hutit, a tributary valley to Wadi

Fig. 5.220 The Hutit deposit with recent buildings and sparse remains from the ancient settlement. Clear traces from an exploited mine are visible (modified Google-Earth image)

Rahama which lies between the valley ends. When the modern mine was put into operation in 1906, a vertical shaft was lowered down to the N side of the vein system. Through utilising ancient building stones and mills for constructing new buildings, the ancient settlement site was virtually completely destroyed. As revealed by their shapes, the round mills originate from boulders gathered up from a slope behind a tailing (Fig. 5.221).

The geology of the deposit is determined by an imbricated zone from basal, volcano-sedimentary sequences and serpentinites with intermediary gabbro layers (Fig. 5.222). The volcano-sedimentary sequences consist of rhyolite tuffs and andesites next to conglomerates (agglomerates ?) and schistic metasediments. The serpentinites are frequently sheared and partially carbonated.

The mineralised quartz exploited first in the Early Arab Period seems to be structured in small, closely grouped Riedel veins (Fig. 5.223). They appear in imbricated border zones between serpentinites and steeply folded metavolcanics. These zones probably also contain silted metasediments, though which, due to the intensive shear foliation, are not clearly identified. The shear zone strikes at an average angle of 120° E. The quartz veins strike approximately in the same

Fig. 5.221 Lower part of an Early Arab Period round mill made from a natural boulder stone, from Hutit

direction at 115° and 120° E while dipping at 80° SW. They thereby generally follow the shear direction of this area and are often interrupted in strike direction or otherwise develop into the small, oblique Riedel veins mentioned above. The reason for this is that the quartz veins follow in isolated, short swarms parallel running and lenticular shear zones. This significantly affected the way in which the mining was managed and had probably also been the reason why it was abandoned, in spite of partly outstanding ore qualities attaining 31 g/t Au (Koshin and Bassyuni 1968).

The exploited vein quartz is blue-grey, at times also white. Inside its sporadically occurring open cavities there is coarse, white calcite. In the paragenesis of the vein quartz one observes ankerite-siderite carbonates. Visible gold was not identified. As it seems, it had been evenly dispersed all over the quartz, which resulted to the removal of the entire quartz vein.

In some parts of the modern mine one may still faintly detect marks dating from the old mine, which reveal that it had once reached depths of at least 30 m.

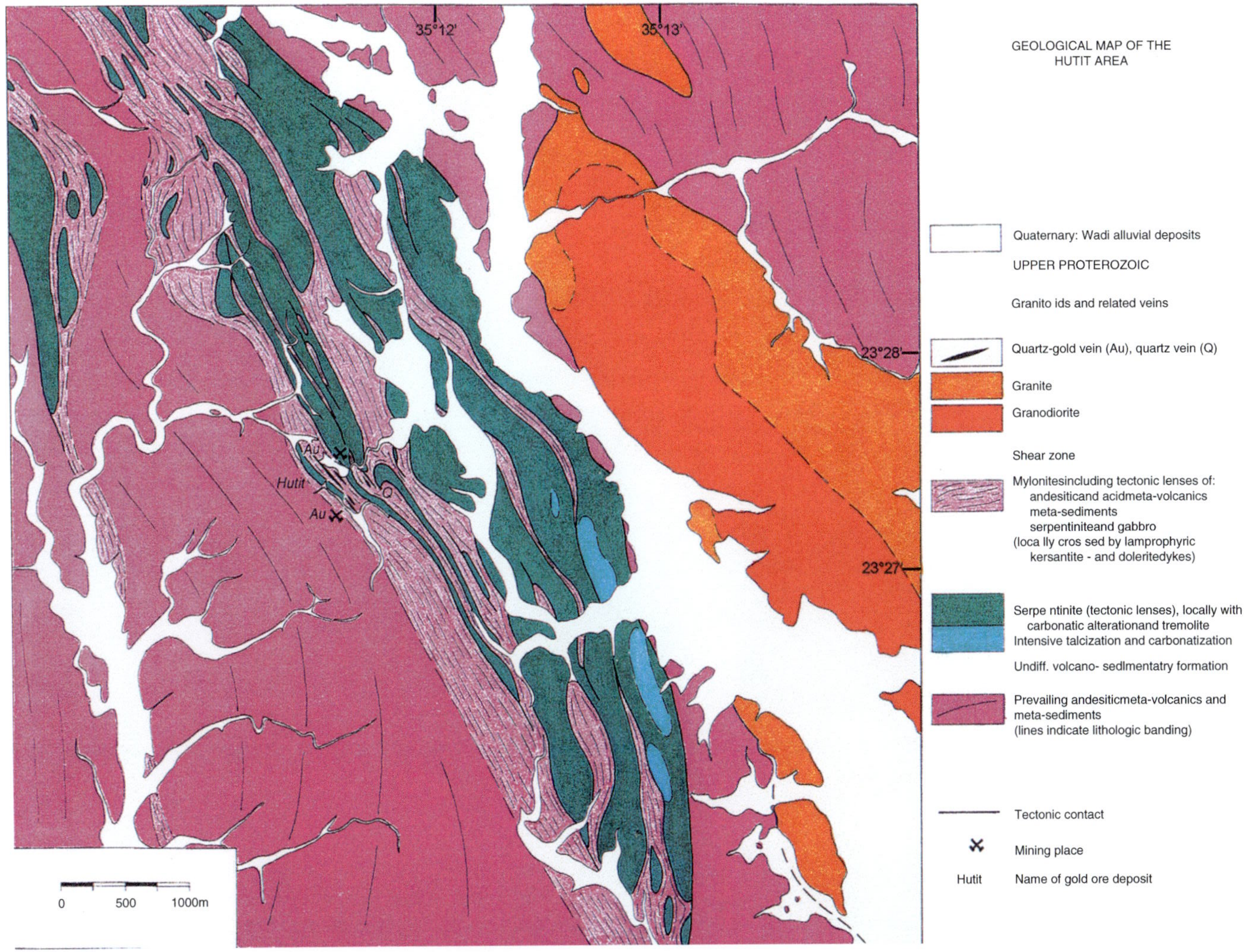

Fig. 5.222 Geological sketch map of Hutit (H. Kräutner)

Fig. 5.223 Ancient intersecting extraction pits at Hutit. Intersecting quartz veins are quite scarse in the Egyptian Eastern Desert. They here form so-called normal Riedel faults

5.5.8 Urga Ryan

Geographic position:	
Mine:	23°21′23″ N, 35°05′0″ E

About 200 m N of Bir Gahlia, in the wadi of the same name there are three noteworthy, single buildings. A closer look reveals their fortified character through massive outside walls, which virtually devoid of openings, were reinforced by corner bastions and wall-protected entranceways.

The walls of the buildings, whose external shapes diverge slightly, are about 60 cm thick and still preserved in heights over 2 m. At their bases they measure between 5 × 5 and 6 × 6 m. Their interiors reveal no further subdivision. The central and southern buildings stand in advanced positions on a mountain promontory. Nearby one can make out an alignment of small, round, and rectangular huts containing round mills and anvils, especially in the northern part, at the foot of a mine corridor.

At the surface in front of the buildings a campsite recently cleared from rubble witnesses modern prospecting work in the area.

The actual ore deposit is a quartz vein which appears at the surface 300 m to the N of the

Fig. 5.224 The mine at Urga Ryan had at first been operated in an open trench pit (largely filled in) and later in an underground mine with its entrance in the foreground

buildings. Few hut remains on the slope indicate mining operations during the Early Arab Period (Fig. 5.224).

Discernible serpentinites at the eastern wadi side are tectonically affected by intensive shearing, which essentially orients parallel to the wadi course. They are imbricated with acid volcanites and heavily sheared gneisses. Within this geologic setting the quartz vein strikes N-S (355°) concordantly with the shear zone system and dips at 80° E. It was exploited in a deep mine down to a level of 15 m below the surface.

Fig. 5.225 A typically filled double-shell wall from an Early Arab Period gold miner's hut at Anbat

5.5.9 Umm Kaliba (also Tundeba or Tunedba)

Geographic position:	23°17′01″ N, 35°09′21″ E

This very remote site is 3 km W of Gebel Tundeba, at the end of a difficult track through Wadi Tundeba.

The geologic surroundings consist of a sequence of metagabbros with transitions to metadiorites, which are crossed by a number of aplite dikes as the only evidence for granitoids. Had it not been for the irrefutable evidence for Early Arab mining, this area would hardly have appeared as a location propitious for searching gold. Virtually no quartz veins are found here, although numerous, large fragments of a milky-white quartz variety scatter in the plain near the huts. As it seems however, these hadn't attracted the attention of the ancient gold miners. Nor had it been possible to extract gold in wadiworkings, and traces pointing to the processing of auriferous quartz chunks from slope rubble weren't found either. The site therefore remains an enigma as to what made it grow to this extent and what qualified it as a gold processing site.

During the latest prospecting work in the mid twentieth century several steep shafts had been lowered into the metagabbro rock to a maximum depth of 15 m, though without striking any productive quartz vein.

5.5.10 Anbat

Geographic position:	23°05′17″ N, 35°20′30″ E

The Anbat occurrence is located in Wadi Hodein, shortly before its entrance to the coastal plain of the Red Sea.

To the E of Gebel Anbat an extensive settlement spreads-out over at least one square kilometre. It consists mainly of single and multi-chambered large huts built in the shell-face technique (Fig. 5.225). They group in numerous, small clusters generally along slopes of low hills.

Fig. 5.226 Spoil heaps from Early Arab Period wadiworkings at Anbat

Gold processing had essentially been based on ores from the wadi sediments, as revealed by small debris heaps in the wadi bed (Fig. 5.226), numerous small, pink tailings, and some loosely scattered washing tables. One of them lies exposed on a protruding rock, surrounded by several round mills and small anvils. The tailing sands have for evident reasons been entirely eroded away. A small, rather well-preserved tailing site was recorded in the S of the district. Its sands are not pink, but almost white and apparently originate from a small, recently exploited mine. A central tailing site seems to be missing at the site.

A large cemetery consisting of at least 60–80 recognisable graves suggests that the local gold production had lasted for a considerable period.

On the whole, the site displays affinities to Wadi Allaqi and other Early Arab sites further to the S, in the Nubian Desert of Sudan. There, places like Tanasheb, Shashuateb, and Techol show among others certain similarities to Anbat. In the same way as at Anbat, the people at these contemporary settlements practiced large-scale gold processing of wadi deposits and ores from even extensive surface trenches on the slopes. Anbat is actually the most northern representative, distinguished by a gradual change of the landscape morphology, by which the valleys become wider and the relief more levelled-out through stronger erosion (Fig. 5.227).

The altered circumstances are indeed reflected by the archaeological findings in terms of mining. In this group the emphasis of gold production is noticeably focused on wadi sediment processing as opposed to underground mining. This led to the development of larger, more extensively exploited districts.

The abundant pottery at Anbat also marks an ethno-cultural transition. New types that are more common further to the S are found here. One of them is painted in the so-called Christian-Nubian style (Adams 1986), another consists of a simple, brown ware with impressed grooves and geometric designs.

Fig. 5.227 The low hills in the Anbat area

Fig. 5.228 Neck and shoulder of an Early Arab Period water pitcher from Anbat

The amphorae are well-burnt with a shimmering, yellowish-red to red-brown, iridescent and grooved slip. They have short necks with three lateral ribs (Fig. 5.228). Inside, the jugs are noticeably light coloured, and the ware itself is almost white.

In the middle sector of the mountains, along the E flank of Wadi Hodein leading to the Red Sea, there is furthermore a row of approximately 20 huts from slate rocks. Some traces from superficial mining can be seen, but for the want finds, there are no safe indications as to the site's age.

A geologic sketch map of the region covering the gold producing district around Anbat was published by Gabra in 1986. It is based on the data from the prospection work by EGSMA and the Soviet Technoexport group.

The units on the Gabra map (1986) referred to as metavolcanics are in fact metasediments, that primarily consist of former arkoses, greywackes and conglomerates much reminiscent to the Hammamat series. Indeed, these southern metasediments contain substantially more arkoses and increasing amounts of marble intercalations, which near Bokari, at Gebel Rokham, just N of the 25th latitude, are already considered adequately suitable of extraction.

The in many cases secondarily carbonated arkoses are tagged in the map by Gabra (1986) as quartz carbonate rocks, which thereby cover large areas around Gebel Anbat. The "alkaline tuffs" marked at its W flank are in reality green-grey metagreywackes. The "acid tuffs" are merely much sheared greywackes and siltstones. Finally, all rocks described as agglomerates are actually conglomerates with well-preserved polymictic boulders.

In the same way as in the Hammamat sequences at the type-site, in-between layers from intermediary and acid volcanites are also discernible here in the metasediments, although at a much lower scale.

The entire metasediment sequence is in a steep isoclinal fold and excessively sheared over long distances, which makes the rock-identification problematic in this area and probably is the principal reason for the petrographic divergencies mentioned above.

Gabra's (1986) diverging classifications exclusively concern the immediate vicinity of the gold producing sites at Anbat and not areas near Wadi Hodein, where basalt-dacite metavolcanics prevail as described by Obeid (2006).

Only one deep mine following a meek quartz vein over approximately 6 m was found, where a mining attempt had been made last century.

Otherwise there is no further evidence for deep mining in the area, apart from small quarry trenches orienting along minor criss-crossing veins and occasionally intersecting, pyrite-yielding arkoses. The Early Arab dwellers in the countless houses clustering over the entire area could therefore with the equipment they had have covered their ore demands from specified areas in the wadi alluvium. The erosion rubble from unnumbered, often only few centimetre-wide quartz veins in the metasediments undoubtedly furnished ample quantities of ore.

The numerous pyrite metaarkoses had certainly not irrelevant either. The fact that they had generally been exploited in trenches at occurring intersections with small quartz veins, suggests that they originally had been auriferous.

5.5.11 Korbiai

Geographic position:	22°47′19″ N, 35°10′02″ E

Due to the tense military situation at the border with Sudan, we were unfortunately unable to visit this site SW of Shalatein.

For the lack of published reports, we resorted to the Google-Earth satellite images to scrutinise the area. Korbiai is thereafter located in the southern, upper stretches of Wadi Kreiga, which bypasses Gebel Korbiai, and more precisely in a small side wadi at the SW slope of Gebel Korbiai. Recent mining has apparently wiped out most traces from ancient mining. About 30 small hut ruins are recognisable approximately 200 m W to this side wadi. Single huts can also be seen further downstream, at the mountain foot as well as on its slope. In most of the side wadis are traces from extensive wadiworkings. Although a dating of the site would seem premature at this stage, its layout, size, and slightly remote location nevertheless strongly suggest a date to the Early Arab Period.

5.6 Southwestern Group

5.6.1 Wadi el-Hudi (Fig. 5.229)

Geographic position:	23°58′54″ N, 33°08′34″ E

Ancient mining in the district of Gebel el-Hudi is especially known for its important amethyst gemstone mines from the Middle Kingdom where altogether five amethyst mines are attested to in the area (Fakhry 1952; Klemm et al. 2002a; Shaw 2002).

We also succeeded in identifying two ancient gold extraction sites in the area.

About 4.5 km SE of Gebel el-Hudi, along the eastern edges of the actual Wadi el-Hudi, an extensive mining settlement is located on the eastern slopes of a sandstone ridge, protruding like an island from the plain (site no. 13 with Fakhry). Whereas some dwellings are lowered into the mountain rock, displaying the occasional masoned porch (Fig. 5.230), others are of a simpler, free-standing kind. The settlement is quite large, and its capacity had doubtless been large enough to hold several hundred inhabitants.

Some of the houses in the more sheltered or elevated areas contain well-preserved and richly painted Nubian pottery dating to a period between the ninth and thirteenth centuries AD In the floodplain on the other hand, simple domestic pottery with coarsely engraved ornaments dominates.

In the settlement area, numerous, for the most intentionally destroyed round mills and pounding stones were found. However, very small, round mills with diameters varying between 20 and 25 cm and unusually small pounding stones suggest that women and children had worked here too (Fig. 5.231).

In the SE, immediately behind the mountain bend, there are more huts plus a large tailing site. In all, three well-preserved washing tables and at least seven tailings can be distinguished inside the settlement area (Fig. 5.232).

Within and beyond the old huts modern detritus (glass, pottery, cans, rusty sieves etc.) had accumulated. Some recent rectangular house ruins are located in the approach of the old settlement in the plain.

The settlement (Fig. 5.233) is associated to a large gold mine beyond the wadi, approximately 500 m NE of it (site 14 with Fakhry). It stands out by light coloured debris heaps on the slopes of a larger hill which is cut by a vertical vein cleft. This ancient, partly collapsed gold mine (Fig. 5.234) had been lowered down to a depth of about 10 m. The quartz ores had undoubtedly been processed in the neighbouring settlement, as revealed by the preserved remains of washing tables, mills, as well as a tailing.

According to the findings, the most significant and intensive mining activity had occurred during the Early Arab Period. A second extraction phase followed only at the beginning of the twentieth century when, by the initiative of British mining companies between 1902 and 1906, an attempt was made to re-open the old gold mines (Schweinfurth 1903). The modern detritus accumulations and well-preserved house ruins go back to this last phase. Sadek (1980) erroneously held this gold mine for an amethyst mine.

Sites 13 and 14 listed by Fakhry are actually closely connected to each other. While site 13 consists of a settlement and a gold mine, site 14 is described as a recent, vertical shaft which was lowered somewhat SW to the old mine at the beginning of twentieth century. We chose to refer to site 14 as the gold mine including the recent shaft.

Contrary to Fakhry, we were unable to detect distinct evidence from the Roman Period at the gold extraction sites. According to our field observations, the Romans had demonstrated little interest in Egypt's natural gold resources. Too many, mostly small gold occurrences scattered over immeasurable territories would have been too demanding on security, if for instance applied with the same standards as those known from the large stone quarries in the northeastern Desert. Though, wherever gold occurrences happened to be located along or near secured trade routes, Roman Period gold mining is nevertheless attested to. On the other hand, according to our knowledge, there is no evidence for the prospection of new occurrences in

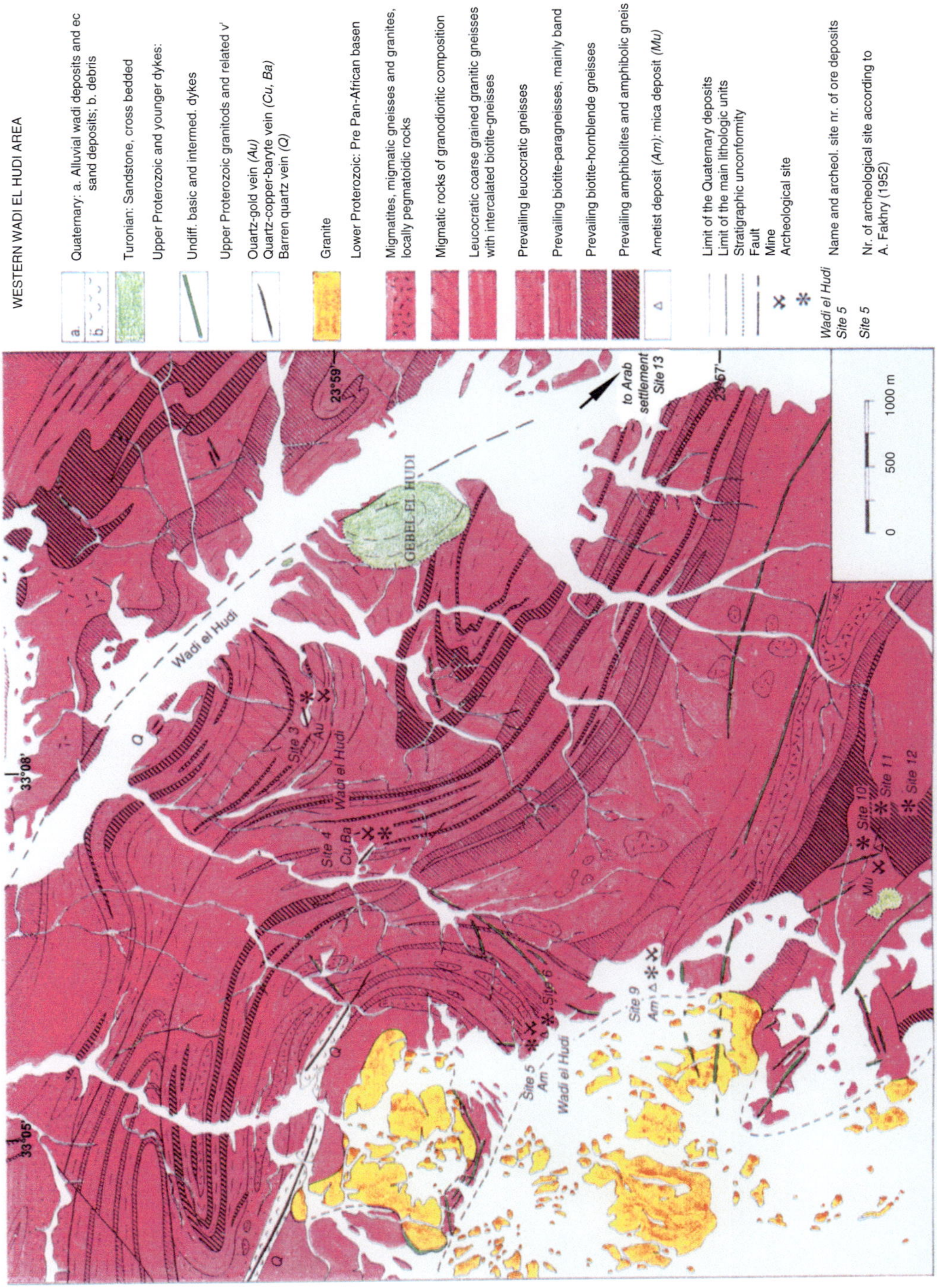

Fig. 5.229 Geological sketch map of the western part of Wadi el-Hudi (H. Kräutner)

Fig. 5.230 Early Arab Period cave dwelling in Wadi el-Hudi. The dry wall belong to a more recent occupation (site 13)

Fig. 5.231 Collection (not arranged by the authors) of Early Arab Period round mills in Wadi el-Hudi (site 13)

Fig. 5.232 Early Arab Period washing table (foreground) and tailing heap in Wadi el-Hudi (site 13)

the Roman-Byzantine Period. After the decline of Ancient Egyptian mining, prospecting activities didn't take up again before the Early Arab Period. In those days, the erstwhile lands of gold, now crossed by pilgrim routes leading from the Southeastern Desert down to Nubia were being rediscovered.

The Aswan region through whose SE district runs Wadi el-Hudi, has ever since the Early Dynastic Period been renowned for its granite, granodiorite, and tonalite, which throughout history have been valued for their qualities in monumental building projects. At first, this had only concerned Egypt, but already by the Ptolemaic Period and especially the Roman-Byzantine era, building stones from Egypt were being traded all over the Mediterranean world. Coarse-grained porphyry granite of a red to reddish-pink colour and commonly referred to as "rose-granite" was by far the most popular stone. Studies on the Aswan stone quarries, including an analysis of the ancient extraction methods, the functional utility of its quarried stones have been presented to the public in detailed reports by the present authors (Klemm and Klemm 1993, 2008b).

The large granite complex near Aswan represents a roof, or at least a slice, of an intrusive granite-granodiorite pluton. In the Wadi el-Hudi area the base is exposed with its root zone. "Root zone" here denotes the area from which granite magma had emerged from its original gneissic metamorphic rocks by melting.

The actual age of the gneisses has not been fully established as it is concealed by extreme metamorphism provoked by very high temperatures. They spread from Aswan towards the SE beyond Wadi el-Hudi into the Precambrian basement of the Egyptian Eastern Desert. To the W and S they are covered by the massive sediments of the Nubian sandstone.

In Wadi el-Hudi the gneisses consist for the most of clearly parallel textured, grey to pinkish-grey rocks chiefly from quartz, feldspar, and varying portions of mica (mostly biotite) and hornblende. This colourful configuration of different rocks reflects the chemical composition of

Fig. 5.233 The Early Arab Period settlement site 13 and gold mine (site 14) in Wadi el-Hudi. The dark hills are inselbergs from Nubian sandstone. The mine surroundings consist of local gneisses (modified Google-Earth image)

Fig. 5.234 Gold mine in Wadi el-Hudi (site 14). The deterioration resistance of its central quartz vein has prevented this inselberg to erode in the same way as the surrounding terrain

the volcano-sedimentary origin. Because the original rocks had initially developed from layered compounds with which they often are intensely folded, the field sections of the metamorphic counterparts too, appear also to be folded (Fig. 5.229).

This describes in a very simplified manner the orogeny of the deeper crustal parts of the Egyptian Precabrian basement that are overlapped by island arch type metavolcanics and metasediments, the Hammamat series that are intruded by the granitoid magmas and the final lava covers of the Dokhan series.

In contrast to the mineralised quartz veins, quartz crystals also formed within the Wadi el-Hudi area in cavities of the gneisses. However, they occured as a result of a slightly radioactive impact in a violet variety, called amethyst. It was especially in the Middle Kingdom that mining for amethyst had reached its apex in large open cast quarries. Extraction had become so effective that today, only meagre residues of a pale-coloured variety is found in the ancient quarry trenches.

Probably attracted by what they thought had been former gold exploitations, Early Arab miners who had been unaware of the amethyst mines, began in their turn to look for gold near these pits and along former prospecting trenches, and with success. Two separate areas were eventually mined. The first was to the W of Gebel el-Hudi, at "site 3" according to Fakhry (1952), and the second at "site 4", which though because of its copper-barium mineralisations, hardly matched the gold yields that can be expected in distinctly deeper zones of the quartz vein.

5.6.2 Wadi el-Hudi, Site 3

Geographic position:	23°57′27″ N, 33°11′18″ E

Site 3 (Fakhry 1952) is located about 1 km NW of Gebel el-Hudi. It composes of a small mine in a hilly plain and about ten surrounding, mostly severely damaged huts. Several round mills and pounding stones scatter between the ruins. Within the mine's immediate vicinity is a well-preserved tailing.

Fakhry dates the site as well as several other ones in Wadi el-Hudi to the Roman Period. The surface finds we made, point however unmistakeably to the Early Arab Period, which in addition is supported by the remains of a prayer site (Fig. 5.235).

Fig. 5.235 Muslim praying space at site 3 in Wadi el-Hudi

5.7 Wadi Allaqi Region

Wadi Allaqi is on one hand known for its gold occurrences, on the other for its famous rock inscriptions (Černy 1947; Piotrovskiy 1964). In archaeological terms though, the region has remained much neglected since the so-called "Nubia Campaign" launched by the UNESCO in the 1960s, mainly as a result of its difficult access since the filling of Lake Nasser.

The wadi runs through the Lower Nubian regions of the Eastern Desert in a wide, sandy, and dry basin from SE to NW. Between the 22nd and 23rd N-parallels it opens into the wide floodplain of the Nile. Today the wadi is flooded by the waters of Lake Nasser over a length of about 60 km.

Since recent years Wadi Allaqi is accessed via an asphalt road from Aswan to the stone quarries located in the S, in the mouth area of its tributary Wadi Heimur.

Our investigation in Wadi Allaqi covered an area between the banks near Umm Ashira, a prominent rock inscription site, and the gold mine at Abu Fas near the Sudanese border, in the SE of the wadi. The gold deposits, however, continue to spread deep into Sudanese territory (Whiteman 1971).

An ancient Egyptian fortress referred to as b3k(j) was erected near the Nubian site of Quban, in the former wadi's mouth area. A corresponding fortress was established on the W side, near Ikkur (Säve-Söderbergh 1941). Both had been built during the Middle Kingdom, although they are known to have remained in use until the New Kingdom (Emery and Kirwan 1935).

In contrast to the fortresses further S near the Nile, where copper smelting furnaces from the fourth and fifth dynasty are known from the Old Kingdom site at Buhen (Emery 1963), the forts at Quban and Ikkur seem to be linked to the gold deposits in the upper parts of Wadi Allaqi. A weight manufactured from limestone with the inscribed cartouche of Sesostris I might indicate that he had in fact been the builder of the Quban fort. Thereby, the Egyptians were searching for gold in Wadi Allaqi anyway no later than the reign of Sesostris I (Emery 1931).

However, the great majority (more than 200) of the rock inscriptions in Wadi Allaqi published by Černy (1947) and Piotrovskiy (1964, 1967), date to the New Kingdom, of which some official titles link them specifically to gold production.

Thus, the well-known stela discovered near Quban from the second reign-year of Ramesses II (Kitchen 1968; Zibelius-Chen 1994) relates to a search campaign and eventual discovery of water for gold labourers. This well had been dedicated by "Ramesses, Meriamun [...]" or "Ramesses, beloved of Amun [...]". The inscriptions in fact suggest that under the New Kingdom gold extraction in the Wadi Allaqi region had been a major concern by the state authorities, and that this policy had been systematically pursued by the beginning of the early eighteenth dynasty.

The well quoted in the Quban stela was finally found in 1962/63 under the Nubian Campaign, about 60 km into Wadi Allaqi, near the Umm Ashira rock with its more than 135 inscriptions. In the meantime, the name of the well has been collated to "The well Ramesses Meriamun [strong in life]".

In the late Ramesside Period gold mining had almost come to a standstill in both the Egyptian heartland as well as the Nubian periphery, probably because of internal economic difficulties. No texts pertaining to expeditions to the gold mines are known to postdate this period. The crisis is well exemplified by the dwindling tributes in the delivery lists in the temples. While under the reign of Amenhotep III, 2.292 kg of gold had been delivered for the construction of the Month-temple in Thebes alone (Helck, Urk. IV, no. 568), only 232.7 kg had been supplied under

Ramesses III to various temples throughout the country (Schädel 1936).

In districts of the Egyptian heartland large-scale gold mining had only resumed by the Ptolemaic Period. In Nubia by contrast, there are no reports of Ptolemaic mining activities to the S of the Dodekachoinos (a 12-mile strip between Egypt and Lower Nubia in Ptolemaic-Roman times, Sethe 1964; Säve-Söderbergh 1941). The work conditions in the gold mines under the Ptolemaic Period as described by Diodorus who quotes Agatharchides (Woelk 1966), are generally ascribed to Wadi Allaqi (Hölbl 2004), though without additional confirmation from the field. Agatharchides's description just as much fits to gold mines now known to have been under exploitation in the southern parts of the Egyptian Eastern Desert in the same period. The most southern Ptolemaic gold processing site known to us today is in Wadi Ghadir (24°47′ N, 34°50′ E), which also happens to be located on the route to Berenice. The well-known emerald mines at Sikeit and its surroundings are even further to the S, between 24°38′30 ″N and 34°49′38″ E, and N of Wadi Gamal. However, whether in Wadi Gamal in spite of several settlements ever gold production took place, remains uncertain.

As certified by the Quban stela, access to the chief gold deposits in the upper part of Wadi Allaqi from the Nile Valley had already existed in the New Kingdom. In particular the location of the Quban fort near the wadi's mouth, emphasises the compound's role as an organising and administrative centre and a base station for expeditions departing to the mines and gold deposits further upstream.

From the evidence in Wadi Allaqi and its tributary valleys, particularly in Nubia, it soon becomes clear where much of the famous Nubian gold had actually come from. The few deep mines found in this region for this matter would on no account have been sufficient. It were rather the wide and open wadies with capacities large enough to occupy simultaneously significant workforces in wadiworkings, where eroded quartz lumps were systematically gathered and thereby producing manifold the amounts a handful of miners were able to deliver from only few narrow deep mines.

According to the information provided by EGSMA even today, 1 g of gold is contained in 1 t of this region's sands.

5.7.1 Hairiri

Geographic position:	
New Kingdom settlement:	22°57′20″ N, 33°27′22″ E
Arab settlement, tailing:	22°57′42″ N, 33°27′27″ E
Main quartz vein:	22°57′42″ N, 33°27′08″ E

Hairiri is a New Kingdom gold producing site that had been revived in the Early Arab Period (Fig. 5.236). Much of this extensive site is totally sanded-in. The landscape evolves in a morphologically soft gneiss-granite over several hills and depressions (Fig. 5.237). Because of the sand drifts, the settlement structure is just barely made out. In protected situations, however, there are also huts with up to 1.80 m high walls. Countless New Kingdom mills and runner stones scatter over the entire area. Many are also found masoned into the walls of Arab huts near the edges of the New Kingdom settlements.

At Hairiri we were able to observe for the first time that New Kingdom mills had been re-used most probably in Kushitic time by deepening the original more flat cavity. Many had been worn down to a degree by which their central depressions had pierced. An additional peculiarity of the mills at this site is the unusually small size of the oval depression, which together with the matching grinding stones vary around 12 cm in

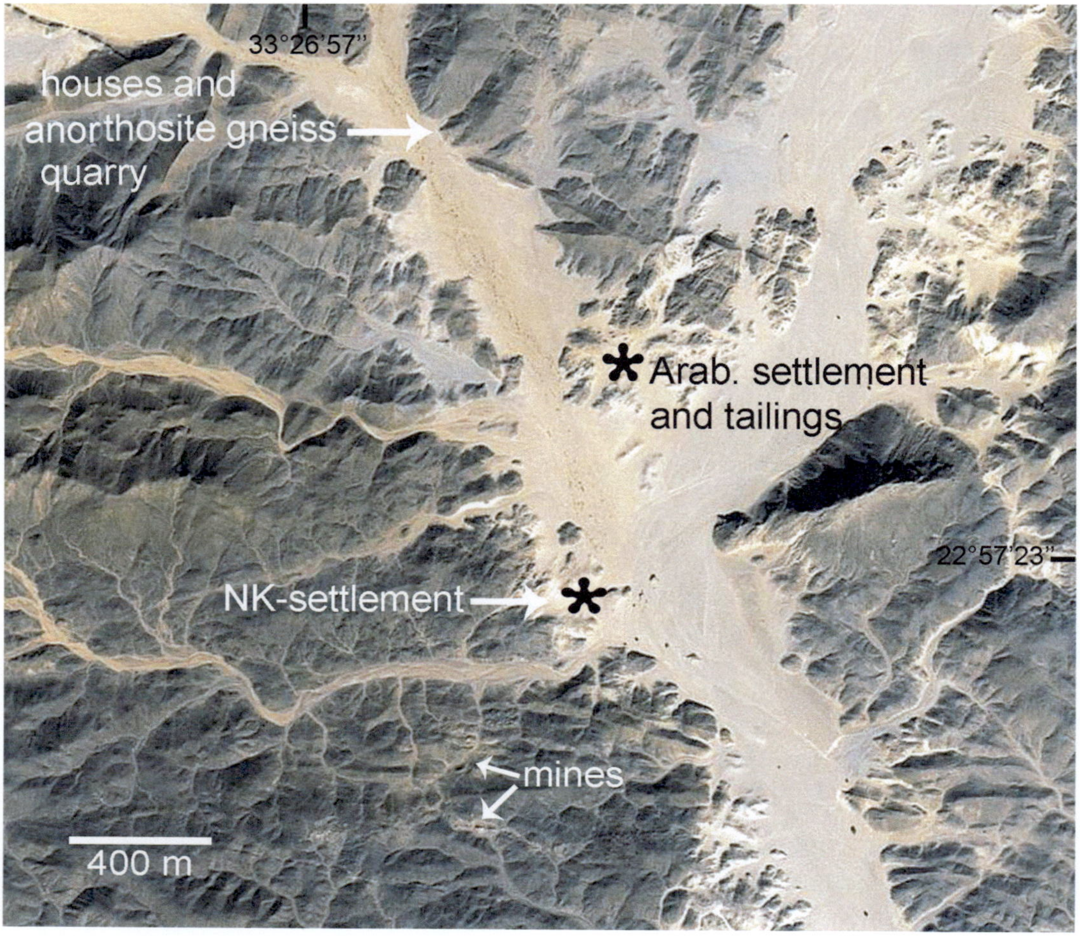

Fig. 5.236 Mines and archaeological sites at Hairiri (modified Google-Earth image)

diameter (Fig. 5.238). The small mills are often made from a purple-brownish quartzite that becomes highly polished from the grinding process. The pounding stones are occasionally cubic with one or several hollows on each side, making them resemble to oversized dices (Fig. 5.239).

Tailings were not recorded, but making up for that, at least five to six washing tables scattered over the entire Early Arab settlement, though under a thick layer of silt. Between the single huts one notices there small prayer sites whose mihrabs are often delimited by vertically standing millstones.

The large size of the settlement indicates that Hairiri had been intensively engaged in wadi-workings, although the associated spoil heaps are lacking, as they probably have been covered by the wadi's thick sand drifts.

There also seems to have been some deep mining in the area. In fact, mines are found approximately 650 m to the S of the main settlement.

Now and then along the house walls, sometimes in isolated locations and sometimes in small waste dumps one notices large, red jasper and translucent carnelian flakes representing bi-products from production processes.

In a side branch of a wadi leading to the N that borders to the large settlement, there are five to six, well-preserved and perfectly circular houses with diameters around 4 m, and wall heights varying between 1.60 and 1.80 m. Round mills were found

Fig. 5.237 New Kingdom settlement at Hairiri covered below a thick layer of sand

Fig. 5.238 One hand rubber stones at Hairiri that were used with New Kingdom mills, sorted according to size. The smallest easily fit into the hand of a woman and even a child

Fig. 5.239 Cubic New Kingdom pounding stone with six impact hollows

Fig. 5.240 New Kingdom mill from local anorthosite hornblende gneiss ("Chefren-gneiss") in Wadi Hairiri

in three of them, and in one there is a still unused, approximately square slab with a central hole.

At the position Anorthisite gneiss quarry (Fig. 5.236) occurs another small New Kingdom settlement of six to eight houses built from anorthosite hornblende gneiss collected from the nearby surface. The tools as well had been manufactured in this rock (Fig. 5.240).

A well-preserved Arab settlement with several tailings is located on the E side downstream Wadi

Fig. 5.241 Early Arab Period house complex with strong double-faced walls

Hairiri. Here as well, the architecture consists mainly of anorthosite gneiss. Two cemeteries and a larger prayer site supplement the settlement in the wadi centre (Fig. 5.241).

Fragments from ostrich eggs were found at all mentioned sites, and even oyster shells at the main New Kingdom settlement.

The geology of this chiefly New Kingdom mine district is determined by a metamorphic sequence of metagabbros and anorthosite gneisses with transitions to hornblende-, garnet-, quartz-, and plagioclase-gneisses. There are even larger amounts of transitions chiefly black-green, medium- to fine-grained amphibolites clearly folded. To the N, the amphibolites go over to quartzdiorite and granodiorite gneisses, which however, contain recurrent, intermediary layers of amphibolites and light-coloured, quartz-yielding hornblende gneisses that resemble anorthosite gneisses. The different varieties of anorthosite gneisses and their hornblende-rich streaks are confusingly similar to the so-called "Chephren gneisses" from Gebel el-Asr, to the W of Abu Simbel, but they also occur in almost identical geologic surroundings.

It therefore is open to debate whether these gneiss sequences after all are part of the genuine African basement, which, if so, must be regarded as a geologic window. This however would contradict Abdel Nabi and Fritsch (2002) who regard the series merely as higher metamorphic parts of the Arabian-Nubian Shield (ANS).

The morphology of the area is also marked by rhyolitic and dacitic dikes, which strike like butte ridges in different directions, although with a leaning to an approximate E-W direction (Fig. 5.242).

The ancient vein mine also follows the general strike of its amphibole wallrock. The supposed main vein swarm may actually consist of only one vein in a multiple, tectonic displacement. The vein reappears on the opposite eastern side of the wadi, though apart from visible tracks from prospecting, displays no traces from genuine mining. At least one of these prospecting tracks stems from the New Kingdom

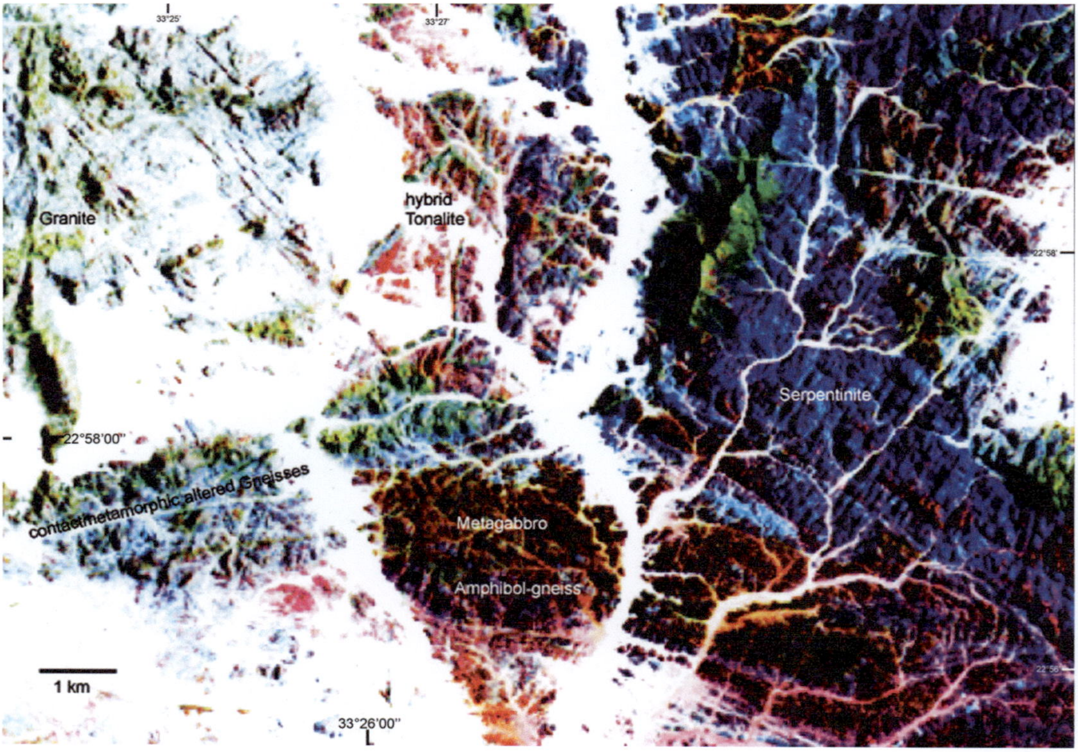

Fig. 5.242 Lithologically processed satellite image of the Hairiri surroundings (TM 174/44, channels 7-4-1)

as is recognised by the grinding mills and pounding stones dating to this period. The extensive New Kingdom settlements in Wadi Hairiri were not able to satisfy their demand for gold ores from such mines only. Traces from wadiworkings in the neighbouring, tributary valleys and hill flanks demonstrate that here too, this method had been applied on a large scale. Extensive floods and sand drifts affecting the entire wadi have evidently distorted and even wiped out much of the evidence.

In both areas the geology is characterised by metasediments (through contact metamorphism) on one hand, particularly former greywackes, and a granodiorite periphery around a granite intrusion on the other, displaying numerous metasediment-xenoliths.

Mined quartz veins are found nowhere in the flanking mountains, which again suggests that the wadiworkings in this generally auriferous area have been shrouded by occasional floods.

5.7.2 Ashira-East

Geographic position:	22°55′20″ N, 33°17′19″ E

Dispersed remains from New Kingdom hut clusters can bee found in Wadi Ashira as well as about 3 km to the E of the road junction between Wadi Ashira and Wadi Heimur.

5.7.3 Umm Ashira

Geographic position:	22°54′04″ N, 33°14′47″ E

This site is located about 5 km S of the bifurcation into Wadi Heimur, near the new asphalt road to the recent foundation of Allaqi City, which is located above the barrage level.

Fig. 5.243 The flooded part of Wadi Allaqi. The inscription rock of Umm Ashira juts out from the water in the background to the left

Relatively large New Kingdom settlements are located along the wadi at this position. On its eastern side, the New Kingdom settlement is largely supplanted by Arab occupation layers. The same in fact, also applies to the southern part of the western settlement. Numerous mills suggest former processing activities of the wadi alluvium.

The surrounding's geology consists basically of contact metamorphism metasediments and metaagglomerates reminding very remotely of the Hammamat formation, although nothing seems to indicate that they represent their southern counterparts.

Several quartz veins running along the surface had been exploited in opencast trenches. In all probability, the settlement had covered its need for ores by supplementing its gold production with ores from the wadi sediments.

The known rock inscriptions at Umm Ashira (22°49′30″ N, 33°10′59″ E) are only a short walk away from Allaqi City (Bedouin settlement of the Bisharin). The top third of the rock still rises from the water and can even be accessed by an embankment during periods of low levels (Fig. 5.243).

5.7.4 Umm Ashira-South

Geographic position:	22°51′55″ N, 33°13′04″ E

This site consists essentially of Early Arab processing workshops containing round mills, washing tables, and small tailings. Much of the site has been damaged by wadi erosion.

At Umm Ashira-South, we were nevertheless able to make a significant observation on a frequently occurring feature, consisting of small stone, box-like cist next to Early Arab, domestic architecture. At Umm Ashira, some are located immediately next to a washing table (Fig. 5.244), which seems to suggest that they had originally functioned as small water reservoirs associated to private gold processing. As a rule, they occur next to single huts and are associated with pans or bowls that may have served for separating heavy minerals from crushed ores. Such washbasins, which usually are made from flat stone slabs or in some cases, former New Kingdom mills, had apparently been covered by a plaster on the inside to make them impermeable. At a large number of sites where they occur, they all appear in Early Arab find contexts. They reveal a basic difference to the gold production in the New Kingdom in that, next to centrally organised processing documented by washing tables and tailings, gold processing was also carried out on a more individual level. Evidence for such decentralised gold production is virtually absent for the New Kingdom Period.

Fig. 5.244 Early Arab Period washing table with flanking stone cist, originally plastered and used as a water container

5.7.5 Neguib

Geographic position:	
Bir Neguib:	22°49′29″ N, 33°43′32″ E
Settlement:	22°50′17″ N, 33°44′00″ E

This gold producing site displays a heterogeneous settlement structure. Settlement remains from the New Kingdom with numerous New Kingdom mills alternate with well-preserved Early Arab houses. The latter are either round or rectangular and sometimes subdivided into several rooms. As a rule, they had been built in the characteristic shell-facing technique.

Only two, hardly used round mills were recorded close to a washing table. In the surrounding area the wadi and the neighbouring heights are otherwise marked by numerous tombs. The ones on the elevations seem to differentiate by their walled, round shapes, which seems to point to a lengthy occupation period at the settlement.

The geologic surroundings consist mainly of granodiorites with transitions to tonalites superimposed in the N by metasediments. In the S of the occurrence too, the magmatites are covered by metasediments, which in this area chiefly alternate with metavolcanics and well developed agglomerates. Here and there, the latter display volcanic bombs in pumice-like lapilli inclusions.

No mined quartz veins were found in the surroundings of the settlement. Gold production had therefore yet again focused probably on selective gathering of gold ore in slope rubble and the wadi alluvium.

5.7.6 Heimur

Geographic position:	22°38′04″ N, 33°18′14″ E

One comfortably reaches Heimur on the new asphalt road from Aswan, just shortly before the Sudanese border.

Wadi Heimur comes from the NE as it flows into Wadi Allaqi, which in its own turn, flows SE-NW. Wadi Allaqi is up to 600 m wide, very sandy, and bordered by flat, hilly flanks.

Fig. 5.245 New Kingdom tailing heap and remains of an inclined washing table within the Heimur mining area

The deposit itself is located in a flat mountain elevation and consists of several adits into the underground mine. However, auriferous quartz lumps had seemingly as well been gathered in wadiworkings in the nearby valley.

The oldest visible mining traces date to the New Kingdom. Close to a tailing site (Fig. 5.245) are several, partly very ill-preserved, multi-chambered New Kingdom houses. The tailing site, which presumably had still been used in later periods, is oriented in a U-shape towards the SW and still accommodates the sparse remains of a washing table.

Many New Kingdom mills and grinding stones are still located in situ in the houses where they had once been used (Fig. 5.246). The houses show the usual classical layout as known from the northern parts of the Eastern Desert. They are approximately 6–8 m long, 4 m wide, and comprise at least three rooms, surrounded by plain walls that sometimes are preserved in heights of about 1 m.

Just below the surface and particularly inside erosion gullies, the entire area is covered by a pinkish-white layer of mylonite powder which is easily mistaken for tailing sand.

In the N, hidden behind the first hill chain next to the valley is widespread Early Arab settlement (Fig. 5.247). Its houses had been erected in the shell-facing technique along the slopes of a valley basin, where one notices a tailing heap and a very well-preserved washing table.

The settlement counts at least 30, relatively well-preserved, round huts built of conspicuously small rocks with the usual gravel fill between two wall-shells.

Associated small mines scatter in the surrounding mountains in a honeycomb pattern.

Small, walled terraces filled with fine wadi rubble had been added to the wealthier houses and had probably served as sleeping quarters.

Sometimes the houses had been equipped with adjoined, small-sized praying platforms. Moreover, in the centre of the settlement there is a mosque with a mihrab pointing to Mecca.

Some of the better preserved houses show signs of short-term re-occupation in the last

Fig. 5.246 Curious New Kingdom mill from Heimur with five separate grinding depressions

Fig. 5.247 Well-preserved Early Arab Period house alignment at the foot of Gebel Heimur

Fig. 5.248 New Kingdom pottery from a small settlement S of Gebel Heimur

Fig. 5.249 New Kingdom bowl from the settlement at the foot of Gebel Heimur

century. In few instances New Kingdom mills had been integrated into the house walls.

A large New Kingdom settlement is located opposite the modern mining area in the open plain. However, it is barely discernible by few mills and flimsy house ruins.

The plain to the W of the Heimur mountain ridge is strewn with small heaps from former wadiworkings. Recent ones too, may be recognised from the nearby rusty sieves from sheet metal displaying holes in different calibres.

No occupation pre-dating the New Kingdom was found at Heimur. The next settlement phase occurs in Early Arab Period. After that, the district then apparently lay derelict until the beginning of the twentieth century.

To the S of the Heimur deposit and on the N side of Wadi Heimur there are two more settlements. One is close to the asphalt road just S of Gebel Heimur (22°37′09″N, 33°18′37″ E), the other is E of it. Both sites are dated by the pottery (Holthoer 1977) to the New Kingdom (Figs. 5.248 and 5.249). The first settlement has been affected by bulldozing during the construction of the road. The second is associated with an extensive area of flat wadiworking heaps that stretch far up the wide wadi.

The geology (Fig. 5.250) around the Heimur deposit is formed by a relatively varied sequence of metasediments, in which at best appear some intermediary acid tuff metavolcanics. In terms of lithology, these series partially remind of the Hammamat formation with its former greywackes, siltstones, and conglomerates, but also of the Igla formation with the typically brown-red siltstones and acid volcanics. In addition, there are substantial metaarkoses, bulky and very ferrous quartzites, graphite-schists (metamudstones) and marbles.

The thickness of this series varies much on a local level within the entire area. It displays an intense, steep, and isoclinal folding, which is inadequately represented in the geologic sketch map owing to the intensity of the alternating and wrinkling folds. The area is furthermore crossed by two large and countless small faults that more or less follow the NE-SW strike of the hinge lines. Up to 500 m wide shear zones as well, generally orient in the same direction. Within the shears, rocks have deformed over several hundred metres to finely powdered and denudated mylonites in lens-like structures. This may even make such areas unsafe for walking and especially for driving as one risks sinking in.

Two, milky-white quartz veins located close to each other and following the shear foliation had been mined in antiquity on the western slope of Gebel Heimur. Within the vein-quartz occur calcite and ankeritic dolomite and now and then muscovite. The ore parageneses consist of haematite, pyrite, chalcopyrite, and very rarely visible gold. Graphite coatings appear recurrently on the dike borders, seldom within the dike itself, as the graphite apparently comes from the wall-rock. The graphite covers supported the deformation and on occasions even the shearing of the quartz veins. Their thicknesses therefore sway between 0.1and 1 m.

The ancient mining traces locate near the ridges in slit-like corridor trenches, which however, have for the most caved in. Although the extraction depths thereby become difficult to determine, depths of up to 15 m are nevertheless likely (Fig. 5.251).

Between 1904 and 1907 the Nile Valley Company Ltd tapped the vein in five stope accesses, an adit and an inside shaft. Moreover, about 500 m further to the SW, three smaller inclined shafts had been lowered into red-brown, sheared siltstones. These however, only accessed small, irregularly arranged quartz veins that apparently turned out to be insufficiently productive. The gold in the main vein apparently has a very irregular distribution because of repeated reports of visible gold. In 1905 Sleeman reported large sectors with graphite (Jakubiak 1989). The gold contents seemingly vary considerably, because according to the recent analyses by the Australian Gippsland Company (2005) these lie between 1 and 70 g/t.

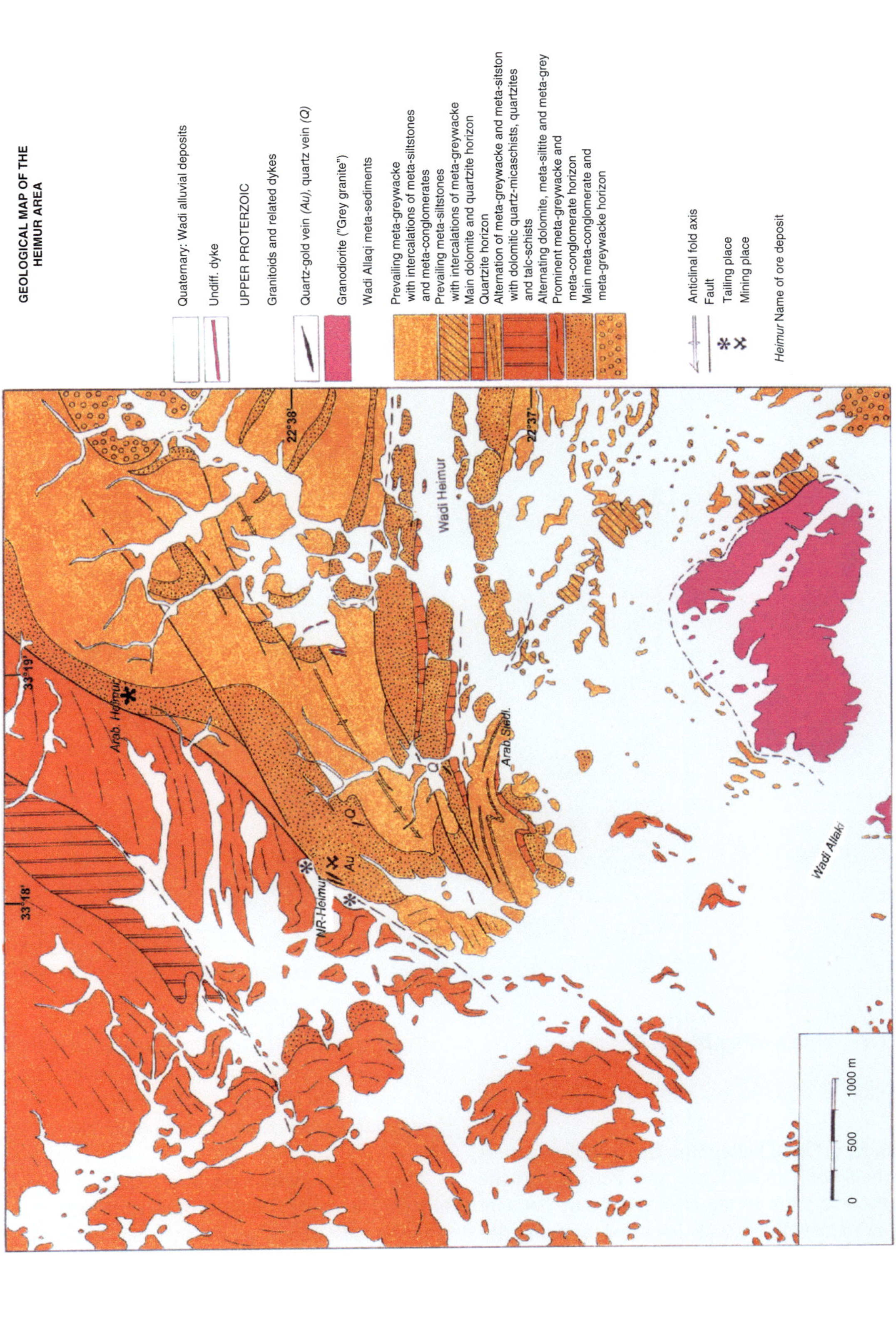

Fig. 5.250 Geological sketch map of the surroundings of Heimur (H. Kräutner)

Fig. 5.251 Collapsed New Kingdom trench pit at Gebel Heimur with recent adit

5.7.7 Nile Valley Block A

Geographic position:	22°35′21″ N, 33°22′20″ E

Mining in "Nile Valley Block A" had begun most probably only in the Early Arab Period. This site is about 1.5 km to the N of the Umm Garaiyat mine and just E of a wide inlet from a small, tributary valley (Fig. 5.252).

Traces from ancient mining have been erased almost completely by modern mining, so the date of the extraction period is merely based on some Early Arab hut remains in the immediate wadi edges.

The mined quartz veins are relatively short and of a bluish-grey colour, not wider than 50 cm at the surface. They appear in metagreywacke, virtually unaffected by hydrothermal activity.

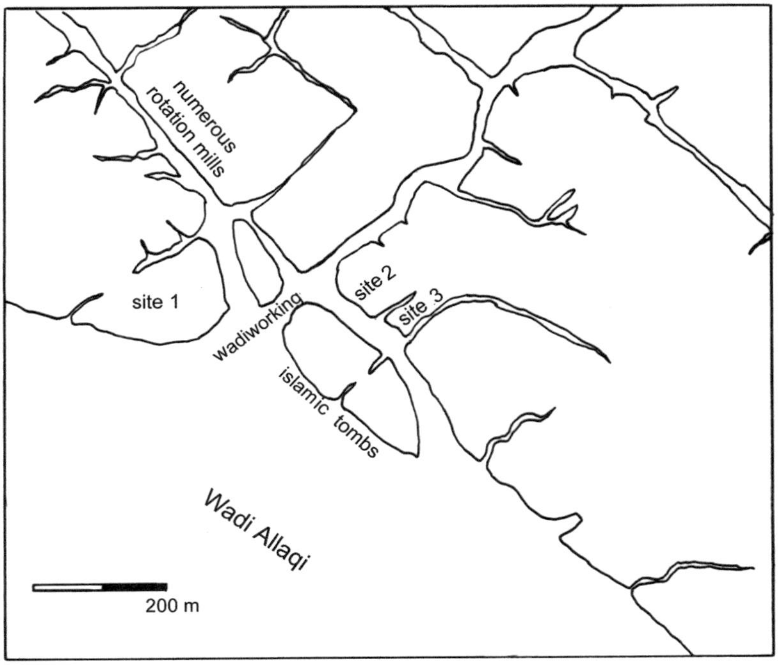

Fig. 5.252 Sketch map of the large Arab settlement "Ahmed City" in NE Wadi Allaqi (R. Klemm)

"Nile Valley block E" (22°36′50″ N, 33°22′33″ E) is about 2.7 km further N and had probably been mined only in the modern era. The mine follows a bluish-grey, pyrite-yielding quartz vein with a 10° strike and an 80° dip W. At 10 m depth it splits up in two. In this case the pyrite seems to contain gold, as the latter appears together with limonite only in the cubic cavities left behind after decomposition of the former. Gold contents average around 18.6 g/t (Jakubiak 1989). Considering the high values, it therefore surprises that the mine had been given up so soon.

5.7.8 "Ahmed Village" (Fig. 5.252)

Geographic position: 22°34′57″ N, 33°21′02″ E

"Ahmed Village" is a settlement area consisting of well-preserved Early Arab houses. Their stonework partly comprises New Kingdom mills. Early Arab round mills are missing, if not extremely rare. As a rule, small terraces had been added to the front sides of the buildings. Some houses are almost circular, others rectangular and divided up in two to three rooms. The area splits up in three sites and each numbers between 20 and 30 houses. Remains from extensive wadiworking activities are noticed over the entire wadi, but it is not clear whether any gold production had taken place here during the Early Arab Period or whether it was exclusively restricted to the New Kingdom Period. An Early Arab house had even been built on top of a small tailing heap from the New Kingdom. Moreover, a significant number of New Kingdom oval stone mills, which had most probably been re-claimed in the Kushite Period had been incorporated to the masonry of the Early Arab Period.

Wide gullies running down the surrounding slopes had been checked for auriferous quartz rubble in so-called colluvium workings, although it is not clear when this had occurred.

The wadi alluvium and much of the slope rubble in almost the entire area to the NW of the Umm Garaiyat mine up to the modern prospecting mine at "Nile Valley Block E" in the N, and Wadi Allaqi in the W had been burrowed

Fig. 5.253 Domestic pottery from "Ahmed-Village"

through. Although the area is thinly speckled by single hut ruins, human occupation clusters essentially within three large settlement complexes in respectively protected and remote locations. Because neither the wadis nor hills in this area have names, we resorted to refer to them collectively as the "Ahmed Villages" in reference to our accompanying geologist from the EGSMA, Ahmed Abdel Moniam, who first drew our attention to this area.

The fact that pottery finds are quite frequent and partially even well-preserved makes this area relatively unusual (Fig. 5.253).

Site 1 (22°34′54″ N, 33°20′57″ E) displays a large cluster of round, attached huts with fairly massive walls built in the shell-facing technique. It strikes that the houses are partly arranged in compounds to form closed, fortress-like barriers facing Wadi Allaqi. A small architectural unit located at the edge of the settlement displaying a genuine dome (approximately 1.30 m high, 2.50 m in diameter) probably represents an ancient stove. A small opening near the base had served for inserting bread after removal of the embers. A small opening in the roof had served as a smoke evacuation (Fig. 5.254).

Here and there one notices modified, re-used New Kingdom mills and grinding stones inside the masonry of the huts. Round mills were not recorded.

For the lack of the usual gold mining and processing tools it may seem possible that the Early Arab settlements had served to accommodate pilgrims on their voyage to Mekka. The comparatively large Early Arab cemeteries in the area may further even seem to suggest the presence of a nearby facility resembling some sort of hospice, or the like.

Site 2 (22°35′00″ N, 33°21′04″ E) consists like site 1 of a narrow valley running from the NE into Wadi Allaqi. On either side it is flanked by up to 1.70 m high house ruins. Their 50–60 cm thick walls from local schist slabs gathered-up from the surrounding surface made the stabilising effect of the otherwise often observed shell-facing masonry redundant. In this case too, the houses are round and their entrances are concealed as they face the slopes on the rear sides.

Fig. 5.254 "Ahmed Village", site 1. Hidden Early Arab Period houses at the edge of Wadi Allaqi, with domed oven (*arrow*)

They thereby again highlight the generally fortified character of the settlement. A separate house at the wadi mouth presumably served as a guard's house. In a narrow side gorge near the northern wadi access, is another building of the same type. To the S is an Islamic cemetery.

Site 3 (22°34′55″ N, 33°21′07″ E) consists of two extremely narrow, parallel wadis separated by approximately only 50 m from each other. The southern wadi has traces from human occupation only along its northern flank, which however spreads slightly to the S at its mouth.

This site too, was equipped with a dome-shaped stove. In addition, there is an approximately 2.50 m long and 80 cm wide cattle trough.

The northern wadi displays buildings aligned on both its flanks and is connected at its upper end to the southern wadi.

The entire settlement is located in a well-sheltered position behind a ridge oriented parallel to Wadi Allaqi. It had moreover been guarded at the wadi entrances by free standing, fortified structures (Fig. 5.255).

Mining tools are again missing. We therefore assume that the settlement had been part of the local pilgrim stop rather than a gold producing site. However, an alignment of clearly distinguishable sand heaps is seen at the edges of Wadi Allaqi indicating former wadiworkings.

In the entire sector of the low hills between actual Umm Garaiyat, at the edge of Wadi Allaqi and Wadi Heimur in the N, traces from intensive wadiworkings are found in nearly all wadi beds, which often associate with hut remains from the Early Arab Period.

A relatively large settlement with more than 20 house ruins (22°37′11″ N, 33°21′40″ E) and a large burial ground about 600 m to the S of the asphalt road in Wadi Heimur is also noteworthy in this respect. Again, this site revealed no gold processing tools. Together with the large cemetery this may again hint the mentioned function related to hospice services.

Fig. 5.255 "Ahmed Village", site 3. Hidden houses with connecting walls, thus secluding the dwellings from the wadi

5.7.9 Umm Garaiyat (Fig. 5.256)

Geographic position (Hafir):	22°34′34″ N, 33°22′33″ E

The Umm Garaiyat gold mine is located about 1 km further N, in a tributary valley to Wadi Allaqi. It had evidently been operated at least since the New Kingdom

At first sight, the site seems of little archaeological interest, due to widespread damage by recent (early 20th century) mining. To the N of some modern house ruins however, one recognises a large enclosing wall near the confluence of a side wadi coming in from the SE. The complex clearly occupies a central position. Its shape is oval and it is made from sand and rocks. As a wall enclosing a surface measuring about 100 × 60 m, the structure's original height must have varied around 1.50 m. In Sudan such structures are still much in use today and referred to as Hafirs, whose function was to retain and to stock water runoff. The Hafir at Umm Garaiyat possibly represents one of the most northern examples (Fig. 5.257). As these Hafirs seem to be a genuine Nubian invention, its existence here may also allude to a Nubian (Kushite ?) occupation after the New Kingdom Period. Inside, the reservoir is divided up into two parts, probably as a consequence to the need for managing different water levels. The complex may date back to the New Kingdom/Kushite Period, though the units inside the enclosure are no doubt later.

A diagonally running retention wall crossing the N-S oriented wadi is just barely made out in the stereoscopic aerial image, though hardly in the field. The wall served to collect water in a cistern, preserved in the form of a large hole to the N of the Hafir. When the cistern was full, the overspill flowed into the Hafir with its two, different-sized basins. One also recognises an inlet at the outside of the southern Hafir wall. Water was thus probably stored as follows: The northern basin and the cistern were filled with water derived by the retention wall of the main wadi, while the southern basin bordering on it was filled with rainwater from the side wadi

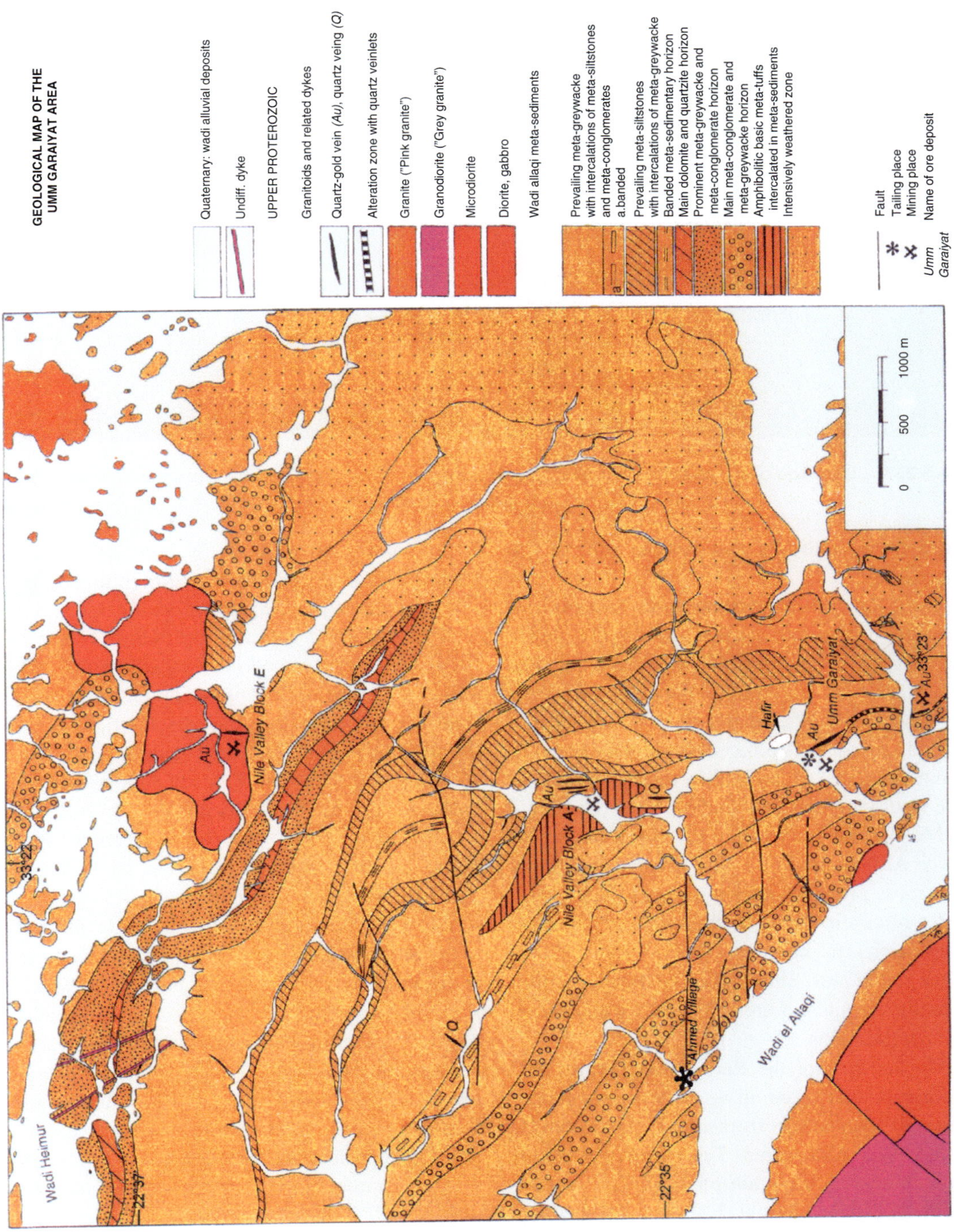

Fig. 5.256 Geological sketch map of Nile Valley Block A and E, Ahmed-Village and Umm Garaiyat (H. Kräutner)

Fig. 5.257 Large Hafir (water basin) at Umm Garaiyat containing several architectural structures from later periods

through the inlet on the eastern side of the compound.

Occasional, still occurring floods today follow the same course as can be told by the sludge deposits inside and outside the compound. The size of the reservoir points to the enormous needs for water engendered by the wadiworkings. Indeed, almost all wadi courses in the district of Umm Garaiyat exhibit visible traces from intensive wadiworkings (Fig. 5.258).

Most of the huts within the Hafir date to the Early Arab Period. By this time it had therefore most probably no longer been in use as a water reservoir. Some of the architecture moreover seems to be associated to more recent mining activities, although more detailed differentiations would require appropriate archaeological investigations.

By the time of the Early Arab Period, gold production had taken up again, which is consistent with the findings from almost anywhere else in the Wadi Allaqi area. The typical processing tools, however, are again missing. This is obviously due to the intensive destruction of the older traces by recent mining.

Modern underground mines seem all to foot on smaller, ancient mines located in the mountains to the E of the wadi. A large Early Arab Period cemetery spreads-out further to the W, some 100 m before the inlet to Wadi Allaqi

The productive quartz vein of Umm Garaiyat occurs in heavily sheared, occasionally kaolinized, silty, and graphite-bearing metasediments with transitions to metaconglomerates. It strikes along the flexures of the shears in an approximate N-S direction and therefore does not follow over its entire length the stratification of the adjacent wallrock. Its dip was recorded between 60° and 80° E. As often observed elsewhere, its thickness varies between few centimetres and 1.8 m in the shear zones and disappears in the S in kaolinised to graphite-yielding and silted wallrock. The gold is dispersed very irregularly in pockets, of which the most extensive one was encountered in depths of 30–40 m below the surface. In addition, vertical enrichment zones had been mined in several levels, whose widths vary between 4 and 15 m. Between them, some sectors are virtually barren (Koshin and Bassyuni 1968).

Modern mining in the district had begun with the intensive interest by the Nile Valley Company Ltd. between 1903 and 1913 and had continued sporadically until 1957 with selective extraction of rich-ore-zones. The mine had been left in a

Fig. 5.258 Heaps from former wadiworkings at Umm Garaiyat

very untidy state, as seen by the abandoned plants and the fairly quaint ruin of a mechanical stamp mill (Fig. 5.259).

The New Kingdom extractions had been lowered into the central quartz vein at various depths up to 30 m, over a distance of almost 350 m. A closer inspection of the yet intact, mined zones revealed that extraction only occurred in selected, ore-rich zones, which resulted to peculiarly forked and chimney-like cavities. Preserved work traces indicate the use of stone hammers and chisels.

The gold grades of the mine vary accordingly, as reported by Koshin and Bassyuni (1968), with high values ranging between 150 g/t and more than 300 g/t. These values may seem somewhat extravagant, considering particularly the fact that the sample sizes are not given. The average content of the mined ores between 1903 and 1913 which varied between 10 and 13 g/t seems therefore more realistic. Today the mine is filled with water right up to the wadi surface and is thus a welcome source for water to the local Bedouins.

Its geologic framework, including the deposits of Umm Garaiyat, is dominated by metasediments more or less intensely affected by contact metamorphism. They are primarily composed of slightly folded, former greywackes alternating with siltstones and arkoses, and recurrently by conglomerates and dolomite marble streaks. To this sequence add former basaltic rocks in the form of amphibolites, which again as a result of contact metamorphism have re-crystallised to coarse-grained hornfels, slightly resembling diorite. It cannot be excluded that some of the hornfelses of this sequence also represent former acid tuffs, although no evidence can confirm this. The necessary thermal energy for the contact metamorphism was undeniably furnished by the large granite intrusion to the NE, as well as the granodiorite intrusion to the SW. The smaller, more recent diorite intrusions around Nile Valley Block E and S of Umm Garaiyat seem on the other hand to have had little influence on the rock alterations.

According to our present-day knowledge, these seem to represent the basic geologic preconditions for the formation of auriferous, hydrothermal quartz veins, provided that the open

Fig. 5.259 Stamp mill at Umm Garaiyat from the early twentieth century

clefts formed by tectonic extension processes supplied sufficient space for the deposit. This was given by the extension of folded metasediments with strike directions adhering to the folded sediment layer structure resulting from the dome-like intrusion of granitoid plutons, which can be seen in the geologic sketch map of the Umm of Garaiyat area (Fig. 5.256).

The extensive wadiworkings in the area around the various settlements within the entire environment N of the Umm Garaiyat mine also reveal gold quartz mineralisations, though which do not originate from large productive veins, but rather fine ones, as observed from the numerous unsuccessful test prospections that had been carried out in the area.

Fig. 5.260 Early Arab Period houses at the edge of Wadi Atshani (background)

5.7.10 Atshani

Geographic position:	22°34′04″ N, 33°32′01″ E

To reach the Atshani deposit a cross-country vehicle is necessary. One gets there only from the N by Wadi Shalman and the large, flatly eroded (granite pluton) Shalman plain. The inlet to Wadi Atshani is not easily found. Furthermore, because Wadi Atshani itself is inaccessible even to cross-country vehicles, one arrives at the site only after walking a distance of 3 km. It leads through a small tributary valley bending-off to the E, where one comes across the building remains originating from mining attempts carried out in the early twentieth century.

In Wadi Atshani however, and particularly along its eastern flanks, slope rubble and wadi sediments had been intensely checked through for auriferous quartz fragments, as testified to by numerous Early Arab hut remains strewn over the entire length of the wadi (Fig. 5.260). Its completely worked-off sediments are also the reason for the appalling driving conditions here.

The observed ceramics consist above all of tall water containers (Arabic: sir) originating from the recent mining activities. The scanty ceramic evidence from the Early Arab Period chiefly shows Nubian geometric motifs.

The geologic environment of the mined vein minerals consist of a sequence of sheared, schist-like and silted metasediments to very finely grained and heavily epidotised metaarkoses, acid volcanites, and prasinites (greenschist: formerly mafic volcanites). Near the main dike in the valley flank we also found heavily sheared diorite.

Although it appears that two main veins follow the border between the schistic metasediments and the noticeably sheared prasinites, we nonetheless cannot exclude that we were here confronted with only one, although slightly displaced vein. The alteration zones next to the veins are stained brown by limonite up to 3 m into the wallrock. The vein thicknesses vary between few

Fig. 5.261 Large stone hammers at Marahig resembling the Pre- and Early Dynastic mining tools from the northern and central Eastern Desert of Egypt

centimetres and 50 cm. The wider parts had mostly been avoided by the miners.

Sticking to the shear direction, the veins strike between 160° and 170° and dip between 65° and 85° E. They can be pursued for more than 700 m. Along the veins, up to 2 m deep pit trenches and occasional, deep mine shafts to depths of about 10 m can be observed. At first sight, the execution of the mines seems rather disorderly and even sub-standard. This however, is first of all due to the partly intensive wallrock alteration and the weathering of the pyrite, which led to murky, brown discolorations and the eventual caving-in of the formerly straight sides of the wallrock.

A recent mining attempt is revealed by 25 m deep inclination shafts, which too, have already partly collapsed. In fact, underground mining here needed to be brought to an end because of safety reasons. About 300 m to the E of these shafts, there is another vein which shows clear mining traces and possibly even ones dating to antiquity.

5.7.11 Marahig (also Marahib)

Geographic position:	22°33′11″ N, 33°28′39″ E

Wadi Maharig forms a wide valley which is much filled with alluvial rubble. It is also extensively littered with countless New Kingdom stone mills, grinding and anvil stones, and very large hammer stones. The mill hollows are clearly deeper compared to the ones at the original New Kingdom from more northern parts of the Egyptian Eastern Desert. The mills here much resemble the ones reported from Hairiri. Hence in this case as well, we are tempted to assume that mining activities had been continued after the New Kingdom, in what is referred to as the Kushite Period, in which the former tools had been reutilised. Some even had pierced from excessive use wear.

The large stone hammers on the other hand much resemble earlier versions of the two-hand hammers known from Predynastic/Early Dynastic find contexts in the northern and central Eastern Desert of Egypt (Fig. 5.261).

Fig. 5.262 Early Arab Period tailing and washing table assembled with New Kingdom stone mills at Marahig

At both wadi flanks there are loose sequences of hut ruins in various states of preservation, some of which are even hardly discernible. Singular washing tables scatter seemingly erratically over the area. Generally, the distinction between the Early Arab and New Kingdom/ Kushite houses is easily made. Early Arab houses tend towards circular shapes and usually are in better states of preservation (with wall heights exceeding 1 m). As a rule, they are also built in the shell-facing technique. Frequently, they have New Kingdom mills integrated to their walls.

A strikingly large tailing site is bordered by two, severely damaged washing tables built from New Kingdom mills (Fig. 5.262). It is also surrounded by Early Arab huts, whose walls also contain built-in, New Kingdom stone mills. Nearby, there are furthermore small basins assembled from New Kingdom mills (Fig. 5.263).

In addition, numerous, minor praying sites with mihrabs made from New Kingdom mills also testify to the re-occupation of this site in the Early Arab Period, which is further testified to by pottery finds (Fig. 5.264).

Seen from the heights of a nearby mountain mounted by an Alam tower, one can still barely make out the remains of the New Kingdom site.

The occupation and mining sequence in this district therefore probably was the following: Tool traces indicate a very first mining activity in Early Dynastic to Old Kingdom times, followed by a New Kingdom gold production period, where most probably only auriferous quartz lumps had been gathered and processed. In the following Kushite Period New Kingdom tools had been reclaimed and re-introduced to the gold production. The site was again occupied in the Early Arab Period. Surprisingly however, no round mills, but merely prayer sites (mosques) and some hut ruins were found.

The geologic catchment area comprises a sequence of partly heavily sheared, acid gneisses which also contain mafic dikes affected by metamorphism, alternating with amphibolite layers. However, further to the S of the settlement the mountains are composed of a metamorphic layer sequence of former conglomerates, sandstones and siltstones, in which recurrently small granite apophyses of an unknown intrusive body are found. On account of the overall geologic setting, gold-quartz mineralisations seem therefore likely in this area.

We noticed some traces from prospecting in some quartz veins in the mountains to the E of the settlement. From their fairly "fresh" appearance, we assumed them to be of a recent date. Although the mountains are crossed by numerous,

Fig. 5.263 Stone cist assembled with New Kingdom mills at Marahig. It had once been plastered on the inside and had served as a water container

Fig. 5.264 Early Arab Period pottery from Marahig

small quartz veins striking E-W, no traces from underground mining or mined trenches were found even in the greater surroundings of the settlement. It seems therefore that the ores had been recovered exclusively from selective gathering of slope rubble and wadi sediments (wadiworkings).

According to Jakubiak (1989), extensive processing of the wadi alluvium had taken place 2.6 km above the settlement and just below the main divide. The rock is reported to consist of tuff schists with bands of violet phyllites and lenses of bluish marble displaying shear strikes from NE to SW.

Another gold processing site is located in Wadi Marahig, approximately 1 km away from its inlet into Wadi Allaqi (22°29′16″ N, 33°26′52″ E). On both sides of the wadi one notices extensive New Kingdom house alignments comprising comprehensive mill assemblages. Some houses are superseded by Arab Period houses, and in a small side wadi, an associated tailing was found together with a washing table. Wadiworking seem to have been the main activity during both occupational periods.

5.7.12 Marahig-East

Geographic position:	22°31′07″ N: 33°30′07″ E

This New Kingdom settlement is located in a valley between Atshani in the NE and Marahig in the SW.

The geologic setting partly consists of a heavily sheared sequence of tonalites with transitions to granodiorites, which makes an exact identification almost impossible. The shearing is oriented NE-SW and corresponds to a large number of quartz veins that reveal no traces from mining.

This relatively large settlement exhibiting a rich assemblage of gold processing tools, seems to have been furnished with ores only through selective gathering of wadi and slope rubble sediments, as visible from the traces left in the surrounding terrain.

The area's lithology is much reminiscent to sheared siltstones of the Hammamat and Igla formation in central parts of the Eastern Desert, where, however, these occur in less sheared environments. There too, the rocks are intruded by an intermediate magma, which noticeably contains much dispersed pyrite. The metamorphic sediments in the surroundings of the intrusions are traversed by numerous, almost horizontal quartz veins, which yet reveal no traces from extraction.

5.7.13 Wadi Murra

Geographic position:	22°31′03″ N, 33°54′13″ E

Wadi Murra is a side wadi to Wadi Allaqi leading N, about 7 km NW of the inlet of Wadi Ungat. Approximately 20 km up the wadi, one notices the ruins of larger houses on both sides of the valley. They were built from heavily sheared and splintery rocks, which bestow on them a very timeworn and devastated appearance. The walls are built of long slabs in a braided pattern masonry by which they had been arranged in alternating positions lengthwise and crosswise to the wall orientations.

Some minor Islamic burial grounds at the wadi's centre and edges seem to indicate that the houses as well, date to the Early Arab Period. Neither pottery nor mills are found here, and it therefore appears that gold production had been based on the processing of wadi sands.

Wadi Murra crosses an elaborately folded, large anticlinal system which strikes southwards to Wadi Allaqi from the N. The geologic location of the wadi sediments' catchment area around the Early Arab settlement is again determined by the intensely folded sequence of metagreywackes, metasiltstones, and metaconglomerates. Especially, the metasiltstones are extremely columnar and dip along their b-axes to the SE. The conglomerates are quite severely sheared and show elongated rubble components. Their weathered surfaces are covered by the debris of their own rubble components. In spared zones the also considerably sheared greywackes reveal primary carbonates.

Partly wide quartz veins cross the entire mountainous backdrop. They strike in two directions, of which the first at 60° with an almost vertical dip, and the second at 90° in an approximate dip of 45° S. Because the quartz veins show no traces from mining, gold production had been probably based on ores from the wadi alluvium.

Contrary to our observations, Jakubiak (1989) reported two small, anciently mined quartz veins and several minor prospecting sites in the upper reaches of Wadi Murra in close contact with the granite of Gebel Abu Brush, about 13 km away from Bir Murra.

5.7.14 Wadi Rylan

Geographic position:	22°26′44″ N, 33°35′41″ E

No evidence hinting to genuine gold mining was found near this wadi. The inhabitants of the three discovered Early Arab house clusters had probably based their gold production on placers from the wadi alluvium by sifting and washing sands as no mills or other processing tools were found.

The geology around Wadi Rylan is made up of alternating layers of metasediments, chlorite-sericite schists, metagreywackes, and metaconglomerates with subordinate occurrences of dacite metatuffs.

5.7.15 Filat

We did not visit this site S of Wadi Allaqi. Apparently it is of little significance.

Other small, Early Arab gold producing sites to the S of Wadi Allaqi are mentioned by Castiglioni et al. (1995). Their registration numbers, geographic positions and short descriptions follow hereafter:

B 28.1: 22°10′42″ N, 33°46′04″ E: "some dozens of round buildings of dry-masoned walls. A large, round millstone from a rotor mill, broken. Traces from ancient mining."

B 31: 22°17′53″ N, 33°39′43″ E: "Numerous dry-wall constructions with round layouts, well-preserved, scattered over the main and side wadis. The buildings are of 4 m in diameter, the entrances are oriented to the N. No millstones. Extensive surface mining in trenches."

B 32: 22°24′54″ N, 33°27′15″ E "Numerous dry-wall constructions with round layouts, partly covered by sand belong both wadi flanks. The position of the mine is indicated by a well-preserved Alam at the top of a hill. Round mills (both parts mill and runner stone). Washing plane with quartz lumps. Two vessels. Area with reddish sand. "

5.7.16 Seiga I (Fig. 5.265)

Geographic position (mine):	22°32′59″ N, 34°07′42″ E

The Seiga I mine is reached via Wadi Allaqi by turning NE into Wadi Seiga. After about 30 km one reaches Wadi Abu Had, which branches-off to the N. Then, about 7 km up this wadi one arrives at a confluence into a valley turning-off to the SE. The mining area is visible from the distance in the shape of a steep incision into the mountain cutting it parallel to the wadi (Fig. 5.266). It consists of a quartz vein system in a NNW-SSE strike which had been mined in a vertical direction from the crest in an opencast technique. Today, the heavily altered wallrocks still impressively flaunt the gap of the extracted vein system (Fig. 5.267). Numerous houses group around the mined hill and at both sides of the wadi.

According to our inspections, this is an exclusively Early Arab mine complex. Similar to Atshani and Neguib, the ores had been processed with round mills found in great numbers inside the house ruins (Fig. 5.268). The houses themselves are well-preserved, particularly in protected locations.

At least six to eight large tailings with washing tables are made out at the wadi edges in the Seiga I district. At the inlets of a number of small side wadis and slope gullies one recognises the remains of minor dams to collect the runoff. On our way to the district, we were already struck by the relative wealth of the vegetation, which evidently reveals a higher precipitation in this region than in the NW parts of Wadi Allaqi.

Fig. 5.265 Settlements and open trench mine in Wadi Seiga (modified Google-Earth image)

Fig. 5.266 Open trench pit of Seiga I and Early Arab Period huts in the foreground

Fig. 5.267 Inside the deep trench pit Seiga I

Among the pottery fragments, amphorae and domestic wares predominate. Here and there we also found shards from a dark-green faïence, more typical for Islamic sites in the S.

Among the round mills that concentrate near the tailings, there are also well-preserved rotor discs. Furthermore there were flat anvil slabs with several depressions, as well as typical, two-hand hammer stones used for crushing the ores before the grinding process.

The geologic setting of the quartz vein consists of a mixed volcano-sedimentary sequence comprising of slightly metamorphic greywackes and sandstones in the immediate vein surroundings.

Fig. 5.268 Wadi Seiga. Base of an Early Arab Period round mill from red felsite, a rotor disc from dolerite, and two pounders

They alternate with prasinites in the NW and in the NE with grey dolomites, graphite schists, greywackes, and conglomerates (Fig. 5.269). The entire volcano-sedimentary series had been exposed to an intense NW-SE tectonic shearing, before being intruded by granitoid magmas, which by assimilation with the prasinites then changed to more diorites, whereas the prasinites in the contact zone transformed to hornfels. As a result of the intrusion, extension faults striking in the shear direction formed, which then filled with quartz minerals of which some were auriferous. Later tectonic faults appeared almost perpendicularly to the extension faults. In the geologic map they clearly exhibit a NNE-SSW strike with lateral displacements of up to 500 m.

The auriferous quartz vein had repeatedly been intersected by adits at the beginning of the twentieth century. No significant extraction, however, ensued. The main vein strikes with a series of smaller, unexploited quartz veins at 160° SE. It dips vertically and is slightly inclined to the W. The mine thereby cuts impressively through a hill with steep slopes. Other important, though barren quartz veins cross the mountains as they strike through the granite in a NE-SW orientation. The Gippsland-Report (2005) mentions clearly fluctuating gold contents of up to 14 g/t.

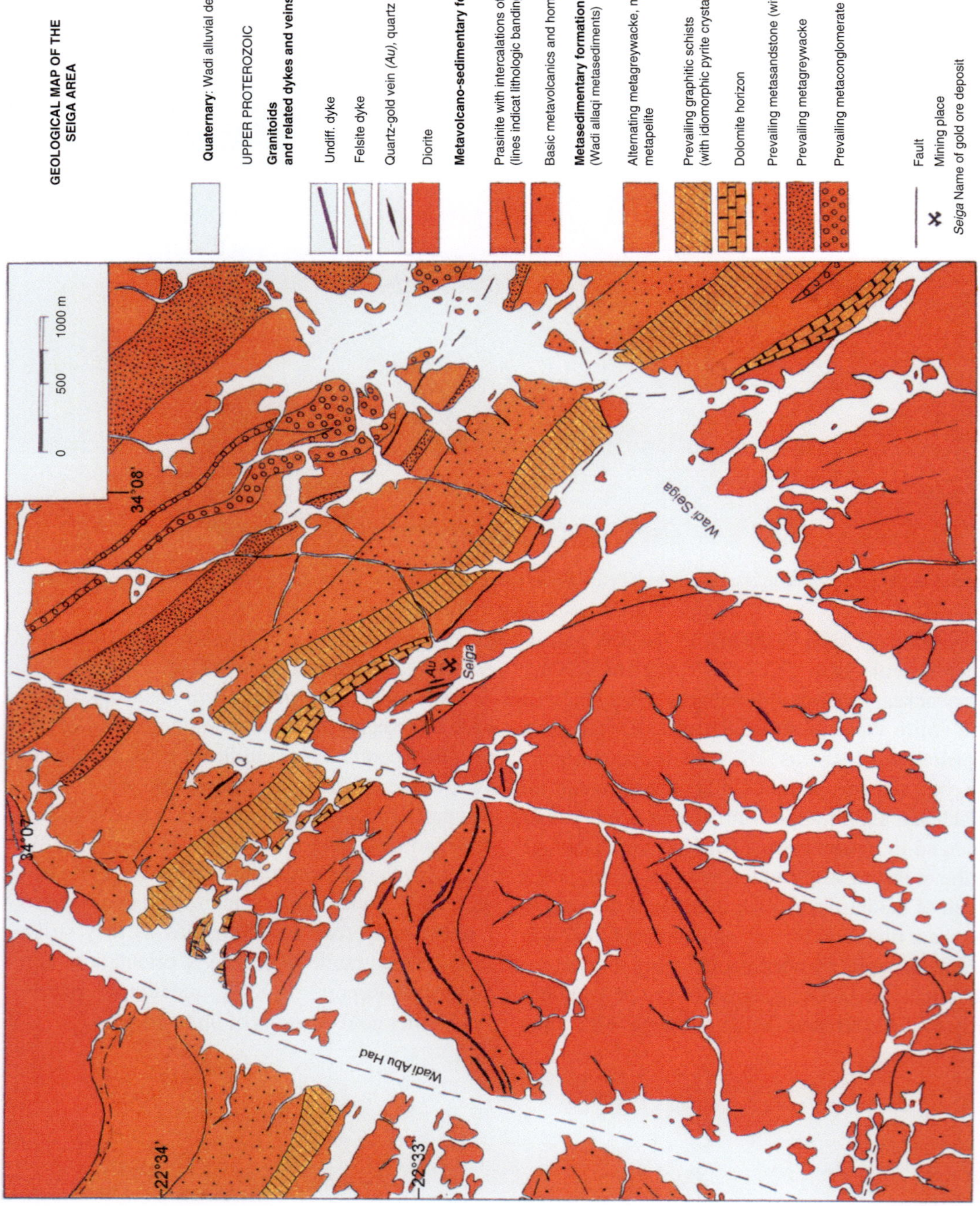

Fig. 5.269 Geological sketch map of the Wadi Seiga surroundings (H. Kräutner)

Fig. 5.270 Well-preserved Early Arab Period (or early 19th century ?) house at Wadi Seiga 2

5.7.17 Seiga II (Fig. 5.274)

Geographic positions:	
Younger settlement:	22°33′26″ N, 34°10′17″ E
Older settlement:	22°33′23″ N, 34°10′21″ E

Two Arab Period settlements are located in a valley basin 5.5 km E of Seiga I and 4 km N of the Shoshoba deposit. The first (younger?) is set within steep-standing metagreywackes and metaconglomerates. Thanks to the stalky building material in the dry walls as well as the site's concealed location, its architecture is still relatively well-preserved. The other site has suffered more and is located in the SE hills of the same wadi.

Although the mountains in the surroundings yield numerous quartz veins, no traces from trench or underground mining were found in the neighbourhood of the settlement. Because of the countless round mills and anvil stones and a vast processing site, which presumably had been established around a former well, this site must have been a major gold processing place. Its needs for quartz ores had consequently been covered by the wadi alluvia in the nearby valleys.

The two differently preserved settlements display the typical Early Arab Period processing assemblages. As in the case of Seiga I, unusually light-coloured amphora shards with robust, stamp-marked handles prevail at both sites.

The first site is quite eye-catching by a number of well-preserved, individual house clusters (Fig. 5.270).

Relatively large Islamic cemeteries cover the surface at the northern access to the valley and the tributary valley to the E of it (Fig. 5.271).

Moreover, there are two water barrages, which are exclusively known from this period's gold mining districts.

Apparently both settlements date to different phases, whereby the less well-preserved houses in the sheltered location seem to be the older ones. The younger site may date to the gold prospecting period under Mohammed Ali, in the first half of the nineteenth century. This period is also attested to in Nubian Sudan (Tanasheb, Tilat Abda), where identical tools to those from the

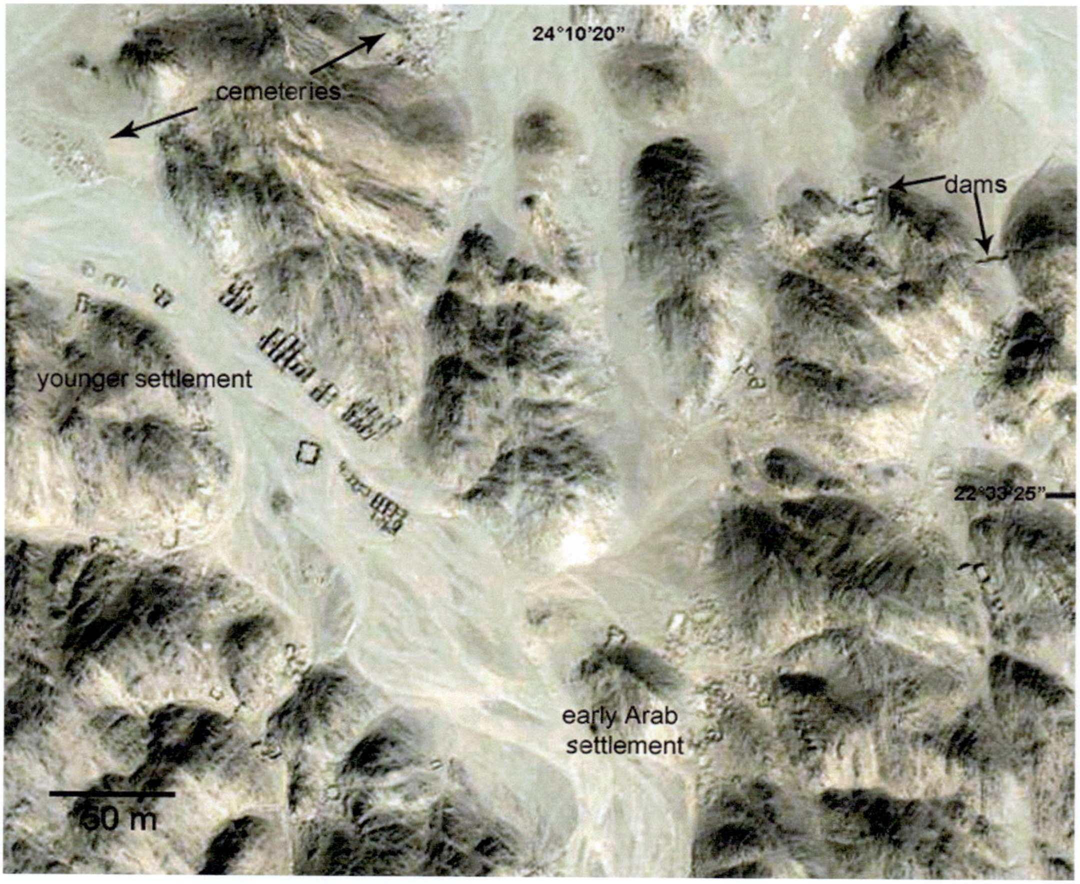

Fig. 5.271 Satellite picture of Wadi Seiga II displaying two Arab Period settlements, dams, and cemeteries (modified Google-Earth image)

Early Arab Period had been used. More detailed data can, however, only be obtained from exhaustive archaeological investigations.

5.7.18 Shoshoba

Geographic position:	22°31′21″ N, 34°01′35″ E

The chiefly Early Arab mining district at Shoshoba is located in a mountain basin with a dense cover of eroded quartz rubble. The surrounding mountains accommodate at least four ancient mines consisting of narrow, partly inclined and sinuous galleries leading down into the depth. At the beginning of the twentieth century, mining had been revived for a short period (Fig. 5.272).

Several tailings, a cemetery, and a group of 20–30 huts are found here. All date to the Early Arab Period, as confirmed by some characteristic painted pottery with animal motifs and banded friezes. According to Adams (1986), they date to a period between the ninth and eleventh centuries. Some glass fragments and stamped amphora handles were also found.

In the rubble at the edge of an erosion gully cutting a tailing site in the Shoshoba plain, an Arab Period round mill was found next to an oval mill from the New Kingdom (Fig. 5.273). This originally during New Kingdom worked stone mill was obviously reused at a period between NK and Early Arab. Best likeliness in this respect seems the Kushitic period. However, to certify this it needs detailed archaeological investigations.

Fig. 5.272 Large, Early Arab Period tailing (*arrow*) at Shoshoba. In the rear to the right, two of the exploited trench pits are to be seen

Fig. 5.273 Original New Kingdom stone mill reused in (?) Kushitic Period (*left*) and rotary discs of an Early Arab Period round mill (*right*) in alluvial debris of the Shoshoba plain

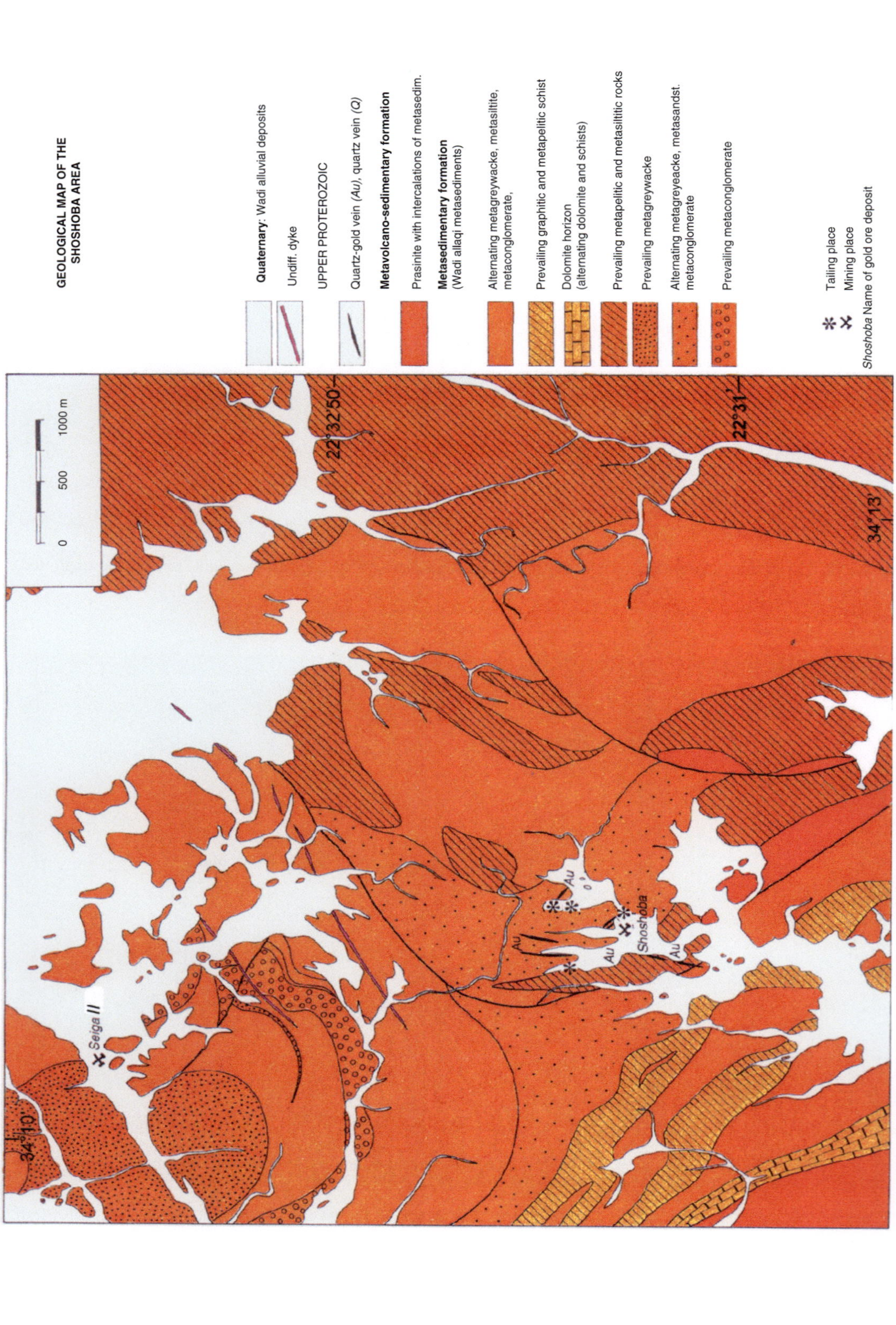

Fig. 5.274 Geological sketch map of the Shoshoba area (H. Kräutner)

Apparently as a result of the large scale at which the wadi alluvium had been processed in the Arab Period as witnessed by the tailings and washing tables, the traces from the New Kingdom had disappeared to a high degree.

The latest occupation is dated to the early years of the twentieth century by glass fragments from purple bottles. Its visible remains consist of a barrack-like building located in the middle of the valley basin, directly next a quartz outcrop. In those days the ancient huts were used as garbage dumps (tins, beer bottles etc.) while rotor mills and pottery vessels were often intentionally smashed.

The geology (Fig. 5.274) of the Shoshoba district consists mainly of metamorphic, partly graphite yielding schists, as well as mostly very heavily sheared metaconglomerates alternating with metaarkoses and metagreywackes. The metaarkoses in general, contain carbonates and are decomposed to such a degree that hand pieces are hardly recoverable. In the vicinity of the productive quartz veins they noticeably contain former pyrite cubes up to 3 cm in size, of which all have decomposed to limonite. Independently of that, metaconglomerates are associated with the quartz mineralisation above all in the north-eastern vein. As a result from shear folding, the metagreywackes had developed to up to 50 cm long spikes and slabs that provided for suitable building material for the adequately well-preserved huts.

The intensive fold structure of the district is less well-discernible in the field than in aerial or satellite images, as the exposures had been covered by debris from decomposition.

The deposit area of Shoshoba is split into several quartz vein systems, of which the two main extraction areas are located around the relatively large erosion plateau of Shoshoba. The SW vein strikes like most others almost N-S in a vertical dip. The NE vein, on the other hand, strikes NW-SE, also in a perpendicular dip. Both should be regarded as vein systems, each consisting of a series of smaller veins that had been mined collectively. At the NE vein prospecting had been carried out at the beginning of the twentieth century, but the resulting exploitation attempts remained fruitless.

The vein mineralisation consists of a milky-white, porous quartz traversed by limonite-filled fissures. Particularly at the selvages there are occasional, elongated concentrations of ankerite-siderite carbonate mineralisation, which both occur together with the limonite. This is foremost observable at the 8 m deep, slot-like shaft in the SW vein. The limonite is in all probability decayed pyrite with which the gold contents are correlated. Prospecting work carried out by Gippsland (2005) evidenced gold contents averaging around 2.2 g/t with maximum values of 4 g/t.

5.7.19 Umm Tuyur

Geographic position:	22°18′16″ N, 34°38′05″ E

We were unable to visit this occurrence during our survey. Thanks to Jakubiak's (1989) study and above all the work carried out by Zoheir (2004), some useful information can nonetheless be presented here.

Traces from ancient extraction indicate that significant amount of ores had mainly been obtained from the central, flat recumbent parts of the mined vein. At the beginning of last century, the mine was run by the African Reefs Gold Mine Ltd until 1925. Six shafts were lowered into the ground to a maximum depth of 82 m. The up to 1.5 m wide quartz vein was mined in several levels and produced gold yields between 25 and 29 g/t (Koshin and Bassyuni 1968). In the deeper levels however, as the vein reaches into graphite schist, it seems to split up into a series of minor streaks of white quartz with significantly decreasing gold contents.

Jakubiak (1989) was able to establish 4.2 g/t Au from some remaining tailings, but it seems as if these are recent.

Unfortunately, noteworthy information as to the ancient history of the mine was not given in the mentioned reports.

The geologic frame of the deposit is extremely complex (Fig. 5.275). In general terms, it is located at the border between the Gabgaba-Terrane of North Sudan and the Gerf Terrane in the south-eastern part of the Eastern Desert of Egypt. This interface is characterised by massive

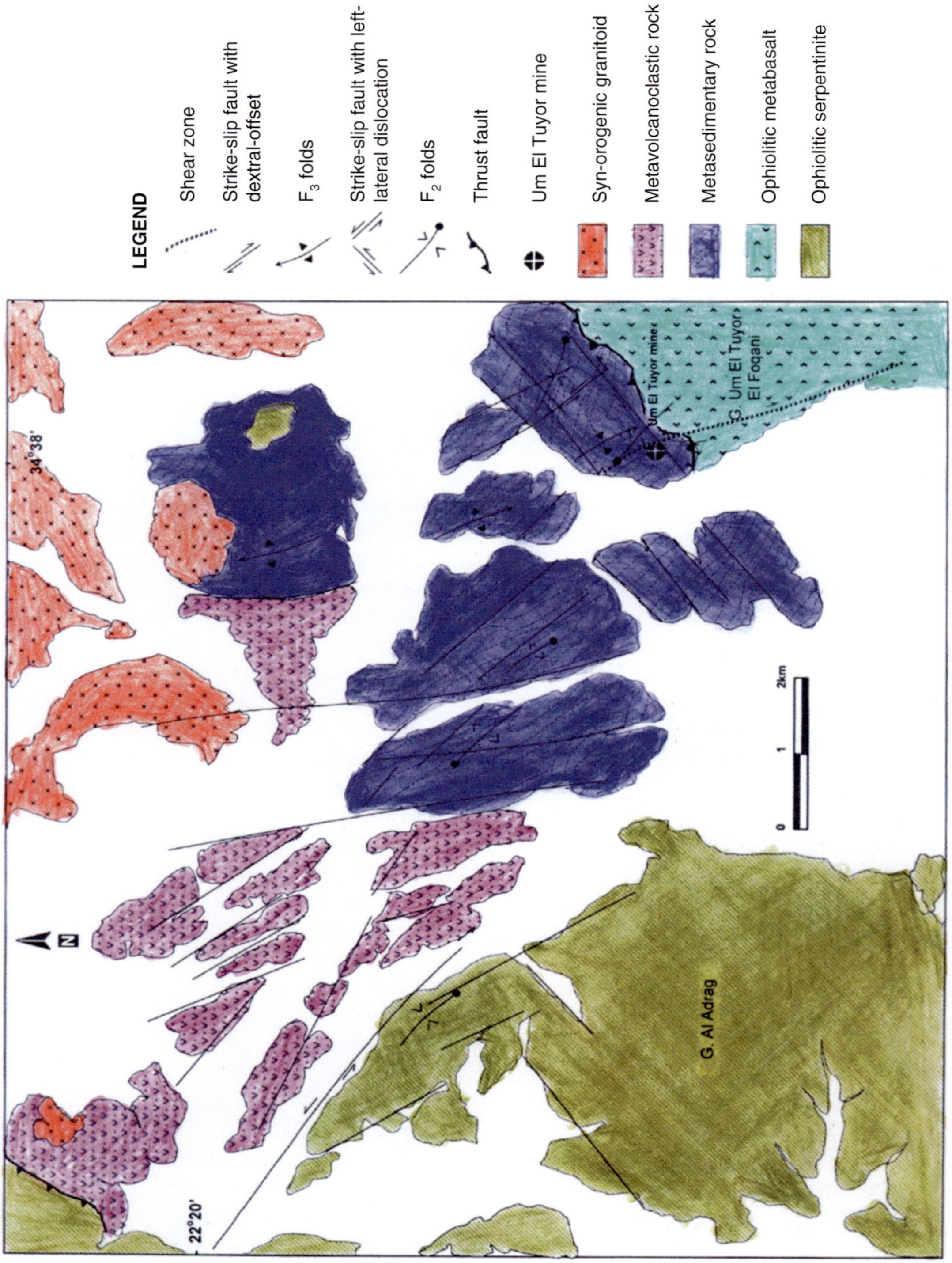

Fig. 5.275 Geological map of the Umm Tuyur surroundings (modified after Zoheir (2004))

overthrust nappes with a diverse sequence of ophiolite serpentinites, gabbros and metabasalts, into which a series of different metasediments are kneaded. These metasediments consist of garnet-biotite schist, metasiltstone and quartzite and represent together with layers of metavolcanic and metagabbro a typical island arc sequence, which together had been abducted onto the continent. The peak of the thereby resulting metamorphism occurred at about 550 ° C and a total pressure of 5.7 kb (Zoheir 2004).

At Umm Tuyur the auriferous quartz mineralisation is embedded in the hostrocks and occurs in several phases along a NW-SE striking shear zone between overthrusted metabasalts and lower lying metasediments. The productive vein irregularly shaped in a dense pinch and swell structure.

The gold contained in them occurs with sulphide minerals, such as pyrite, arsenopyrite, and subordinately sphalerite, galena and chalcopyrite. In general, it then holds between 13 and 23 % silver, in some carbonate-rich quartz parts even up to 30 %. Only the gold contents primarily enclosed in sulphide minerals, in this case mainly the pyrites, and subsequently released during the decomposition of the host mineral lose some of their original silver and relatively increasing their gold content to 88–94 %. Gold is reported to be visible in the blue-grey quartzes. The lode dips at 45° SW in its middle section, but at 80° near its ends where it modifies its strike.

Thanks to thorough investigations by Zoheir (2004) the formational conditions of the main mineralisation have been determined at about 400 ° C for the earliest, and 350 ° C for the final stage at pressures between 1.3 and 2 kb.

5.7.20 Betam

Geographic position (modern mine):	22°16′42″ N, 34°30′35″ E

The occurrence is situated in the upper part of the northernmost of the two tributary valleys of Wadi Umm Tenedib, about 10 km SW of Gebel el-Adrak, and just to the N of the border between Egypt and Sudan.

We had not the chance to visit this occurrence during our survey in the region in January 1993. We here therefore refer to the information given by Koshin and Bassyuni (1968), Jakubiak (1989), and in particular to Zoheir and Qaoud (2008a) with respect to the ore geneses.

Traces from ancient settlements are spread loosely over a wide area. They are often covered with rubble and sand drifts and are appear within a stretch of more than 2.6 km from NE (22°16′41″ N, 34°30′30″ E) to SW (22°16′04″ N, 34°29′16″ E). Curiously, most of these house structures occur definitely mainly south of the mined vein system, and here only to low percentage within the ore-bearing metasediments but in metagabbro. According to Koshin and Bassyuni (1968), the area only reveals minor exploitation traces. It consequently seems that this large and dispersed settlement had functioned less as a mining site than a halt or resting place for pilgrims on their way from Egypt to Mecca. By its general layout, anyway, the architecture seems to hint to an occupation in the Early Arab Period. Numerous, small dam constructions designed to collect the occasionally occurring runoff were identified in Google-Earth images. Further, especially in the SW peculiar round structures between 5 and 10 m in diameter seem to represent a line of tombs, which partially remind the isolated round tombs in the Nubian Desert.

Inside this widely scattered and unorganised settlement area is a noteworthy, fairly large compound whose architecture much reminds that encountered at the seemingly planned settlements Umm Eleiga in Egypt and Derahib, Bir Kiaw, Terfawi, Uar, and Omar Kabash in Sudan.

These sites have all in common that their layout had been arranged according to a standardised, almost military principle, thereby disclosing the advanced organisation of the building authorities. This notion is supported by the aspect that these settlements had not necessarily been directly involved in the gold production sector. Their central roads and various repository buildings seem to indicate to their function as supply stations for the gold mining sites dispersed within the close and distant surroundings. The gold seekers, in their turn, who may for the most have been com-

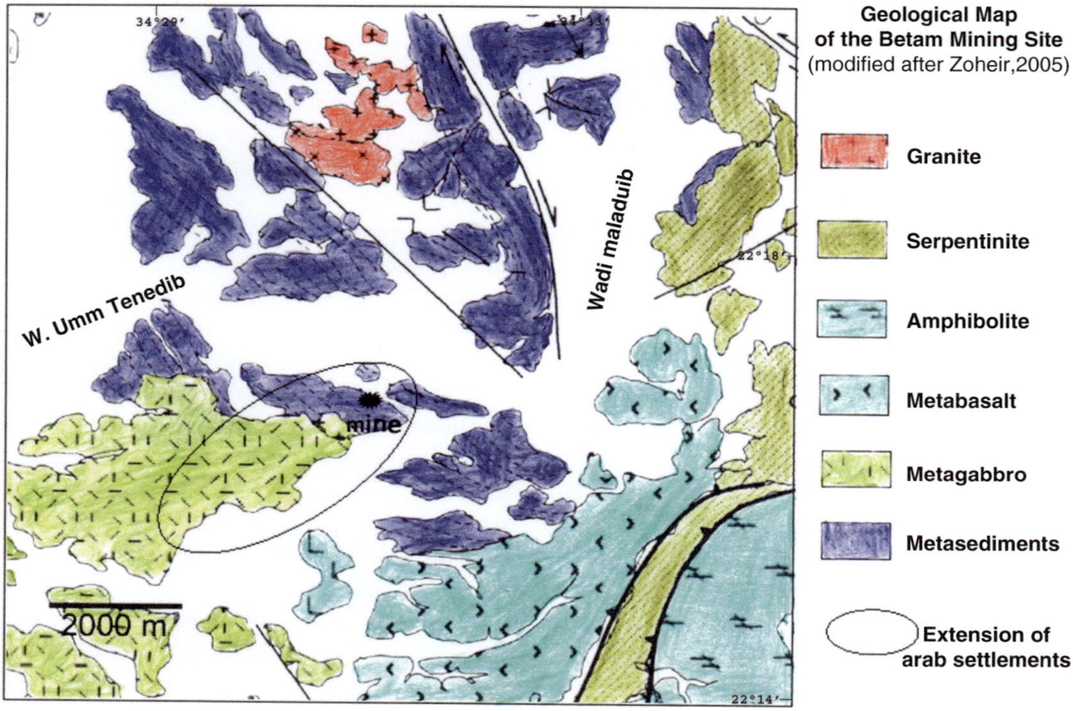

Fig. 5.276 Geological map of the Betam deposit (modified after Zoheir (2004))

posed of seasonal and migrant workers from the Arab Peninsula, seem by their settlement patterns to have been organised merely in loose, uncoordinated groups. By contrast, because of their paramilitary nature, historically identified work gangs of mainly Egyptian origin acting under the command of el-Omari during the reign of Ibn Tulun had probably been more hierarchically organised. It is therefore tempting to consider the mentioned settlements in light of that historical context. However, such deliberations would require much supplementary archaeological evidence from detailed investigations.

The geologic surroundings of the area consist of extremely disturbed, dark, rusty-brown schists, and occasionally graphite yielding metasediments. The latter are marked by a series of NW-SE striking shear zones running parallel to the foliation, into which later granite has penetrated. The metasediments are bounded by ophiolite metagabbro and metabasalt (Fig. 5.276).

The formerly mined gold quartz vein strikes parallel to the foliation in a NW-SE direction and dips at 45° NE. Its width may reach 2.2 m. More parallel veins within the silicified schist of the wallrock also seem to have been mined at the same time. The ancient extractions apparently had been relatively shallow, although there are possibly more vestiges deeper down, buried under the extremely fragile wallrock schist. Therefore, the determination of the gold contents of the ancient ores is also quite problematic. Samples from debris heaps produced values between 1 and 22 g/t Au (Jakubiak 1989).

Beside the genuine ore mineralisations in the quartz vein system consisting of pyrite, arsenopyrite, galena, and subordinately of chalcopyrite and gold, Zoheir and Qaoud (2008) describe three hydrothermal alteration zones in the wallrock close to the quartz mineralisation. They distinguish between an external chlorite-calcite zone, a middle sericite-chlorite zone, and an inner zone of pyrite-sericite. The latter may be highly auriferous, especially if traversed by small, sulphide-rich quartz veins.

Fig. 5.277 Ruin of an early twentieth century mining building and a dam at Abu Fas

5.7.21 Abu Fas

Geographic position:	22°08′79″ N, 33°52′50″ E

Three buildings from recent mining activities at the beginning of the twentieth century as well as a retention wall running straight through the wadi constitute the most conspicuous mining relics (Fig. 5.277). An alignment of round and rectangular huts containing round mills from the Early Arab Period is located in a wadi to the S (Fig. 5.278).

A small, wall-lined well in the middle of the main wadi has dried up (Fig. 5.279).

The geologic surroundings of the mined quartz lodes consist of amphibolite to sericite gneisses, partly with remarkable portions of carbonate, which strike approximately NW-SE and dip in the deposit area between 65° and 85° NE. The sequences are clearly folded. To the NW of the main ridge Jakubiak (1989) was able to make out acid metavolcanics.

Mining attempts had been made at three quartz veins, which all strike WSW-ESE. The northern and middle veins dip 45° NE and the southern one 75° NE. The veins are at about 1 m thick, occasionally increasing to 6 m. The examined quartz was translucent to opaque.

Around 1919, a contractor named Alfred Nahman prospected the veins in four short shafts and an intersecting face, although apparently without any ensuing genuine mining.

The small Early Arab settlement in the valley to the S of a ridge however, confirms all the same that gold had been extracted in the area.

Fig. 5.278 Early Arab Period mining site in Wadi Abu Fas just below the mined hill

Fig. 5.279 Sanded-in well in the Abu Fas mining district

Fig. 5.280 The Bedouin cemetery at Ungat and the two distinct peaks of Gebel Ungat

5.7.22 Ungat

Geographic position (Bir Ungat):	22°06′55″ N, 33°45′37″ E

On the plateau at the foot of the twin summit of Gebel Ungat lies an impressive Bedouin cemetery consisting of N-S oriented graves with the burials in each case pointing S. One of them is bordered by vertically deposed, narrow stones as well as a mihrab pointing E (Fig. 5.280). On some graves storage vessels for water and cereals had been placed.

At the foot of Gebel Ungat one recognises some house ruins with New Kingdom stone mills of red granite (Fig. 5.281). Some of them show a secondary grinding depression indicating a phase of reuse. This suspects a witness of a later Kushitic activity at the site. As we failed to discover any mine, we presume that the site was specialised in wadiworkings, today eroded away by occasionally occurring, rough water erosion.

Schweinfurth (1903) mentions graffiti of animals near Bir Ungat and, somewhat further away, a rock inscription by a scribe called Amenhotep. In addition to the mills, this may be another indication to gold production during the New Kingdom in this area.

In terms of geology, the site is set on a S-facing terrace of a wide and apparently frequently flooded plain. The catchment area consists of schistic metabasalt, intruded by the granite of the twin mountain Gebel Ungat. The granite has parallel flow structures at its margins and is partly marked by shearing and partly by a matrix without any particular orientation. The high-rising summits of Gebel Ungat consist of the latter.

Since no traces from ancient trench or underground mining were found in the area, and the wadi alluvium is basically judged as auriferous, gold production had probably been based on wadiworkings. The intensive erosion in the wadi, however, vetoed all observation of any such evidence.

Fig. 5.281 Scarce remains of a New Kingdom settlement at Ungat with oval mills (front right)

5.7.23 Romit

Geographic position (median position of the settlement):	22°19′31″ N, 35°47′00″ E

During our survey, we were neither able to visit this occurrence nor the one at Umm Egat. We therefore here render the short descriptions for both given by Koshin and Bassyuni (1968):

"It is an extreme south-eastern gold occurrence situated 5 km to the south-west of Gebel Ti Keferia on the left hand side of Wadi Bagawai. Numerous quartz veins occur in a crushed and decomposed diorite. The veins are composed of smoky quartz with calcite and siderite."

In the Google-Earth image one recognises an exceptionally extensive, apparently Early Arab settlement covering several hills and partly stretching into partly hidden positions in the southern wadi openings and consisting of several hundred huts. In spite of the images good resolution, mined vein systems as cited above were not identified in the mountains, though wadiworkings are clearly visible in the valleys.

Ball (1912) described Romit is an Arab processing site.

5.7.24 Umm Egat

Geographic position:	22°01′41″ N, 34°55′07″ E

We had no chance to visit this site directly at the Sudanese-Egyptian border, thus we again refer to Koshin and Bassyuni (1968):

"The occurrence of Umm Egat also known as Alfawi, it located in Wadi Allaqi, 5 km NW from the state border. There are three distinct lines of foliation of the schist of north-western trend, along which lenticular bodies of quartz occur, up to 3.6 or more meters in thickness. The ancient evidently worked this gangue and the softer parts of the vein. The presence of gold was found in material resulted from the oxidation of pyrite."

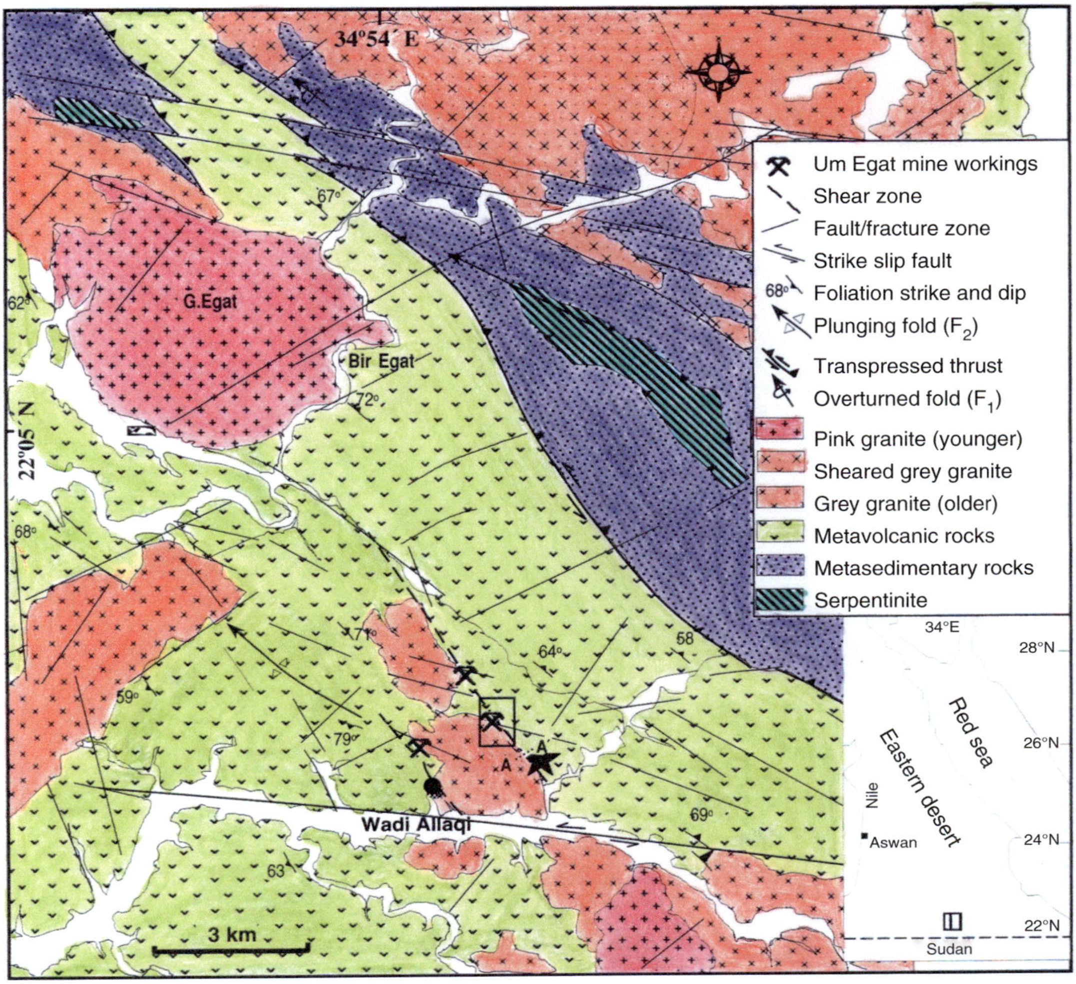

Fig. 5.282 Geological map of the area around the Umm Egat deposit (modified after Zoheir et al. (2008b))

A detailed geologic description of the occurrence was presented recently by Zoheir et al. (2008b). Thereby, the minerals worthy of extraction appear in clear boudins and lenticular quartz bodies within a heavily sheared sequence of metavolcanics. These are imbricated with metasediments and serpentinites in the W edge of the Hamissana zone (Fig. 5.282). The main ore minerals are arsenopyrite and pyrite next to low contents of chalcopyrite and galena. The sulphide minerals are limited to the margins of the quartz bodies. According to different geo-thermometers, they had formed between 270° and 340 ° C. Especially gold flitters smaller than 50 µm are associated with both first mentioned sulphide minerals in these marginal areas. Finally, gold also appears in altered wallrock parts where it concentrates in small, spongy globules during pyrite oxidation.

6 Gold Production Sites and Gold Mining in Nubia/Sudan

Contrary to the above discussion on the gold production sites in the Egyptian Eastern Desert, for which we were able to follow a rough sequence from N to S, this is presently no longer the case for the sites in Sudan's North-Eastern Desert (Nubia) or its hills along the Red Sea. The sites are therefore hereafter grouped according to either geographic or geologic sectors, within each of which, however, the same sequence from N to S is maintained (Fig. 6.1)

6.1 Group: The Red Sea Hills

6.1.1 Shishaiteb or Shishutaib

(Whiteman 1971: "Shashitaib")

Geographic position: 21°58′54″ N, 36°02′45″ E

This site is most probably an Early Arab mining attempt along a quartz vein running over some small hills (Fig. 6.2) at the given position. In the first half of the 20th century the site was intensely explored with several vertical shafts by M. Bishop, the miner of Oyo further south, but it seems as real gold production never started due to his all of a sudden early death. The excavated quartz vein at Shishaiteb itself strikes NNE-SSW and dips at an angle between 50° and 70° SE. It occurs in a granite and hornfels wallrock in an almost E-W (80° S) striking shear zone, whereby the hornfels probably represents former silt-stones. The vein's N-S striking Riedel fractures are all highly carbonated. A modern shaft located further S, which probably dates to the time of Bishop, is 20 m deep with a drift gallery that ends after 10 m. The thickness of the opaque to white quartz hardly exceeds 1 m, but the vein displays many clefts that are filled with partially well-crystallised carbonates.

The ore paragenesis consists mainly of pyrite, which for the most has decomposed to limonite at the surface, of chalcopyrite (partly malachite) and traces of gold.

However, a graveyard with seven graves is located near the modern mine. Tailings pointing to active gold production were not found.

6.1.2 Shishaiteb-South

Geographic position: 21°58′50″ N, 33°03′03″ E

Here are about 15 circular, relatively well-preserved Early Arab huts of about 60 cm high with double-shell walls.

R. Klemm, D. Klemm, *Gold and Gold Mining in Ancient Egypt and Nubia*, Natural Science in Archaeology,
DOI 10.1007/978-3-642-22508-6_6, © Springer-Verlag Berlin Heidelberg 2013

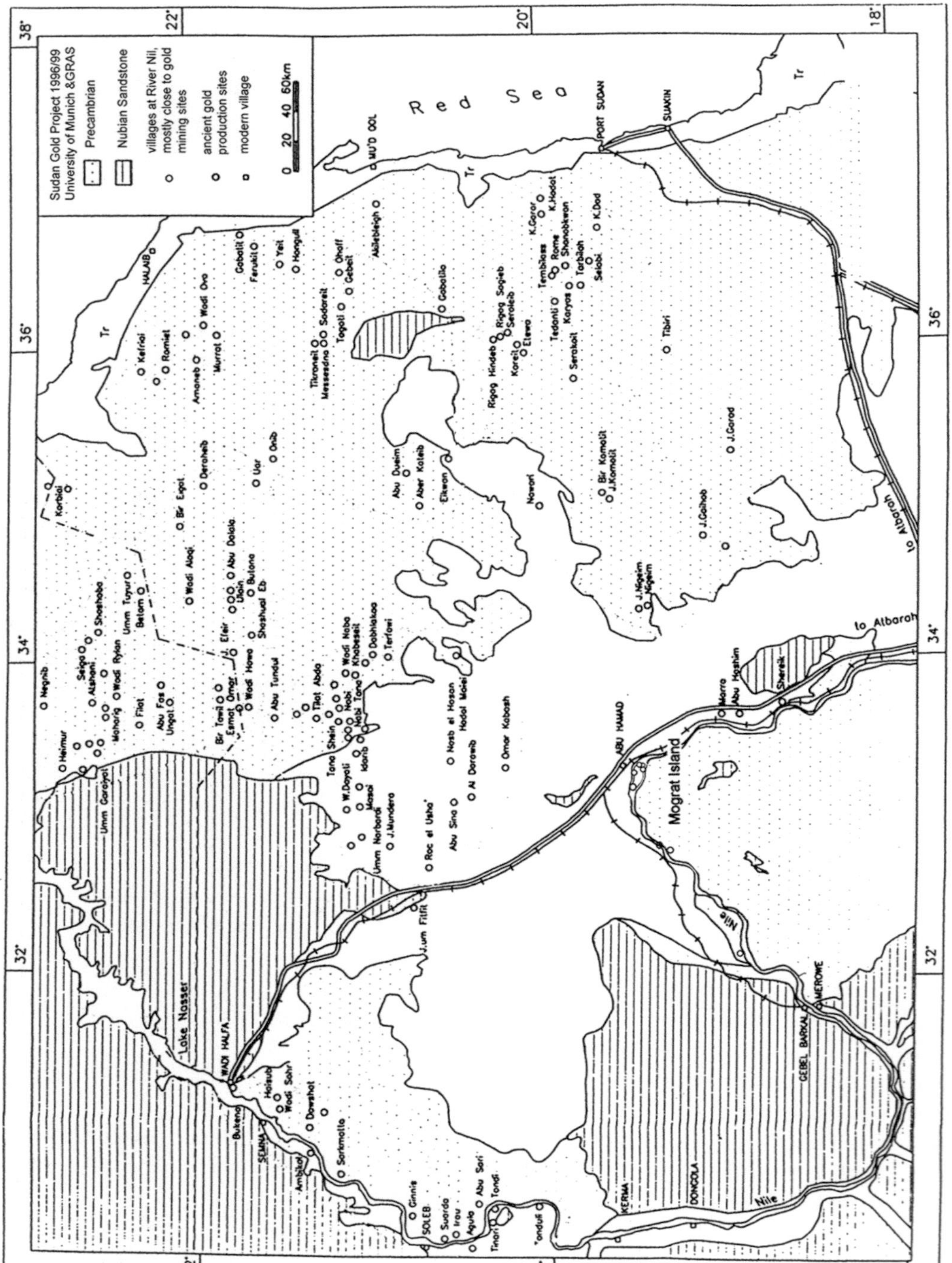

Fig. 6.1 Distribution map of known exploited gold occurrences in Nubia/Sudan

Fig. 6.2 Early Arab Period trench pit at Shishaiteb with spoil heaps covering the slopes

The geologic setting of Shishaiteb represents the hybrid margins of a large granite intrusion complex, which has assimilated large quantities of Nefirdeib metasediments that in terms of petrography are very similar to the Hammamat series. They are partly assimilated as hornfelses but partly also completely dissolved. The assemblage is viewed as particularly propitious for the formation of auriferous, hydrothermal quartz mineralisations (Fig. 6.3).

Fig. 6.3 Modern, approx. 20 m deep shaft at Shishaiteb, following a 1 m wide quartz vein

Fig. 6.4 Arab huts (foreground) in a side wadi of the relatively lush and well-hydrated Wadi Oyo

6.1.3 Oyo

Geographic position (modern tailing):	21°56′00″ N, 36°07′37″ E

The mining district at Oyo is accessed relatively easily once one has passed the difficult conditions in Wadi Diib. At position 21°48′21″ N, 36° 04′35″ E one bends-off into the wide wadi coming in from the NE and then at position 21°52′10″ N, 36°08′50″E into the smaller, tributary valley to the NNE where the track leads directly to the mine.

The more difficult route leads to Wadi Migraff via Wadi Diib from the SW. From position 21°55′35″ N, 36°05′39″ E, Oyo is attained in a 2-h walk through a boulder-covered gorge. Shortly before the actual deposit one reaches a more or less flat valley. Early Arab Period huts expectedly exhibiting round mills scatter along both its flanks as well as in the side valleys and gullies (Fig. 6.4). In between are recent house foundations from concrete and recently re-occupied huts from the 1930s and flimsy shacks built from steel barrels and canisters (Fig. 6.5).

Oyo is for the most a relatively recent site, at which both wadiworkings and underground mines have been extensively operated. Between 1920 and 1944 a notorious British miner called Michael Bishop had been in charge of the mine. Unfortunately he had managed to destroy virtually every trace left behind from earlier mining periods.

Fig. 6.5 Dwelling hut built from oil barrels and cans from the Bishop-era at Oyo. To the right, a round mill from the Arab mining phase

There are three wells in the main wadi. All were filled with fresh water during our stay in the area (Fig. 6.6).

The modern mines are located in a wadi leading S and mainly consist of a large cyanide tailing just outside the shaft and other associated facilities. The premises are scattered with hardware and car wrecks dating back to the 1930s (Fig. 6.7).

The quartz vein system at Oyo occurs in metasedimentary rocks of mainly conglomerates, greywackes, and siltstones apparently belonging to the Nefirdeib sequence (Fig. 6.8). The components of the conglomerates are sometimes quite sharp and reminiscent of the "Breccia Verde" in Wadi Hammamat, Egypt (Klemm and Klemm 2008b). Occasionally, the granitic rock components may also have rounded shapes and may measure up to 20 cm in diameter. The sedimentary series strike in a mean NE-SW direction, which corresponds to that of the series in the satellite image (Fig. 6.8). They are regularly interspersed with lenticular granite apophyses that possibly originate from the Shishaiteb-granite in the NW, which actually represents a clear indication for the proximity of a thermal energy source behind the productive quartz mineralisation.

Tectonically, the productive, hydrothermal quartz vein mineralisation is located in an approximately 7 km wide NE-SW oriented syncline, which essentially composes of the mentioned metasediments and which repeatedly has been displaced up to 100 m by NW-SE striking fault tectonics.

In a wider geologic perspective, the Oyo syncline represents an elongated massif, approximately measuring 15 km from NE to SW with an average width of 2 km. It consists of hornfelses from conglomerates, greywacke, and siltstones with intermediary layers of mainly acid volcanics that had sunken into the roof of a granite intrusion. In the periphery the granite magma had thereby contaminated by assimilation to granodiorite.

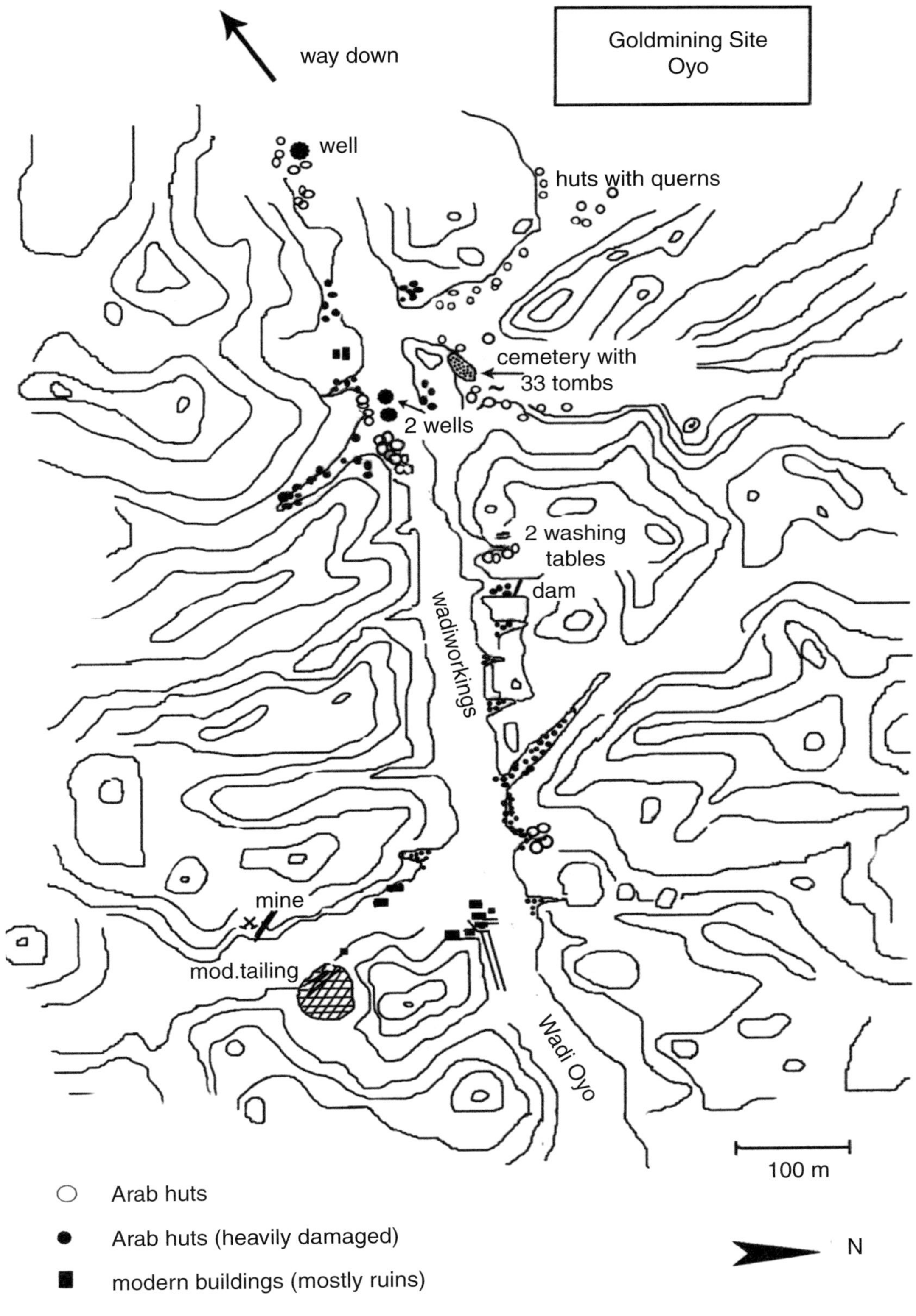

Fig. 6.6 Archaeological sketch map of Oyo (R. Klemm)

Fig. 6.7 Mining remains consisting of enormous tailings and retention walls from the first half of the twentieth century, in a southern side valley of Wadi Oyo

This intrusion may have emplaced simultaneously to the little less elongated and oval Shishaiteb granite, located about 9 km further W (Fig. 6.8).

The Oyo syncline transits into an anticline in the SE, whose core probably displays the same sequence of sedimentary rocks as the Oyo syncline. Its limbs however, are composed of an acid to intermediary sequence of magmatic, effusive rocks markedly reminiscent of the Dokhan volcanics of Egypt's Eastern Desert.

The Oyo folding is apparently disrupted by a major fault to the NW that runs through Wadi Oyo-Migraff. Further W it is cut by the Shishaiteb complex. Although at the surface the latter consists of metasediments evocative of the Hammamat sequence, these strike ENE-WSW and are extensively modified through contact metamorphism to hornfelsic rocks by the intruded Shishaiteb-granite.

Three quartz veins running through the upper reaches of Wadi Oyo had been mined. They are located at the SE flank of the valley and are accessible from a small tributary valley turning-off to the S. They occur in a slightly sheared zone oriented roughly parallel to Wadi Oyo with a predominant EW strike (but at places up to 70° E) and display an average dip of 45° S (Fig. 6.9).

The main shaft is flooded to the third face. A detailed recording of the drifts could not be carried out, and reference was therefore made to the historic plans.

The ore paragenesis in Oyo is relatively varied, consisting mainly of pyrite, idiomorphic arsenopyrite, chalcopyrite, some secondary

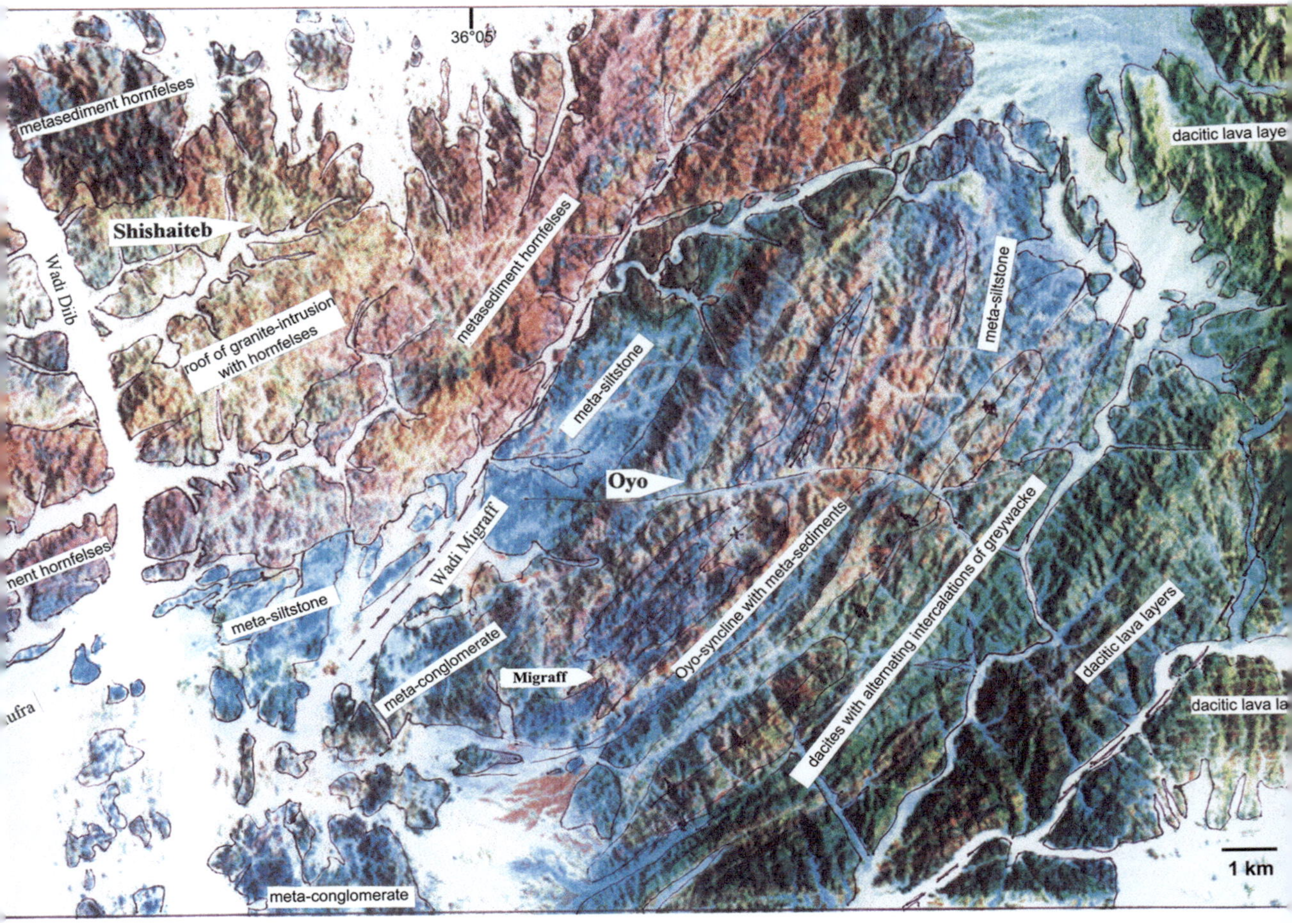

Fig. 6.8 Lithologically processed satellite image of the area around Oyo and Mikraff (TM 172/45, channels 7-4-1)

copper minerals, galena, sphalerite, and gold, of which some visible grains were observed.

Mining in the Early Arab Period had probably mainly been limited to wadiworkings, but possibly also operated within the upper oxidation zones of the mineralised quartz veins. Unfortunately though, most traces from ancient mining, especially near the vein mineralisation have been obliterated to such a degree that soleley detailed and time-consuming archaeological investigations would have to precede any reconstruction attempt.

This widespread damage is probably also the reason why no traces from the New Kingdom were recorded at Oyo. Considering the size of the deposit as well as especially the presence of visible gold in an almost ideal geologic environment, Oyo certainly had not escaped the highly experienced eyes of the New Kingdom prospectors.

Fig. 6.9 View into the mined quartz vein system at Oyo after modern exploitation. The auriferous quartz vein is only few centimetres thick

6.1.4 Mikraff

Geographic positions	
Twentieth century mine buildings:	21°54′28″ N, 36°06′17″ E
Arab wadiworkings settlement:	21°54′36″ N, 36°06′06″ E

In the actual valley of Wadi Mikraff there is a small, Early Arab settlement and a well. Both, the position of the settlement and that of the well at the edge of the wadi indicate that the settlement had primarily been specialised in wadiworkings.

From here a small wadi branches-off to the SE. After about 350 m one arrives at a deposit, which apparently had already been processed in antiquity. Unfortunately, the ancient traces have suffered so much through recent mining that only few huts from the Early Arab Period can be discerned (Fig. 6.10)

A quartz vein had been exploited in several shafts at the slopes of the SW valley flanks. The vein strikes NW-SE and dips 60° W. At its surface outcrops one notices a number of ancient extraction pits.

Another quartz vein, which probably also had been mined in ancient times, is located at the NE flank of the valley. It has a NNW-SSE strike and a 70° dip E.

The occurrence at Mikraff is located within an oval, granodioritic intrusion dome measuring about 700 m in diameter at the extreme SW end of the Oyo massif described above. However, intermediate to acid volcanic rocks seem to be predominate over the metasediments here.

Fig. 6.10 The Mikraff mining district with Arab Period settlement remains (modified Google-Earth image)

6.1.5 Hangul

Geographic position: 21°17′57″ N, 36°28′52″ E

The occurrence at Hangul is an Early Arab settlement consisting of about 50 round huts, which had apparently focused on wadiworkings. According to Boyle et al. (1986), a number of additional trial pits are still weakly discernible on both sides of the valley. Their locations are recognised by the absence of desert varnish. Furthermore, some prospecting had taken place at the surface of a number of quartz veins. The quartz veins run sub-parallel to the mainly NE-SW striking, layered rock sequences of the volcanosedimentary series. These series are largely identical to those of Gebeit and its neighbourhood, of which they represent the northeastern offshoots.

Fig. 6.11 Early Arab Period settlement remains at Ohaff with fractured round mills and pounding stones

6.1.6 Ohaff

Geographic position:	21°05′55″ N, 36°22′39″ E

In Wadi Ohaff is a small gold mine that had been run by Bishop in the 1930s.

The mine, together with its surroundings is much affected by this recent development, which greatly handicaps the detection of remains from earlier occupations. Even so, the following extraction periods were identified:

Early Arab Period (iron chisel and round mills).

Early twentieth century (two stone houses, one oven, purple bottle shards of English origin, terraced camp platforms along the mountain slopes).

Mining phase attributed to Bishop (tent platforms (?), beer bottles, cans, parts of machines at the mine itself, cyanide leach basin).

The small terraces were probably built because of the steepness of the slopes on both sides of the wadi and for protection from flooding.

The round mills are located at the mouth of a small side wadi near the mine (Fig. 6.11).

According to the MINEX report, there are more hut ruins in another side wadi.

A circular forge area with charcoal residues and a furnace are located on a small terrace in the Wadi.

The mine at Ohaff itself had been lowered into considerable depths. Outside the sealed-off adits is a large tailing from cyanide leaching processes (Fig. 6.12). There are also other large piles of untreated, apparently low-grade gold ores.

Next to these relics, no traces from earlier mining could be detected.

The mined, auriferous quartz vein occurs in massive to laminated tuffs and strikes between E-W in the S of the deposit, and SW-NE to the N. It dips between 30 and 70° S to SE.

Fig. 6.12 Large tailing at Ohaff from the first half of the twentieth century (Bishop-Period)

6.1.7 Gebeit

Geographic position (Bishop's house):	21°03′01″ N, 36°19′00″ E
Eastern New Kingdom settlement, wadi centre:	21°04′30″ N, 36°20′10″ E
Early Arab settlement:	21°03′52″ N, 36°18′58″ E

Only few data are available that might shed more light on the different extraction periods at most gold deposits in the Eastern Desert of Egypt and Sudan. The mining archives in both Egypt and Sudan only reveal little on the re-launching of the gold mining industry at the beginning of the twentieth century. On the other hand, for the deposit at Gebeit, Harrison et al. (1980) have published a summary report on its recent mining history. The following, brief discussion hence mostly draws on their work.

Thereby Theodore Bent from the Royal Geographic Society had identified the mine in 1896 as an economically viable gold deposit. Subsequently, a prospecting license covering an area of 800 square miles was handed out in 1903 to a certain Noel Griffin, a British geologist from the so-called Gebeit Mining Syndicate. The rights were then acquired by the "Sudan Mines Ltd." at the end of 1904, but which was forced to cede them in 1908 as a result of financial shortcomings.

In 1911 an Australian smith called J.W. Ashworth who had some mining experience, started work at Gebeit and began systematic extraction in 1913. By March 1914 he had already recovered 93 kg of gold produced at an average grade of 62 g/t Au. In 1915 already 110 kg of gold could be retrieved at a total production of 1733 tons of crude ore, which corresponds to an average grade of 63.5 g/t. With the introduction of cyanide leaching in 1917, a new chapter in the site's gold production opened. In 1922 gold production reportedly increased to 269 kg. This series of successful years, however, came to a sudden halt in 1928 with the abrupt departure of the director, Mr. Clarke, and other European staff members.

The "Sudan Goldmines Company" took up production in 1933 and continued operations

Fig. 6.13 Recent mining industry at Gebeit, abandoned in 1956

with variable, but yet good results until 1937. In that year, Michael Bishop took over the management and continued its operation with much success until 1942, when it had to shut down because of the war.

Mining operations were then resumed with much effort in 1948. Due to water leakage resulting from the temporary shut-down of several levels, productivity decreased to only about 7 g/t Au. With the sudden passing of Bishop in 1956, production was stopped. The total yield at Gebeit until Bishop's death thereby added up to at least 4,200 kg of gold (Fig. 6.13).

Eventually, after a feasibility study carried out during a year by Robertson Research International Ltd., the British Minex Mineral (Sudan) began successful production in a joint venture with the Sudanese Government and Greenwich Resources PLC until its unexpected halt in 1990.

The oldest mines in Gebeit are within the extraction area operated by Bishop. He extended old galleries, which by their typical sinuous course are believed to at least date to the New Kingdom. Bishop drove between 10 and 15 declines and around 20 vertical shafts into the ancient gallery network. Today, they are partly buried and have transformed the ancient mines, especially the ones from the New Kingdom, beyond recognition (Fig. 6.14).

A large tailing next to the miners' shacks still bears witness to the former mining operations.

Today the miners live in a makeshift village from sheet metal that has formed around the modern mine run by the Minex Company until the late 1980s (Fig. 6.15). They continue to extract ores inside the former under archaic conditions. As they probably leave their mining gear inside the mine, they disappear barefoot and equipped with only a rope into the depths to reappear 1–2 h later with half a bag-full of ore chunks. They then immediately hurry back into their sheds to continue their officially illegal activity in hiding. There the ore is pounded in mortars made from metal tubes with round bases (approximate diameter: 12 cm; height: 40 cm). Massive iron bars, about 5 cm thick and rounded ends, act as pestles. The pounded produce is then sieved and washed in tin bowls and pans.

Fig. 6.14 Ancient mines at Gebeit, to a great extent destroyed by modern mining in the first half of the twentieth century

Fig. 6.15 Sheet-metal shanties in the Gebeit mining district, seen from the old mining area

Subsequently, the gold is amalgamated with mercury. The mercury is then evacuated from the obtained alloy simply through heating, in which highly toxic mercury vapours are released and likely to be ingested by the workers. The raw product is then sold to various dealers or used as currency in the local shop. This in fact, was the alleged reason why none of the gold was shown to us. In fact, with respect to the activity's illicitness, we were denied gathering exploitable information, especially such as taking pictures of the different production stages.

The old Bishop tailing is reserved for the poorest of the poor as it too is panned before being processed as just described.

Around the old mines of the Gebeit settlement no ancient tools were found. This hardly surprised us considering the intensive processing that is still oing on here today.

6.1.7.1 Gebeit Surroundings

Next to the large Early Arab settlement mentioned above there are more, fairly small settlements in the vicinity. The largest is about 800 m to the N of the metal shacks, at the intersection of Wadi Gebeit with several tributary wadis (21°03′50″ N, 36°18′49″ E). Unfortunately these sites too, have suffered from widespread destruction, although their initial site can still be made out at the surface. Schweinfurth (1903) writes that T. Bent had counted between 700 and 800 ancient huts in the entire area at the beginning of last century.

Further settlement remains stretch along the eastern margins of the wadi as well as in the eastern tributary wadis.

One of the huts contains a round flour mill from vesicular lava as well as opaque-green glass shards often found at Early Arab Period sites in the region.

Pottery is completely missing. Among the tools are the usual assemblages consisting of round mills, pounding stones and stone hammers.

6.1.7.2 Gebeit Sharq (East Gebeit)

Geographical position:	21°04′30″ N, 36°20′15″ E

A large New Kingdom settlement to the NE of the main mine at Gebeit was found in Wadi Gebeit-Sharq.

About 15 rectangular, single and multi-room house ruins line-up along the side terraces on either side of a relatively wide tributary wadi oriented approximately parallel to Wadi Gebeit. Such houses usually occur in context with evidence from wadiworkings.

The settlement, however, is most probably also associated to at least two quartz veins mined in several pits over a distance of up to 300 m at the lower, western ridge. Traces from subsequent mining in later (Early Arabic?) periods, especially recent ones, however, have to a great extent obliterated the earlier ones. It is assumed that particularly the western part of the settlement was linked to the extraction activities at the vein.

Large quantities of small, oval grinding mills and fist runners were found. Moreover, there are anvil stones, of which some come in the shape of cubes with wear marks at several sides. They lie scattered within the houses, but sometimes also near the mentioned, ancient extraction sites. All tools have been blackened by desert varnish (Fig. 6.16).

The buildings average between 10 and 12 m in length (Fig. 6.17) and usually have five to eight rooms arranged in consecutive suites. The walls are built in a single shell and are preserved at average heights of only 50 cm. In this case too, there is a marked lack of pottery.

Since no round mill was found, it is possible that the site remained uninhabited during the Arab Period.

Two grave mounds are located near the site. The framing wall around one contains a number of oval, New Kingdom mills (Fig. 6.18). It is therefore at least contemporary or most likely later than the settlement. The walls of the second showed no datable artefacts.

Fig. 6.16 New Kingdom pounding stone at Gebeit-East, with several impact hollows

Fig. 6.17 Severely disturbed New Kingdom settlement remains at Gebeit-East, with numerous grinding mills

Fig. 6.18 Burial mound at Gebeit-East partly covered by New Kingdom mills

6.1.7.3 Tagoti and Tigranit

Geographic position:	21°07′47″ N, 36°11′ 38″ E

Both sites were described as "ancient workings" by Fletcher et al. (1984a).

In Tagoti there are two huts (one circular, one rectangular) in a small side wadi running next to Gebel Arrad. A small, probably related trench, though buried by drift sands is located at the mountain top.

Tagoti is a minor site, which had probably been briefly occupied in the Early Arab Period. This is for the most based on a fragment from a round mill, which interestingly had been made from a grinding mill dating to the New Kingdom.

The site of Tigranit was not found, despite a relentless search by one of our local teammates.

6.1.7.4 Walati

Geographic position:	21°07′40″ N, 36°14′ 19″ E

About 15, partly severely damaged house ruins from the Early Arab Period are located on a gravel terrace of a side wadi along the southern flank of Gebel Arrad. We found relatively large quantities of round mills, which though, had been destroyed intentionally. No pottery was found at the surface.

The mine, from which significant quantities of quartz chunks are still found inside the huts, is located at the top of the slope. Its blue-grey quartz however, seems to have been retrieved mainly for trial purposes.

The gravel banks to the W of the site contain some old tombs and are probably associated to the settlement.

6.1.7.5 Gumarob

Geographic position:	21°06′12″N, 36°17′00.2″ E

Gunarob is situated in a side wadi of Wadi Arrad, at the end of Khor Adarib, and consists of countless, small rock piles from former wadiworkings, which spread from the wadi up to the gently rising slopes. In the adjacent, relatively low mountains there are numerous outcrops of visible, light-colored quartz. Eroded lumps may have been picked up in the plain and the slopes and subsequently processed. However, neither tools, nor pottery, or even ruins from buildings were found at this location.

6.1.7.6 Khor Massesana

Geographic position: 21°09′59″ N, 36°20′11″ E

A small occurrence in the form of a narrow trench was located in Khor Massesana in a relatively flat mountainous landscape. Several round mills were observed right next to it, thus pointing to a date in the Early Arab Period. Other small quartz outcrops can be made out at the surface of the decomposed granodiorite in the surrounding mountains.

At the northern edge of Wadi Hudayo near its confluence with Wadi Direigat are the ruins of a sizable caravanserai next to a well (21°10′30″N, 36°20′41″E).

Here we found three fragments of green glass originating from a specific pilgrim flask destined to hold holy water from the Samsam well in Mecca. Such bottles were first published by Llewellyn (1903) and are frequently found at Early Arab mining sites.

6.1.7.7 The Geology of Gebeit and its Environment

The geologic environment of Gebeit includes the individual deposits of Gebeit and its surroundings as well as the occurrences at Tigranit, Tagoti, Ohaff, Walati, Gumarob, and Khor Massesana. In the following, they are therefore all discussed together.

The area consists of volcano-sedimentary sequences with an apparent prevalence for volcanic rocks (Fig. 6.19). In the actual deposit district of Gebeit the latter chiefly consist of rhyolite to dacite lava with partial feldspar inclusions measuring up to 7 mm, as well as agglomerates, and pyroclastics. Additionally occurring andesite lava and tuff generally seem to be located at the distal ends in relation to the actual mineralisation.

Through numerous core drillings carried out in the area of Gebeit during recent operations by the Minex Company, Sudan, a well-ordered drill core archive was established and made available to us. It proved extremely helpful for our understanding of the local geology.

Both, in the field, but especially in the drill cores numerous sediment variants are found that initially consist of tuffites and greywackes, but which also exhibit transitions to carbonate variations. Often these sediments contain some graphite-bitumen substances, and especially finely disseminated, mostly idiomorphic pyrite. Microscopic investigations however, revealed additional framboid structures indicating at least a partial, primary sedimentary genesis for pyrite. This finding is in so far relevant as the entire volcano-sedimentary sequence apparently emerged in a reductant marine environment, by which the so-called, recently much discussed VHMS (volcanic hosted massive sulphide) deposits occur by preference on the sea floor associated with intermediate to acid volcanics. Through hydrothermal leaching of the slightly auriferous VHMS deposits and through subsequent mobilisation of these fluids into shear zones and cleft systems, mineralisation of auriferous quartzes with low sulphide contents as observed in the surroundings of Gebeit, become conceivable. However, this provides the presence of a thermal anomaly supplying the necessary energy. Numerous magmatic dikes in the close and more remote surroundings of the deposit, as well as close by granodiorite apophyses and minor intrusions advocate the proximity of a magmatic heat center.

The primary quartz veins are much weathered and cavernous. Their thickness does not exceed 1 m but usually only measures less than 10 cm. They strike SW to NE and dip steeply between 30° and 75° S.

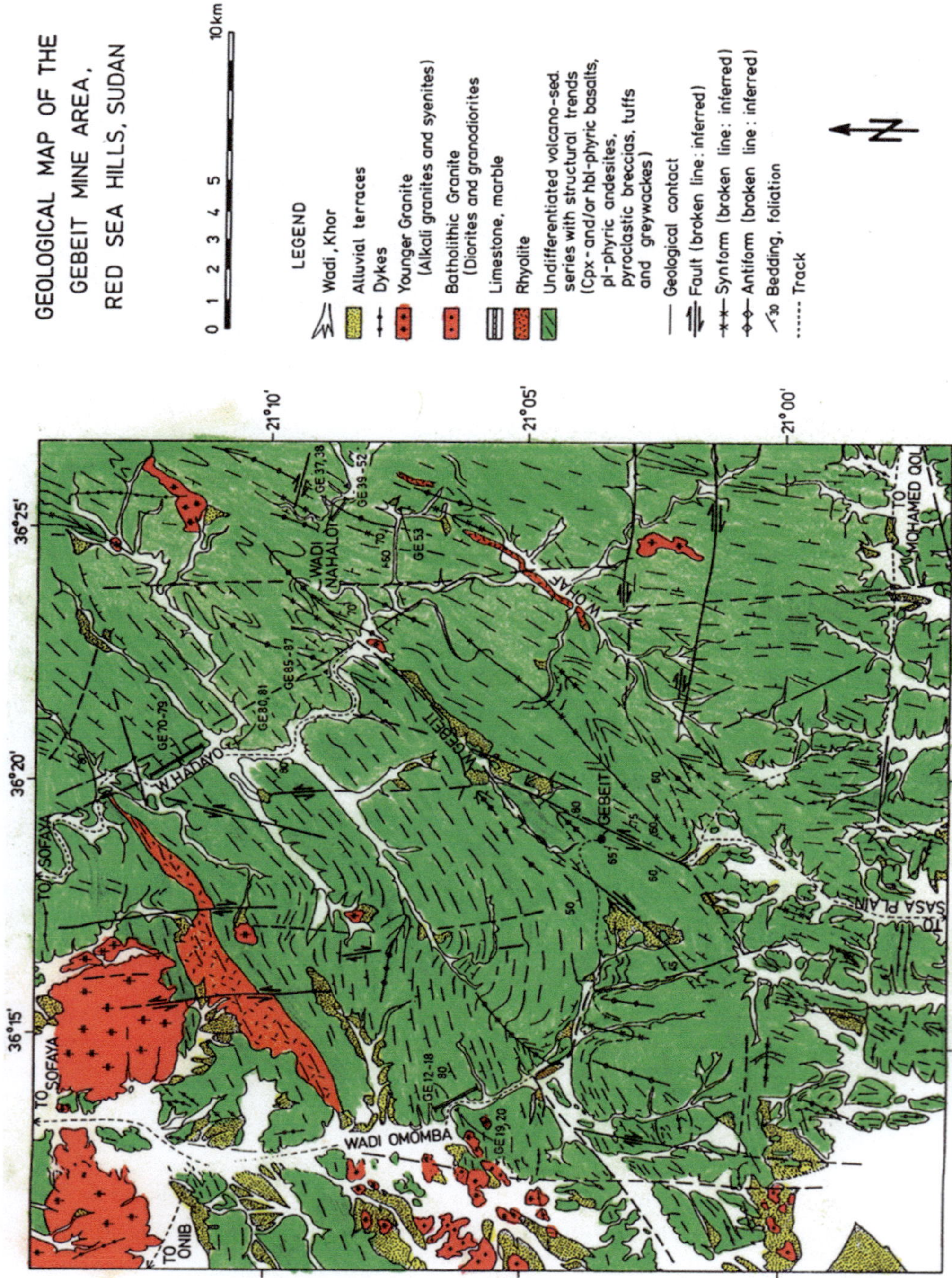

Fig. 6.19 Geological sketch map of the Gebeit area (modified after Reichmann and Kröner 1994)

Fig. 6.20 Early Arab Period house ruins in Wadi Gabatilo

6.1.8 Gabatilo

Geographic position: 20°31′39″ N, 36°12′43″ E

Wadi Gabatilo is very sinuous and accommodates on either side a vast Early Arab settlement with partly, yet well-preserved houses. They consist mainly of angular, shell-facing walls with entrances facing the slope sides. There are also detached, round huts, which often contain round stone mills (Figs. 6.20 and 6.21).

The building material consists of flat slate rocks gathered from the surrounding surface. Their relatively good state of preservation is probably explained by the flat shapes of the rocks, as the thereby increased friction inhibits impending displacement of the masonry.

Several, inclined washing tables assembled from gathered stones are found loosely distributed over the entire area. The water backflow systems are still partly distinguishable by the remnants from small basins. One also often observes small, cist-like units, usually just outside the house walls. The tool assemblage corresponds entirely to that of the Early Arab Period, which here in Sudan also often entails the absence of pottery.

As opposed to Fletcher et al. (1984a), who record 200 houses from the "pharaonic era" at Gabatilo, we counted only about 100 Early Arab huts. Evidence from older periods was not detected.

The geologic structure of the mountains surrounding Gabatilo reveals close similarities to that at Gebeit. It is therefore viewed as the former's southern extension displaying an interruption at the Sassa plain.

The rock sequences consist of 10 m (to about 100 m) thick series of andesite, dacite and rhyolite lava flows, tuffs and agglomerates in a dense, isoclinal fold structure with amplitudes ranging between 200 and 300 m. This in turn, is superseded by a relatively large fold structure. As to their stratigraphy, the units are referred to in the geology of Sudan as part of the Nefirdeib series.

Fig. 6.21 Early Arab Period round mill with rotor disc in Wadi Gabatilo

Minor as well as tiny quartz veins run parallel to the foliation structure ubiquitously through the adjacent mountains. Among them, only the blue-grey variety is auriferous. Evidence from scant mining activities is preserved in the form of two small mines. Together with the wadi deposits, these were probably the reason for the establishment of this settlement in Wadi Gabatilo.

6.1.9 Nafirdeib (Lakobdog)

Geographic position:	20°30′32″ N, 36°24′49″ E

This site, which also is referred to as Nefardeib, is located on a hill in a tributary valley to Wadi Nefardeib. It consists of ten huts and a 20–30 m long, Early Arab mining trench in a shifted vein. Whereas a number of round huts may have served as sheltered workshops, a single, rectangular building with several rooms may have been the dwelling house. In all, it consisted of three rectangular rooms and two, fairly large and round annexes.

Since the round mills from the local andesite rocks show little use wear, the extraction period at the site must have been comparatively short and was probably contemporary with that at Gabatilo. Yet again, there was no pottery.

In the gorge leading from the higher lying deposits down to the side wadi of Wadi Nafardeib we located an inclined washing table along with a small stone cist.

The rock sequences around this small mine consist of andesite lava and tuff. Similar to the evidence from Gabatilo, they are folded with dacite to rhyolite sequences. Despite the slightly different spelling, this site seems to be the name-giving location for the Nefirdeib series, which determine the geology of the entire surrounding mountains.

6.1.10 Rigag Sageib (Rigag Hindeb or Ragaghundab)

Geographic position: 20°11′42″ N, 36°01′38″ E

In the 1930s a British Army Sergeant Major called Foley reactivated this ancient mine by driving four shafts into it.

Just S of the mine is a fairly large settlement from the 1930s consisting of well-preserved, round houses. Their foundations are partially made from round mills that had been brought here from a neighbouring Arab Period site. The latter consists of very severely damaged, round to rectangular hut ruins revealing one and two rooms. At least two inclined washing tables are still barely discernible.

In samples from debris heaps there were traces from visible gold. Though this was evidently secondary, relocated gold, as it was found inside the cavities left behind by dissolved pyrite in the vein quartz. Probes of this gold revealed a purity of up to 93 % Au, while that of the primary gold bound in the pyrite had lower values between 85 and 87 % Au.

The gold mineralisation is bound to a quartz-mineralised shear zone in an andesite wallrock. Several quartz generations were identified.

Extraction had been carried out in four shafts of which however, three probably originate from the most recent mining under Sergeant Major Foley. The apparently ancient shaft in the S of the vein was readjusted and lowered to considerably deeper levels. Today, the shaft is filled with groundwater below 50 m, thus obstructing any examination of the vein beyond that point.

A systematic sampling of the rock sequences in the upper parts of the mine was undertaken. With an additional methodical sampling of the debris heap, we were able to identify ore parageneses between specularite hematite and pyrite, galena, chalcopyrite, sphalerite, and carbonate (ankerite/siderite) within the quartz. No visible gold was detected in the parageneses dominated by sulphide, but primary gold particles were detected under the ore microscope.

The immediate wallrock in Rigag Sageib is a porphyritic andesite with macroscopically well-visible hornblende inclusions. But in the outer periphery andesite tuffs also alternate with carbonate layers and apophysis-like, small, fine-grained diorites. To the NE of the deposit a fairly large granite pluton with a granodiorite aureole intruded the described sequence above (Fig. 6.22).

The entire district was intensively combed through during 2 days, in search for the sites Rigag Sageib 2 and Rigag Hindab, but unfortunately without success. It eventually turned out that Rigag Sageib was identical with Rigag Hindeb. The survey at least had served for identifying a reasonably wide range of different rock varieties whose determination would otherwise have proven problematic in processed satellite images. Subsequently, they were coded in the aerial photos. The coarse-grained granodiorites and diorites finally exemplify that the geologic structure of the entire region is much more complex than the dominant volcano-sedimentary series as initially suggested.

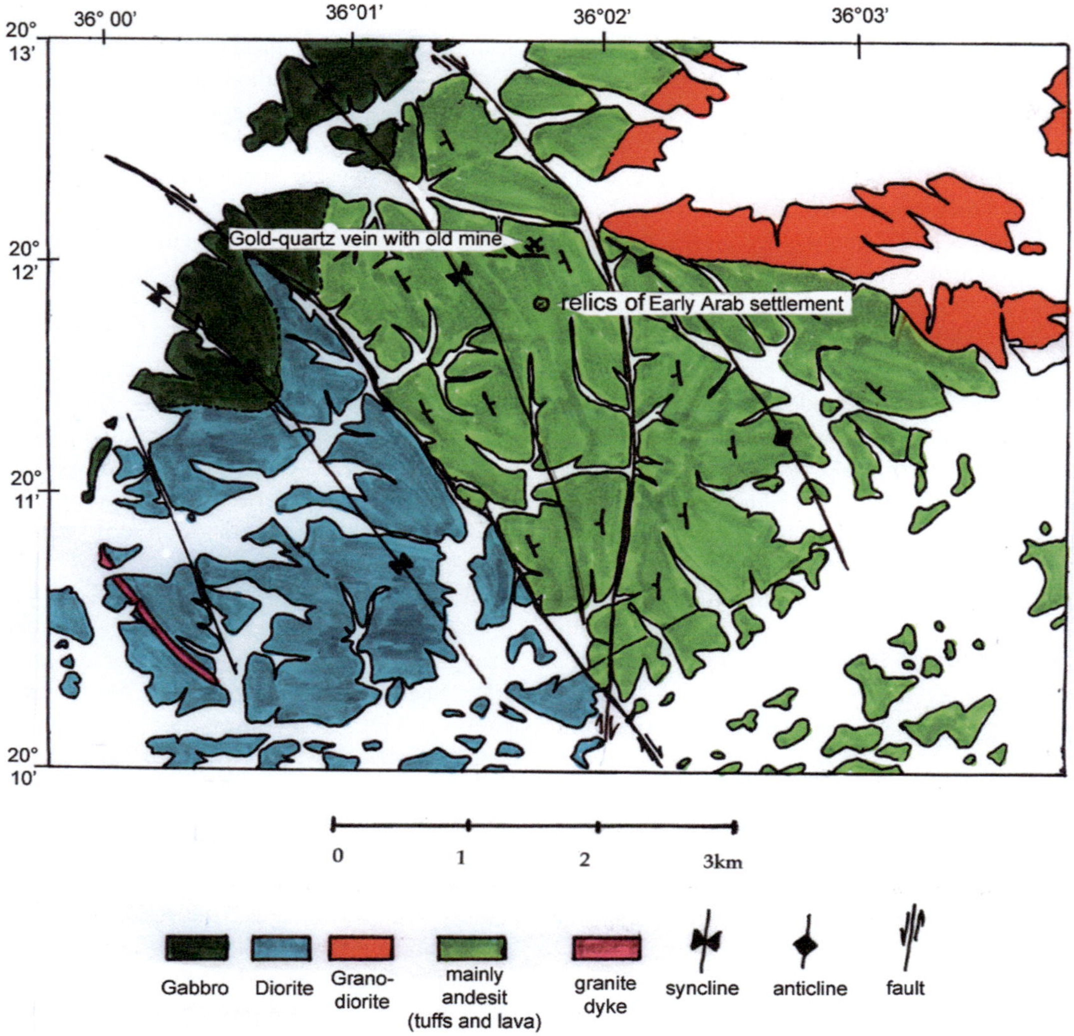

Fig. 6.22 Geological sketch map of the Rigag Sageib area (T. Kirchbichler)

6.1.11 Shikryai (also Serakoit)

Geographic position:	20°00′46″ N, 35°38′58″ E

The name of this deposit is Shikryai. In some reports it is also referred to as Serakoit, so called after the relatively distant well at Bir Serakoit.

This ancient gold mine, which was still functioning in the early days of the twentieth century, came under scrutiny by the Soviet-Sudanese joint "Techno Export" project in the 1970s. This virtually resulted to a total destruction of about anything that could have been of interest to archaeologists.

The mine follows an elongated vein with irregularly arranged extraction drifts that seem to adhere to a selection of rich-ore-zones. Thereby short, lateral drifts also came under excavation, which occasioned the seemingly disordered appearance.

In front of the shaft adit, which was only built under the Soviet-Sudanese prospecting, are the remains of some New Kingdom grinding mills and runner stones, but in a slightly smaller version and with a deeper use wear than otherwise usually

Fig. 6.23 Beadly damaged Early Arab Period settlement site at the foot of Gebel Shikryai displaying intentionally broken round mills

observed in Egypt. The depressions in the original New Kingdom grinding mills indicate a secondary re-use at some later stage, which we strongly suspect to have occurred in the Kushite-Meroë Period (cf. below).

Our hopes to discover occupational remains from the New Kingdom Period were disappointed in spite of our efforts to locate them.

An extremely ruinous, Early Arab settlement consisting of scattered huts, a prayer site, and associated small-finds spreads-out along both sides of the main wadi, at the foot of Gebel Shikryai (Fig. 6.23). Small, cist-like containers made from stone slabs were often noticed next to the huts. Llewellyn (1903) thought them to represent special receptacles destined to hold gold dust. We prefer to think that they had a more general storage function.

To the N of the mining site, there is a relatively large processing platform with an adjacent, well-preserved building from flat slate slabs. Its two, interconnected rooms were accessed from the NW corner. The width of the double-shell walls averages around 80 cm, while the narrow cavity in its center is filled with fine gravel to achieve extra stability as well as protection against the wind.

Quartz ore lumps obliterated for processing are preserved in sizes ranging between a fist and a melon. All mills at the site had been deliberately broken. No remains from former tailings or washing tables were observed.

Because of several recorded, narrow trench pits, which are viewed as characteristic for as early as Middle Kingdom workings in Egypt, we decided to systematically search for typical fist hammer rejects in the slope debris, but in vain. In some areas however, we were able to identify chisel marks that seem to suggest a date to the New Kingdom.

At this site too, the exploited quartz mineralisations follow approximate, N-S striking shear zone systems, which in their turn are largely determined by the strike of the volcano-sedimentary wallrock sequences (Fig. 6.24). Apparently, this fault zone was still active after the mineralisation of the gold-quartz, because even at the deepest,

Fig. 6.24 New Kingdom pits at Shikryai in intensively sheared volcano-sedimentary wallrock

still accessible mine face about 20 m below the original outcrop accessed through a recent shaft, only oxidised ores were recorded. Identifiable ore parageneses are therefore very meagre. Next to the dominant quartz, which occurs in several generations, only limonite, hematite, and occasionally some siderite were found. Only the debris heap that had seemingly formed during recent mining attempts by the Soviet-Sudanese prospectors, produced residues of chalcopyrite, arsenopyrite and pyrite. An analysis of this debris material also revealed tiny gold sequins in this paragenesis.

The wallrock consists predominantly of andesite lava with occasional tuff interstices to tuff lapilli and slightly metamorphic agglomerates. In addition, there are also rhyolite layers and dikes as well as occasional, intermediate marble layers. Several displacements suggest an intense imbrication of the sequences from E to W.

6.1.12 Romi

Geographic position: 19°44′32″ N, 36°33′03″ E

In Fletcher et al. (1984a, b) Romi is described as an old "processing center". Indeed, a spacious Early Arab Period processing platform is found here, though without any recognisable traces from associated mines. Therefore the site may have served for processing quartz chunks collected in the wadi bed.

The site is most of all marked by a rectangular complex built around a central courtyard with a now silted-up well. The complex was originally equipped with corner bastions, and along the inside of the walls, there were long room alignments. The walls are generally preserved in only one single masonry layer. The individual rooms measure almost invariably 4 × 5 m and consist of double-shell walls of about 50 cm width. The facility was accessed through a central gateway in the S. A path of the same width as the entrance leads up to the well in the centre of the complex. Outside there are more ruins, and even further out one notices more hut remains, burials, as well as a prayer site. At some locations along the wadi margins are the partly well-preserved remains of a wall that probably served as a protective flood barrier.

Silhouettes of round tombs are seen near the hilltops and in the plain are grave mounds scattered between hut ruins. These mounds had probably been amassed with stones formerly used for the huts and consequently would post-date the settlement period of the site. Here and there, one comes across round mills from either light granite or the local quartz porphyry or rhyolite. Pounding and hammer stones are found as well. We even discovered a round flour mill from coarsely-pored, vesicular lava whose origin has yet to be identified in Egypt and in Sudan. The sparse pottery finds consist of a simple and coarsely tempered, local ware and grooved amphora fragments of a red ware dating to the Early Arab Period.

In spite of thorough investigations, no evidence dating to the New Kingdom periods was found outside the well complex. However, within its rubble wall masonry some New Kingdom runner stones, one grooved mallet, and an old fist hammer clearly displaying traces from use wear were found. We did not find any oval grinding mills of the New Kingdom period which would have complemented the runner stones.

There was no tailing heap near the well (19°44′33″ N, 36°32′57″ E), but this was hardly surprising considering the relatively frequent rainfall in this area relatively close to the Red Sea coast.

Here at Romi as well, the round mills had been deliberately destroyed.

The geology of the site consists of an alternating sequence of metasediments and metavolcanics.

6.1.13 Sheiteb

Geographic position: 19°44′19″ N, 36°31′25″ E

Except for four ancient trenches, there is no other archaeological evidence at Sheiteb. Fletcher et al. (1984a, b) designate the site as a "prospect", which most probably is the case.

The trenches had been dug into fine-grained rhyolite rocks, which had been intruded by a microdiorite.

6.1.14 Shanobkwan 1

Geographic position (shaft):	19°42′57″ N, 36°34′06″ E

This site is referred to by Fletcher et al. (1984a, b) as "ancient workings".

In the midst of a slightly profiled plain rises an inselberg topped by a so-called Alam, a tapering stone cairn to mark the landscape. It is flanked by an ancient mining trench. Approximately 300 m to the N is a house ruin dating to the Anglo-Egyptian condominium, as evidenced whisky bottles and other shards from the modern era. The house has two rooms, and its 1–1.5 m high walls are plastered both inside and outside.

This most recent occupation had probably contributed significantly to the further destruction of the heavily damaged hut ruins in the immediate vicinity of the mine. Compared to the mine's size, the number of huts seems too low. The shell-facing walls, the small hammer stones with wear marks at both extremities, the round mills (of which, many had been deliberately broken), and a very large praying space date the site to the Early Arab Period. Another prayer site is located in a shallow depression, about 300 m E of the huts. It consists of an approximately 20 m long and slightly curved wall preserved in heights between 30 and 50 cm and whose centre reveals an approximately 1 × 1 m wide mihrab pointing to Mecca.

A typically oval, New Kingdom grinding mill was also recorded at the site. It so far forms the most convincing evidence that mining had already taken place on a moderate scale during this early period, although no occupational traces have been revealed.

The mine itself is about 150 m long and has at least three, partially caved-in shafts. The westernmost shaft possibly dates to the time of the condominium. An underground inspection led to the discovery of ancient, narrowly winding galleries connecting the shafts to each other. In the slope debris of the western part of the mine a worn stone tool very similar to a hand axe was found.

The 60 cm thick quartz vein with ancient mining traces reveals at least three quartz generations (45° E/70° N, 55° E/80° N, 45° E/72° NW). It is bound to a dextral shear zone system in a medium-grained granodiorite, which itself had intruded a sequence of acid tuffs and compact volcanics. Close to the vein it is extremely altered by hydrothermal activity.

Noteworthy as well, are the partly quartz-mineralised Riedel faults (64° E/82° N, 80° E/8NW, 80/85 N, 75° E/90°).

The three quartz generations consist respectively of a white, a probably older grey (auriferous?), and of another white variety in the Riedel faults. Fletcher et al. (1984a, b) were able to detect 16.5 g/t Au, with an average content varying between 7 and 10 g/t. Ore mineralisations within the samples were not visible under the magnifying glass.

6.1.15 Shanobkwan 2

Geographic position:	19°42′57″ N, 36°33′02″ E

This small mine cuts across a valley, and today, its trench is largely filled-in, but probably hadn't been very deep to begin with. To the W in the plain, which is strewn with slope rubble, fragments of round mills scatter all over. Pounding and hammer stones are abundant, but no potshards were found.

The mine follows a quartz vein, which is visible at the surface for a distance of about 500 m (strike 27° E/dip 88° E). Now buried shaft holes are located at its northern and southern ends. The quartz vein itself follows a shear zone in the granodiorite.

6.1.16 Selobit 1

Geographic position:	19°39′50″ N, 36°37′26″ E

Selobit is a relatively small vein occurrence at some distance up the slope of the mountain of the same name. The evidence from some 12–15, although severely ruined huts about 200 m away seem to indicate that the mine had been exploited for a comparatively short phase in the Early Arab Period. This is also reflected by the insignificant use wear observed at some of the round mills.

An investigation resulted to an average of one round mill per hut (Arabic: "raha"). All round mills had been deliberately broken, in many cases

Fig. 6.25 Boudinaged quartz vein in heavily sheared and altered quartzdiorite at Selobit 1

only through marginal chipping at one spot. This phenomenon has been described repeatedly in previous published reports. In addition, there are the typical small hammers and about 25 × 30 cm large, amorphous stone slabs on which the ores had been crushed to a pea-sized grain.

The shaft of the ancient mine follows the dip of the vein. The shaft can be accessed to a depth of about 6 m. Actual extraction occurred on two quartz veins with widths between 10 and 45 cm in respective strike/dip rates of 24°/60° NW and 26°/63°NW. It had formed through hydrothermal circulation in a quartzdiorite hostrock. It is emplaced in mylonite shear zones and even displays significant boudinage (Fig. 6.25). The hanging wallrock shows clear to significant alteration through silicification and sideritic carbonation.

The original depth of the shaft is no longer determinable due to caving-in. The drift gallery, which is also collapsed, had been extended only towards the N.

The vein quartz itself is milky-white and reveals no distinctive mineral generations. Macroscopically visible sulphide minerals or even gold particles have been identified neither in the rock material inside the mine nor in the sparsely existing debris material just outside and in the settlement area.

6.1.17 Selobit 2

Geographic position:	19°39′29″ N, 36°37′11″ E

About 1 km to the SW of the Selobit 1 mine are three small workings. The westernmost one consists merely of a single, deep shaft, probably from prospection work carried out at the beginning of the twentieth century. No findings of archaeological significance were observed in its immediate vicinity.

The mined quartz vein strikes 5° W and dips steeply W (due to inadequate exposure, its dip angle couldn't be measured). The hostrock is a clearly altered microdiorite. The three pits evidently represent the result from prospecting work only, which had only produced little debris material.

6.1.18 Miradaab

Geographic position:	19°39′17″ N, 36°37′18″ E

The mine at Miradaab is an ancient, mined trench running tangentially along the slope. It was partly opened at the surface in a 20–30 cm wide gap, which raises some technical questions. Here and there, the trench broadens, at other places it is buried. The debris heaps and the surface in the surrounding vicinity of the mine revealed neither tools nor pottery. In contrast, approximately 500 m to the E of the mine were five rooound mills, but no architectural remains or tailings. The mine at Miradaab is therefore assumed to date to the Early Arab Period, which then would also account for the narrow gap, which for lighting purposes could have been hammered out by means of long iron chisels.

The largely collapsed and slit-like trench follows a quartz vein, whose thickness is not well determinable. It is bound to a shear zone and strikes 46° E with a SE dip. As it is affected by metamorphism, the identification of the wallrock is not well identified on a macroscopic level. It is either a microdiorite, or else a re-crystallised, metamorphic andesite. On the other hand, residues of diorite are detectable in the sheared parts of the hanging wallrock.

The quartz chunks gathered from the debris heaps are milky-white and revealed apart from oxidation traces from iron, neither sulphide nor gold mineralisations.

6.1.19 Khor Dat and Miteb

After a several hour-long march through ravinous terrain, which is even impassable for cross-country vehicles, we did not succeed in finding the positions of Khor Dat and Miteb as indicated by Fletcher et al. (1984a, b).

According to Boyle et al. (1986) the occurrence at Khor Dat is located at 19°38′52″ N, 36°44′38″ E. A high resolution optical scan of a Google-Earth image of the area around the indicated position revealed no evidence of a possible mining site.

The coordinates given for Miteb were equally erroneous, which is particularly unfortunate, because according to Boyle et al. (1986), extremely high gold values of up to 100 g/t had been registered there.

The investigation as to the reasons for these important inaccuracies showed that the specified coordinates resided on older measurements. These may diverge considerably from data obtained with modern GPS measurements. Even the slightest deviations of only few seconds may render the location of a given site impossible, especially as it often is likely to consist of small trial pits, which even in the field may barely be visible.

6.1.20 Abidoidib

Geographic position:	19°37′35″ N, 36°36′16″ E

Abidoidib is a small, ancient mine, which consists of two parallel trenches running horizontally just below the surface through schistose rock. A shaft leads down to one of them, from which a drift gallery branches-off. The second is filled-in. No artefacts were found near the mine. Just as the sites mentioned above, this mine seems to have been relatively insignificant and wouldn't have required any specific settlement.

The workings are bound to several, only few centimetre-thick quartz veins that run through a shear zone system (strike: 10° W, dip: 63° W) in a greywacke and siltstone sequence.

It is important to note that neither pronunciation nor spelling of place names are standardised in this region. Consequently, we were for instance unable to tell out whether or not "Abidoidib" and "Abidoididi" designated the same place, and we didn't find out either which of the two localities under the cited as "Miradaab" is referred to in the published reports.

6.1.21 Romeib

Geographic position:	19°37′20″ N, 36°29′ 25″ E

This small mine trench measuring only between 4 and 5 m in length, is easily overseen in the plain. Only fragments from one or two round mills dating to the Early Arab Period were found in the immediate vicinity of the mine. Consequently it is highly probable that it only represents a small trial mine.

The deposit is located in a much altered sequence of metasediments and marbles, into which a comparatively large serpentinite nappe had been tectonically thrusted. This is the area where the round mills concentrated. They had presumably been used during a short period for the processing of ores from wadiworkings.

6.2 Group: Wadi Amur-Ariab

6.2.1 Tibiri

Geographic position:	19°20′49″ N, 36°03′33″ E

The Tibiri deposit is located within a flat, undulating and mountainous landscape. It consists of parallel quartz veins in a NW-SE strike through terrain covered by light, shimmering quartz. They show traces from several, shallow pit workings today partly filled-in by sand (Fig. 6.26). Due to high gold concentrations within very restricted zones revealing visible gold in handpieces (Fig. 6.27), the sector was thoroughly explored in investigations by the French Geological Survey (BRGM) and the British Minex Company that also involved deep drilling.

In the eastern part, at the edge of a hill are a number of huts and a few weakly discernible, round mills. Surface pottery was not found.

Tibiri, Igariri and Hashai are located within a large prospecting area warranted to MINEX, which was classified as economically promising. From a standpoint of archaeology though the area seems to be less interesting.

The wallrock of the quartz veins in the area of the main workings is a coarse-grained gabbro, which changes to diorite further E. To the W, the mined quartz vein is located within a metadacite.

Fig. 6.26 Strikingly red ore spoil heaps at Tibiri resulting from oxidised pyrite. In the foreground, the badly damaged remnants of an Early Arab Period settlement

Fig. 6.27 Tiny gold particle (*arrow*) in an originally pyrite-rich cavity of a quartz ore chunk from Tibiri

The main quartz vein is up to 2 m wide. Next to it are other trenches oriented in parallel direction. The primary ore paragenesis consists of pyrite and chalcopyrite. The clefts within the weathered zone contain hematite and occasionally visible gold (Fig. 6.27). Again, the gold seems to occur primarily in sulphide minerals, and only in a secondary, re-arranged and enriched state together with hematite. Other secondary minerals are malachite and limonite.

6.2.2 Aliakateb

Geographic position:	19° 20′19″ N, 35° 55′24″ E

In the midst of the glistening, quartz-covered flats of Khor Takhwan rises a low hill crossed by several quartz reefs that protrude in butte-like ribs. We recorded three vein directions NNW-SSE, ENE-WSW, NE-SW.

A large, New Kingdom time settlement site spreads-out between this hill and a granite inselberg in the NE (Fig. 6.28). It consists of an estimated 50–60 houses built in the typically elongated, rectangular architecture, which has been well documented in the Eastern Desert of Egypt for the period in question. The relatively well-preserved walls are from rounded granite blocks and covered by drift sand to a point that only the top masonry layers stick out from the surrounding sand layer (Fig. 6.29).

One of the largest houses measures 30 × 8 m and comprises a suite of five rooms and interior passageways. Usually, the houses only have one or two rooms and average around lengths between 8 and 12 m.

Fig. 6.28 Much sanded-in, large New Kingdom settlement at Aliakateb, Wadi Amur

Fig. 6.29 Aligned, multi-chambered New Kingdom houses at Aliakateb

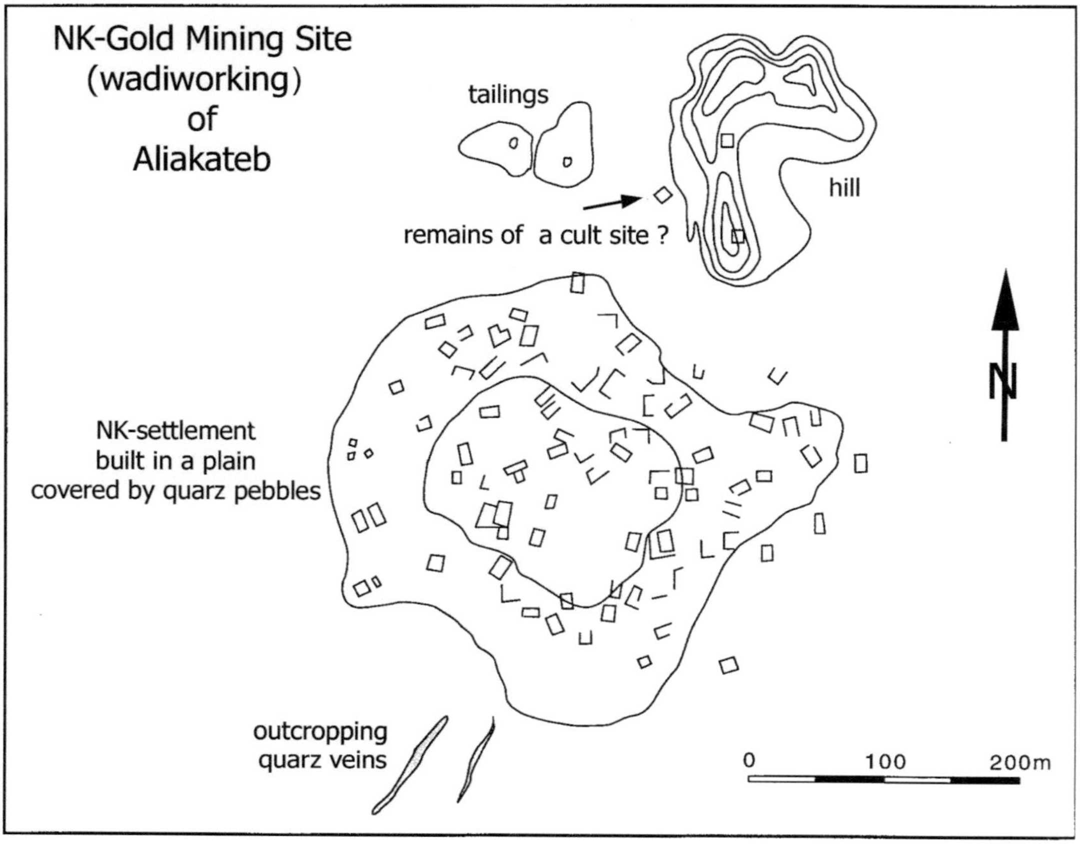

Fig. 6.30 Sketch plan of the New Kingdom settlement mound at Aliakateb (R. Klemm)

Few intact, but many deliberately smashed, New Kingdom oval mills scatter over the site. Their grinding surfaces are partly in an advanced stage of corrosion, although some fragments from a light-grey, granite rock are free from weathering and have smooth surfaces. The runner stones are slightly smaller than usually observed in Egypt. The same is also valid for the mills, which here too, had been re-used and accordingly exhibit a more advanced use wear than those from the New Kingdom, possibly pointing to a later re-occupation of the site.

Within the settlement area, at least three major debris heaps were counted. They are from sandy, brown earth interspersed with charcoal fragments, of which several samples were taken for radiocarbon dating. Tailings were not identified inside the settlement. Approximately 100 m to the N of the settlement, two flat and very light-coloured sand mounds probably represent the sagged residues from former tailings (Fig. 6.30).

Both mounds are interspersed with Nubian pottery fragments, some belonging to ceramic phase II b (after Bitak 1968) dated to mid of the second century BC, but most of them are younger and date to the late first millenium AD.

Interestingly, this pottery has so far been thought to be mainly restricted to the Nubian Nile Valley region only. Hence, its occurrence in a far distant desert area is a new finding (Fig. 6.31).

Fig. 6.31 Nubian pottery fragments from Aliakateb

6.2.3 Wadi Amur

Geographic position:	19°19′01″ N, 36°03′57″ E

South of Gebel Malifay numerous New Kingdom stone mills scatter at a site between timeworn Romibs (pie-shaped graves) and a couple of grave mounds. The mills are covered with desert varnish, and are therefore difficult to detect in this rubble-dominated landscape.

The mills from local granodiorite rocks are particularly noteworthy. They at first bring to mind the New Kingdom oval stone mill type, but they also are more deeply hollowed-out (Fig. 6.32), similar to the ones seen at Nubt (c.f. Nubt). Because this very hollow mill at Nubt represents a later re-use of the former New Kingdom type, it therefore only seems reasonable to assume that here too, this mill is a more recent version, maybe dating to the Kushite Period.

The site is dominated by several large, circular graves called Romibs. It is conceivable that there is some connection between them and the processing site indicated by the mills. To determine this link, it would be necessary to carry out archaeological excavations at the graves and elsewhere at the site.

Fig. 6.32 New Kingdom ore mill, displaying considerable use wear, probably from secondary utilisation later in the Kushite Period. Wadi Amur

6.2.4 Hadanaib

Geographical Position:	19°17′40″ N, 35°51′21″ E

In a dark wadi filled with gravel, which also contains quartz chunks, is a small, Early Arab site consisting of five to six round huts barely distinguishable at the wadi's edge. Some well-preserved, round mills along with typical anvil stones are found too. In between, there are small heaps of quartz rock. Because no mine was detected in the vicinity, this site had probably been specialised in wadiworkings.

Two quartz generations can be distinguished. One is red and identified by our Bedouin guide to be the auriferous one. The second is light-grey to white and is supposedly barren.

The red colour originates as usually observed, from secondary limonite-hematite on clefts inside the quartz. The wallrock consists of a distinctly sheared (shear direction: 20° E) quartz porphyry, and an equally sheared microdiorite.

6.2.5 Igariri

Geographic position:	19°36′09″ N, 36°10′17″ E

Due to erroneous coordinates, and especially to the poor driving conditions in Khor Igariri, we initially failed in finding this site. Finally, one of our geologist colleagues managed to get there with the help from a local Bedouin.

6.2.6 Hashai

Geographic position:	19°11′32″ N, 36°09′31″ E

This site at the narrow and gorge-like, rear end of Khor Ambufeki is only reached by foot. The Early Arab settlement is located in a bend just before a steep ascent to the surrounding mountain ridges. An associated mine in the immediate vicinity was not found.

Fig. 6.33 The Early Arab Period gold mining settlement at Hashai

Up to 12 m long, relatively well-preserved houses had been built in thick double-shell walls from local, flat rock slabs with the entrances facing the slopes (Fig. 6.33).

A delimited praying site displays a mihrab in its centre which as at Nubt, is oriented towards the NE and not towards to Mecca (Fig. 6.34). This may represent the orientation towards Al Quds (Jerusalem), which in the Early Arab Period had actually claimed the prayer direction.

A massive quartz vein crosses the site, and several smaller veins can be observed in the surrounding mountains. Tailings or washing tables were not observed.

Unfortunately, no pottery was found. Only a single round mill including a disc-shaped rotor stone points apart from the mosque to the Early Arab Period. A second mill was located in the rubble about 100 m further down the wadi, where on a left-hand-side wadi terrace one barely just discerns the potential remains of another settlement.

A Hashai-mining site occures about 4 km away east of the settlement along the flank of Gebel Dangag where one comes across a gossan within a red-brownish area coloured by limonite. According to Newall et al. (1985) the site displays an "adit and shaft". It occurs within a heavily foliated and displaced sequence of acid, chloritised metavolcanics (rhyodacite) and andesites. Immediately at the ore outcrop the footwall sequence consists of a much kaolinitised metavolcanic, which opposite to the actual ore body is formed like a butte. The working focused on a 1–5 m thick and almost perpendicularly plunging limonitic gossan zone. Small, centimetre-large quartz lenses are found here, as well as a rock with intense micro-folds. Newall et al. (1985) described this rock as a "shale" in close association with the sulphide body. The near-surface alteration of the sulphide thus led to the formation of the gossan and thereby to a relative enrichment of gold. The mine itself follows the strike of the gossan. The southern mine consists of a small shaft, which is already blocked after 2 m. The main mine is also filled-in after few meters. The drift follows the quartz-hematite zone.

Fig. 6.34 Mihrab oriented towards Al Quds (Jerusalem) in a mosque at Hashai

Both, the weathered lenticular sulphide bodies (mainly former pyrite) and the geologic setting with the intensely folded metavolcanic streak, which continues its course to Ariab, display striking similarities with this remarkable gold mine.

6.2.7 Nubt

Geographic position:	19°01′41″ N, 36°10′43″ E

Not listed by Fletcher et al. (1984a, b), our attention was drawn to this site by Sudanese geologists. We found it after long reconnaissance work and much help from a local Bedouin.

The site at Nubt catches the eye primarily as a burial ground. It might be referred to as a "central cemetery" that apparently had developed in connection with some other important Bedouin sites. It may for instance have grown over a lengthy period around the burial of a high dignitary.

The oldest burials are certainly linked to a group of at least 30, so-called Romibs resembling enormous stone pies amidst the flat, hilly terrain (Fig. 6.35). Between them are a collapsed building from light-coloured sand bricks and an irregularly distributed number of smaller square or round structures of schistose rock, which probably are graves too, although more recent ones than the Romibs.

In the close periphery of the graves lie scattered fragments of tombstones from a pale, streaky marble. Some are framed by incised, decorative bands, often topped by a frieze composed of hexagram motifs, so-called Solomon's seals. The framed area cites the lines of the first sura in the Quran, and often gives a date and the name of the deceased individual (Fig. 6.36).

Because of the dates on these tombstones, of which some are being kept in the National Museum in Khartoum, the occupation of Nubt falls within a period between 750 and 914 AD (Glidden 1954; Sanders and Owen 1951). Furthermore, the inscriptions often quote women, which may suggest that the settlement was occupied on a perrennial basis.

Fig. 6.35 Different grave types at Nubt. The ones in the foreground are single burials framed by stone circles. Behind them (*arrow*) is a large, cake-shaped grave (Romib), most probably of an older date

Fig. 6.36 Marble tombstone from Nubt with Arabic inscription and a frieze consisting of Solomon's seals (*arrow*)

In this respect, it is however remarkable to observe that the site and the surrounding area is virtually devoid of domestic architecture. It is therefore conceivable that Nubt had exclusively consisted of tents and huts from organic materials according to Bedouin traditions.

According to Y. F. Hasan (1967), Nubt had been the first Arab locality in the country of the Bedjas and also a camel trading post with important connections to the ports of Suakin and Badiya. He further reports that, up to 60,000 camels were used for the supply of the miners working throughout the Nubian Desert. The camels had primarily been reared by the Bedjas and had been much valued in the region.

The ruins of a large complex, apparently a mosque, are located on top of a bordering hill. Its semi-circular mihrab is again located at the north-eastern side of its wall. In addition, a spacious stone wall delimits the entire site, including also the surrounding mountain ridges. Presumably, this zone had once corresponded to the Early Arab Period trading post.

According to our accompanying Bedouins, so-called Red Sea Towers are located to the S, behind a mountain crest, dividing the site in two. We were able to locate such towers on several occasions during our visit to the area. Our own impression is that these towers are actually graves built for important personalities, a view which is supported by their actual location within the great cemetery of Nubt (Fig. 6.37). The towers cluster along the pilgrimage and trade route leading to the port cities of the Red Sea (Hasan 1967).

In the southern part, although before one arrives at the mentioned crest, one distinguishes at least three mounds of fine rock debris pervaded with pottery fragments. Between them lie decayed vestiges of domestic architecture. The debris contains mills reminiscent of the ore mills from New Kingdom in Egypt, with the difference also observed elsewhere that they are smaller and more worn-down (Fig. 6.38).

Because the entire area around Nubt is devoid of round mills and its wider surroundings visibly lack traces from mining, it is quite likely that the settlement had not been involved in the gold processing industry during the Arab Period. The evidence from the oval grinding mills, however, does indicate that Nubt had indeed been a gold mining site during the New Kingdom. Their secondary wear hollows show that they also had served at some later stage, probably the Kushite Period, again for gold processing. Since there apparently had been no deep mine in the area, the gold ores had probably been recovered in wadi-workings. The site's development to a trading post had only occurred about 2,000 years later.

The name "Nubt" as well as "Nubia" may be associated with the ancient Egyptian word "nub" which means "gold". However, since Nubia is referred to in Ancient Egypt by a variety of different names, and because the name "Nubia" does not appear before the Christian Period, this presumption remains conjectural. A. Grimm (1988), however, discusses a list fragment from the Temple of Djedkare-Asosi from the end of the fifth Dynasty (Old Kingdom) tallying "exotic animals, plants and cities", and quoting a country referred to as "ta nebu" or "Land of Gold". He identified this land with Nubia, a region for which the list gives three cities as the dispatchers of the gold deliveries. This is so far the only reported evidence of an etymological connection between "nub" (gold) and the geographical denotation "Nubia". But more than anything else, it demonstrates that Nubian gold occurrences had already been known as early as in the Old Kingdom, which is only confirmed by our observations in the field.

Fig. 6.37 One of the so-called Red Sea Towers, concentrating especially in the Wadi Amur area. These isolated, funerary towers are found along the route leading to the Red Sea and are linked to pilgrimage and trade rather than to gold mining

Fig. 6.38 Originally evenly worn New Kingdom stone mill from Nubt, probably reclaimed and worn down in the Kushite Period

6.2.8 Ariab

Geographic position (western pit entrance):	18°41′52″ N, 35°23′34″ E

Ariab is the name of an E-W oriented wadi and that of a gold mine, which is jointly managed by the French BRGM and the state of Sudan (GRAS) under the supervision of the "Ariab Mining Company". It is thereby the only gold mine in the study area currently under exploitation. It was discovered only a few years ago with modern prospecting methods (Fig. 6.39). Today, the district consists of seven separate, opencast mining operations. Since no traces from ancient mining are known from here, the area is interesting only from a geo-scientific perspective.

The gold mineralisation is bound to a much leached-out, secondary decomposition zone of a massive sulphide mineralisation in former marine rhyolites. The gold itself occurs inside a fine, porous quartz residue matrix, in an extremely fine grain. It had therefore been undetectable to the ancient prospectors.

This type of mineralisation is crucial for our understanding of the genesis of hydrothermal gold-quartz mineralisations connected to vein systems and shear zones within a volcanogenic environment, because it may represent their primary ore type.

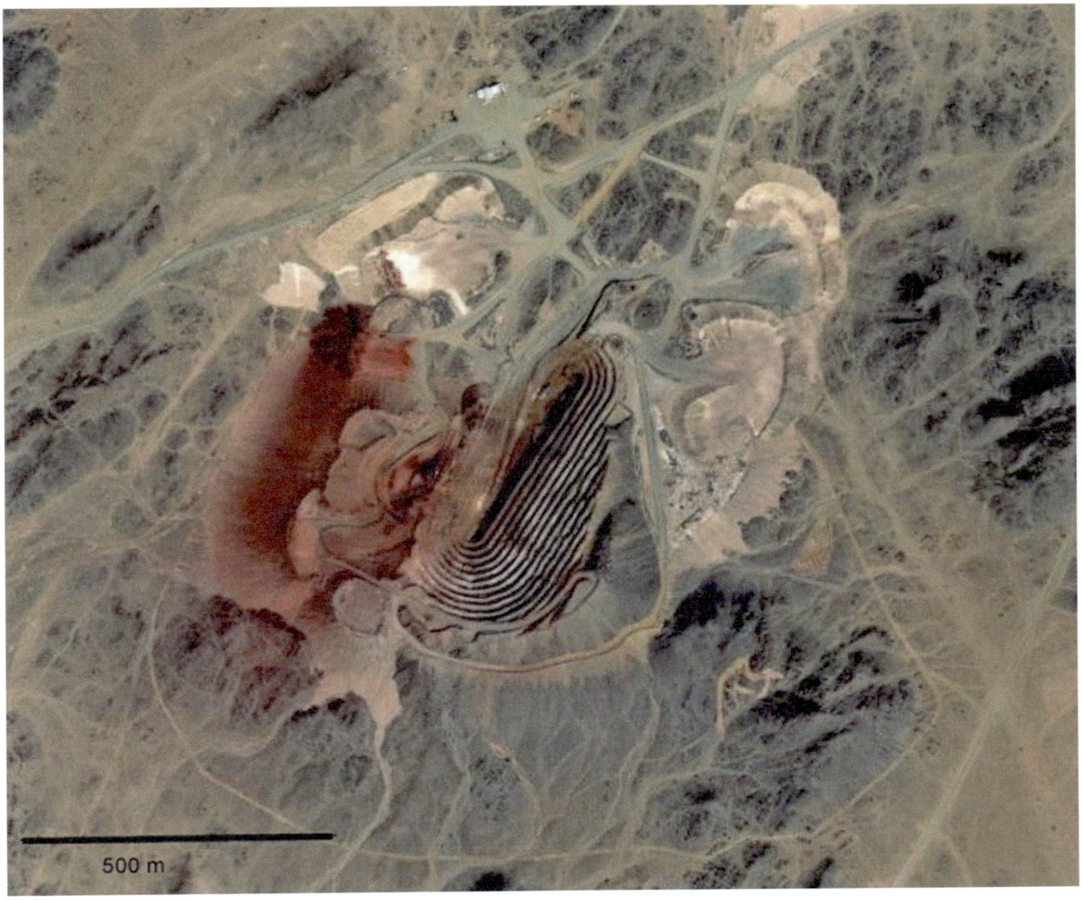

Fig. 6.39 Large, open-cast gold mine at Ariab (modified Google-Earth image)

6.2.9 Mishalliet Gurad

Geographic position:	18°55′20″ N, 35°13′37″ E

This site was pointed out to us by an employee at the Ariab mine. It is located in volcano-sedimentary hill chains, about 30 km NW of the Ariab mine and W of Gebel Gurad.

This Early Arab Period site is marked by a very wide area of wadiworkings to the NW of Gebel Gurad. It covers approximately 1 km^2 in a plain dominated by low hills, where typical double-shell houses scatter in a loose distribution. Several, inclined washing tables with still visible water basins can be discerned. Tailings were not recorded (Fig. 6.40).

Furthermore, there are countless potshards, for the most a light-coloured ware in a relatively good condition. The shapes consist mainly of amphorae and cooking pots. Coarse and relatively thick oven walls are also found and masses of greenish glass shards.

The occurrence was prospected by GRAS, whose traces are still visible everywhere.

Fig. 6.40 Relatively well-preserved washing table with (sand-covered) collecting basin and a round mill at the far end at Mishalliet Gurad

6.2.10 Ganait

Geographic position:	18°43′56″ N, 35°15′45″ E

The Ganait mine is located high on the mountain and strikes by its red-brown colour (Fig. 6.41). Its narrow structure and slightly sinuous course evokes the mines of the New Kingdom era, although most of the mining activity probably occurred during the Early Arab Period. According to our accompanying French geologist, within the framework of a decade-old, lead and zinc prospecting program carried out in a joint venture by BRGM and GRAS, gold was accidentally discovered. It was only then that mining started in neighbouring Ariab.

In the lower terrain and on the gentle slopes of the surrounding hills near the mountain there are isolated huts displaying the typical mining inventory from the Early Arab Period. In a wadi located S as well as immediately W of the mine, several huts had been built in an alignment. In all, they add up to a number between 40 and 50 units. It struck that many of the therein contained round mills displayed only little use wear. This finding alone would imply a relatively short extraction period, although the appearance of the mine contradict this. It seemed therefore reasonable to look for older occupation traces in the area.

A thorough inspection of the close and distant surroundings of the mine eventually led to the discovery just to its SE of several New Kingdom oval mills and their respective grinding stones near two large, round graves and even included to their stonework (Fig. 6.42). Again, the mills are smaller than the ones in Egypt and especially more worn down. Quartz chunks lie

Fig. 6.41 Severely sheared quartz veins at Ganait, probably exploited already in the New Kingdom, but predominantly in the Early Arab Period

Fig. 6.42 Large circular tomb at Ganait covered by numerous New Kingdom mills (*arrows*), indirectly hinting to older gold mining activities at the site

scattered around, indicating that next to genuine mining such loose ores had been processed here too. Unfortunately, hardly any remains from architecture were preserved, and probing the adjacent wadis in the hope to encounter typical complexes from the New Kingdom remained without conclusive results. Consequently, an estimation of the scale at which mining was operated in this period is problematic. Only the mine itself seems to hint to intensive extraction activities in this period, especially considering that genuine deep mining had only rarely been practiced in the Early Arab Period, as it had been usually restricted to wadiworkings or shallow quarry trenches rather than underground galleries.

On the other hand, Dr. Merani, historian at Khartoum University pointed out to us, that “Arabs” are generally preferred to the Bedja Bedouins when it comes to excavating holes for sub-surface constructions, such as wells, foundations etc. The Bedja on the other hand, are allegedly still reticent by their nomadic traditions to “dig” the earth. From a cultural perspective, mining skills were therefore only prone to develop among “Arabs”. Especially within a desert environment, this cultural distinction between both groups is enhanced, which partly also accounts for the reference to an explicitly “Arab” mining.

A lode-like ore body running parallel to the rhyolites and rhyolite tuff layers had been mined at the site. In addition, approximately 100 m N of it is a dark grey quartzite, presumably containing graphite, which suggests an origin within an anaerobic environment. The productive vein mineralisation consists mainly of barite, which through mineralisation of hematite has become bright red.

The vein itself was accessed through two main shafts. Samples were taken from the seven first meters from the southern shaft. Because of substantial sand deposits however, its original depth remained undetermined.

6.2.11 Derbeikwan

Geographic position:	18°43′10″ N, 34°55′54″ E

Approximately 50 km in a straight line to the W of Ariab one arrives at the large wadiworking site Derbeikwan, in the middle of a white and glistening, quartz-covered plain next to a flat ridge marked by heavy sand drifts.

The entire area has been rummaged through, which has resulted to a characteristic landscape of innumerable, flat quartz rock piles. We also found elongated trenches, again bordered by low heaps in the same way as occasionally observed at wadiworking sites in Wadi Allaqi.

The strikingly few and carelessly built huts of this site show by their double-shell walls nevertheless the typical features of Early Arab Period architecture. As a rule, they have a round to rectangular shape with only one room. The finds consist of round mills. Remarkably large numbers of washing tables were recorded between the huts. In the southern part of the site a large ore processing site at the edge of a long trench was recognised. It was covered with at least 10–15 round mills and numerous anvil stones. The remains from several washing tables can be distinguished here, although due to strong erosion, the associated tailings have disappeared.

The installations are not easily identified in aerial and satellite images as the entire area is intermingled with steeply dipping panes of andesite and andesitic agglomerates that may easily be confused with house ruins. Moreover, the surface of the area is marked by important sand drifts, which also tend to cover the ruins (Fig. 6.43).

Auriferous quartz veins occur in highly sheared, approximately N-S oriented volcano-sedimentary sequences of the Nefirdeib formation. They are comparable with the mineralisations near Wadi Gebeit and Wadi Amur.

No genuine mine was found. The processed ore material consists of quartz with some pyrite and malachite and traces from visible gold.

Fig. 6.43 Early Arab Period gold mining site at Derbeikwan covered by sand drift. The entrances to some of the huts are marked with large stone slabs (*arrow*)

6.3 Group: East of the Hamisana Suture

The Hamisana suture separates the Gebeit terrane from the Gabgaba terrane in the W. It therefore constitutes a major tectonic boundary zone.

6.3.1 Kamoli

Geographic position:	21°45′20″ N, 35°08′51″ E

The site of Kamoli is represented by an Early Arab Period mine up in the western mountain slope of Wadi Kamoli, a small northern side wadi of Wadi Rak, and surrounded by a number of huts built from globular granite nodes.

Small quartz swarms concentrate in a highly sheared graphite schist mined in an elongated trench (Fig. 6.44). As a result from boudinage of a formerly massive vein, these swarms have in addition disintegrated into single, lenticular segments. Few meters further up the eastern side of the wadi one notices the beginning of a highly stretched granite lens, which is post-intrusively heavily deformed. In its own turn the granite is penetrated by a yet, slightly deformed gabbro. This gabbro is the wallrock to the small vein deposit at Bir Rak discussed hereafter (Fig. 6.45)

The relatively meager observation conditions at the partially collapsed trench nonetheless reveal a steep dip E and a 5° E strike. Due to comprehensive oxidation, the original ore paragenesis cannot be adequately determined.

Fig. 6.44 Long trench running down the W slope of Wadi Kamoli

6.3.2 Wadi Rak and Bir Rak (Fig. 6.45)

Geographic position:	21°45′14″ N, 35°09′14″ E

The site is located in the upper section of Wadi Rak, not far from Kamoli. One recognises a number of quartz-veins along the northern margins of the wadi. Numerous Early Arab hut ruins are hidden away in the neighbouring tributary wadi. A small settlement is located near quartz outcrops on the southern mountain plateau. Throughout the northern part of Wadi Rak there are numerous additional house ruins, of which some reveal stone enclosures. The ones in better conditions of preservation are built from schist, the rest from granite rocks. They had probably been associated to former wadi-workings (Fig. 6.46).

This small deposit, whose economic importance had definitely not been very momentous, is most of all remarkable for its location within an intrusive gabbro. It had intruded an elongated granite body by deformation and was subsequently deformed in its own turn. Thereby, a relative dating of the deformation phase within the Hamisana zone becomes possible. It occurred approximately simultaneously with the intrusion events, in which first one portion of the intrusive bodies was deformed and then to a lesser extent the other (Fig. 6.45).

The macroscopic ore paragenesis of the slightly folded quartz vein consists of pyrite and chalcopyrite. Visible gold was not found.

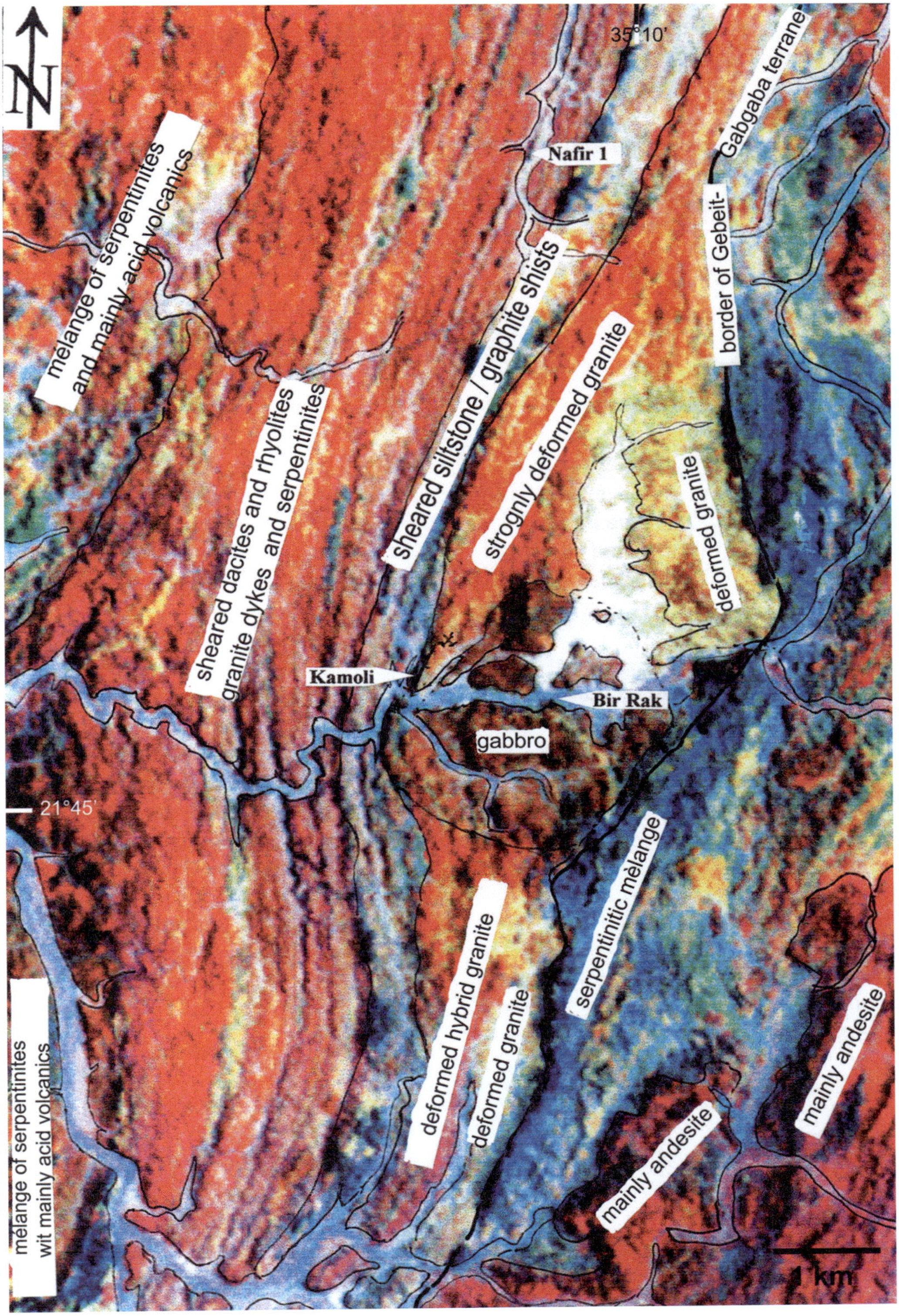

Fig. 6.45 Lithologically processed satellite image of the Kamoli and Wadi Rak area (TM 172/45, channels 7-4-1)

Fig. 6.46 Large house complexes built from different rock types in the northern part of Wadi Rak

6.3.3 Camel 1

Geographic position:	21°44′56″ N, 35°07′44″ E

This site is defined by several dozens of rock drawings on a large cliff that consist of dromedaries, ostriches and animals that seem to represent canidae (Fig. 6.47). These graffiti are undoubtedly not very old and likely to have been executed by passing Bedja. The rock wall is appropriately located near a thoroughfare, which probably accounts for the concentration of drawings at this site. According to Mustafa Khazim, the geologist who accompanied us, ostriches were still seen in the Nubian Desert as late as1983.

6.3.4 Camel 2

Geographic position:	21°50′51″ N, 35°10′33″ E

The Camel 2 site is defined by another group of rock drawings representing camels (dromedaries) and cattle on an upright rock surface. An Arabic inscription reads "Ibn Hassan Ibn Ali Ibn N. Homeira, bismillah al ahram al ahrim". According to Mustafa Khazim this inscription is old, due to the antiquated expression "Ibn". Unfortunately, there is otherwise no indication as to its date (Fig. 6.48).

Fig. 6.47 Probably recent rock representations of camels, ostriches and canidae at Camel 1

Fig. 6.48 Rock drawing and old inscription in Arabic at Camel 2

6.3.5 Uar (Alaar) (Fig. 6.49)

Geographic positions:	
Mine (shaft):	21°41′33″ N, 35°08′16″ E
Large Arab complex:	21°41′42″ N, 35°08′21″ E
Arab settlement:	21°41′05″ N, 35°08′13″ E

Uar is a mine district with a large settlement displaying a well-organised layout, being its most striking feature (Fig. 6.50). It is located in the valley of Wadi Uar and had been built around a broad, central street of about 152 m length. At its N end the complex is about 40 m wide. Rectangular, architectural units line-up along either side of the central street. Because their sizes vary, the settlement's outer contour is irregular. The southern end is characterised by a large, open space situated next to workshops, small tailings, amphora fragments and round mills (Fig. 6.51).

Two types of rock had been used for the construction the complex. The central and W sectors consist of metasediment rocks from the surface in the western mountains. The eastern part on the other hand, is built from globular granite nodes collected in the eastern hills and is therefore in a worse state.

Alike similar settlements mentioned earlier, there is actually little doubt that this site had mainly functioned as a supply post with storerooms as well as a caravanserai. This is even indicated to by the name-giving word "uar", which signifies "to distribute" in Arabic. Some rooms are 15 m long and only 2.5 m wide (inside dimensions) and are therefore interpreted as store rooms. In its central part, the site bulges out on its western side. At regular intervals of about 15 m along that western facade, 5 m long annexes seem to have been added to the central structure.

In this western side of the central part, some walls from flat schist rocks still rise to about 2 m above the plain.

Because some mills were found inside and near the western annexes, it is conceivable that miners lived here too at some stage, while the older, central buildings had probably already been forsaken and left to fall into decay.

The N-S oriented, central street is 4–5 m wide and had probably once been paved. The eastern side consists of elongated, rectangular rooms measuring 2.5–3 m by 5–8 m (inside dimensions). Along the street their dry stone walls consist of schistose slabs, and towards the E of globular, granitoid nodes. No mining tools were found inside the buildings. In the N there is an open space reserved for prayers. Around it there is an accumulation of the usual litter from the early twentieth century, consisting of broken bottles, cans etc.

At the well close to the actual mining sector, which is situated in the neighbouring wadi to the S, several hut ruins group along the mountain slopes. Associated to them are round mills, including the rotor stones. The rotor discs, which for the most are made from granitoid rocks, had been transported from distant quarries, whereas the lower sedentary stones often consist of local material.

The accurate date of the complex within a particular phase of the Early Arab Period needs yet to be determined. It belongs to a type documented in NE Sudan at Bir Kiaw, Derahib, Omar Kabash and other sites with well-preserved walls. In Southern Egypt Samut, Sukkari and Umm Eleiga may also be counted to this group. These sites have in common that they are not hidden away in the mountains, but set in clearly visible locations in the middle of the wadis. They once had belonged to integrated supply systems for possibly both, pilgrims and especially gold miners who settled in their wider surroundings.

At the inlet to the tributary wadi branching-off to the actual mining area in the SW are some recent house ruins from the Anglo-Egyptian condominium. They are associated with elongated, grey tailings, which still clearly reveal the tracks for the rail trolleys. Furthermore, on the N-side is an open mine with a shaft whose debris heap reflects the sunlight in a spectrum of different colors. To the SE of a water-bearing well in the wadi, a myriad of small processing sites and tailing heaps plus an artificial amassment of rotor stones testify to the former mining activities in the area (Fig. 6.52).

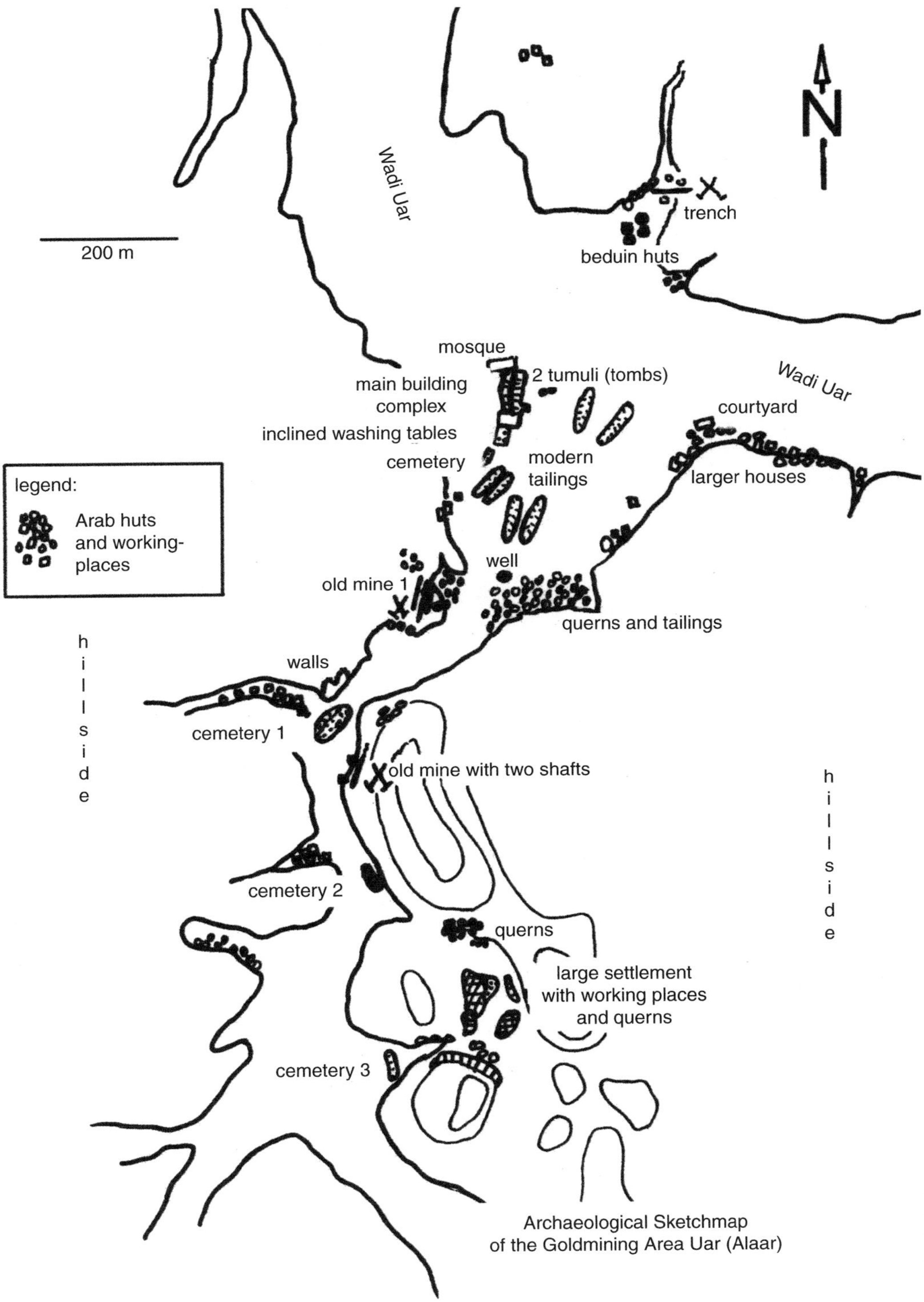

Fig. 6.49 Archaeological sketch map of the mining district at Uar (R. Klemm)

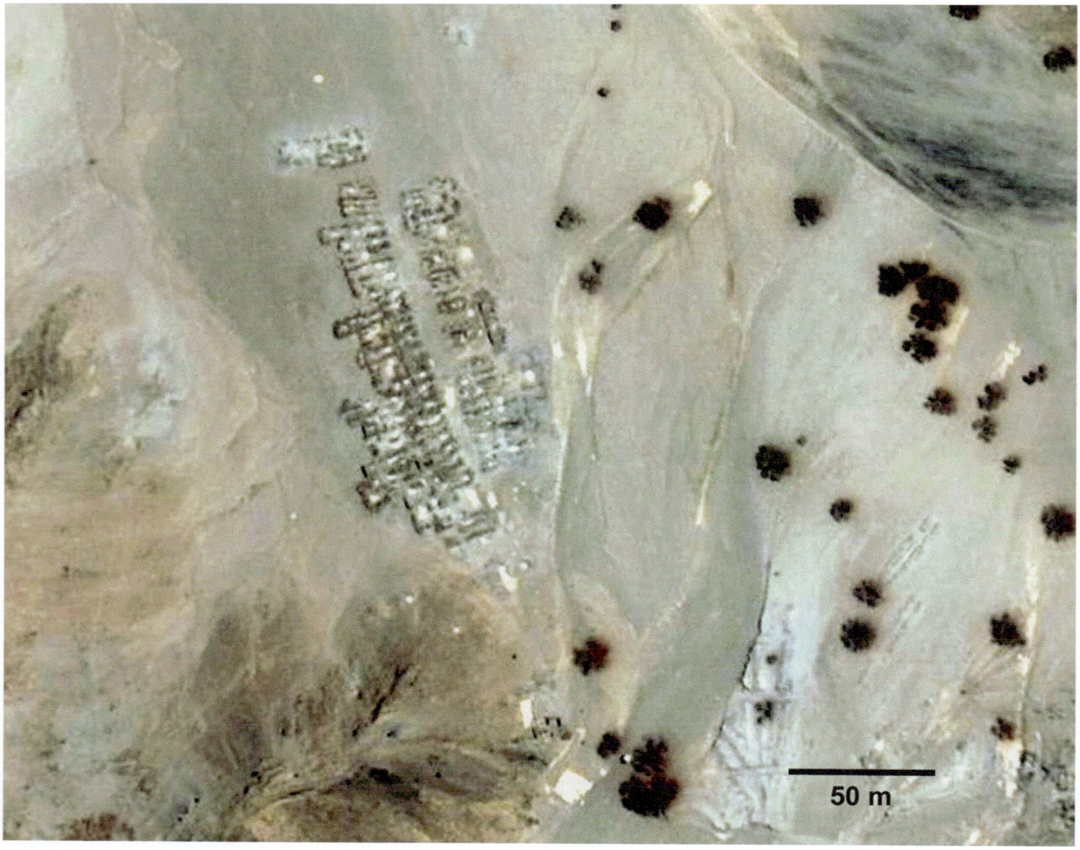

Fig. 6.50 Large Early Arab Period complex at Uar (modified Google-Earth image)

Further up this actual mining wadi, more processing areas as well as small dwelling houses align along both its sides. To the W is an Arab cemetery with about 50 well-preserved, very flat graves. To the S is another old mine, which had partly been operated in an open pit, partly as a sub-surface mine. In the early twentieth century, it was reopened again. Further up, on the N side of the wadi, there are more houses and another cemetery. In turning-off behind a ridge to the S one yet again comes across a large number of house ruins with countless round mills. A third cemetery is found here as well, whose severely dilapidated graves are oriented in a N-S direction.

Within the settlement areas we found two stamped amphora handles (hexagrams), green faïence shards, and obliquely grooved shards from dark-red amphorae.

Despite our efforts, we didn't succeed in finding any evidence whatsoever testifying to the New Kingdom Period, neither in the district nor at the mine itself.

Fig. 6.51 Large Early Arab Period complex in Wadi Uar laid out around a central alley

Fig. 6.52 Collection of Early Arab Period rotor discs from various, locally unknown granites found at the SE entrance to the mining wadi at Uar

Fig. 6.53 Remains of relatively well-preserved New Kingdom houses at the wadi edge of Uar. Architectural evidence from central sectors of the wadi has largely been lost

6.3.6 Uar, New Kingdom Complex

Geographic position:	21°42′32″ N, 35°07′36″ E

Approx. 2 km to the N of the large Uar-complex are the remains of a New Kingdom settlement including a gold processing site (Fig. 6.53). It is squeezed into the mouths of two side wadis, just SE and NE of a projecting hill with a quartz outcrop emerging at the top. Conjectured ruins continuing into the main wadi have been washed away. In the rear part of the southern wadi there are additional ruins of six round huts whose green shards, however, indicate to the Early Arab Period.

This relatively large New Kingdom complex at the entrance to a tributary wadi has an approximately rectangular shape. About 20 New Kingdom oval mills were counted here. Approximately the same number was counted regarding anvil stones with central cavities and typical grinding stones in various sizes and materials. The largest building measures 16 m in length and was built with the schistose slabs from the surrounding surface. The slabs had been placed in alternating layers, in parallel and perpendicular orientation to the walls. A barrage-like wall had been built at the top of a small side wadi to the SE to decelerate and control through-flowing water. No dwelling remains were found here. The date of the wall is unknown, but it probably is contemporary with a number of Early Arab huts at the inlet to the wadi. The predominantly New Kingdom site yet again yields very little pottery. Also typical is the open distribution of its dwelling units and processing sites. Within this area one also discerns the structures originating from washing tables.

The ores had probably been retrieved from the quartz rib and its eroded rocks in the wadi, thereby suggesting the former existence of wadiworkings. Other quartz veins that lead from the adjacent mountains into the wadi had presumably also been mined sporadically, though without leaving any apparent traces.

No traces from human activities were found on the opposite S side of the valley, even though mill fragments were seen in the central rubble gully.

Due to its location at the extreme eastern border between the Allaqi-Hamisana zone and the Oyo-Onib zone, the geologic position of the Uar deposit is quite remarkable. Apparently, the gold mineralisation had formed independently from the influence of both tectonic zones, thus after their collision. It seems that they had formed exclusively in connection with their geochemical and petrographic environment and the thermal vicinity of post-orogenic magma intrusions. Such intrusions are not necessarily granitoid, but may as in this case also be gabbroid.

The geologic and lithological framework of the Uar deposit is, insofar as the Hamisana zone is concerned, made up of severely sheared, partly extremely pyrite-rich graphite schist in the immediate vicinity to the vein mineralisation at the main shaft. Moreover, the wide spectrum of rocks in the area is composed of metagreywackes, metaarkoses, as well as metasiltstones next to metaandesites, and 1 km further N, of serpentinites and quartz carbonates. The rock units of the Oyo-Onib zone in the deposit area consist on the other hand of amphibolites and metarhyolites.

Whereas the partly intensely sheared series of the Hamisana zone strike chiefly in a consistent manner from N to S, the amphibolites and the metarhyolites of the Oyo-Onib zone display an average NNW-SSE strike (Fig. 6.54).

The productive quartz vein mineralisation too, largely follows the general N-S strike of the Hamisana zone and dips with the wallrock of black shale at about 45° E. An accurate measurement could not be carried out because a potentially primary quartz vein had been folded and transformed to small, lenticular quartz boudins by later shearing. The b-axes of the folds (crenulations?) plunge gently N.

The main shaft, which inclines at 45°, can be followed only over a short distance, as it is heavily silted-up and flooded at about 20 m depth. A second inclination shaft about 500 m further S on the eastern slope of a small, S running side wadi, which branches-off from the main shaft in Wadi Uar, was followed and sampled over a short distance. An adit located about 50 m further S, on the other hand, turned out to be caved-in to such an extent that any exploration inside the gallery would have been unwise.

Next to traces from gold, the ore paragenesis consists primarily of pyrite, copper sulphides probably a little arsenopyrite, for besides malachite there is also some azurite that usually is indicative of primary As-mineralisations. From the primary Fe-sulphides too, only the limonitic oxidation products remain.

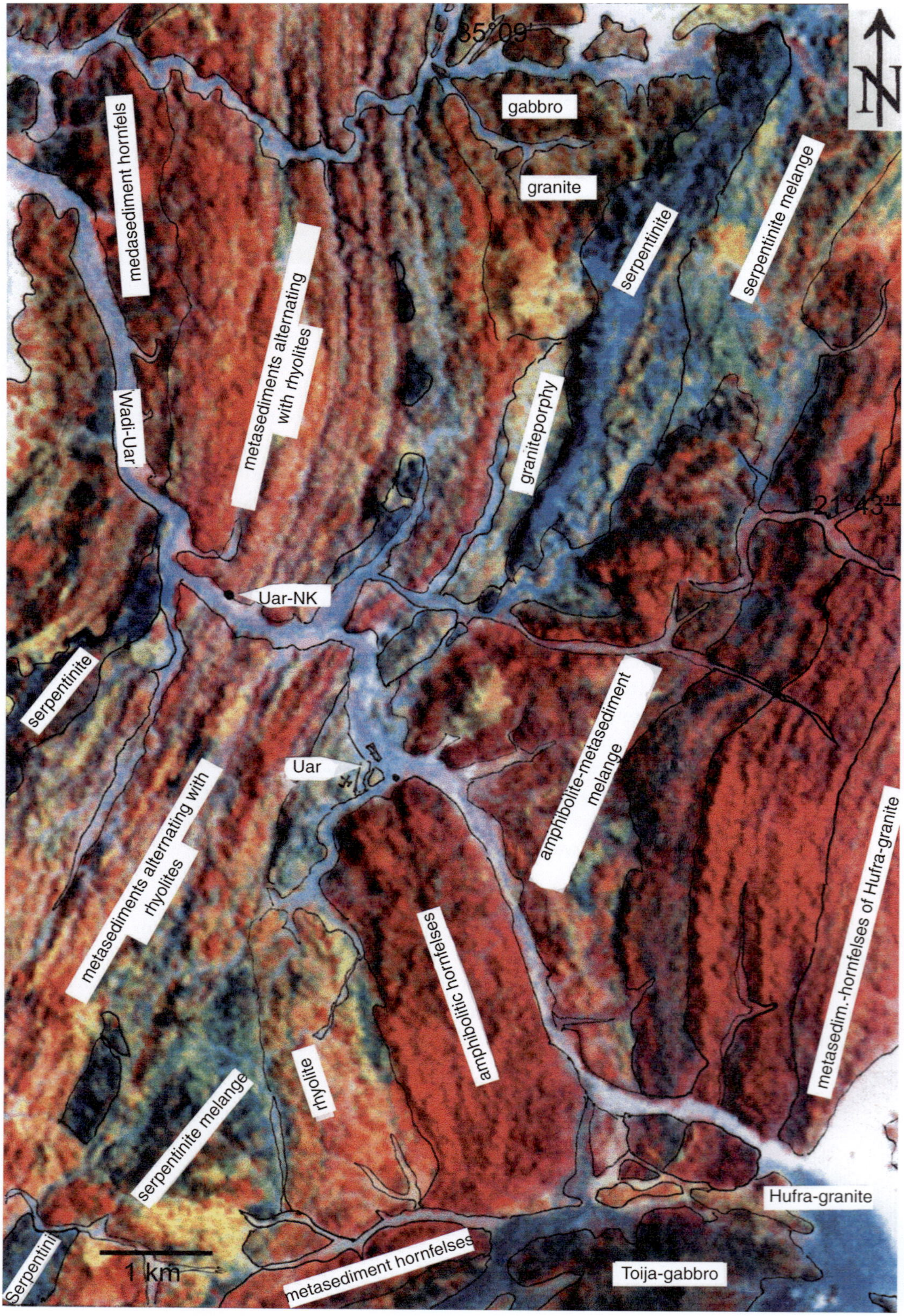

Fig. 6.54 Lithologically processed satellite image of the area around Uar (TM 172/45, channels 7-4-1)

Fig. 6.55 Large, old cake-shaped tombs in a recent Bedouin cemetery at Hufra. New Kingdom mills lie scattered over a site at the edge of the wadi. Some of the circumferential stones around the graves are former New Kingdom mills

6.3.7 Hufra

Geographic position:	21°31′51″ N, 35°11′44″ E

Just N of the junction of Wadi Onib into the Hufra plateau a New Kingdom settlement with well-preserved rectangular floor plans stretches out over about 150 m along the wadi edge. In front of the houses are at least ten old Romibs, many younger Bedouin burials (Fig. 6.55), and numerous tumuli graves of which one is a recent tomb of a sheikh. It was partly assembled with New Kingdom oval mills, which had also been integrated to the stonework of the Romibs. Furthermore, there are typical, flat anvil stones with central depressions. The establishment of this cemetery had probably been viable only with the ample rock supply furnished by the ruins of the New Kingdom site. Parts of it are therefore severely dilapidated. Particularly noteworthy are the remains of washing tables and their basins. They seem to be slightly narrower and longer than usually observed for the Early Arab Period. Although no traces have survived, a nearby well once undoubtedly provided the gold processing operations with the needed water.

Geologically, the site lies within the clearly hybrid SW margins of the extensive Hufra granite intrusion, about 1 km N of its contact with the mafic series of the western Onib nappe.

For the lack of evidence from mines, the settlement had probably been specialised on wadiworkings, but as the wadi floods would have washed away all traces from such wadiworkings no statements as to the origin of this ore can be made. Neither were there any signs of nearby quartz veins, in spite of random prospecting in the area.

Pottery finds are extremely meagre. Some Nubian shards were recovered, as well as ostrich egg shells and a cowrie shell.

6.3.8 Onib

Geographic positions:	
Onib Major mine:	21° 28′54″ N, 35°16′26″ E
Large New Kingdom settlement:	21° 29′32″ N, 35°16′11″ E
Early Arab settlement:	21° 29′19″ N, 35°15′43″ E

6.3.8.1 Onib Major

At least four long mines run parallel to each other from NE to SW, across three hill groups. The sketch map shows that the vein continues towards SW where yet another, though recent shaft is discernible (Fig. 6.56).

Two Early Arab Period occupation areas cluster respectively in the NE and SW of the district. The area in the NE consists of about 50 houses and a large number of round mills, which here and there are also found integrated to the stonework of the nearby houses dating to the British mining phase in the early twentieth century. The concentration in the SW is close to another quartz outcrop and consists essentially of a large courtyard with numerous mills and a processing site, as well as a washing table (Fig. 6.57).

The main vein itself shows a distinctly protruding, untouched and thus ostensibly barren quartz outcrop. Instead of excavating the entire vein, only the primary and therefore more convenient generations at its margins had been removed (Fig. 6.58). To the SW of the vein is a series of British constructions amidst purple water and beer bottle shards. Nearby is the tomb of an Italian miner by the name of Felice Lenzi, who died in 1907 (Fig. 6.59) the year in which the British mining concession at Onib expired.

Under inclined sunlight one notices the faint elevations that witness bygone wadiworkings in the main wadi and along its gentle slopes.

Apart from a few remarks dropped by Dunn (1911) concerning the uniqueness of the vein lengths at Onib, the site has otherwise been completely omitted in published reports. Although only few New Kingdom mills lie scattered around in the wadi from erstwhile floods, the evidence from the sinuous courses of the mined galleries as well as the rock pillars left standing expose the typical features of New Kingdom mines as known from Egypt. The beginning of mining at Onib therefore, may well date to the New Kingdom.

The details of the geology around Onib are rendered in a geologic sketch map of a lithologically processed satellite image (Figs. 6.60 and 6.61).

It shows that the exploited quartz veins are located in a complex sequence of metasediments essentially following the general NE-SW strike of foremost greywackes and conglomerates. Furthermore, there are marble layers, mostly bituminous graphite schist, siltstone, and interstices of dacite and rhyolite volcanics.

A small diorite intrusive stock occurs about 5.7 km away, to the NE of the main mine. Approximately 3 km to the E, there is another large diorite complex, which partly has altered the volcano-sedimentary margin to hornfels through contact metamorphism. Both diorite intrusions may have represented the thermal source of the ore mineralisation.

About 3 km to the W, there are magmatogeneous sequences with diorite, gabbro, pyroxenite, serpentinite, and amphibolite completely integrated to the NE-SW tectonic elongation of the entire zone. They show no direct connection to the formation of the deposit itself. They also seem to belong to a separate tectonic nappe (Stern et al. 2004). Moreover, auriferous quartz mineralisations no longer occur in these zones.

The whole area is marked by an intense isoclinal folding, whose b-axes in general strike NE-SW. At least two tectonic fault phases dis-

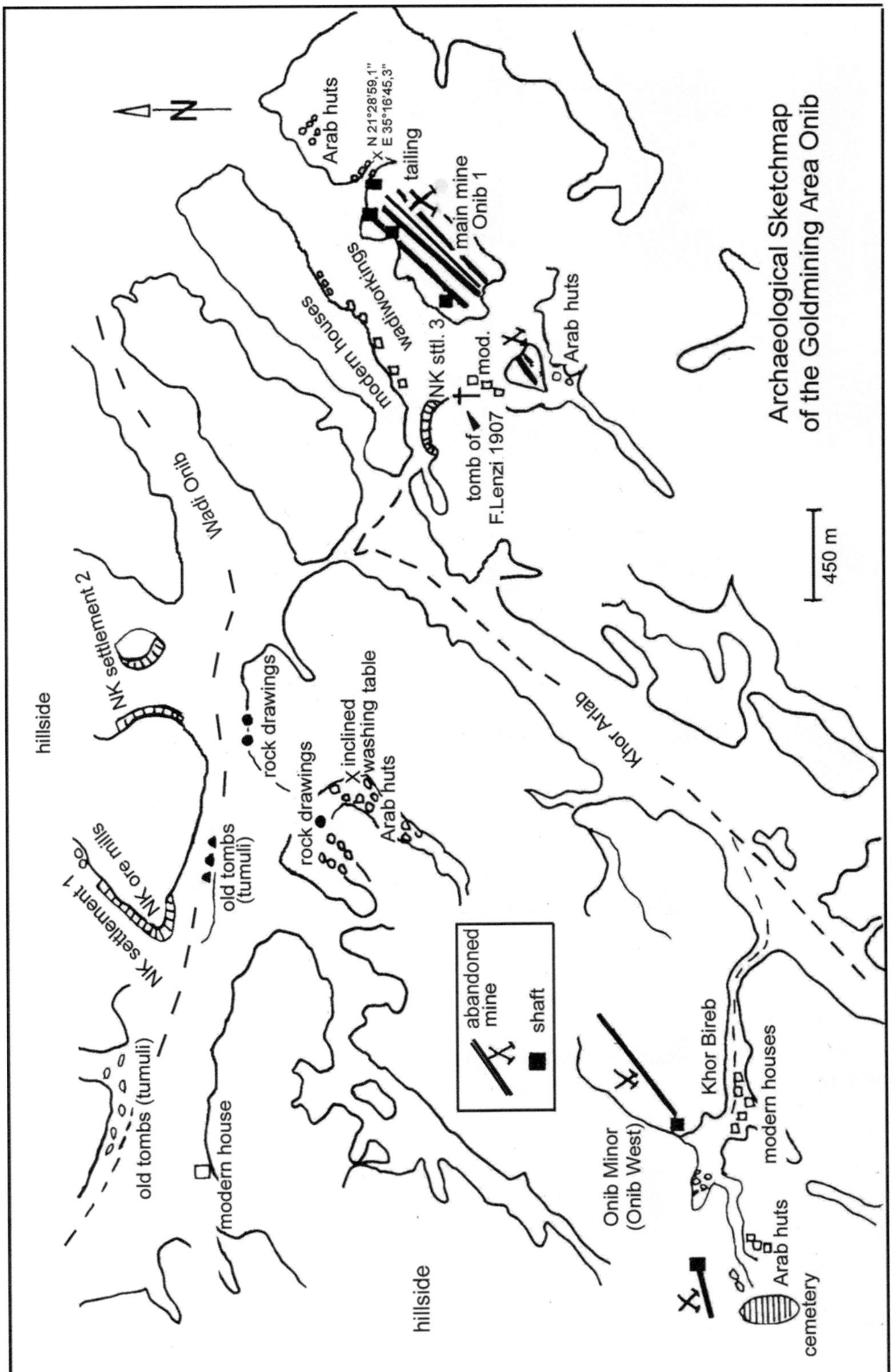

Fig. 6.56 Archaeological sketch map of the Onib district (R. Klemm)

Fig. 6.57 Early Arab Period settlement remains at Onib including the remnants of a washing table

place the series up to 500 m in a roughly NW-SE as well as E-W direction.

The productive quartz vein within the main mine consists of at least two generations, of which only the marginal zones degraded by limonite were extracted during the New Kingdom. The milky-white quartz core was left standing. The adjacent wallrock is also extensively altered by hydrothermal circulation. The most conspicuous alterations consist of pervasive formations of carbonates and pyrites, which at the surface are only recognised by limonite pseudo-morphoses. Especially the thin satellite veins are much folded, the main veins, due to their higher competence less so.

The folding occurred in several phases, which is well-discerned in the crenulations and chevrons at the area's outcrops. The conglomeratic parts of the wallrock are also significantly sheared (Fig. 6.62).

An outline of the chronology of the ancient mining phases emerges according to the so far identified occupation periods. At first, the veins had been mined at the surface in simple pit trenches during the New Kingdom. In doing so, only the uppermost, well-oxidised rich-ore-zones had been exploited. In the Early Arab Period these areas had been considerably expanded and deepened further into the ground with depths reaching more than 20 m. During modern mining, which ended in 1907, the pits were lowered to about 40 m depth. Because of disrepair, the shafts are no longer accessible without special equipment.

Fig. 6.58 Main quartz vein at Onib. Mining operations had probably taken place in the New Kingdom. Only the alteration zones along the wallrock and the first quartz generation had been removed for processing. The remaining quartz vein is only productive in lower depths

Fig. 6.59 Tombstone of the Italian miner Felice Lenzi who died in 1907 at Onib

Fig. 6.60 Sheared and folded quartz-chlorite-gneiss at the Onib mine

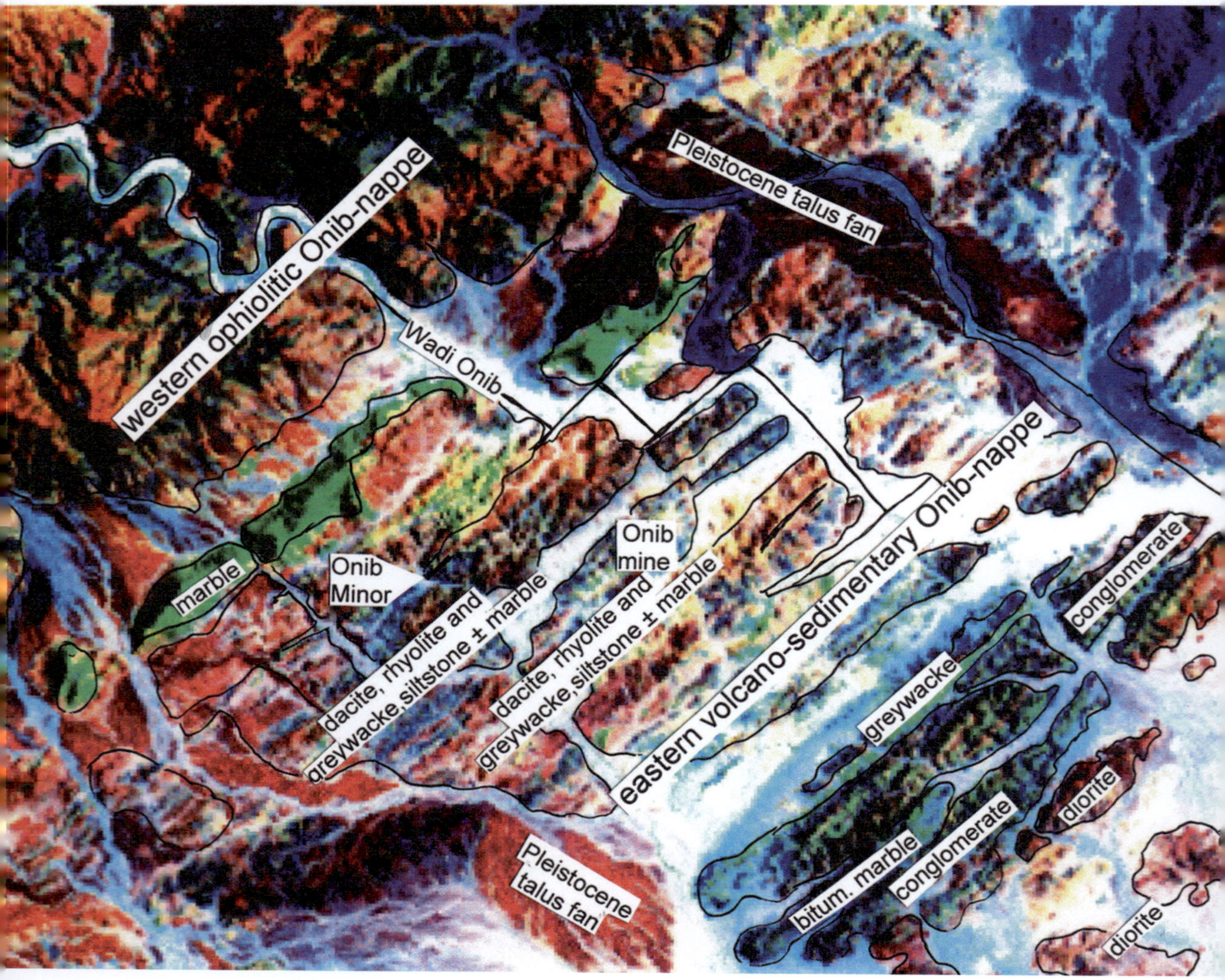

Fig. 6.61 Lithologically processed satellite image of the Onib area (TM 172/45, channels 7-4-1)

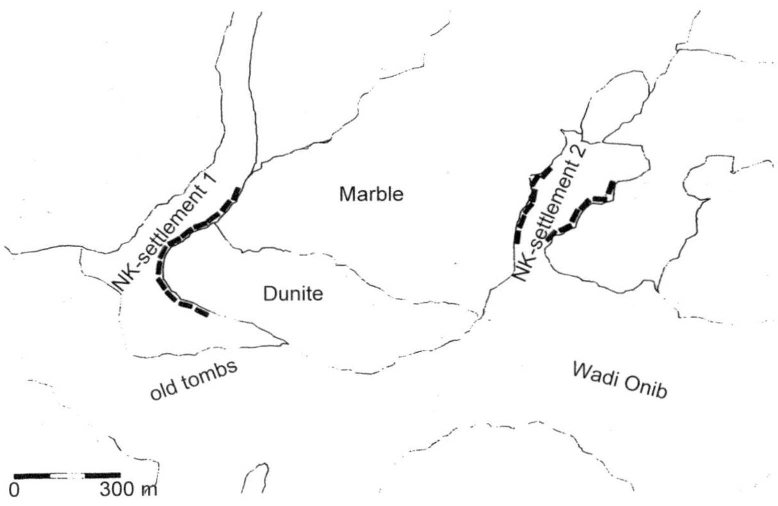

Fig. 6.62 Sketch map of two large, New Kingdom settlement sites to the N of Wadi Onib

6.3.8.2 Onib New Kingdom Settlements

Geographic positions	
Site 1:	21°29′40″ N, 35°15′37″ E
Site 2:	21°29′36″ N, 35°16′00″ E

Further to the W in Wadi Onib, about 2 km away from the main mining site, are the flimsy remains of two, yet relatively large New Kingdom settlements on the northern slope of Wadi Onib (Fig. 6.62). Numerous rock drawings along the southern wadi margins mainly exhibit elongated animal figures with horns formed to long arches (21°28′0 4″ N, 35°15′36″ E; Fig. 6.63).

A small Arab Period settlement with a relatively well-preserved washing table is located across the New Kingdom settlement on the southern edge of Wadi Onib.

6.3.8.2.1 Site 1

There are eight round and tumulus graves near this site. They had partly been built from the building stones and some artefacts gathered from the New Kingdom settlement.

The remains of the settlement are for one located along the edge of a wadi terrace on a stretch of 30–40 m, and for the other at the foot of the mountain to the N. Some oval stone mills were found near two pits filled with sand halfway up the mountain. The rectangular houses had been built in the usual arrangement forming long alignments. As a rule, the rooms still contain New Kingdom mills and grinding stones. Corresponding specimens were found integrated to the masonry of more recent tombs in front of the site. With the houses typical location along the margins of the wadi their occupants had probably specialised on wadiworkings. As they had probably been built in a continuous alignment around a hill, parts located at the very edge of the wadi have in the meantime been eroded away. The length of the settlement is about 180 m (Fig. 6.64).

Noteworthy is the enormous size of the oval grinding mills of which some are 60–70 cm long and 60 cm wide. The pounders consist of cylindrical stone pestles weighing several kilos. They generally match with the depressions found on flat stone slabs that had served as anvils. Unfortunately, only few shards were found.

Fig. 6.63 Rock carvings representing longhorn cattle at Onib

Fig. 6.64 New Kingdom settlement (site 1) at Onib with cake-shaped -and tumuli tombs in front

Fig. 6.65 New Kingdom settlement remains (site 2) in Onib at both wadi edges

6.3.8.2.2 Site 2

Site 2 consists of two opposite house alignments at the respective eastern and western foot of mountain and a small hill (Fig. 6.65).

The western alignment is about 250 m long, the eastern one is somewhat shorter. The alignments run parallel to the wadi and consist of houses which by majority are divided up in groups of five to eight square rooms, each measuring between 12 and 15 m² in surface. Seven such houses were counted on the E side. Their ruins are yet again covered with remarkably large grinding mills. In addition, there are cubic pounding stones with deep cavities and measuring about 40 × 40 cm.

On the western side is a washing table whose orientation is perpendicular to that of the houses. It is located next to a noteworthy accumulation of large oval mills. As expected, no remains from tailings were found here as this region receives comparatively large amounts of rainfall. One of the washing tables measures 2.50 m at the base. Due to the ill state of preservation of the other, its dimensions could not be established.

Apparently, there were two more washing tables at the NW side, although these were not identified with certainty.

Otherwise, the site displays the usual finds, ranging from small grinding stones, oval grinding troughs, anvil stones and matching pestles.

The settlements seem rather to have been associated with the local wadiworkings than with the nearest mines at some distance away. It is also conceivable that the ores had been brought here from the mines, due to the availability of water at the settlement. In Egypt, this has in fact been observed more frequently for New Kingdom sites than for Early Arab settlements, the latter being usually located in close vicinity to the mines. A nearby, central well in the plain served all neighbouring settlements. It may have sanded up not that long ago as suggested by only recently abandoned Bedouin huts in the vicinity.

Under altered lighting conditions we noticed at least four or five additional washing tables near a damp, green vegetation strip running parallel to the settlement. Apparently there was enough water here (and there still is) to operate several washing tables simultaneously.

Both settlements are built from dolomite rocks with fine karstic grooves at the surface. This rock series occurs only half way up the eastern slope where they go over to conglomerates. The majority of the stones is therefore likely to originate from the western mountain slopes.

There is almost no pottery from both sites, and if so, then only in the form of very small, heavily eroded shards.

Here again, the evidence points to three main occupation periods:

1. The first dates to the New Kingdom, when house alignment had been built along the wadi edges. In this case again, the location of the settlement had primarily been determined by the availability of water and ores in the wadi and less by the distance to the mines. However, New Kingdom mills also occur scattered around the tomb of Felice Lenzi indicating New Kingdom operations at the main vein itself.
2. In a second phase an Early Arab village is established in a random arrangement immediately at the mine, consisting of few minor installations and singular huts and an ore processing site. Its location has no apparent link to natural water resources.
3. The third phase occurred only in the early twentieth century.

In Onib there is still plenty of water in a well, which apparently had been dug already in antiquity. The district is therefore traditionally also much frequented by mobile populations, which is also reflected by the large number of nearby funerary structures from various periods.

6.3.8.3 Onib Minor

Geographic positions:	
Mine:	21°28′42″ N, 35°15′02″ E
Southern Early Arab huts:	21°28′24″ N, 35°14′45″ E

The Onib Minor mine is located 2.8 km SW of Onib Major (Fig. 6.66). Today it has the appearance of a long crack running through the mountain though it has collapsed underground (Fig. 6.67). A trench had probably been cut in the New Kingdom, and further exploitation had taken place in the Early Arab Period and the early twentieth century. To the W the terrain is predominantly marked by Early Arab settlement remains. There are also many recent house ruins dating to the Anglo-Egyptian condominium, of which some display the remains of fireplaces. The ruins are well-enough preserved to visualise the original roof supported by long, wooden shafts covered with large slate tiles that had been retrieved from the surrounding rocks. In the 1930s or even 1940s there was a second extraction phase represented by important accumulations of industrial waste.

At the foot of the mine is a settlement from the early twentieth century whose house walls partly contain built-in round mills. The square and rectangular houses were built partly on top of Arab huts, whose vestiges are still discernible.

To the S, the settlements date mainly to the Early Arab Period with mill inventories consisting of different rock types.

Isolated in the main wadi is an Arab cemetery with about 250 different-sized graves in a NE-SW orientation. The graves probably also include burials of women and children due to there small size (Fig. 6.68).

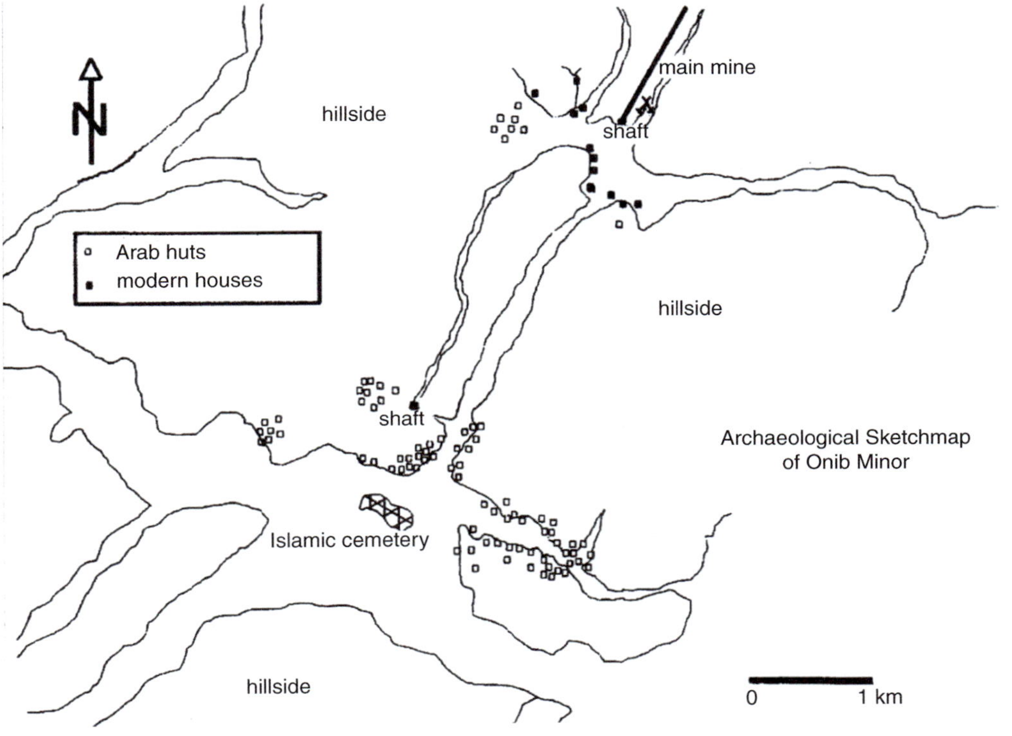

Fig. 6.66 Sketch map of the mining district Onib Minor (R. Klemm)

Fig. 6.67 Vein quartz deposit at Onib Minor. Mining had already taken place in the New Kingdom but its main period is attributed to the Early Arab Period. A short and intensive exploitation phase had also taken place at the beginning of the twentieth century

Fig. 6.68 Large Islamic cemetery to the S of Onib Minor

The geologic setting of the Onib Minor vein mineralisation is akin to that of the main deposit. A slight difference is that the small marble lenses are more frequently observed here. Towards the NW they then culminate to a large marble band which has been exposed to high tectonic stress. Generally, Onib Minor seems to be less productive, which hardly surprises, as this mineralisation is located further away from the thermal energy source.

6.3.8.4 Settlement SW of Onib Minor

Geographic position:	21°25′42″ N, 35°10′33″ E

A relatively large, but badly ruined and windblown settlement that had probably been specialised on wadiworkings is located on the north-western slope of a wide NE-SW striking, granitoid dike. The site is 9.5 km SW of Onib Minor at the indicated position. We did not visit the site and were only able to scrutinise it on the satellite image after its discovery on Google-Earth. To judge by its ruinous state and the architecture, the site seems to date to the New Kingdom.

6.3.9 Tabon-North

Geographic position:	21°17′08″N, 34°04′46″ E

According to Boswell (1984a) this site is a copper deposit. It is actually a poorly preserved site in the wadi consisting of five huts probably built in connection with gold prospecting. The very adverse driving conditions make this an extremely difficult site to visit.

6.3.10 Tabon-East

Geographic position:	About 21°13′58″ N, 37°06′51″ E to 21°13′49″ N, 35°06′56″ E

This Early Arab settlement spreads over several hills and small side valleys. There are countless round mills and much medieval

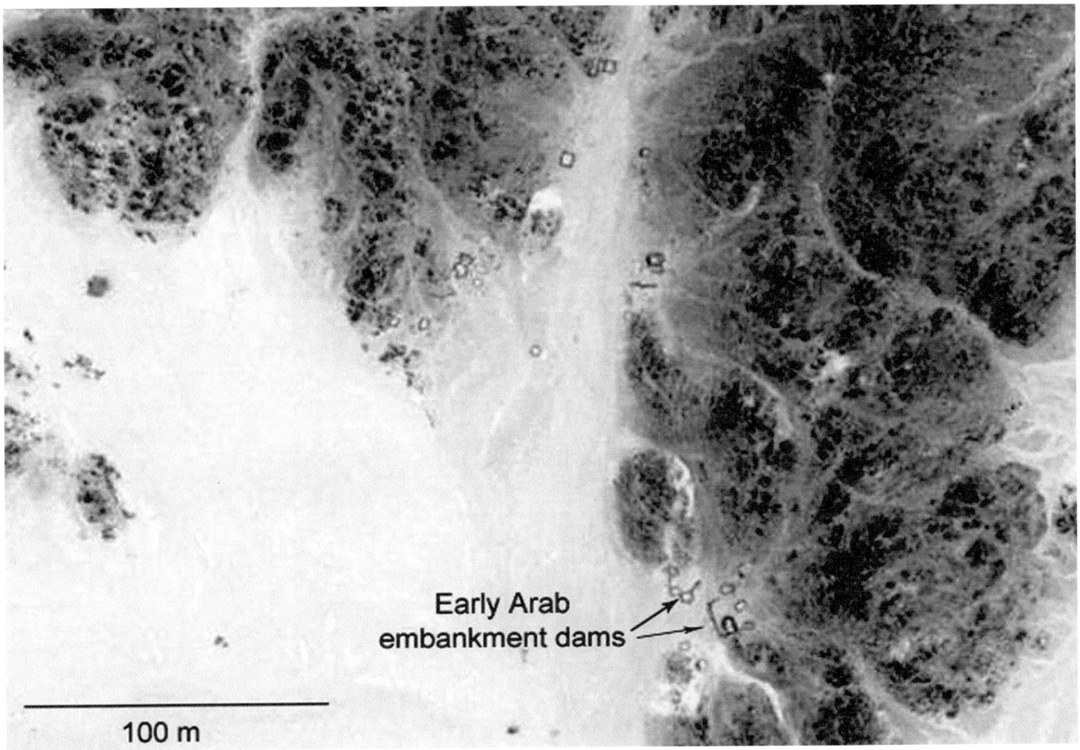

Fig. 6.69 Tabon-East and its water retention walls close to some Early Arab Period huts (modified Google-Earth image)

pottery. The huts, which had been built in the characteristic shell-facing technique, are partly relatively well-preserved. They consist of the severely sheared, local granite. The stones are strikingly spiky, which actually contributes to the stability of the walls. To the W of the first site there is another one in an area near quartz workings in the mountains. It consists of about 15 houses and work platforms. Next to shards from green glass and green glazed pottery there is also a pottery variant with an apparently local pattern, but also larger fragments of dark-red amphorae.

As commonly observed elsewhere at Early Arab Period mining settlements, small dams had been built at the inlets of small tributary wadis along both eastern and western halves of this elongated settlement (Fig. 6.69).

Notwithstanding the presence of some trial pits (21°14′06″ N, 35°07′05″ E), the entire settlement complex was perceptibly specialised on wadiworkings.

Geologically, the site is located in the northern part of an elongated and partially severely sheared granite stock (Fig. 6.70). It contains numerous xenoliths from rhyolitic hornfelses, into which the granite had originally penetrated. Together with its wallrock it had then been exposed to a N-S oriented tectonic shearing. This also corresponds to the strike direction of the mineralised veins, which equally to the entire shear zone are slightly folded.

Fig. 6.70 Lithologically processed satellite image of Tabon-East and Adarmo-East (TM 172/45, channels 7-4-1)

Fig. 6.71 Early Arab Period washing table with lateral annex at Tabon-South

6.3.11 Tabon-South

Geographic position: 21°12′38″ N, 35°04′13″ E

An Early Arab settlement site with an estimated 20 hut ruins lies hidden away in a tributary valley to Wadi Tabon. Although a small quartz vein striking 40° E (60° SE dip) reveals traces from mining, the site's livelihood had concentrated essentially on the wadiworkings whose remains spread along the N-S oriented wadi. Apart from a few huts near hillside, most of the architectural remains are severely dilapidated.

A washing table attached to a small room near the Early Arab huts was an unusual observation. The huts had been built from the schistose and spiky rock material gathered-up from the surrounding terrain, but which for the most are poorly preserved (Fig. 6.71).

Together with the occurrence discussed hereafter, Tabon-S is situated geologically in a zone of severely sheared rhyolite and quartz porphyry, close to the western border with the ophiolite series of Onib-Sol Hamed suture, which is sheared into the Hamisana zone. The quartz mineralisation is slightly folded in accordance with the intense tectonic stress in this zone. Macroscopically, the ores contained in the quartz as well as the altered wallrock consist exclusively of pyrite (Fig. 6.70).

6.3.12 Early Arab Village to the NE of Egait

Geographic position: 21°06′59″ N, 35°10′10″ E

This is the location of a presumably small, Early Arab Period settlement whose individual huts tend to spread into the adjacent northern and southern side valleys. The site had visibly been specialised in processing ores from the wadi alluvium. We were only able to survey the site via Google-Earth images.

6.3.13 Egait- North

Geographic position:	21°06′30″ N, 35°04′39″ E

Minor mining attempts had been made on a protruding quartz outcrop just below the summit of a hill. In a neighbouring gorge one still discerns two severely damaged, round huts where a fairly unworn rotor stone from granitic rock was found. This finding may indicate to a relatively short-term occupation of the site. At the foot of the hill one recognises additional hut ruins.

The few collected shards belong to grooved amphorae of a dark ware from the Early Arab Period.

Another small, Early Arab Period settlement consisting of about 15 hut units is well hidden in a valley, approximately 1.3 km W of this site (21°06′38″ N, 35°03′54″ E).

6.3.14 Egait-South

Geographic position:	21°03′25″ N, 35°04′20″ E
N end of the settlement:	21°03′39″ N, 35°04′21″ E

Two small mines in two quartz veins are located just N of a settlement site composed of about 25 huts and work platforms in a winding side valley of Wadi Egait. Apparently however, mainly wadi-workings had been operated here, judging by the clearly visible heaps of sorted-out wadi debris (Fig. 6.72). A very well-preserved washing table including its drain channels and a collecting basin is set slightly off the huts and a number of work platforms (Fig. 6.73)

Geologically, the veins occur in a N-S extending and much sheared sequence of amphibolite, alternating with rhyolite at the immediate margins to the equally, significantly sheared Adarmo granite stock.

6.3.15 Adarmo-East

Geographic position	
Adarmo East 1:	21°01′47″ N, 35°09′36″ E
Adarmo East 2:	21°01′12″ N, 35°10′47″ E

Adarmo-E 1 consists of two large, Early Arab Period houses, each with two rooms. In between are a barrage wall and a very well-preserved washing table. Furthermore, slightly hidden away in a side wadi branching-off at a right angle from Khor Adarmo are several small huts. The site's economy was probably based on wadiworkings (Fig. 6.74).

There seem to be several trial trenches along a vein striking between N-S and 30° W within a highly sheared, granitoid wallrock. The quartz still contains traces from pyrite and arsenopyrite, although these sulphides seem to be fully decomposed to limonite near the surface.

Adarmo-E 2 also consists of an Early Arab settlement at the N edge of the eastern part of Khor Adarmo, again in a hidden location. It consists of about 15 houses and huts.

A NNW-SSE (steeply dipping E) extraction cavity is located approximately 300 m E of the settlement. The up to 2 m thick quartz vein occurs in considerably sheared and hybrid granite. It displays several, though quite shallow extraction trenches distributed over a distance of 300 m.

Geologically, the vein mineralisations occur in highly sheared metasediments within the contact area of a granite stock. The granite itself is sheared lenticularly in a N-S strike (Fig. 6.75). In the deposit area it is much contaminated by the metasediments. Further S it again becomes intensely sheared.

Fig. 6.72 Spoil heaps from Early Arab Period wadiworkings at Egait

Fig. 6.73 Very well-preserved washing table at Egait, with water collecting basin and backflow channel

Fig. 6.74 Part of the gold production site at Adarmo-East with huts and washing table

Fig. 6.75 Thin section of a severely mylonitised granite, which had recrystallised in a shear zone, thereby displaying parallel oriented quartz and feldspar components. The original granite structure is barely discernible

6.3.16 Bir Kiaw

Geographic position:	20°50′43″ N, 35°01′31″ E

Bir Kiaw consists of an open space around a large, central well, which is frequently used by Bedouins today (Fig. 6.76). To the N is an Arab fort (geographic position: 20°50′47″ N, 35°01′30″E), which with its round corner bastions and small embrasures had been built from roughly mortared, flat stones (Fig. 6.77). Judging by the type of walling, the fort is relatively recent and dates probably to the Ottoman Period.

On a wadi terrace to the E of the well is a large settlement from the Early Arab Period. Unfortunately however, many single house foundations are no longer fully preserved. Its centre had probably once been the location of a now collapsed tower. The settlement itself clearly divides into three sectors, of which two are on either side of a broad, central road. The third forms a 120 m long house alignment in the SW (Fig. 6.78). From a more general perspective, the site much resembles those at e.g. Uar (Alaar) and Derahib and had probably developed at this location because of the well.

Unlike the Ottoman fort, the structures at the settlement had been built of dry stone walls.

To the W and S of the Ottoman fort are several round mills and processing sites as well as remains of severely dilapidated walls. Washing tables were not found, but near a number of small quartz veins observed in the mountains, one seems to have processed wadi rubble.

The mine attributed to this small processing site consists of various minor, exploited quartz veins striking in an E-W direction. They occur in a highly fractured and friable rhyolite-dacite wallrock with an approximate NNW-SSE strike. Although due to its location at the southern foothills of the Oyo-Onib zone, its isoclinal folding is less steep than the corresponding series of the Hamisana zone to the W.

Fig. 6.76 The well at Bir Kiaw, fenced- in by branches to prevent animals from falling into it. The water is hauled in hides attached to long ropes

Fig. 6.77 Small Ottoman (?) fort at Bir Kiaw

Fig. 6.78 Early Arab Period settlement and well at Bir Kiaw (modified Google-Earth image)

6.3.17 Listi

Geographic position:	20°50′03″ N, 35°00′05″ E

The predominant listvenite, which had formed by the surrounding, carbonated serpentinite rock, inspired us to name this New Kingdom settlement as Listi. The settlement consists of a small number of houses with several, almost square rooms. The find assemblage composes of numerous, oval New Kingdom mills, which by their shapes, i.e. their wide, flat, and evenly worn grinding hollows, are identical to the mills from the Egyptian heartland. The same goes for the other tools, being pounders in the shapes of small cylindrical pestles and fist grinders (Fig. 6.79). Pottery finds were, on the other hand, very sparse.

The New Kingdom settlement is located in a geologic environment, characterised by highly carbonated serpentinite that have altered to N-S striking, elongated bodies of lenticular quartz carbonates (listvenites).

As serpentinite may occasionally contain up to 200 ppb Au (measured at el-Sid in Egypt), it is generally regarded as a pre-enriched gold source (proto-ores). Through hydrothermal leaching and subsequent concentration processes the resulting quartz carbonates may in some regions develop to productive gold deposits (e.g. in the Caucasus). A similar enrichment process can be assumed for the present district. It is therefore quite surprising to see this fairly unconventional gold deposit identified by the Egyptian prospectors, whose

Fig. 6.79 New Kingdom anvil stone and pestle in Listi

expertise yet again deserves our full approbation.

If in fact the listvenite quartz carbonates had constituted the basis for the mining activities here, they would manifestly have been extracted in open pits. Such opencast pits are actually found in large numbers close to the New Kingdom settlement as well as along the nearby serpentinite ridges. They have different and irregular sizes but measure in average about 5 × 10 m and are only few meters deep.

Fig. 6.80 Fort Shashuteb, Early Arab Period huts and work platforms near a wadi mouth. Several circular tombs (Romibs) scatter in the surroundings (modified Google-Earth image)

6.3.18 Shashuteb

Geographic position:	
Early Arab settlements:	20°49′16″ N, 34°59′51″ E
Fort:	20°49′20″ N, 34°59′52″ E

Shashuteb comprises several Early Arab settlements dispersed over several wadis. In the following discussion they are referred to respectively by their geographic coordinates.

On the E side of the main wadi, which flows into Wadi Kiaw from the S, there are at least four washing tables. At the inlet are the remains of a tower-like fort from the Early Arab Period, which undoubtedly had served for the protection of the settlements further up the wadi (Fig. 6.80).

We found some painted shards of the Nubian-Christian type as well as round mills and the base of a green glass vessel.

A small mine limited to one quartz vein is located to the W in a side valley. The fact that it is associated to a large settlement, which in addition seems to have benefited from the protection of a fort, obviously point to a major Early Arab gold production in the district, which then would mainly have been based on wadiworkings.

Another Early Arab settlement (20°49′15″ N, 34°59′24″ E) is located on the western slope of a 400 m long, very carbonated serpentinite body (listvenite). It consists of 30 small huts that spread over two large erosion gullies. It too had possibly been, specialised in processing auriferous erosion ores.

Further to the W, there are at least two more Early Arab settlements, of which one counts about 40 housing units (20°49′36″ N, 34°58′49″ E.).

Lastly, another settlement associated to extensive wadiworkings is found at 20°49′38″ N, 34°58′18″ E.

The immediate geologic catchment area of this gold mining district is composed of a limited zone of alternating gabbros and serpentinites striking NNW-SSE. However, in the field we were not able to establish whether the serperntinite had emerged from the gabbro, as potential residue structures had become too distorted from slight carbonation. Small granitoid apophyses from the western granite stock intrude these serpentinites and, in this tectonically highly stressed border environment between the Gabgaba and Gebeit terranes, may very well have mobilised gold contents from the basic to ultrabasic series in hydrothermal system (Fig. 6.81).

6.3.19 Eikwan 1

Geographic position:	20°38′31″ N, 34°59′53″ E

The Early Arab settlement at Eikwan1 was not among the sites visited by us in the field. We only discovered the site later on in a satellite image provided by Google-Earth. The settlement consists of more than 70 house units and huts. The largest is a house comprising several rooms that measure about 9.5 × 6 m. The average size of a house varies around 6 × 4 m only, and that of a hut around 2 × 4 m (Fig. 6.82).

As a rule, these Early Arab settlements comprise prayer sites delimited by stone alignments with mihrabs facing E.

6.3.20 Wadi Eikwan

Geographic position:	20°36′42″ N, 35°03′23″ E

This site comprises some Early Arab houses with a well-preserved washing table (Fig. 6.83).

6.3.21 Eikwan 2

Geographic position:	20°35′16″ N, 35°03′55″ E and 20°35′27″ N, 30°03′50″ E

This large site is somewhat hidden away in the mountains and is made up of about 100 small houses (approximately 4.5 × 3.5 m) and above all small, singular huts (2 × 2 m). There is also a poorly-preserved washing table, but whose surrounding drains are in a relatively good state of preservation. Prayer platforms bordered by stones of which some are poorly-preserved are discerned just about everywhere.

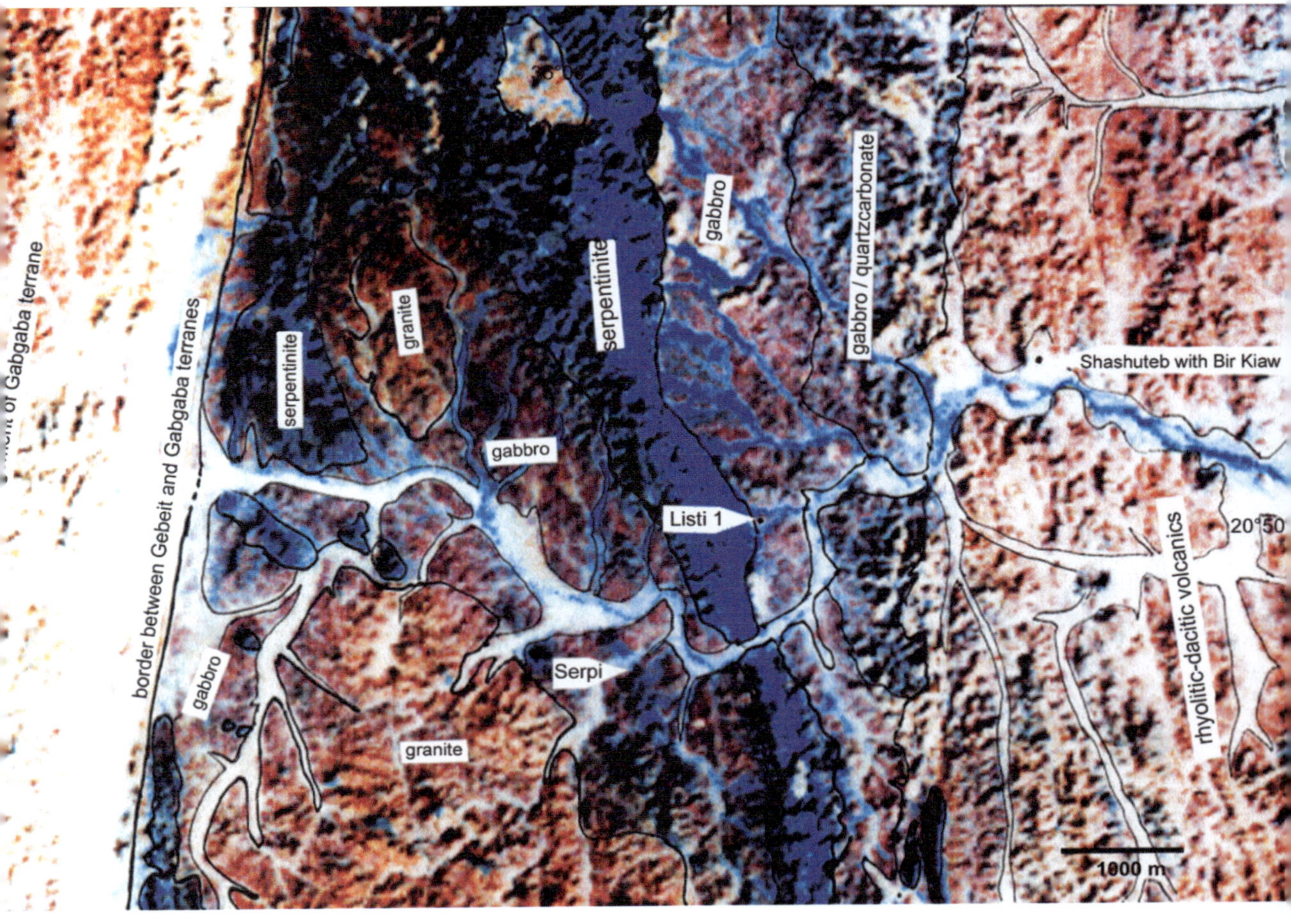

Fig. 6.81 Lithologically processed satellite image of the area around Shashuteb (TM 173/46, channels 1-4-7)

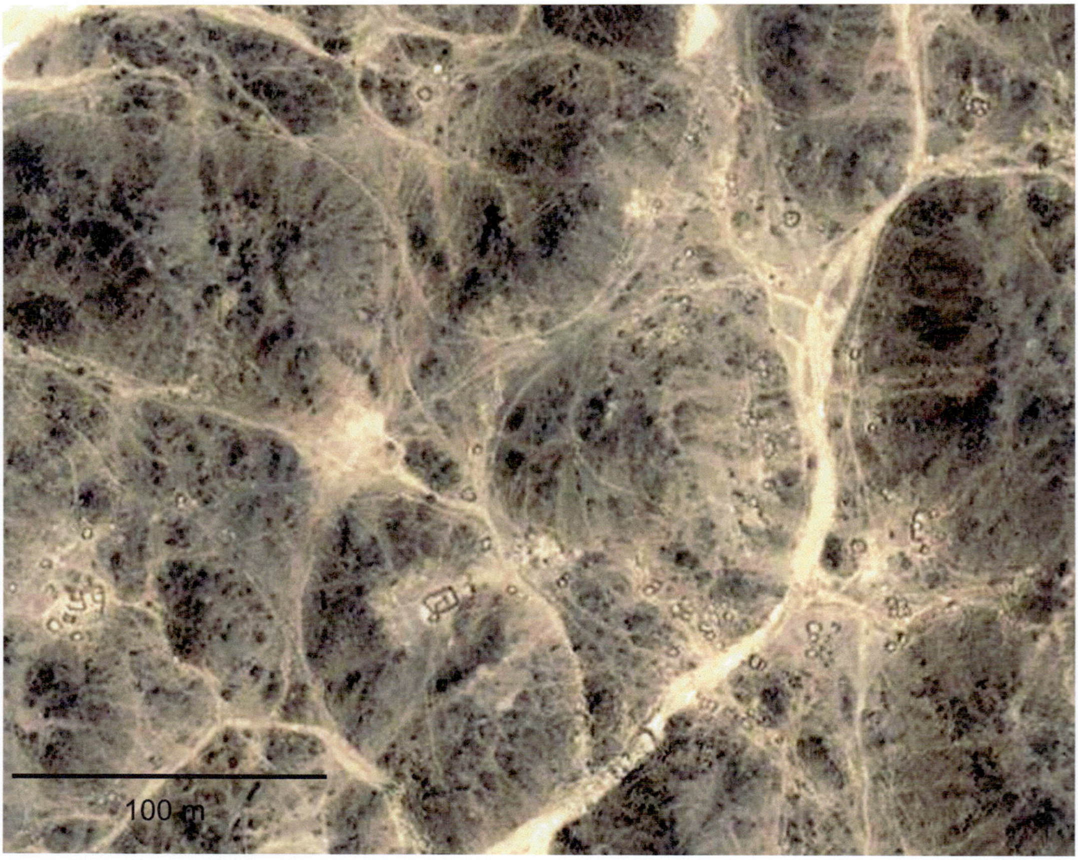

Fig. 6.82 Early Arab Period settlement at Eikwan 1 with numerous huts (~2.5 × 3 m) and a large house (overseer?), measuring 10 × 6.5 m (modified Google-Earth image)

Fig. 6.83 Early Arab Period washing table with sand-covered water collecting basin in Wadi Eikwan

6.3.22 Eikwan 3

Geographic Position: 20°31′44″ N, 35°04′30″ E

Eikwan 3 also represents a relatively large gold processing site located at the foot of a mine that is oriented parallel to the rock foliation (Fig. 6.84). It contains large quantities of anvil stones, some round mills, and a large prayer enclosure. Two inclined washing tables oriented in opposite direction to each other, and at least three more, singular ones are grouped together at this processing site.

The round mills are apparently not manufactured from the local rock, which might explain why many sites are completely devoid of mills. It is quite imaginable that they had been removed when the sites had been abandoned.

The prayer sites at Eikwan 1–3 are delimited by stone alignments, into which also broken anvil stones had been integrated. A large rock in upright position often served as a prayer niche.

6.3.23 Eikwan 4

Geographic position: 20°31′18″ N, 35°05′43″ E

This site counts about 30 rectangular houses with fairly massive walls. Some have small terraces on the outside. Between them one also observes round huts, consisting of single rooms. A small underground mine occurs some 600 m south of the settlement in a small tributary wadi just across the main wadi (Fig. 6.85). However, no direct ancient traces could be detected here, and it seems to be an early 20th prospection shaft.

Invariably broken round mills scatter about in the open terrain. A small group of houses is located in the background near the settlement. Small dams had been erected across lateral gullies to assure the water supply. The settlement is located in two side valleys that flow from the N into the main wadi. At this location, the main wadi is distinctive by its vegetation cover, notably by a group of trees, where one notices the traces from a recently abandoned Bedouin campsite.

Fig. 6.84 Adit of an exploited, auriferous quartz vein oriented parallel to the foliation at Eikwan 3

Fig. 6.85 Lowering a Jacob's ladder into the mine shaft at Eikwan 4

All sites in Wadi Eikwan display the typical mill inventory, including flat pounding and grinding slabs from the Early Arab Period.

The geologic environments around all Eikwan sites are, in spite of some local disparities, very similar to each other (Fig. 6.86). They all occur in the periphery of predominantly andesite to rhyolitic hornfels and a mostly coarse-grained, hybrid granite, which is tectonically elongated in a NNE direction. Wadiworkings had presumably been exclusively operated at Eikwan 1, 2 and 4, which might account for the large size of these settlements. The ones at Eikwan 3 and Eikwan mines are on the other quite small, probably as a result from the reduced workforce capacities within the operated mines.

However, there seems to be no link between the productive gold processing sites and the basalt and basaltic hornfels in the SE of the area.

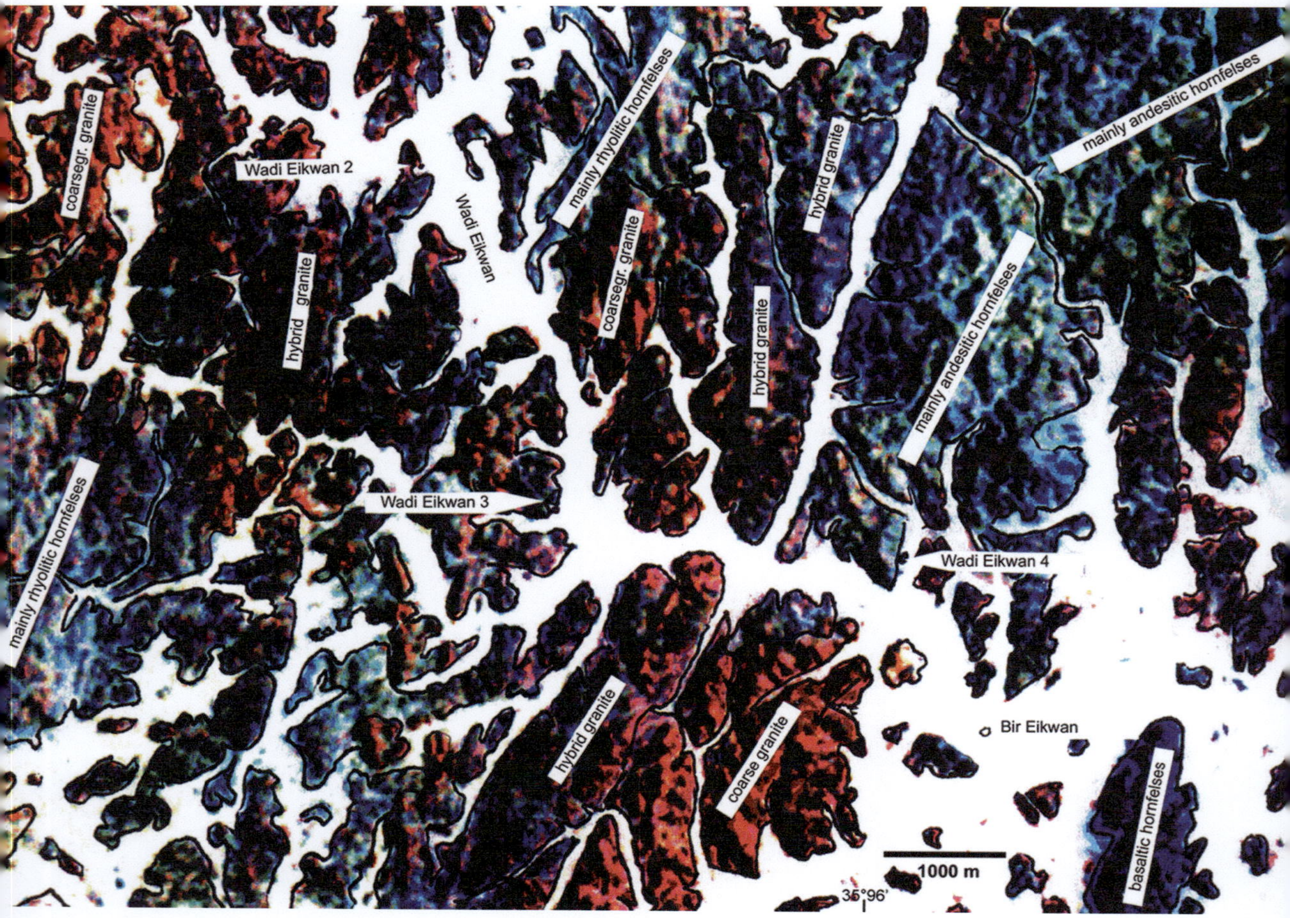

Fig. 6.86 Lithologically processed satellite image of the area around Eikwan (TM 172/46, channels 7-4-1)

6.3.24 Eikwan-South

Geographic position:	20°28′35″ N, 35°04′36″ E

This site is characterised by a mined trench in the mountains and several quartz outcrops in the hills. Closeby are a number of prayer and processing sites. Again, no mills were recorded. It is possible that the site had been abandoned during its prospecting. Insufficiently worn, round mills were therefore likely to have been kept and taken along to the next working site. In this context, about 12 house plans are located in a small, tributary wadi to the N, and the surrounding terrain is dominated everywhere by the heaps of former wadiworkings.

The overwhelming part of the mining operations in this district seems to have been focused on wadiworkings. We discovered however two quartz veins running parallel to each other with a 10° E strike that had been mined in small trenches. They still contained some pyrite, though which had largely oxidised to limonite. One sample even revealed visible gold.

The wallrock of this deposit consists of partially porphyritic rhyolites with an average NNE-SSW strike and a foliation at around 10° E. Noteworthy is the occurrence of a microdiorite dike containing marginal xenoliths of a granite body, located about 2 km to the NE.

The following three settlements were detected with the aid of Google-Earth scan only. We unfortunately did not visit them.

6.3.24.1 **Settlement 1** (20°31′20″ N, 34°57′32″ E)

A presumably Early Arab settlement of more than 30 huts is located in a concealed position at the mouth of a southeasterly side wadi into a main wadi, oriented WSW-SSW. Apart from two insignificant mining attempts on the northwestern slope of the main wadi, no additional mine trenches were recorded. Gold mining had therefore probably been limited to wadiworkings..

6.3.24.2 **Settlement 2** (20°28′42″ N, 34°58′02″ E)

This settlement is located 4.9 km further S, yet again in a slightly hidden position on both slopes of a tributary valley, to the N of the main wadi. It consists of about 12 small, square huts whose sides average around 3.5 m. By their arrangement and geographic position, they seem to date to the Early Arab Period.

6.3.24.3 **Settlement 3** (20°27′18″ N, 34°56′48″ E)

About 3.5 km SSW to the above mentioned settlement is again a slightly hidden site in a northern tributary valley of the same NNE-SSW main wadi. The about 30 round and square huts measure between 3.5 and 4.3 m in diameter. In the vicinity are distinct traces from wadiworkings.

6.3.25 Abu Mereim Salalob

Geographic position:	20°25′11″ N, 33°01′12″ E to 20°25′11″ N, 33°01′54″ E

The Early Arab Period settlement area at Abu Mereim Salalob stretches out over a distance of about 700 m through Wadi Salalob. It spreads from its western flank over a low hill into a small tributary wadi coming in at its eastern side. The site at the western edge of Wadi Salalob reaches into small side valleys while counting more than 100 small huts. There are some larger houses measuring about 14 m in length and usually containing three rooms. At least one prayer site is accounted for. The central settlement is located on an intermediary hill in the wadi itself and consists of about 80 huts. The 25 huts of the third settlement complex are aligned along a side wadi oriented in southeasterly direction from the E flank of the main wadi (Fig. 6.87). As often observed at many other Early Arab settlements, these sites too, are concealed from the main wadi.

Whereas the two western sites had probably been predominantly focused on wadiworkings, a quartz vein was exploited in the East. It runs from a group of hills across a side wadi to another small hill group. In addition to the few work platforms that were identified in this valley, only one washing table and one mill were recorded.

The mine in question is a 50–80 cm thick quartz vein, which strikes 40° E and dips 45° NW. It consists of two quartz generations, of which only the displaying limonite decay was mined in widths between 10 and 30 cm at the footwall, while the more massive, milky-white quartz was left intact at the hanging wall.

Next to gold, the ore paragenesis was determined as consisting of pyrite or arsenopyrite (?), some chalcopyrite, and cleft hematite.

The geologic environment of the vein occurrence consists of a foliated rhyolite, although at

Fig. 6.87 Early Arab Period settlement at Abu Mereim Salalob

the surface a large part of the actual vein runs into a severely sheared hybrid granite, which is to be viewed as a contaminated, marginal zone caused by melting of the rhyolite wallrock through a large granite stock of about 2.5 km to the NE.

6.3.25.1 Settlement (20°24′42″ N, 35°00′28″ E)

This location is defined by an apparently Early Arab settlement discovered in a Google Earth scan. It is located about 2.5 km to the SE of Abu Mereim Salalob. The settlement had evidently been specialised on wadiworkings restricted to the alluvial fans coming in from the NE into a wide valley.

The wallrock of this deposit consists of partially porphyritic rhyolite with an average NNE-SSW strike and a foliation at around 10° E. Noteworthy is the occurrence of a microdiorite dike containing marginal xenoliths of a granite body located about 2 km to the NE.

6.3.26 Techol

Geographic positions:	
Techol West:	20°12′32″ N, 34°53′07″ E
Techol East:	20°13′05″ N, 34°54′09″ E
Wadiworkings:	20°13′00″ N, 34°52′15″ E
Trenches:	20°12′56″ N, 34°52′50″ E

Coming from the W, the Arab gold producing site at Techol begins with a wide, sandy plain and extensive wadiworkings covering the entire area and especially the central wadi. Clearly perceptible heaps of gathered quartz debris from the wadi marks the landscape over several kilometres (Fig. 6.88).

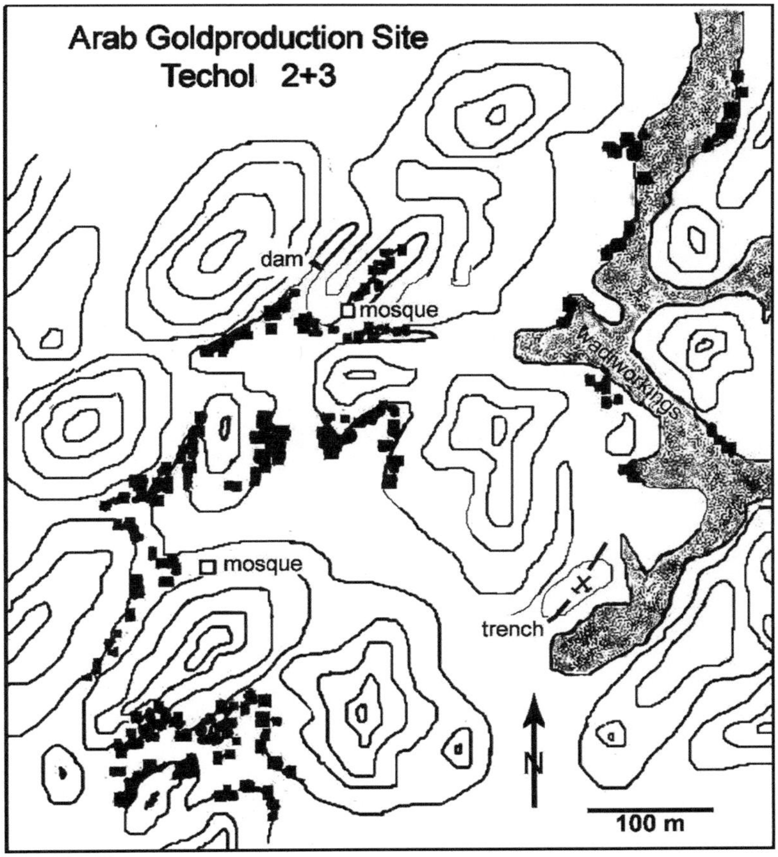

Fig. 6.88 Sites 2 and 3 in the Techol area with expanded field of wadiworkings, mapped by aid of a satellite image (R. Klemm)

Techol consists of three large settlements that had been built along the flanks of the main and side wadis flowing through a scenery of low hills (Fig. 6.89). In addition, one recurrently comes across huts that are loosely arranged in smaller clusters.

At settlement 1 the mostly rectangular houses are usually subdivided into several rooms. Their walls consist of flat and elongated slabs, so that occasionally they are preserved in heights up to 1.60 m (Fig. 6.90). At settlements 2 and 3 the huts have a more rounded appearance and are built from gabbroid boulders, which have left them in a much worse state of preservation.

At least three fully preserved washing tables were seen at settlement 2 (Fig. 6.91). All major house groups had prayer sites attached to them (Fig. 6.92).

According to Adams (1986), the recovered painted pottery belongs to the "Late Christian" Period. Nubian pottery shards with grooved, rhombic patterns were also found.

In its western part, the gold mining area at Techol had presumably exclusively aimed at the processing of alluvial quartz rubble. Because of the unusually large size of the settlements, the question arises as to whether or not certain quarters also specialised in activities other

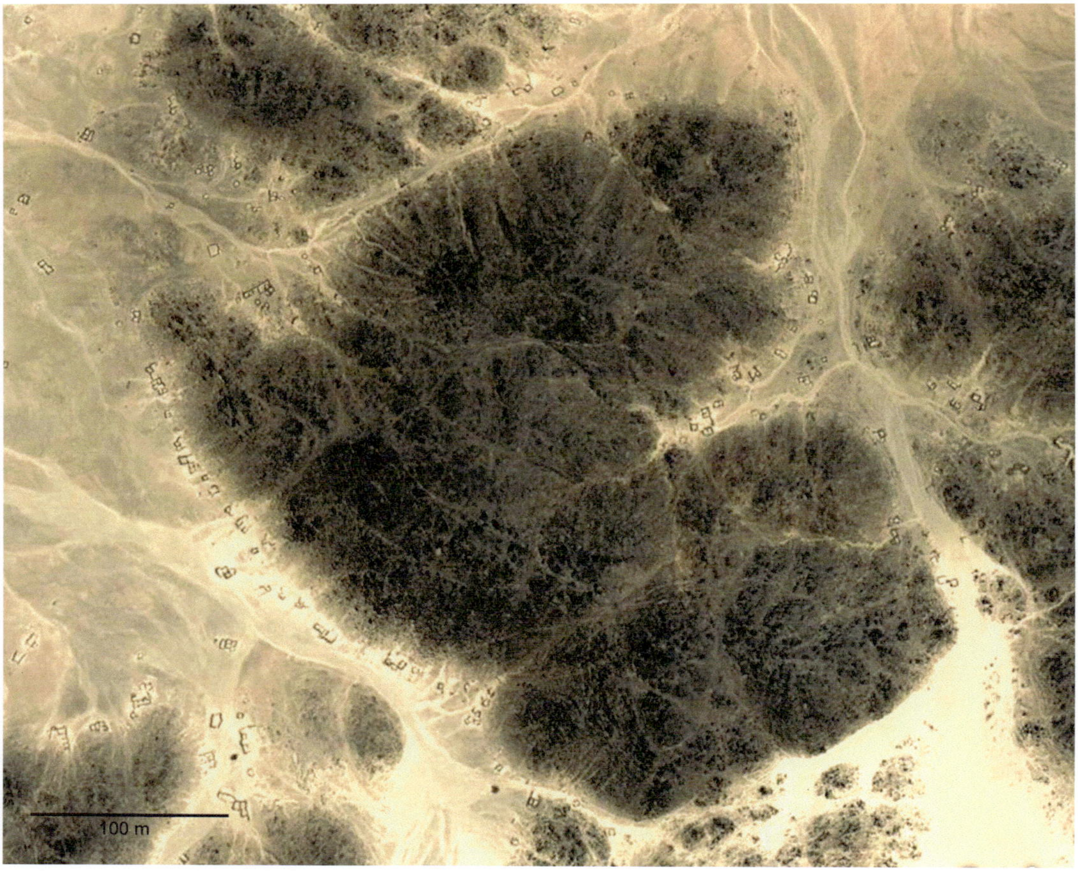

Fig. 6.89 Techol-West with countless huts along the wadi edges (modified Google-Earth image)

than the mining industry. Because of the mentioned scarcity of gold processing tools in the residential areas, one is tempted to assume that certain quarters had been linked to the supply of the settlement and to trade (caravanserai). Such specialisation would in fact account for the relatively large amounts of pottery in certain areas and the disproportionate number of praying sites.

Independently of that, the eastern settlement areas at Techol are to be regarded as genuine gold production areas, and even more so by the fact that mined quartz veins were observed in the vicinity. Understandably though, these mines were yet again not large enough to account on their own for the size of the neighbouring quarters.

Sporadic finds of New Kingdom oval stone mills, incorporated into the masonry of Arab

Fig. 6.90 Multi-roomed Early Arab Period house in Techol is well-preserved because of its flat building stones. Some New Kingdom ore mills and pounding stones are integrated to the masonry. To the right one recognises a washing table oriented towards the slope

Fig. 6.91 Well preserved washing table at Techol 2, whose water collecting basin and backflow channel are buried under the sand

Fig. 6.92 Large praying site at Techol 1. The mihrab is formed by standing stones

architecture, indicate indirectly former activities within this area.

In terms of geology, the Techol area (Fig. 6.93) belongs to the southern part of the Gebeit terrane and consists of NE-SW striking series of hornfelsic meta-qartzdiorites, -diorites and -rhyolites, which most probably are underlaid by a large granitoid pluton occurring in the erosion plain in the NW. The productive quartz veins and the wadi bed with traces from wadiworkings are exclusively bound to the rhyolitic-quartzporphyric sequences.

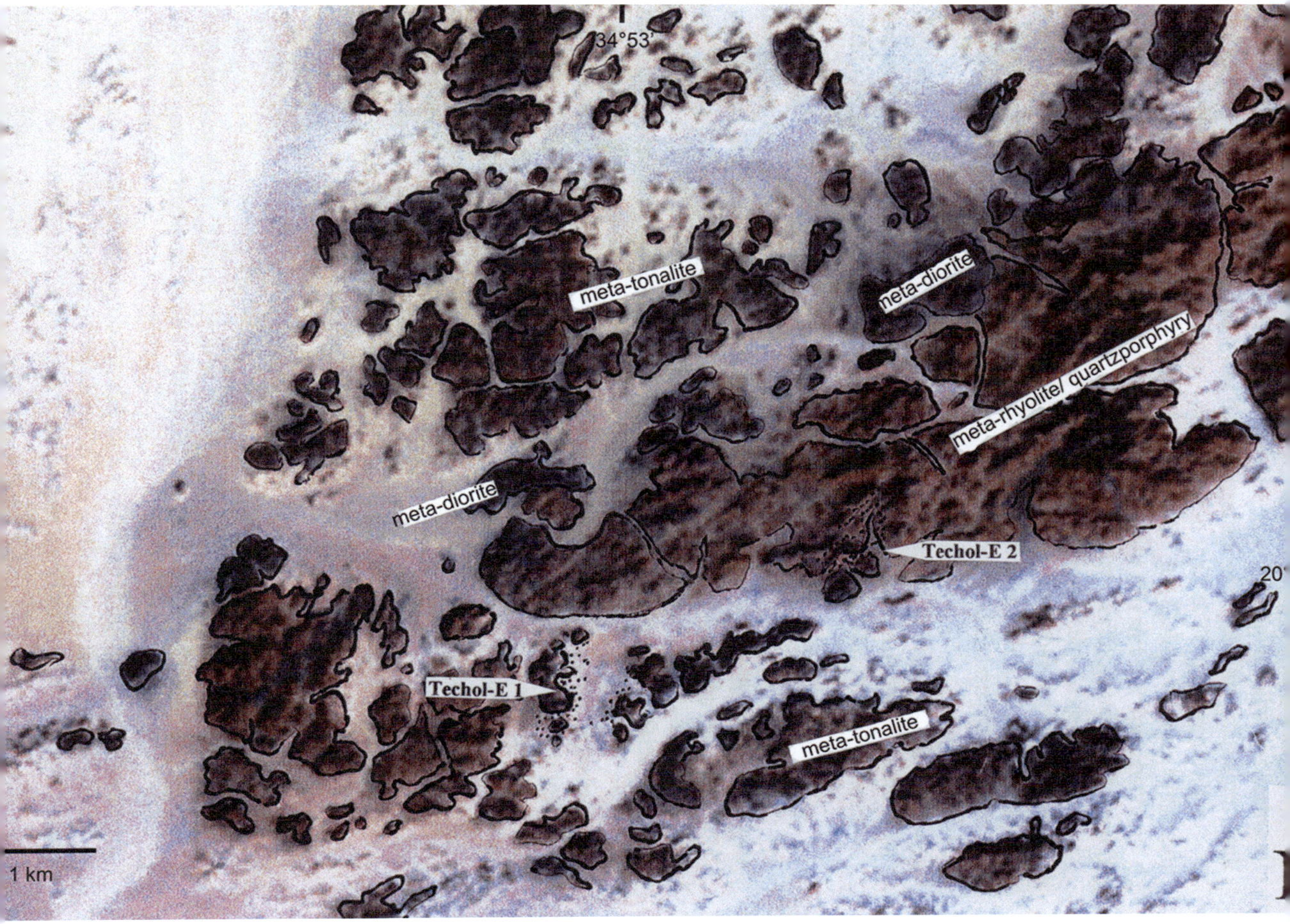

Fig. 6.93 Lithologically processed satellite image of the Techol area (TM 173–46, channels 7-4-1)

Fig. 6.94 Sand-covered Early Arab Period settlement at Nawarai with mostly round huts and working places

6.3.27 Navarai

Geographic position:	20°11′53″ N, 34°42′44″ E

A spacious, Early Arab settlement is located in a gravel plain consisting mainly of gabbro (Fig. 6.94). The huts are predominantly circular and usually have two rooms between 2 and 3 m in diameter. Next to a hut we found an intact round mill (quern), an exceptional find, as querns at some moment either have normally been willfully destroyed or are missing altogether. Well-preserved round mills consisting of both lower and upper parts were liable to be taken to other mining sites, as they often consist of rocks not available where they were used (Fig. 6.95).

We also found the rudimentary remains of four washing tables that stand out by their relatively short (3 m) sluice surfaces. A prayer platform pointing to the NE, however, is still well-preserved.

To the E of the settlement, about eight to ten Romibs are distributed over two hills, and on the western slope near the margin of the settlement, an approximately 50 cm long infant's grave was recorded, which again proves that families too, lived at the Early Arab mining settlements. Halfway up the slope some minor extraction trenches were noticed. Larger mines do not seem to have existed here. Apparently, gold production, or rather its attempts, seems to have been based on the gathering of abundantly available quartz fragments at the surface. The yield must nevertheless have been quite inconsequential, because the gold producing site at Navarai is located within a mountainous massif constituted of gabbro and diorite, which according our experience doesn't represent the most promising environment for gold production.

Fig. 6.95 Well-preserved Early Arab Period ore mill and two different rotor stones in Navarai

6.4 Group West of the Hamisana Suture

In addition to the large Early Arab site at Derahib, a number of contemporary settlements are located further N, on the Sudanese side of Wadi Allaqi. The following listed sites are quite small and are therefore only referred to by their geographic positions:

21°58′42″ N, 35°06′46″ E (small settlement in a southern side wadi)

21° 58′32″ N, 35°07′38″ E (13 m long complex with a large, central room and seven smaller ones)

21° 58′02″ N, 35°08′19″ E (small settlement)

21° 57′37″ N, 35°08′34″ E (three large buildings with several rooms)

21° 57′36″ N, 35°08′03″ E (small settlement in a southern side wadi)

21° 57′32″ N, 35°08′12″ E (small settlement in a southern side wadi)

6.4.1 Derahib

Geographic positions:	
Free space in large settlement complex:	27°57′02″ N, 35°08′27″ E
Northern fort:	21°56′51″ N, 35°08′32″ E
Mine:	21°57′10″ N, 35°08′20″ E
Southeastern settlement:	21°56′36″ N, 35°08′42″ E

From a distance, in approaching Derahib one makes out the silhouettes of what one might think are tall buildings rising from the middle of the desert landscape (Fig. 6.96). However, they soon turn out to be the ruins of nonetheless massive fortifications whose walls and corner bastions are still preserved in impressive heights. The complexes are situated on the W side of Wadi Allaqi and separated by about 50 m from each other (Figs. 6.97 and 6.98). In the surroundings and especially the area between the forts one notices

Fig. 6.96 The two, well-preserved Early Arab Period Forts at Derahib, unique vestiges in the Eastern Deserts of Egypt and Sudan

the ruins of house clusters covered by thick layers of drift sand.

On the eastern side of the wadi, to the NE of the forts, there is an approximately 400 m long compound consisting of agglutinating, architectural units that group around both sides of an approximately 10 m wide, paved street. The rooms of the buildings form relatively elongated rectangles and are arranged perpendicularly to the central axis. Here and there some taller ruins jut out from the rest of the remains (Fig. 6.99).

The building material of the settlement consists of the two prevailing rock types in the area. One is a flat metasediment slab originating from the W side of the wadi; the other is a more globular, granitoid type from the E side. The distribution of both rock types within the site in fact reflects that of their origin. While the quarters to the W of the central road had been built with the former, the latter had been used for the sector to its E. Owing to the higher stability of the walls built from the flat slabs, the W sector is in a comparatively better state of preservation. Its resemblance to the compound in Uar is therefore perceptible (see above).

An intersecting side road is clearly discerned in the central part of the complex. At least four, large rock heaps have formed along the western part of the settlement. A cemetery is located at its northern end, and at the southern end bordering an eastern side valley is an 18 × 13 m large, Islamic prayer site (Fig. 6.100). Further E, one is struck

Fig. 6.97 Tower gate of the northern fort at Derahib

Fig. 6.98 Vaulted passageways inside the northern fort at Derahib. Note the clay mortar and remains from wall plaster

Fig. 6.99 The large settlement site at Derahib displaying partly well-preserved walls. The entire complex is crossed by a central alley, a common feature at Early Arab Period sites in the Eastern Deserts of Egypt and Sudan

Fig. 6.100 Large Islamic prayer site at the S end of the large settlement at Derahib

by a large building whose two rooms are enclosed by very high walls, which may possibly represent a former prison. Numerous Islamic tombs are grouped at the inlet of a small side wadi to the E (Fig. 6.101) and further up in the main valley.

The former location of a possible well is marked by a partially walled depression surrounded by several buildings about 100 m W of the northern fort-tower. It may even have been the main well of the settlement, as we had not been able to identify any corresponding structure within the main settlement area. The well may be older than the settlement itself, which would also explain its location.

Among the pottery, fragments from amphorae are relatively common. Green glass shards and green to brown, glazed plates and bowls from the Early Arab Period are scarcer. The British in their turn, have left behind purple glass shards from mineral water bottles.

Everywhere in the wadi one comes across for the most severely dilapidated, architectural remains. Along its southern edges and especially in a wadi branching-off to the S are more settlements as well as a heavily damaged cemetery. In isolated positions, certain tool finds unquestionably designate some of the architecture as miner's dwellings.

Although we carefully inspected all the wadis in the district, we found no traces dating to either the New Kingdom or the Ptolemaic Period. This is perplexing, in that Castiglioni (1992, 1998) claim to have identified the site with the lost Ptolemaic city of Berenice Panchrysia, a Ptolemaic gold mining city founded by Ptolemy II Philadelphus in 270 BC, and already quoted by Pliny in his Natural History

We unfortunately didn't find any evidence that might substantiate this identification, considering especially that the gold mines are by far too small to justify a settlement of this extent.

These mines are located directly above the site, in the western mountains (Fig. 6.102). In the Dunn report (1911) it is noted that they were chockfull with hundreds of tons of bat-guano, an observation we are unable to confirm. To us, the mining complex did not seem to have been very important, as its exploitation had been unsystem-

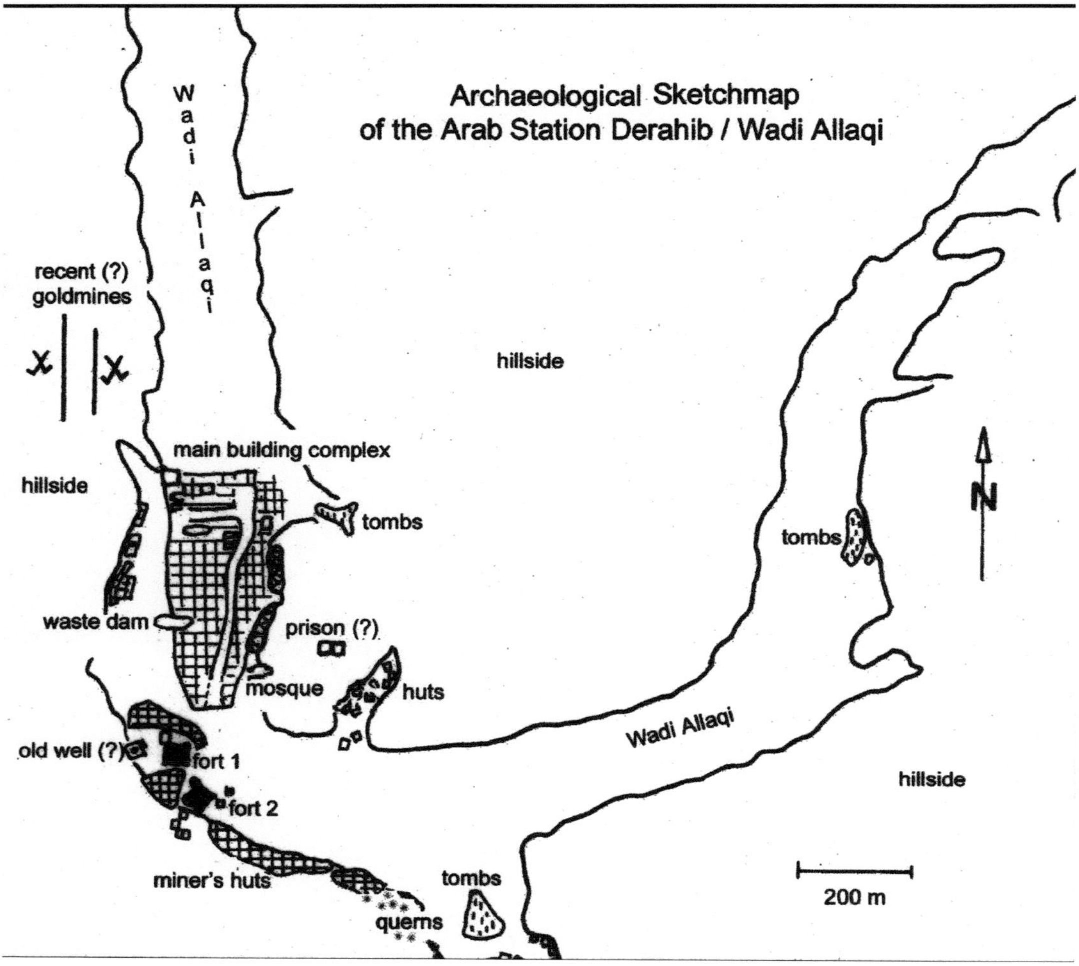

Fig. 6.101 Archaeological sketch plan of Derahib (R. Klemm)

atic and its size generally quite reduced. In the early years of the twentieth century however, the mines were explored in new shafts by a British mining company, though without subsequent mining on a significant level.

In our opinion the entire settlement area was centered around an important caravanserai serving the region's allegedly large camel herds, which according to written sources apparently added up to altogether 60,000 heads (cf. A. Al-Maqrizi, quoted by Y. Fadl Hasan 1967). Such large herds may well have been necessary to assure the supply of the numerous miners disseminated throughout the Nubian Desert during the goldrush under the rule of A. Ibn Tulun and Al-Omari. Commodities like food and water had been certainly stored here as corroborated by the mentioned abundance of amphora fragments and the elongated, depository-like architectural units along the central road of the large settlement. In this connection, the two fort-towers may have functioned as the central coffers for the payable gold coming in from the numerous gold mines in Wadi Allaqi and its adjacent areas to the S. At the same time the site most certainly too, operated as an administrative centre for all these activities.

Similar units with apparently related functions are also known at Umm Eleiga in the southern

Fig. 6.102 The relatively small mining district at Derahib shows inspite of the settlement's large size only few traces from gold mining in the Early Arab Period. This may however, be partly due to the mining attempts carried out at the beginning of the twentieth century

part of the Eastern Egyptian Desert, furthermore at Uar, Omar Kabash, Shashuteb at Bir Kiaw (also with a fort), as well as in Wadi Terfawi. All of these sites are located in Nubian Sudan and date to the Early Arab Period. They are also located centrally in relation to the countless gold mining sites in their respective hinterlands.

According to our observations, Derahib's importance as an actual gold mining centre may on the other hand have been relatively limited. To date, solely the site's western mines testify that gold was actually mined here. Ore processing on the other hand, occurred chiefly near the huts of the southern settlement where only few round mills have been found. The diversity of rocks that had served for manufacturing the rotor stones to some extent reflects the mobility of the individual miners during this period. Evidence of an occupation older than the Early Arab Period was not recorded at Derahib.

It is nevertheless likely that wadiworkings had been operated here, like almost everywhere else in Wadi Allaqi. Traces thereof are however very hard to spot.

We did not detect any indication as to gold melting activities. However, some gold assaying crucibles from the British mining company have in fact been found.

Derahib is strategically located at the main transit route from Aswan to the Red Sea, and in particular at a junction with a road leading S to the gold-rich province of Onib.

The geology around Derahib is dominated by the Allaqi-Hamisana zone with its multi-coloured rock sequences that consist of greywacke, siltstone, conglomerates, and repeated marble intercalations alternating with serpentinite, amphibolite, rhyolite, dacite, and agglomerates (Fig. 6.103).

In Wadi Allaqi these persistently much folded series display an average NNW–SSE strike but

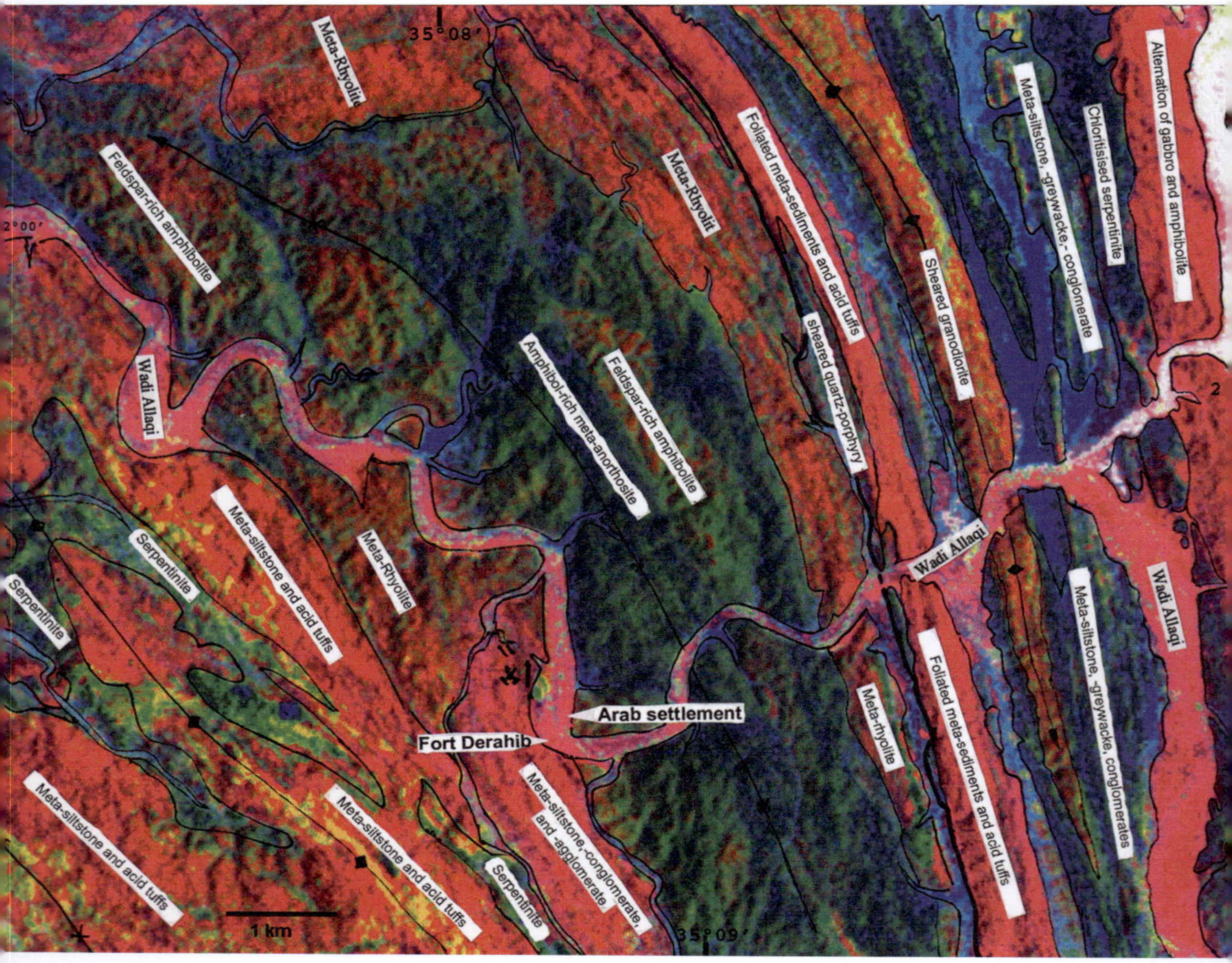

Fig. 6.103 Lithologically processed satellite image of the area around Derahib (TM 172–45, channels 7-4-1)

then change to a chiefly N-S direction east of Derahib and continue towards the Hamisana zone in the S. In a prominent and common shear zone system their eastern margins attain the Oyo-Onib zone, which turns-off to the NE

Partly enormous granitoids intrude both zones in a syn-kinematic order (e.g. the Hufra complex with 35 km longitudinal diameter). Sufficient thermal energy in their chiefly granodiorite border areas and surroundings they had been supplied to prompt hydrothermal gold quartz mineralisations. Under continuing tectonic stress such occurrences subsequently had sometimes transformed completely to boudins (cf. Uar, Kamoli).

However, as far as hitherto known from the ancient mining traces, most auriferous vein quartz mineralisations are bound to the numerous post-kinematic granitoid intrusions, whereby these in some instances as by assimilation of their host rocks may also be composed of diorite an even gabbro.

Extractible gold concentrations required however a geochemical gold pre-enrichment of at least of about 20 ppb inside the intruded rock sequences. These preconditions are often given in mafic to ultramafic, former ophiolites and their serpentinites, basalts or amphibolites as well as the numerous intermediary and acid volcanics, plus the resulting products such as greywackes and conglomerates.

According to this model, especially in a zone like that of Hamisana, in which all of these series occur in steep isoclinal folds and with the necessary thermal energy from intrusive complexes one may expect extensive hydrothermal gold enrichment. Through sustained compression over lengthy periods, the thicknesses of the vein occurrences may decrease substantially, if not even reduced to boudins through shearing processes. The result in both cases are small quartz veins or even lenses, that only rarely allow for cost-effective exploitation in modern mining. With the ancient wadiworking method however, fragments of such small deposits, which previously had been transported into the wadi grounds by erosion could be gathered selectively and subsequently processed in an economically viable manner.

According to the archaeological findings, gold mining at Derahib during the Early Arab Period occurred only at a limited scale, mainly in wadiworkings and to a lesser extent in deep mines in the hills. Significant gold mineralisations along this much frequented route would certainly not have escaped the attention of the New Kingdom prospectors, as they after all had managed to identify reliably just as little opulent mineralisations within the gold mining districts located further S.

6.4.2 Miri

Geographic position: 21°56′07″ N, 34°52′52″ E

This location is defined by a number of small trenches and a dozen small huts on a gravel terrace and furthermore, at least two prayer sites and some tombs.

6.4.3 Wadi Ward Miriyam

Wadi Ward Miriyam runs about 6 km N to Khor Nesari, at first in a roughly parallel, E-W direction before bending-off to the SE. The wadi hosts a number of sites attributable to the Early Arab Period as well as the New Kingdom. The following discussion treats the larger sites which we tagged with numbers1–6. Because this nameless but yet interesting wadi also needed to be labelled, we decided to name it "Ward Miriyam" (Ward = Rose and Miriyam = Marie).

6.4.3.1 Wadi Ward Miriyam 1

Geographic position: 21°55′14″ N, 34°53′29″ E

Investigations at a wadi heading E, in an area N of the mouth of Khor Nesari, led to the discovery of an extended New Kingdom settlement spreading-out on both sides of the wadi in the vicinity of a hill marked by an Alam tower. The houses consist of large, rectangular rooms with massive, single-shell walls and contain numerous, relatively large, New Kingdom oval mills as well as only few round mills from the Early Arab Period.

The New Kingdom houses at "Wadi Ward Miriyam 1" (Fig. 6.104) are well fortified and may have served for the protection against raids from desert populations. Their walls are unusually well-preserved, probably as they are also set

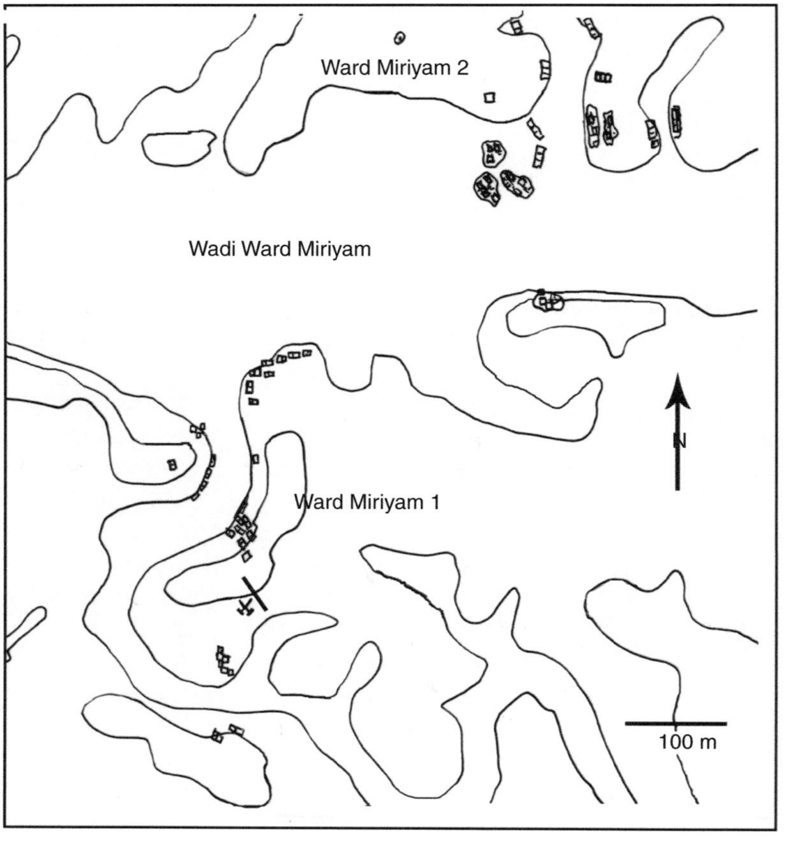

Fig. 6.104 Sketch plan of the gold mining sites Wadi Ward Miriyam 1 and 2 (R. Klemm)

in a protected situation on a wadi terrace, in a remote valley recess. The site consists mostly of agglutinating, rectangular rooms in the N sector, but also round huts whose walls integrate natural rock protrusions. The houses could not be accessed from the valley side.

Particularly large grinding mills were observed inside the round huts at the N side of the site. Their oval grinding surfaces are 40–60 cm long and 30–40 cm wide. They display various stages of their gradual hollowing-out from use wear and consist partly of the local, coarse porphyry rock. In addition, there are numerous fist grinders and cubic anvil stones as well as pestles (Fig. 6.105).

The fist grinders, which measure up to 20 cm in length and 10–12 cm in width, are quite large. Sometimes they are worn on both sides. Some display a depression resulting from pounding, which hitherto hasn't been observed. In the case that the quartz chunks had been too bulky for the grinding process, they had apparently first been crushed on the mill surface with the grinding stone (Fig. 6.106).

Fig. 6.105 New Kingdom pounding stones and broken New Kingdom mill (*top*) from Ward Miriyam 1

The mine associated to this settlement is located in a partly internally split quartz vein system. The wallrock is a granitoid, intensively altered by hydrothermal circulation and weathering (Fig. 6.107). The quartz mineralisation consists of at least two generations, of which the white one had not been mined. Only the thinner quartzes situated in the partly split-up, recumbent zones of the vein structure, near the highly altered and to some extent pyritised wallrock, had been extracted. The decomposed wallrock had probably supplied the iron necessary for the pyritising process, which subsequently led to the spontaneous decay of gold-sulphide complexes and the precipitation of gold and pyrite.

Fig. 6.106 New Kingdom one hand runner stones (*front*) in Ward Miriyam 1

Fig. 6.107 Partly excavated quartz vein in Wadi Ward Miriyam 1. As most of it is barren, much was left standing, thereby indicating the original level of the surrounding terrain

6.4.3.2 Wadi Ward Miriyam 2

Geographic position:	21°55′19″ N, 34°53′38″ E

At the northern side of the main wadi, about 650 m further E, we came across an unusually large, New Kingdom settlement. The site is located at the mouth of a small side wadi and is divided up into at least four large and clearly defined areas, each consisting of a cluster of interconnected rooms. The rooms themselves are quite large and contain several New Kingdom mills (Fig. 6.108). At the eastern edge of the settlement are the remains of a relatively well-preserved, inclined washing table. It is about 4 m long and 80–90 cm wide. Its original height is no longer determinable since only its contours are visible at the surface. Although it is similar to the

Fig. 6.108 Typical New Kingdom house clusters at Ward Miriyam 2

ones from the Early Arab Period, the find context duly dates it to the New Kingdom. The discovery only adds to the evidence that such appliances were already used at the time of the New Kingdom to concentrate and separate the gold from the ground quartz powder (Fig. 6.109).

The buildings themselves had apparently been used partly for dwelling and partly as workshops. Terrace-like extensions suggest that the low-lying ones had been dwelling units, whereas the narrow structures on higher ground seem to indicate some sort of storage facilities. Tools are found only rarely. Further up the mountains a number of buildings with massive walls scatter along the slopes. None yielded tool inventories. They may therefore have served as observation posts used for supervising most of the settlement, as some houses are hidden behind small neighbouring hills.

Gold production had certainly also been based on ores from the wadi sediments, because the small mine located to the SW had by no means been able account for a settlement of this size, and especially its large number of oval stone mills.

These in turn, are strikingly large and may measure up to 60 × 80 cm large stone slabs whose 40–50 cm long grinding depressions are more round in shape than oval. The grinders too, are particularly large but otherwise come in the usual shapes. Also typical for New Kingdom sites is the scarcity of pottery finds. At Wadi Ward Miryam 2 they have been preserved in only very small fragments.

Former Alamat towers are discerned along all surrounding heights.

6.4.3.3 Wadi Ward Miriyam 3

Geographic position:	21°55′16″ N, 34°54′49″ E

Totally 12 house groups from the Early Arab Period were counted here. An Arab Period settlement is located some 100 m further up the wadi, on its N side.

Fig. 6.109 Remains of a New Kingdom washing table and oval mills at the edge of a New Kingdom settlement in Wadi Ward Miriyam

6.4.3.4 Wadi Ward Miriyam 4

Geographic position:	21°55′08″ N, 34°55′06″ E

This is another Arab settlement, again at the wadi's N side and consisting of two house groups. It had probably been the processing site for ores coming from a mine to the SE (on the opposite S side of the wadi).

6.4.3.5 Wadi Ward Miriyam 5

Geographic position:	21°55′03″ N, 34°55′14″ E

A narrow and rectangular stone structure measuring 31 m in length, whose function we were unable to determine is located in front of a mine, on the southern flank of a side valley. It may possibly have served as a cattle line-up.

The mine is the largest one we recorded in Wadi Ward Miriyam. It is at the rear end of the slope to the S of the settlement site (Fig. 6.110). It had probably already been under intensive exploitation in the New Kingdom and undoubtedly supplied the downstream processing sites with ores. Different locations for settlement and mine had been common for gold mining communities in the New Kingdom but no longer in the Early Arab Period. In other words, if an Early Arab settlement is the only evidence of human occupation in the immediate vicinity of a mine, this does not automatically certify its exploitation exclusively for that period.

The partially folded quartz veins, which exhibit a slightly chaotic arrangement within the mining district, occur in a sheared, granitoid wallrock where they probably are bound to an ancient fault system. Consequently, the thicknesses of the veins vary considerably, revealing a so-called pinch-and-swell structure. Only the quartz-containing wallrock interstices had been extracted between massive, white, sometimes brecciated quartz veins.

Fig. 6.110 Important deep mine in Wadi Ward Miriyam, in which abutments have been removed and replaced by large boulders

The ore parageneses consist next to (invisible) gold, of pyrite and remains of primary copper minerals, notably malachite.

6.4.3.6 Wadi Ward Miriyam 6

Geographic position:	21°54′51″ N, 34°56′16″ E

This position marks the end of a sequence of isolated Early Arab huts on both sides of the wadi between Ward Miriyam 5 and 6. A small mine is recognised on the mountain near Ward Miriyam 6. In addition, a widespread area of wadiworkings was identified in the valleys behind the mountain ridge of Ward Miriyam 6.

Smaller hut concentrations are distributed over the entire distance, preferably at small inlets along the N side of the wadi. They all are similar to the settlements Ward Miriyam 3 and 5. In general, the huts are well-built and offered sufficient shelter.

The geologic structure of the entire area covering Khor Nesari and Wadi Ward Miriyam much recalls the orogeny in the area of Wadi Gabaideb, because here too, formerly volcano- sedimentary units of the Allaqi-Hamisana zone had been intruded by a vast granitoid pluton, whose hybrid roof is exposed at the surface. In fact, within a wider geologic context this area for the most forms one unit with the Gabaideb complex, of whose large boudinage zone it represents the northernmost part. This northern complex is crossed in its centre by Wadi Ward Miriyam in an approximate E-W direction. Nevertheless, significant differences remain. Among them the circumstance that unaltered granitoid magma areas are never clearly attained here, but merely the exposed hybrid edge and roof zones, as seen in the lithologically processed satellite image (Fig. 6.111).

From this it becomes evident that the closest proximity to the granite is located in the S of the Ward Miriyam complex, where it is crossed by the wadi in a NW-SE direction. Towards the N the hybrid zone increasingly becomes richer in wall-rock components that eventually end up by dominating in a hornfels formation in Wadi Allaqi.

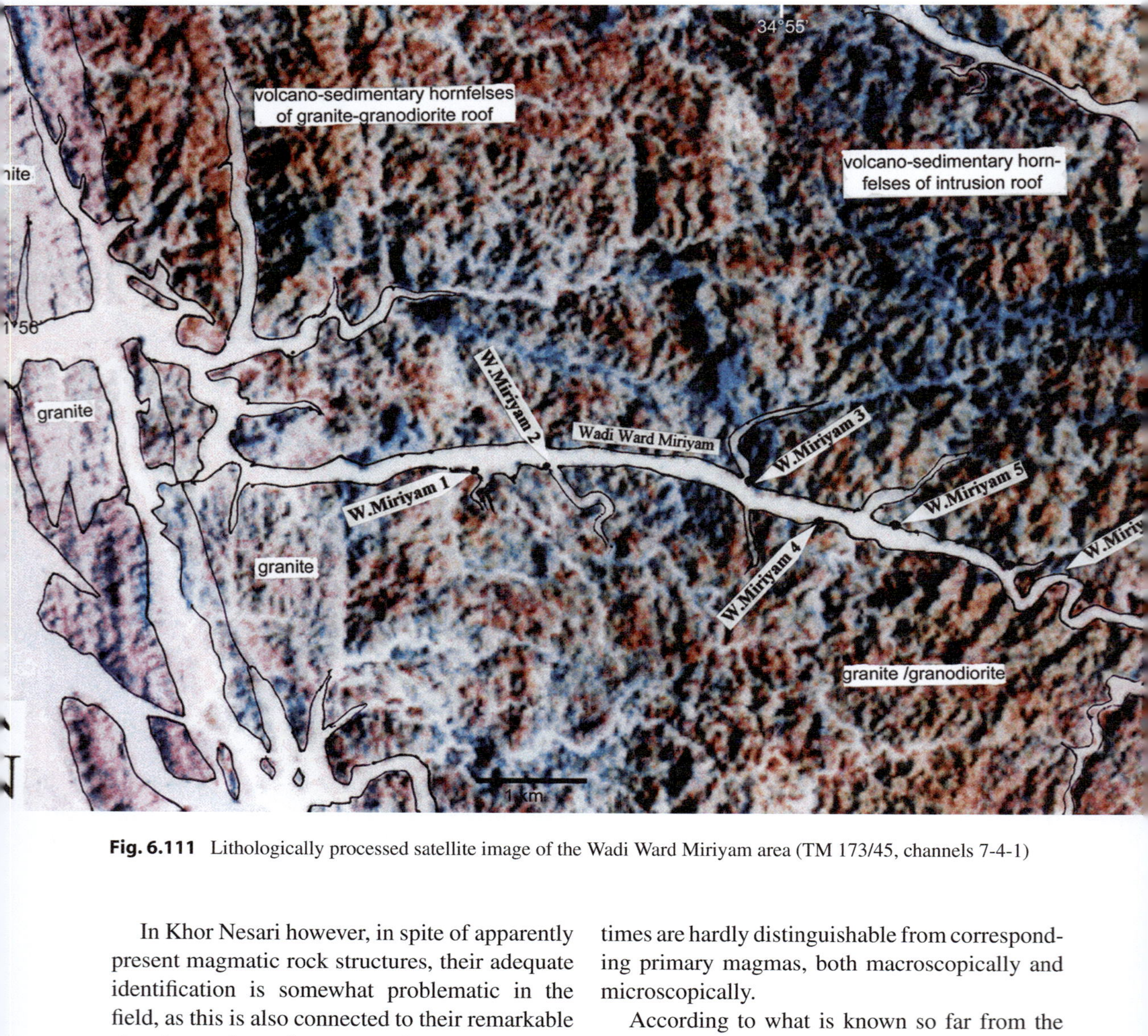

Fig. 6.111 Lithologically processed satellite image of the Wadi Ward Miriyam area (TM 173/45, channels 7-4-1)

In Khor Nesari however, in spite of apparently present magmatic rock structures, their adequate identification is somewhat problematic in the field, as this is also connected to their remarkable shearing and ensuing susceptibility to weathering, which again is typical for such hybrid rocks.

In Wadi Ward Miriyam already, unmistakable relics from hornfelsic wallrock can be observed, especially at the wadi mouth. Depending on the chemistry of the source rocks digested to hybrids by the granite magma, granodiorite and even diorite rocks occur here in large streaks, which at times are hardly distinguishable from corresponding primary magmas, both macroscopically and microscopically.

According to what is known so far from the Eastern Desert of Egypt and Sudan, such hybrid zones are ideal target areas for auriferous hydrothermal quartz mineralisations.

It is hence not surprising that numerous, ancient gold processing sites are found along both valleys, despite a possible link between a concentration of Early Arab sites in the thoroughfare of Khor Nesari and their location on the pilgrim route to Mecca. One may therefore expect more ancient sites in areas both to the N as well as S of this area.

6.4.4 Wadi Nafiryam

6.4.4.1 Nafir 1

Geographic position:	21°48′14″ N, 35°09′31″ E

A comparatively large settlement with Early Arab assemblages is located on the E slope of a ridge in Wadi Nafiryam ("good water") (Fig. 6.112). Some huts even spread right up to the ridge itself. Traces from wadiworkings, as well as two, well-preserved washing tables with tailings can be discerned over a distance of approximately 500 m in the wadi (Fig. 6.113). Some ruins from large houses containing a complete round mill and some anvil stones are seen not far from the washing tables. We furthermore found painted shards of the so-called figurative Classic Christian ware (Adams 1986) as well as Nubian pottery with incised patterns near the southern washing table.

6.4.4.2 Nafir 2

Geographic position:	21°50′57″ N, 35°10′06″ E

The small settlement of Nafir 2 is reached after about 700 m into a side wadi that branches-off to the E from Wadi Nafiryam. Here, on the southern side of the wadi are about 10–12, poorly-preserved round huts. Between 20 and 25 additional structures are seen on the slope of a tributary wadi, some also in a gorge, of which one reveals wall heights up to 70 cm. Apart from a couple of anvil stones, no tools were found. The site probably represents an Early Arab wadiworking site. A small dam, of which one half has survived, is still standing in the bed of a small wadi to the N of the settlement.

6.4.4.3 Nafir 3

Geographic position:	21°51′51″ N, 35°09′40″ E

Nafir 3 is chiefly represented by a Bedouin cemetery. Furthermore, there are approximately eight Arab Period huts on the southwestern slope. In the wadi itself are the abandoned remains of a Bedouin campsite suggesting a possible, nearby presence of a well.

6.4.4.4 Nafir 4 (Including Bedouin Cemetery in the N of the Wadi)

Geographic position:	21°51′39″ N, 35°09′01″ E

Some fairly large and well-preserved huts displaying a round mill from the Early Arab Period are located in a side wadi leading N from Wadi Nafiryam, behind an Arab cemetery. Further N in the same valley is another group of 15 to 20 huts. The wadi turns-off N from the plain with the Bedouin huts. To the SW, behind Nafir 4 there are more Early Arab huts on both sides of the Wadi. Apparently, intensive wadiworking activities had taken place in this district.

6.4.4.5 Nafir 5

Geographic position:	21°52′03″ N, 35°09′01″ E

Nafir 5 consists of a relatively large settlement site exhibiting two phases of occupation. The first with its characteristic tools and mills dates to the New Kingdom. The second occurred in the Early Arab Period, when small New Kingdom grinding stones and New Kingdom mills had been included to the stonework of the Arab house walls. The houses reveal the usual assemblage from the Early Arab Period.

Fig. 6.112 Early Arab Period houses in Wadi Nafiryam displaying a peculiar masonry of long, spiky stone slabs

Fig. 6.113 Remnants of an Early Arab Period washing table in Wadi Nafiryam with an unusually large water collecting basin

The New Kingdom site Narfiryam 5 (and possibly Narfiryam 4) testifies to the highly sophisticated skills of the New Kingdom prospectors, who recognised the productive gold concentrations in the listvenite quartz carbonates inside a serpentinite lens running through this area. By the Early Arab Period, this knowledge had gone lost, since according to the ore residues in the huts, only wadiworkings had been operated, while the admittedly slim prospects from deep mining had not been recognised.

6.4.4.6 Nafir 6

Geographic position: 21°52′48″ N, 35°08′34″ E

The small, Early Arab settlement is located about 1.5 km N, at the western edge of the serpentinite swathe of Nafir 4 and 5 in a small, parallel side wadi of Wadi Nafiryam. Since no traces from mining were found, probably wadiworkings or mining attempts only had been carried out here.

Wadi Nafiryam runs over a distance of approximately 7.5 km in a parallel direction to the NNE strike of the Hamisana western branch. It then meanders-off to the NE, transversely to the series. The geologic sequences consist of a colourful series of predominantly acid volcanics and clastic, sedimentary rocks, which despite their mostly intense deformation remind of the Hammamat series. In Sudan however, these rocks are referred to independently as the Nefirdeib formation and indeed contain much more volcanics than their counterparts in Egypt. Serpentinite and amphibolite ophiolites are intrinsically imbricated with these units. They are often transformed to listvenite quartz carbonates and already belong to the Gabgaba terrane. These sequences are all recorded in a steep, isoclinal fold and are repeatedly intruded by round gabbro stocks (Fig. 6.114).

Through overthrust and intense shear processes within this zone, the older marginal volcanosedimentery series of both terranes as well as their pre-and syn-kinematic, granitoid intrusions had been severely deformed in the shear direction (Fig. 6.115). Close to Wadi Rak the zone divides into a western and eastern branch. The former merges into the Haiani-Allaqi zone, while the latter forms the border between the Gerf terrane, already belonging to Egypt, and the Gebeit terrane. Nafir-Derahib is positioned within the furcation of the Haiani-Allaqi zone and the Gebeit terrane, confining the Onib-Sol Hamed zone.

6.4.4.7 Nafir-Derahib

Geographic position: 21°52′36″ N, 35°11′11″ E

This position refers to a widely dispersed, Early Arab settlement consisting of several clusters, each numbering between five and ten individual huts, and occasionally houses of up to 10 m in length. They are located mainly along the edges of the numerous, shallow wadis running parallel in a N-S direction. In some cases they also spread beyond the low, elongated hills between the wadis. The sparse finds consist of round mills, few anvil stones, and even less so of Early Arab Period pottery.

No evident traces resulting from quartz vein mines were found, which suggests that the local mining industry had concentrated exclusively on wadiworkings.

The geologic environment consists of similar petrographic sequences to those observed in Wadi Nafiryam. Only the acid volcanic rocks occur in larger amounts here. About 700 m E, there is an elongated serpentinite ridge. Both sequences are potential sources for primary gold, provided the necessary presence of hydro-

Fig. 6.114 Lithologically processed satellite image of the Wadi Nafiryam area (TM 172–45, channels 7-4-1)

thermal anomalies able to set-off circulation in order to dissolve and transport the primary gold traces. The closest, though severely deformed granite massif, however, only occurs at the southern foothills, about 4 km E of the settlement complex and is therefore an unlikely source for the hydrothermal formation of gold quartz veins.

Scattered settlements are located close to the pass leading to Derahib. At least 30 houses containing round mills spread beyond a hill and far into the wadi. This settlement is perhaps identical to that described by Castiglioni and Vercoutter (1998) "to the SE (of Derahib) is a settlement similar to Derahib". This widely scattered settlement area is however not comparable with Derahib itself.

Fig. 6.115 Intensively sheared part of the western Hamisana zone, in which singular rock components have formed to lenticular boudins

6.4.5 Khor Nesari 1

Geographic position: 21°49′46″ N, 34°59′27″ E

This location is defined by a fairly large number of huts on the N side of the wadi with Early Arab find assemblages. An Alam tower rises on the most prominent hill of the area. Continuing on the S side one furthermore notices some additional huts in the NW.

6.4.6 Khor Nesari 2

Geographic position: 21°49′46″ N, 34°57′41″ E

On the southern side of the wadi, where wadiworkings had probably been operated, there are also Early Arab huts extending further W into the recesses of this large wadi. Clearly visible traces from wadiworkings and deep holes from the surveying work under the Anglo-Egyptian condominium mark the wadi floor. Some greenish glass shards, the remains of a round mill, and some light-coloured pottery fragments from amphorae represent the meagre archaeological finds.

6.4.7 Khor Nesari 3

Geographic position: 21°50′47″ N, 34°55′44″ E

This site consists of approximately 15 poorly preserved, small, round huts from the Early Arab Period. Small mines can be made out near the top of a mountain on the S side of the wadi. On the N side, just above a small settlement, there is another exploited mine. Yet another ancient mine, about 200 m further W displays transverse pillars in a wide quartz vein on the S side of the wadi (Fig. 6.116).

Compared to the relatively small settlement the exploited quartz veins seem considerably developed. They strike 35° E and dip SE. The large vein to the W is up to 2 m wide and occurs in a granodiorite rock with a slightly schistose texture. It in fact represents a ghost structure of the formerly dissolved wallrock through hybridisation provoked by the granite intrusion.

No more than a third of the approximately 2 m wide quartz vein was extracted along its hanging wall. Externally, this part is almost indistinguishable from the remaining white quartz of the vein (Fig. 6.117).

Two additional trenches at the Khor flank to the NE of the settlement reveal superficial mining. They occur in a seemingly sheared, granitoid wallrock. They respectively strike 40° E and 10° E with a SE and E dip.

6.4.8 Khor Nesari 4

Geographic position: 21°50′54″ N, 34°55′25″ E

About 500 m NE of Khor Nesari 3 another Early Arab settlement lies hidden away in the wadi behind an elongated gravel heap. It includes the ruin of a strikingly large house measuring about 20 m in length and surrounded by several smaller houses. They are associated to a mine distinguished on the mountain to the NE. Khor Nesari 3 and 4 should be regarded as one site, whereby Khor Nesari 4 being fortified, had possibly been established under Al-Omari. According to Hasan (1967) 30,000 Bedjas (Bedouin tribe) had been forced within the Nubian Desert to socage services in the mines here and coerced to convert to Islam. Two large, outdoor spaces surrounded by walls inside a compact complex, as well as a washing table located next to the ruins of a large building in the plain seem to reflect a master-planned organisation behind the reconstructed sequence of work processes in gold production. The complex also includes three large cemeteries, of which two are placed on a gravel terrace parallel to the wadi. The small, individual huts further E date probably to a later occupation period (Fig. 6.118).

An inscribed shard quoting "Allah" was found in addition to the usual assemblage of round mills, green glass and painted ceramics. It strikes that many buildings have open courtyards surrounded by high walls. In most courts we observed a washing table and several round mills. The massive walls suggest that work had been carried out under supervision.

In the narrowing valley to the E of the large complex at Nesari 4, one also notices the remains from wadiworkings.

The outcropping rock here is essential for the comprehension of the geologic setting in this district. Rhyolitic xenoliths also occur next to the ghost structures described above in the diorite-like bedrock. Because of their structural dissolution through the hybridization effect of the granite magma, these appear in a schistose structure. In addition, there are especially dark xenotliths in small to several meters-thick fragments of a yet largely undigested amphibolite, which of course here occur in a hornfelsic structure. By partial assimilation of these basic fragments from the intrusion roof the original granite magma had seemingly adopted the characteristics of diorite (Fig. 6.119).

6.4.9 Khor Nesari 5

Geographic position: 21°54′51″ N, 34°56′16″ E

More Early Arab Period houses are located on either side of a side wadi, to the S of Wadi Nesari, though without exhibiting any mining gear.

A possible explanation for the differences between Early Arab settlements, specifically those that contain mills and mining tools and those that don't, may be that ones usually consisting of loosely distributed huts represent

Fig. 6.116 Ancient mine following a winding quartz vein and secured by buttresses at Khor Nesari 3

Fig. 6.117 Apparently, only partly productive quartz vein at Khor Nesari 3. The barren (white quartz) areas had therefore been disregarded

Fig. 6.118 Building complex at Khor Nesari 4, fortified with strong walls

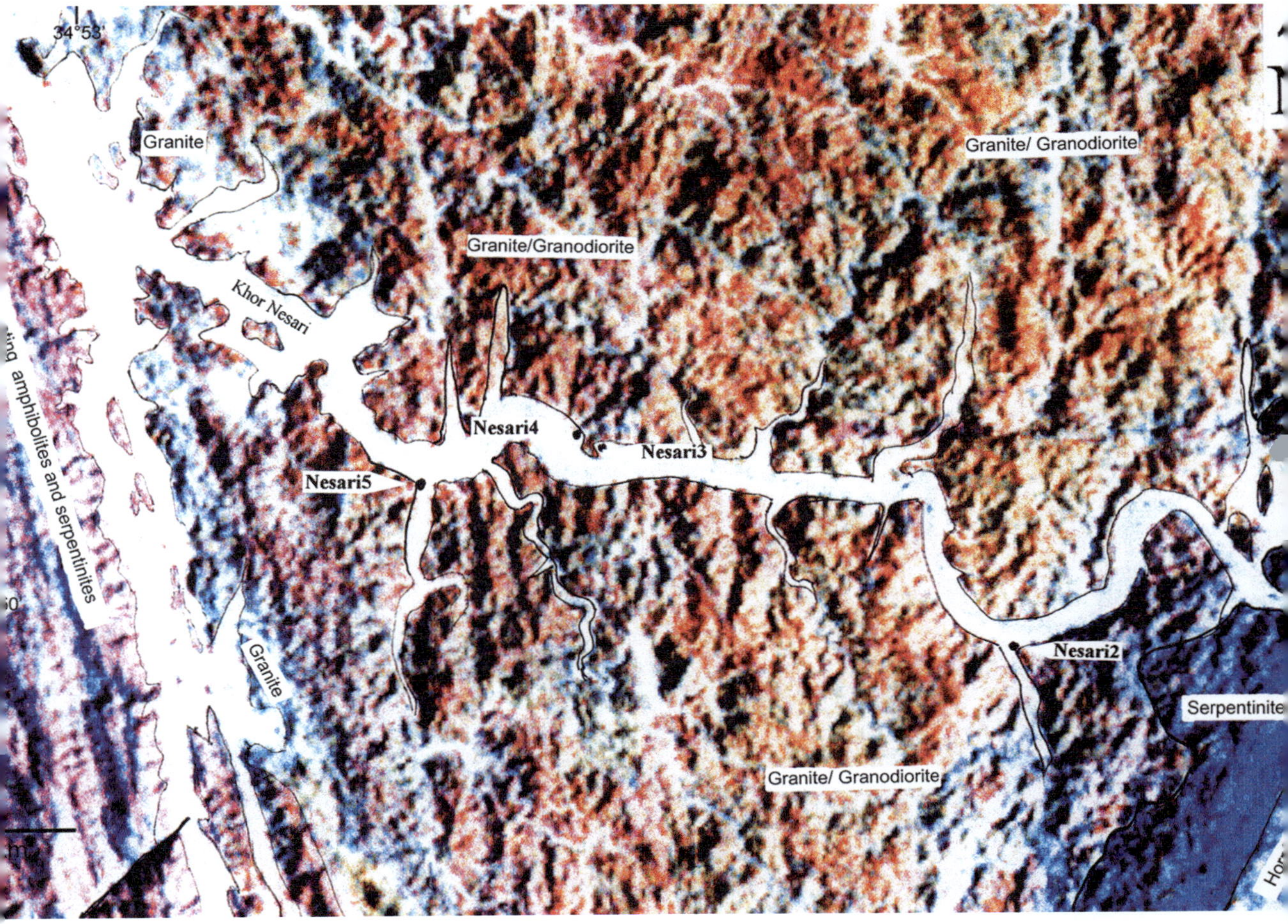

Fig. 6.119 Lithologically processed satellite image of the Khor Nesari area (TM173/45, channels 7-4-1)

dwellings of late-coming, "free entrepreneurs", who depending on a given goldrush had been liable to move and take their gear along with them to wherever they decided to settle. Such mobile miners would first have sampled the ores before deciding whether to stay or to move on. Many settlements consisting of only few, rudimentary, round huts in relatively disorganised arrangements might reflect such a scenario. They contrast to the more fortress-like and structured settlements, which together with their processing sites and tools may belong to an earlier gold mining phase, including that organised under the occupation of Al-Omari (Hasan 1967).

6.4.10 Hamisana-North

Geographic position: 21°38′36″ N, 35°00′20″ E

This site is located about 2.5 km N of the Hamisana mine. It consists of approximately 30–40, predominantly rectangular houses, as well as plain, round huts located further up the slope. The ruins of an imposing, relatively well-preserved, and rectangular building with eight rooms were recorded in the central part of the site (Fig. 6.120). Its dimensions are 26.5 × 8 m, and its walls are 80 cm thick. An isolated, circular room in its middle with particularly solid walls may have served as a vault.

Fig. 6.120 Central building in Hamisana-North with sturdy walls, surrounded by huts and round work platforms

Because there is no evidence from mining, neither in the form of tools at the site itself, nor that of any nearby quartz extraction, this site may also have functioned as some sort of storage or administrative facility for the gold processing sites in the wider district. The round huts would then probably represent simple dwelling places.

In the wadi debris below the huts as well as two other locations, we discovered objects resembling to New Kingdom grinding mills. In addition, more New Kingdom mill fragments were found in the wadi centre, which may point to a disappeared, small, New Kingdom, wadiworking settlement in the area.

6.4.11 Hamisana Mine

Geographic position:	21°37′26″ N, 35°00′54″ E

About half-way up the eastern slope, just S of Hamisana-North, a mine is followed over a distance of 1,5 km parallel to the wadi (Fig. 6.121). A characteristic washing table and some huts with round mills are located at the S end of the mine. No other tools and no pottery were found here. The number of huts is relatively low, compared to the length of the mine. They consist of spiky fragments from metasediments and spread loosely along minor rock outcrops, especially in the southern sector, which also contains an Islamic cemetery of about 15 burials.

Everywhere in the wadi there are noticeable traces from wadiworkings (Fig. 6.122)

No evidence predating the Early Arab Period was found at the mine.

The vein mineralisation appears at the southern end of the Hamisana granite stock. According to its external shape, which is located in the centre of the Hamisana zone, it belongs to a synkinematically deformed granite type, even though deformation structures are hardly perceptible in its few, central rock outcrops (Fig. 6.123).

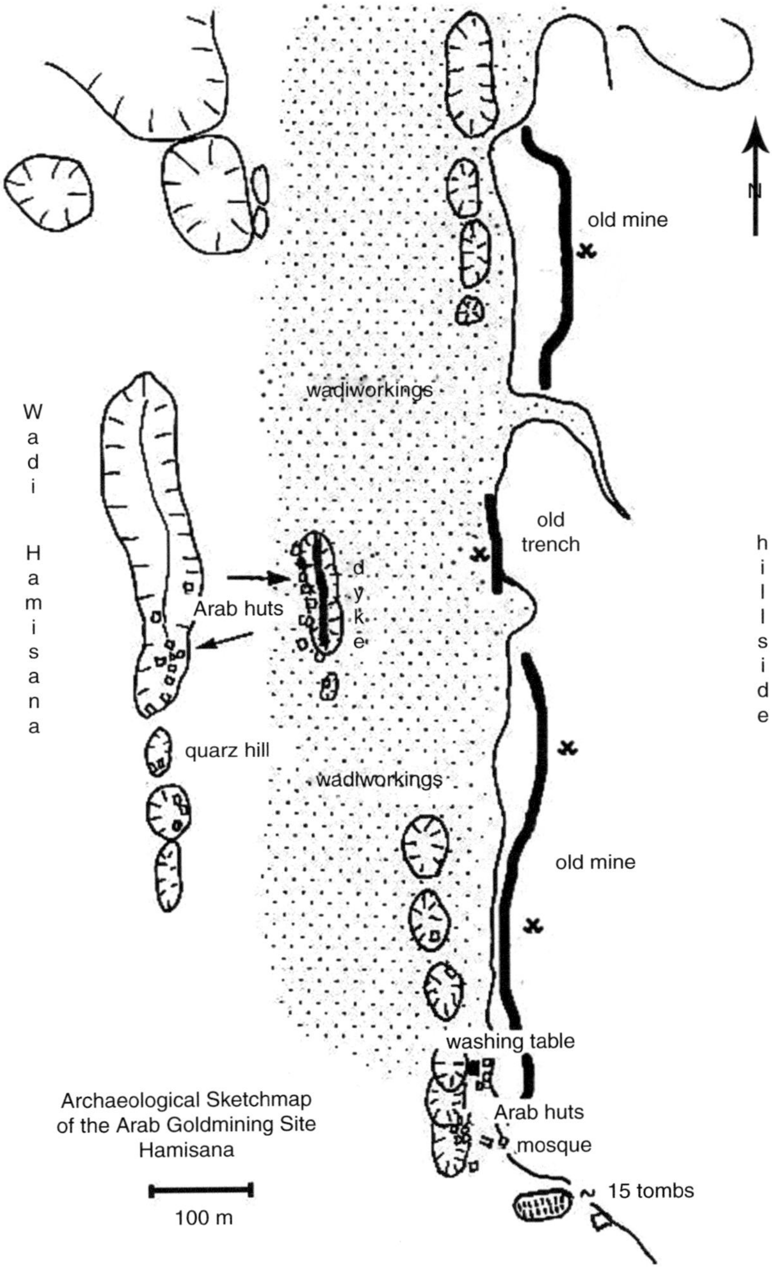

Fig. 6.121 Archaeological sketch map of the Hamisana mining district (R. Klemm)

Clearly deformed portions of the granite around the vein, located about 11 km further S of the Hamisana granite pluton, possibly go back to a first, syn-kinematically intrusion surge. The main intrusion, however, is post-kinematic.

The actual auriferous quartz vein mineralisation in Hamisana appears in the border zone between highly deformed sediments consisting of possible greywackes, well-identified conglomerates, and slightly less deformed granite-apophyses. The vein had been mined at the

Fig. 6.122 Small heaps in Wadi Hamisana witnessing former wadiworkings. In the right background, part of the old mine is to be seen

surface in shallow trenches over a distance of 1.5 km. It orients parallel to the deformation foliation in a NNE-SSW strike and a 45° E dip, which corresponds with the discernible anticline structure intruded by the granite. The sporadic trenches appear to match with suspected boudins along the quartz vein, but this cannot be confirmed, because of poor observation conditions at surface profiles. Extension during mineralisation of the vein seems to have led repeatedly to sudden fractures in which subsequently idiomorphic quartz crystallisation could take place.

Mineralisation had therefore occurred in several generational phases of which a rather wide, grey one had apparently been neglected. The same applies to finely mineralised Riedel veins, which seem to be devoid of gold. A milky, white quartz variety with some pyrite and clearly discernible malachite coatings had on the other hand been extracted.

6.4.12 Khazim-City

Geographic position:	21°35′14″ N, 35°02′36″ E

A large settlement of mostly round huts is set within in a hilly landscape in the mountains to the S of the Hamisana mine (Fig. 6.124). We named it after our colleague Mustafa Khazim, who had discovered the site. Since there is no evidence to suggest former mining activities here, the reason for the site's existence is unknown.

The building materials of the settlement consist of carbonated serpentinite. However, it is noticeable that the settlement is concealed between two mountain ranges, in a remote location, slightly off the main wadi. It also strikes that almost all houses had been built on small outcrops and natural rock protrusions, which make them look like under-dimensioned fortresses. Some huts are accessed through

Fig. 6.123 Lithologically processed satellite image of the Hamisana area (TM 173/45, channels 7-4-1)

Fig. 6.124 Early Arab Period settlement at Khazim-City, extending far into the hills. No evidence suggestive of gold mining was found here. Because of its relatively well-preserved mosques, it may have played a role in activities connected to pilgrimage and trade

very narrow doorways (0.50 m). From a distance, it is hard to distinguish the houses from the natural rock protrusions, which only adds to the generally concealed character of the site.

On the W side is a well-preserved mosque measuring 2 × 4 m with its mihrab at the long side. More praying sites are found in the surrounding area.

6.4.13 Large Mining Settlement

Geographic position:	21°24′44″ N, 34°35′38″ E

Castiglioni and Vercoutter (1998) mention a large mining settlement with well-preserved, but silted up washing tables at this position. They also report a large cemetery with some round graves. Round mills as well as abundant pottery fragments are described to date from the "Islamic Middle Ages". Some shards are described as originating from Central Nubia.

A number of funerary structures excavated by Castiglioni and Vercoutter (1998) in the vicinity produced a date of 1500 BC, which would thus be consistent with the early New Kingdom Period.

6.4.14 Wadi Gabaideb

Traces from ancient gold mining in Wadi Gabaideb can be recognised over a distance of 7 km. They include several settlement clusters referred to in a numerical sequence from S to N. Their locations are plotted in the lithologically processed satellite image (Fig. 6.125).

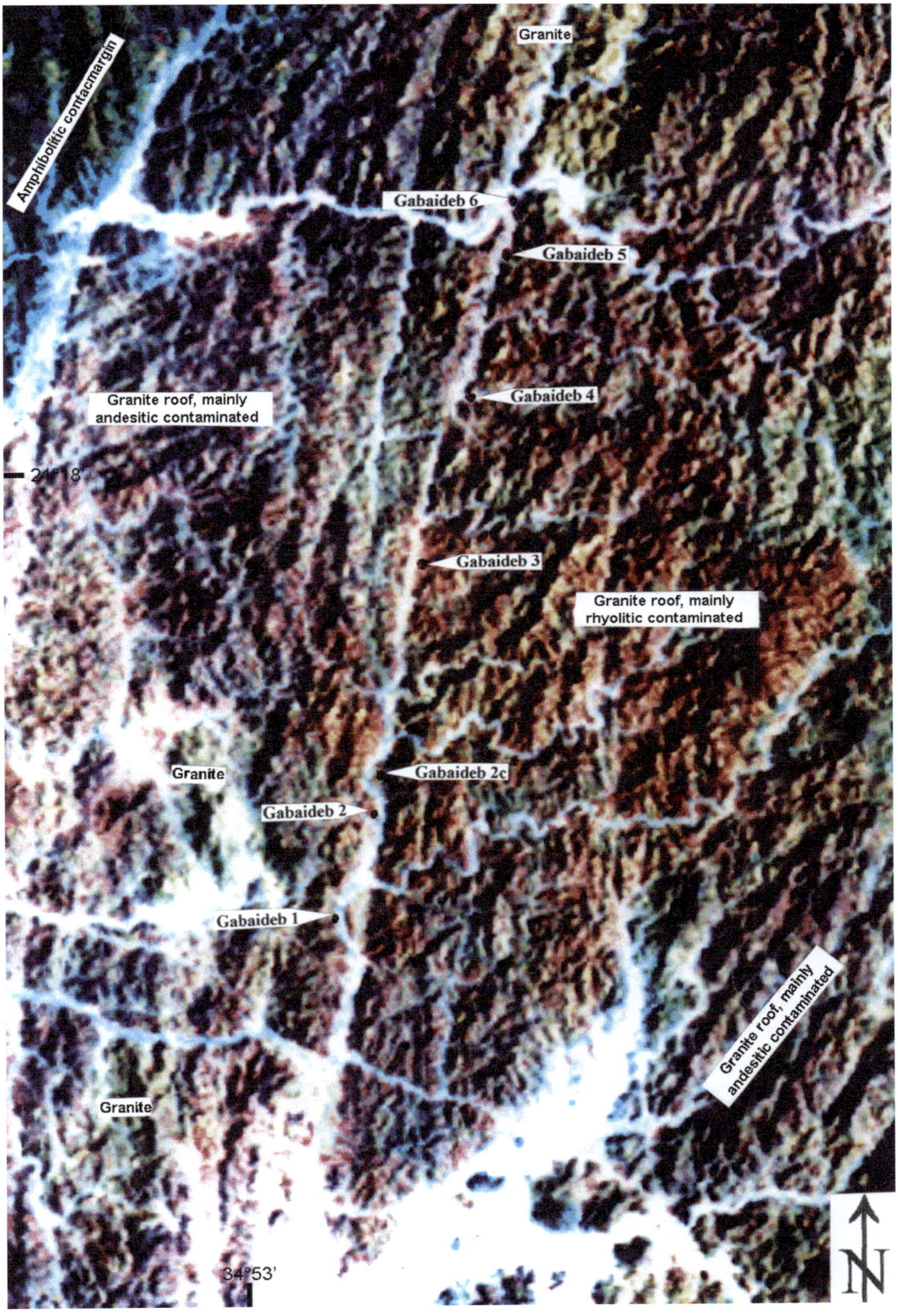

Fig. 6.125 Lithologically processed satellite image of the Gabaideb area (TM 173–45, channels 7-4-1)

Fig. 6.126 Early Arab Period settlement at Gabaideb 1 with clearly visible traces from wadiworkings

6.4.14.1 Gabaideb 1

Geographic position:	21°15′47″ N, 34°53′31″ E

This Early Arab site consists of a small number of huts at both sides of a wadi oriented N-S with visible wadiworkings in the E (Fig. 6.126).

6.4.14.2 Gabaideb 2a+b

Geographic position:	21°16′12″ N, 34°53′45″ E

This relatively large settlement is located on a wadi terrace and consists of large houses aligned in the typical manner of the New Kingdom. The tools also date to the same period. After an interval, more house ruins are visible to the N.

6.4.14.3 Gabaideb 2c

Geographic position:	21°16′22″ N, 34°53′40″ E

Numerous, large, multi-chambered houses from the New Kingdom (Fig. 6.127) are located on two, flat hill groups along the wadi's eastern terraces. In accordance with the settlement sites from this period, this site too, has very little pottery at the surface. Washing tables covered by debris material are barely discerned. By contrast, there are many large, New Kingdom oval mills, one cubic stone anvil, and numerous fist grinders.

6.4.14.4 Gabaideb 3

Geographic position:	21°17′30″ N, 34°53′56″ E

This site consists of 12 hut ruins. No surface finds were recorded here.

6.4.14.5 Gabaideb 4

Geographic position:	21°18′18″ N, 34°54′12″ E

Twenty large, rectangular houses measuring between 6 and 8 m in length, and with up to two interconnected rooms are located here in slightly elevated positions on the gravel terraces along both sides of the wadi. Unfortunately, no tools

Fig. 6.127 Rectangular New Kingdom house ruins cluster at Gabaideb 2 amid wadiworkings

were found. According to their layout however, the ruins may date to the New Kingdom. Significant traces from wadiworkings spread from here into the neighbouring side wadis.

6.4.14.6 Gabaideb 5

Geographic position: 21°19′05″ N, 34°54′26″ E

This site consists of six to eight, fairly large and rectangular houses that contain no tools. Another site labeled as Gabaideb 5a is about 250 m further to the NE (21°19′17″ N, 34°54′27″ E). About 20 small, round huts date to the Early Arab Period. They have one or two rooms, are partly equipped with double-shell walls, and group around an imposing quartz outcrop. Further hut ruins scatter over 3 km at the E side of an intersecting wadi in an E-W orientation. Rock art representations are also found nearby.

Wadi Gabaideb runs through an homonymous geologic complex, which on the map resembles a large lens stretching over 32 km from N to S and 15 km from E to W.

Numerous sites are found in this valley, where driving conditions are very difficult. The sites seem primarily to have been focused on wadiworkings and less on the exploitation of auriferous quartz veins.

The lithologically processed TM-image (Fig. 6.125) reveals that the entire complex rests on a syn-or post-intrusively deformed, granitoid intrusion, partly visible at the surface in the S and NE, but otherwise only detectable by its hybrid roof position. Numerous relics occur from the primary volcano-sedimentary series with hornfels structures, whose former nature as greywackes, siltstones, or various rhyolite rocks are not only blurred macroscopically, but also under the microscope. Only former conglomerates can still be safely determined.

Consistent with their hybrid character, the magmatogenous rocks in this complex appear in a relatively wide variety. They consist mainly of different granodiorites yielding hornblende and biotite, but also of quartzdiorites with transitions to diorite, and not less importantly, of an array of granitic variants, depending of the assimilated volcano-sedimentary rock types in the intrusion roof.

Such hybrid rims around large granitoid intrusions form ideal areas for hydrothermal gold quartz mineralisations, provided the intruded series display adeaquate, primary Au-anomalies of at least 20 ppb. Such preconditions are often given for the volcano-sedimentary series in the Egyptian-Sudanese bedrock and confirmed by the countless, ancient gold mining sites.

Under given circumstances, it is however somewhat perplexing that the ancient prospectors had only exploited the relatively modest and scarce, primary gold quartz deposits. Possibly the New Kingdom prospectors, who after-all had been successful in finding virtually all significant gold deposits in NE Sudan, had preferred to limit their prospecting activity to this easily accessible valley only. Due to its potential therefore, a more thorough, modern gold deposit survey would seem judicious within this complex.

At Gabaideb 1 the small gold quartz veins exploited in antiquity strike in a N-S and NNE-SSW direction and dip 55° W. At Gabaideb 2 and 5a the strike is N-S (no dip values). Otherwise, all the other sites in Wadi Gabaideb as well as the just mentioned ones are wadiworkings, of which primary vein mineralisations are still unknown.

6.4.15 Caravanserai

Geographic position: 21°5′44″ N, 35°02′16″ E

This building complex consists of schistose stone slabs and is very well-preserved. Since the site delivers very little finds at its surface, we can only assume that it had represented some sort of supply station or caravanserai. No signs relating to gold production were detected here.

6.4.16 Mining Settlements W of Khor Adarmo

At least five mining settlements, probably all specialised on wadiworkings are located in the two adjacent parallel, N-S running wadis W of Khor Adarmo, and in a large, E-W oriented, transversal valley. None of the sites were known to us prior to our field campaigns. We therefore didn't visit them. We only managed to identify them in a systematic screening survey on the computer in high-resolution Google-Earth images.

6.4.17 Settlements in the E-W oriented Wadi

6.4.17.1 New Kingdom settlement

Geographic position: 21°12′24″ N, 34° 51′36″ E

This site probably dates to the New Kingdom and consists of around 30 discernible house ruins along the northern wadi edge and another 14 house and hut ruins on its southern side. The settlement remains partly spread from the wadi edges into the mouth areas of two small, tributary wadis.

6.4.17.2 Early Arab settlement

Geographic position: 21°12′33″ N, 34°51′52″ E

About 800 m further E, a possibly Early Arab site of about 35 huts and house ruins is located in a secluded situation behind a small hill at the inlet of a northern side wadi. The side was discovered only by Google Earth images but its location and the layout of the architecture furnished the dating criteria.

6.4.18 Settlements S of the E-W oriented Wadi

6.4.18.1 New Kingdom settlement (?)

Geographic position: 21°11′28″ N, 34°53′56″ E

This and the following site are located in the next wadi, running parallel to Khor Adarmo. It extends from the indicated position over about 500 m to the N. The original settlement appears to have been much larger because only higher lying sectors near the slopes and inlets of the side wadis appear to have survived severe inundations in the past.

6.4.18.2 Early Arab settlement (?)

Geographic position:	21° 10′39″ N, 34° 53′59″

This settlement spreads chiefly over various small hills and is slightly oriented away from the wadi. It consists of about 45, small huts and five large, multi-room house ruins apparently dating to the Early Arab Period.

6.4.18.3 Early Arab settlement (?)

Geographic position:	21°10′12″ N, 34°53′06″ E

This large site is composed of about 90 hut and house ruins and was localised only by highly magnified Google Earth image. Between five and seven houses were counted to contain several rooms. Its slightly hidden position on the slopes of small side wadis is a typical feature of settlements dating to the Early Arab Period.

6.4.18.4 Khor Adarmo-West

Geographic positions:
21°06′22″ N, 34°57′17″ E (about 15 huts and houses)
21°06′28″ N, 34°57′13″ E (30 huts)
21°06′32″ N, 34°57′24″ E (about 20 huts and a funerary enclosure)
21°06′41″ N, 34°56′39″ E (about 20 huts and houses)
21°06′45″ N, 34°56′47″ E (25 huts)

Together, the above sites distribute within an area no larger than 2 km long. The relatively high occupation density is exemplified by the comparatively large number of architectural units given within the brackets.

The sites date predominantly to the Early Arab Period, and for the most spread in hidden positions in small side wadis along the western side of Khor Adarmo. The tool assemblages consist of characteristic round mills, small anvil stones and stone hammers, as well as more or less well-preserved washing tables, whose basins as a rule are covered by sand. Together with the medieval pottery the entire assemblage points to the Early Arab Period. In general, the dwellings along the valley rims are associated to wadiworkings, whose small heaps are especially visible in the numerous side wadis.

Attempts at small-scale, deep mining had been made in open pit trenches following auriferous quartz veins.

Adarmo-West is an Early Arab site extending along both sides of the wadi. At the southern side some houses group inside a tributary gorge, where wall alignments crossing two gullies indicate to the former presence of dams. Ore processing in adjacent work platforms enclosed by low walls had therefore benefited directly from the collected water in the dams. Potential washing tables were not recorded in this area.

On an opposite foothill below a mountain ridge pointing directly E, a large, open space is marked by two rounded wall structures in its centre. At some distance one recognises a small number of huts. The round wall enclosure seems to mark a funerary precinct apparently restricted to outstanding members of the community. Similar structures are also known from other sites (e.g. in Hamisana and Wadi Dom 5). Archaeological investigations within such precincts in many aspects would contribute significantly to our understanding of these ancient communities.

6.4.19 Khor Adarmo Fort

Geographic positions:	
Fort:	21°08′33″ N, 34°55′43″ E
Settlement:	21°08′27″ N, 34°55′40″ E

The site consists of a cluster of about 100 huts and some larger house ruins along extensive wadiworkings. The large, reddish quartz chunks, which cover the desert surface, are a relatively uncommon occurrence. Numerous working sites scatter in the surrounding district. They concentrate along the wadi edges, whereas the huts are mainly located on the slopes.

Opposite the main settlement in the N are the ruins of a fortress-like complex (Figs. 6.128 and 6.129). It was safeguarded by at least one tower and had probably served to repel Bedouin raids.

Similar fortifications are known from Derhaib and Bir Kiaw. They may have served for the protection of the local Arab mining populations and

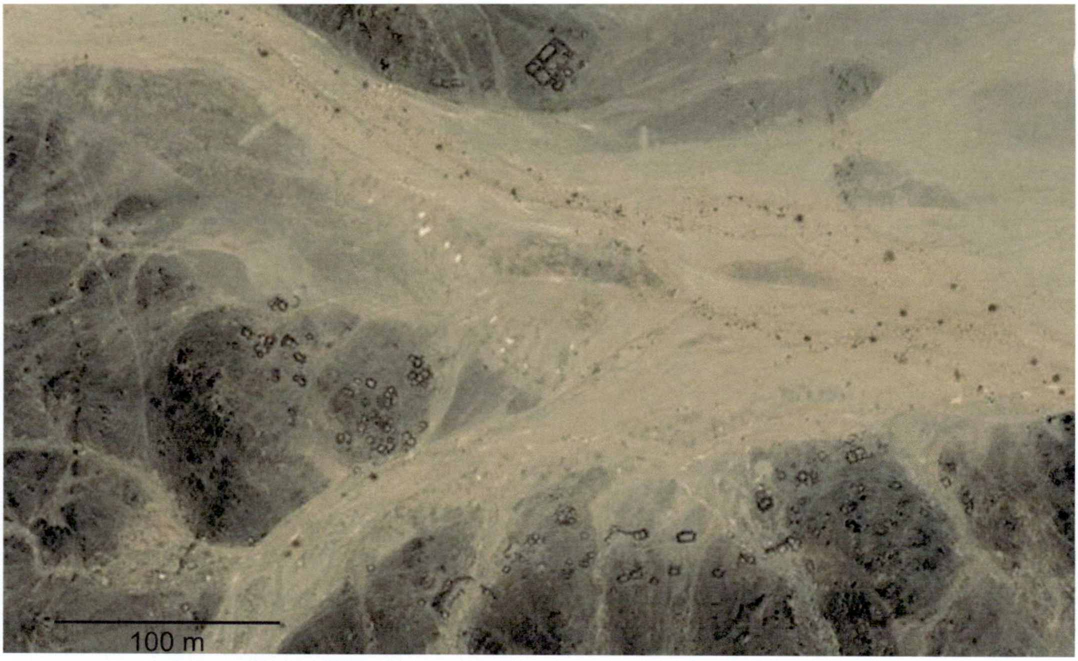

Fig. 6.128 Large Early Arab Period settlement in Khor Adarmo and a fortified structure in the upper part of the image (modified Google-Earth image)

thereby of their gold. The small finds consist among others of a round mill, which strikingly enough exhibits severe use wear on both its sides. Typically shaped pounders had been shaped from the ubiquitously available quartz chunks.

Genuine mining had been carried out in a trench lowered into the wadi floor, in a NW-SE orientation. Another small extraction pit is located in the hills to the NE of the settlement.

The few recorded shards display painted "Late Christian" motifs described by Adams (1986). Moreover, we found green, glazed pottery shards as known from sites in Southern Egypt. We also noticed some pale-green glass shards thought to belong to pilgrim flasks (so-called "Samsam" flasks according to Llewellyn 1903).

The geologic situation at Adarmo-West is characterised by a zone much affected by intense tectonic activity. The zone is in fact identified as a large suture between the Gebeit and Gabgaga terranes. In this area virtually all geologic units are highly sheared.

Consequently, the surface granite is only recognised as such on a macroscopic level. Under the microscope it reveals to be a severely shredded mylonite, whose feldspar crystals appear as brecciated lenses, while the matrix is ground to such an extent that singular components are no longer correctly perceptible, even under high magnification.

Before its tectonic disruption this granite had penetrated into a sequence of predominantly rhyolite volcanics that in the border area recrystallised to hornfels before being sheared to mylonites by tectonic activity.

The auriferous quartz mineralisation had then formed in a parallel strike to the predominantly N-S shear shortly after the tectonic stress in an extension phase resulting from the thermal aftermath of the granite intrusion.

6.4.20 Wadi Eleij Area

Until recently, no evidence from ancient gold mining was known from the districts to the E and W of Wadi Eleij. This is in so far unusual, as its geology shows the distinctive sequences of andesite to dacite volcanics with subordinate conglomerates (Nefirdeib series) intruded by granites to

Fig. 6.129 The fort in Khor Adarmo with its strong exterior walls

diorites, which are so characteristic for areas with mineralised gold deposits. In this district the Nefirdeib series are particularly affected by diorite intrusives. According to our experience concerning gold deposits in late-Proterozoic bedrock, we therefore expected to find corresponding mineralisations in the area around Wadi Eleij.

In order to verify this hypothesis, large-scale geologic maps were edited covering an area between 20° 55′ – 21°06′ N and 34°32′ – 34°44′ E. Furthermore, the different rock sequences were exposed to thorough geochemical and petrographic tests (Tichatschke 2001) and embedded in the local and regional tectonic deformation structures (Schmid 2001). During the therefore necessary field work, additional attention was given to possible evidence from ancient gold mining activities.

Preliminary results showed that the whole area represents a side branch of the large suture in the Hamisana zone, whose Nefirdeib series had been subjected to intense folding and deformation by shearing. The volcano-sedimentary sequences of the Nefirdeib series containing ophiolite residues, just as the Gebeit and Gebel Gerf terranes, are viewed as island arc units that had prompted primary gold enrichment up to two digit values on the ppb-scale. Although no major auriferous quartz mineralisations were found in the field, numerous mining settlements and unambiguous traces from former wadiworkings distinctly indicate that gold production had in fact been viable in this area over lengthy periods.

These exclusively Early Arab settlements, which all contain more or less large, square prayer sites, are listed in the following with their geographic locations in a N-S sequence:

Settlement 1 (21°07′04″ N, 34°35′55″ E)

Settlement consisting of about 40 small huts measuring rarely more than 2.5 × 3 m. They are spread over several hills in a side wadi to the W of Wadi Eleij.

Settlement 2 (21°06′38″ N, 34°35′57″ E)

About 60 huts and some slightly larger houses located on the southern slopes of a fairly large valley basin.

Fig. 6.130 Stone slab in a mihrab of a prayer site at position 21°04′47″N, 34°34′09″ E, bearing a Kufic inscription dating probably to a period between the tenth and twelfth centuries. It may initially have served as a tombstone, as instead of "Allah" it reads the name of an individual "al-Khidr (?) ibn…" (pers. comm. St. Heidemann, Jena)

Settlement 3 (21°06′22″ N, 34°35′35″ E)

About 1.5 km SW of the first settlement 35 huts group at the SE flanks of the same wadi as well as in hidden positions in small side wadis.

Two neighbouring settlements (21°05′37″ N, 34°33′58″ E and 21°05′31″ N, 34°34′05″ E)

Both are located in a narrow tributary wadi, approximately 300 m apart from each other and consist respectively of about 30 and 15 huts.

Settlement 4 (21°05′33″ N, 34°35′53″ E)

Consisting of about 60 huts and two large, multi-roomed houses (16 and 12 m long), mainly on the north-eastern slope but also on a flat hill, this is the largest settlement to the W of Wadi Eleij

Settlement 5 (21°04′58″ N, 34°34′28″ E)

This settlement is located in a small, tributary wadi and stretches over 330 m on both flanks. It consists of about 40 huts and a few slightly larger houses with two rooms.

Settlement 6 (21°04′47″ N, 34°34′09″ E)

About 15 huts divide into two distinctive groups. Between both is a praying platform with a stunning Kufic inscription on a stone slab of the mihrab (Fig. 6.130).

Two small settlements (21°05′14″ N, 34°42′22″ E and 21°05′01″ N, 34°42′13″ E)

Both consist of small huts. The southern one has a large house. Both sites are in a hidden valley east of Wadi Eleij.

Settlement 7 (21°04′49″ N, 34°42′33″ E)

Thirty small huts, some of which are built closely together are found in the centre and at the margins of a wide area of wadiworkings.

Settlement 8 (21°04′07″ N, 34°40′20″ E)

A small settlement of about 50 small huts hidden in the hills.

Settlement 9 (21°04′02″ N, 34°40′35″ E)

About 50 small huts hidden away in the hills and the side arms of a tributary wadi.

Settlement 10 (21°03′52″ N, 34°40′29″ E)

In its present state, the settlement consists of more than 50 huts and three, slightly larger houses. Many of the huts are located on alluvial islands in the wadi, where they seem to have escaped

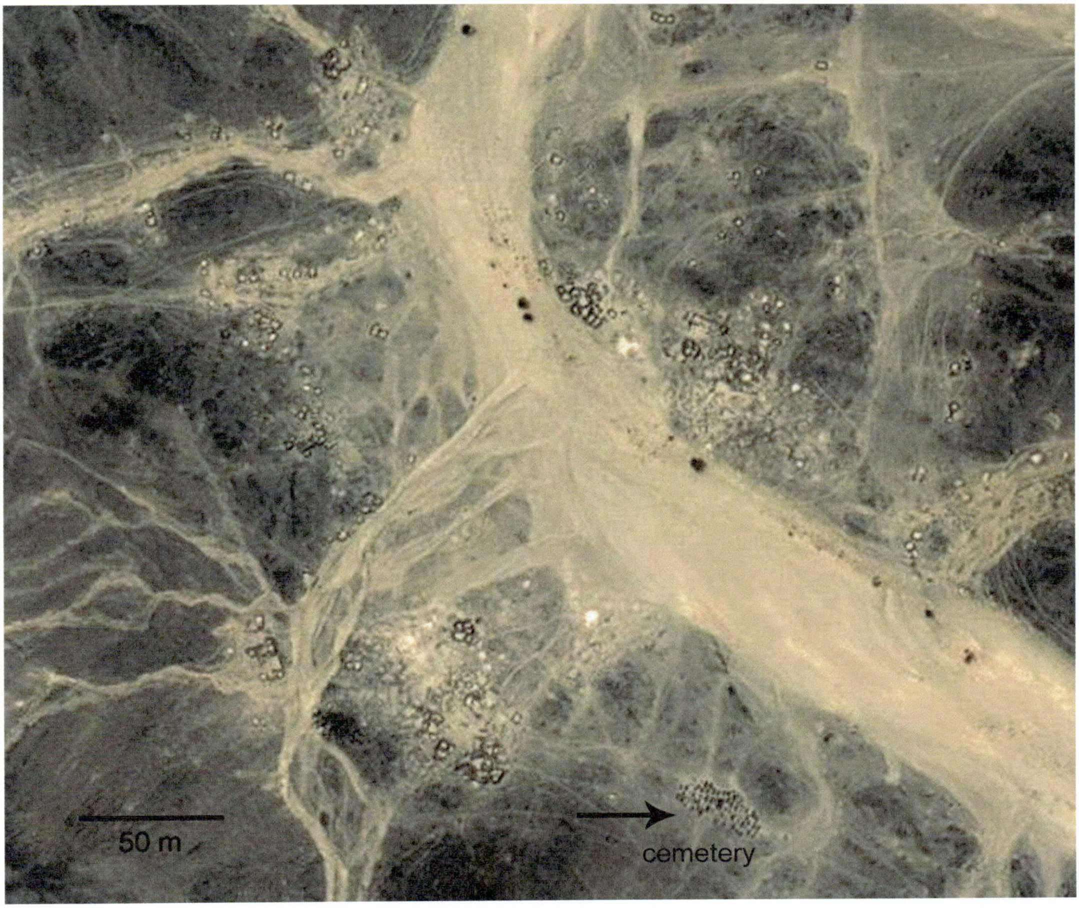

Fig. 6.131 Large Early Arab Period gold production site (21°03′50″N, 34°40′15″E) to the W of Wadi Eleij (modified Google-Earth image)

destruction from floods. It is therefore conceivable that the settlement had once been much larger and that many huts have been washed away.

Settlement 11 (21°03′50″ N, 34°40′15″ E)

This is the largest settlement in the area of Wadi Eleij with over 150 huts and houses. They spread over several hills and tributary wadis, and there is a large cemetery of about 60 graves (Fig. 6.131).

New Kingdom (?) **Settlement 12** (21°03′36″N, 34°41′03″ E)

Only 25 houses of varying sizes stretching along the wadi margins with a possible date to the New Kingdom.

Settlement 13 (20°59′16″ N, 34°37′07″ E)

About 30 huts in the middle of a small tributary wadi with wadiworkings.

Settlement 14 (20°56′30″ N, 34°37′47″ E)

A settlement with 50 huts on slopes and hills hidden in tributary wadi. It is one of the larger settlements in the area.

Settlement 15 (20°56′05″ N, 34°42′55″ E)

More than 30 small huts in concealed locations in several small tributary wadis to the E of Wadi Eleij.

Settlement 16 (20°55′29″ N, 34°40′34″ E)

More than 30 huts loosely distributed and hidden away in a tributary wadi.

Fig. 6.132 Trench pit in Wadi Sukai and associated Early Arab Period mining settlement behind a mountain ridge (modified Google-Earth image)

6.4.21 Sukai

Geographic position:	20°49′14″ N, 34°46′24″ E

The site at Sukai (Fig. 6.132) consists of an elongated mine in the wadi floor. Today it is for the most filled with sand and scrub (Fig. 6.133).

In a parallel oriented, gorge-like wadi, there are four, large contiguous house complexes and about 30 round work platforms, of which 11 are grouped around a well-preserved washing table in the wadi's NW extension.

Round mills including the rotor stones, small pounding stones and flat grinding stones are found in large numbers. The secluded location of this settlement has probably contributed to the fact that so many tools are still found here. Noteworthy as well, is a profusion of amphora shards.

The small size of this settlement is characteristic for sites located near mined quartz veins as opposed to ones specialised on wadiworkings.

Due to the lack of any older archaeological evidence, the mine at Sukai dates exclusively to the Early Arab Period.

The productive quartz vein at Sukai extends for about 120 m in a N-S direction and had been exploited to a depth of about 6 m in a trench measuring approximately 30 m, the remaining parts being only superficially excavated.

Fig. 6.133 Trench pit at Sukai with burning shrubs

The vein itself is about 1.5 m wide and consists of at least two quartz generations, of which the barren one is massive and white. Only the other (older?) generation had been exploited. It is split up into several, few centimetre-thick lodes within a rhyolite wallrock marked by intensive limonitic alterations. The limonite alteration zones themselves had not been mined.

A petrographic sample traverse through the deposit area showed that the immediate surroundings predominantly consist of rhyolite (chloritised quartz porphyry) and to a lesser degree of layers of rhyolitic agglomerates. To the E this NNW–SSE striking series passes over into equally oriented porphyritic andesites.

A small, divided diorite intrusion with a noteworthy NE-SW striking andesite dike sequence is located approximately 3 km W of the mined deposit (Fig. 6.134). It had penetrated along the general strike line into a basaltic sequence and the rhyolite hostrocks of the productive quartz mineralisation.

The andesite dike swarm mentioned above clearly breaks through the diorite stock and the metavolcanics further SW, but not the rhyolites or andesites further to the E. In the valley however, it is affected by a NNW-SSE oriented fault, which seems to indicate an initially larger geographic distance between the productive quartz vein and the intrusion stock. The granite intrusive stock SW of the basalt band contains no andesite dikes and thereby post-dates the diorite.

Fig. 6.134 Lithologically processed satellite image of the area around Sukai (TM 173/46, channels 7-4-1)

Fig. 6.135 Relatively well-preserved Early Arab Period house at Abu Duneim, built with flat stone slabs

6.4.22 Abu Dueim

Geographic position:	
Large house S of the wadi:	20°50′44″ N, 34°56′50″ E

This fairly large settlement of about 70 huts is located in a tributary valley branching-off to the W of the N-S oriented wadi, separating the Gebeit and Gabgaba terranes. The huts consist of flat, schistose walls, often enclosing several rooms and preserved in heights up to 2 m (Fig. 6.135). In between, on the south-western slope, there are also simpler huts that had been built at considerably higher altitudes. They are both round and rectangular, their room dimensions ranging between 2 × 2 and 3 × 3 m. One of them is equipped with an adjacent rubble terrace measuring 2 × 2 m.

Neither mining tools nor washing tables were found here. Some quartz chunks are nevertheless still located in situ, in a small heap near a house. The pottery consists of simple cooking ware, and there are some shards from green glass. A so-called Romib is located in elevated terrain near the eastern border of the settlement (Fig. 6.136), and at the highest point a now collapsed Alam had once towered above the site. On the other side of the wadi there are silted-up trenches and distinct traces from former wadiworkings, next to small piles of quartz rubble. We identified a thoroughly burrowed-trough hill with the ore deposit at Abu Dueim. Nevertheless, even here, no tools were found.

Fig. 6.136 Large round tomb (Romib) at Abu Duneim

6.4.22.1 Karaibitar 1

Geographic position:	20°39′46″ N, 34°4′35″ E

Karaibitar represents a New Kingdom settlement of about 40, mostly rectangular houses typically arranged in alignments. Some houses are relatively large. They are located in the eastern extension of the main wadi (Fig. 6.137).

Here, we found specific variants belonging to the oval New Kingdom mills, which had been reclaimed at some later stage and worn-down in deeper and narrower furrows (Fig. 6.138). At other sites (e.g. Nubt and Ganait), where this mill type is also found, we called it the "Kushite" type.

The houses are rectangular and built from the, local spiky rocks, partly masoned in a braided pattern. Some walls are preserved in heights between 50 and 80 cm. The houses are located on a wadi terrace and extend slightly onto the slope. Some houses are associated with cist-like stone structures on the outside, a finding that has also been observed at various other sites. To the N, a small but massive structure supplanted by a dome had been attached to one of the houses.

For the lack of evidence from genuine mines, gold production had probably for the most been based on wadiworkings. In the evening light one can actually make out the remains of numerous quartz chunk heaps left behind over the entire area to the S of the settlement (Fig. 6.139).

The region to the E and W of the Hamisana fault consist mainly of volcanogenic, mostly dacite to rhyolite rock sequences, which here and there also display intermediate andesite

Fig. 6.137 The site at Karaibitar with a house alignment typical for the New Kingdom Period (modified Google-Earth image)

layers (Fig. 6.140). All sequences strike approximately N-S, but beyond Karaibitar with an increasing tendency towards NNE-SSW. Regardless of this however, in all sectors the series reveal more or less well-developed folds, in some cases isoclines, with the b-axes generally adhering to the general strike. Granite intrusions had often penetrated into the anticline structures. In the field the intrusions are nonetheless only visible, often as minor apophyses or indirectly through hornfelsic rocks, and because of their low spatial distribution, they cannot be identified in processed satellite images. This is all the more regrettable, as productive gold mineralisations in the contact aureoles of intrusions are not uncommon, although they remain almost undetectable in preselecting procedures in remote sensing surveys.

Because of the similar geologic environments around Karaibitar and Abirkateib and their common link to the wider tectonic unit of the Hamisana group, both sites are treated in a common 1:50,000 scale geologic map.

Fig. 6.138 Broken New Kingdom grinding mill at Karaibitar with a secondary, central use wear depression, suggesting a reclaim (?) in the Kushite Period

Fig. 6.139 Spoil heaps from wadiworkings appearing in the evening sunlight of the Karaibitar plain

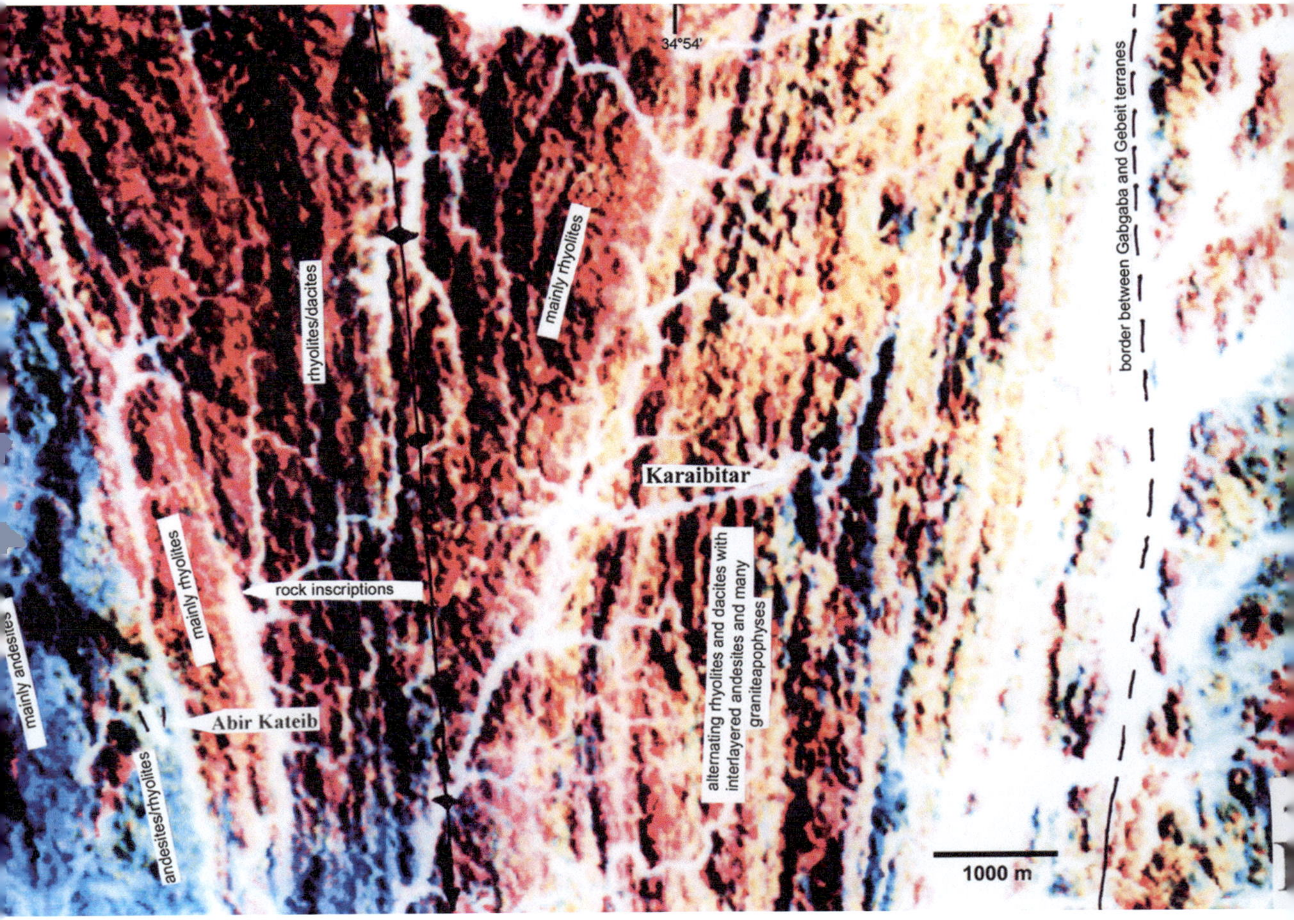

Fig. 6.140 Lithologically processed satellite image of the area around Abirkateib (TM 172/46, channels 7-4-1)

6.4.22.2 Karaibitar 2

Geographic position:	20°39′48″ N, 34°53′43″ E

The scarcity of the Early Arab hut remains and its round mills, as well as the sparse extraction traces at two small quartz veins bearing marks from recent prospecting, suggest that mining had never exceeded the trial stage and that the site had been abandoned soon after the inspection of the veins.

Here the veins occur in prevailing quartz porphyry, and there are no signs in the geologic surroundings, which might suggest a gold mineralisation at this location.

There are no direct references to hybrid borders with granitoid intrusions, and the quartz porphyry reveals no formation of hornfels. The granite stock at the surface about 4 km to the S seems to plunge at a steep angle. Neither this nor another granite stock, which is indicated by a contact aureole approximately 7 km to the N, are considered as potential energy sources for thermal gold mineralisation processes here.

A striking aspect about this area is that it discloses a palpable lack of experience among the Early Arab gold prospectors. In fact, the finding at Karaibitar 2 fits in well with the general idea that Early Arab mining had usually been operated at, or in very close proximity to the New Kingdom mines, and that wadi deposits had been exploited according to a random "trial and error" method. In this they were able to distinguish unerringly between auriferous and barren quartzes. Certain skills had nevertheless been required, as only in the rarest cases is visible gold found in the quartz fragments recovered from the wadi beds.

Contrary to the mistake committed by the modern prospectors at Karaibitar, mining traces from the Early Arab Period are therefore under no circumstance to be taken as reliable indicators for potentially exploitable gold deposits.

6.4.23 Abirkateib

Geographic position:	
Modern mining buildings:	20°38′37″ N, 34°51′23″ E
Bir Kateib:	20° 36′50″ N, 34°52′27″ E

The Abirkateib (Fig. 6.141) mine was run by the Basheer family from Port Sudan until 1994, when the license was withdrawn. The archaeological remains have unfortunately been largely destroyed during this latest mining phase.

A survey around two shafts in the centre of Abirkateib as well as in a neighbouring wadi between Bir Kateib and a recent water drill-hole, led to the initial conclusion that there had only been one Early Arab and one recent mining site in the area. At first we found no traces from the New Kingdom here. Even extended investigations into a parallel wadi with several wells remained unsuccessful as to our hopes of finding more ancient traces. Intensive overbuilding through recent structures and tailings in the immediate vicinity of the old mines, whose layouts are yet much reminiscent of those from the New Kingdom, may have largely contributed to the eradication of associated remains.

Among the amassed stones forming the boundaries of a recent sports field on a terrace at the eastern edge of the main wadi, we were also able to find a number of New Kingdom mills that further represent an indication of such an earlier occupation at Abirkateib.

A thorough inspection on the eastern side of the main wadi near the modern mine revealed about 40–50 hut ruins belonging to an Early Arab settlement. The rectangular huts had been built in carefully layered walls measuring about 50 cm in width. There are round huts too, whose masonry had been executed with flat rocks, though less neatly. The northern part of the settlement has badly suffered from the recent occupation and today much resembles a landfill. As often observed elsewhere, round mills lying scattered about on the ground are only preserved in fragments.

We found a washing table in a small side wadi to the W, and some more Early Arab huts to the N in a small wadi incision. Two 1 × 1.20 m large stone cists were recorded nearby. They are made of large schistose slabs whose corner joints had been fitted with a waterproof plaster (Fig. 6.142). The bottoms of the cists consist of single slate slabs.

In the parallel wadi to the E are at least 30 more huts (Fig. 6.141). In the following parallel wadi further E is a rock slab with prehistoric petroglyphs.

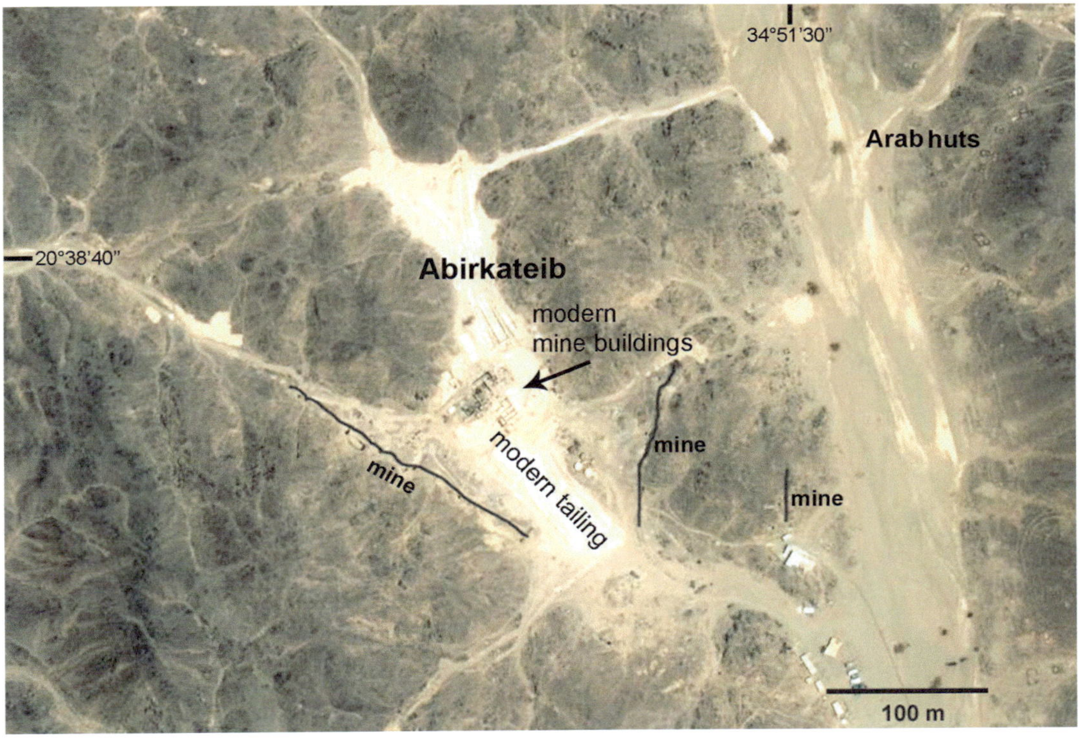

Fig. 6.141 Mining district at Abirkateib with ancient trench pits (*black*). Modern mining installations are located in the centre, and parts of the Arab Period settlement are distinguishable at the eastern wadi edge (modified Google-Earth image)

The deposit in the area of Abirkateib had been chiefly mined in three, still easily accessible quartz lodes (Fig. 6.141).

They occur in clearly foliated sequences with slight internal folds, consisting predominantly of andesite (sometimes significantly amygdoidal), dacite and rhyolite, each with fine-grained, intermediary layers of tuff. Coarse-grained hornblende-granite containing much K-feldspar had penetrated into these sequences. Consequently, the volcanogenic series in the deposit area had been converted to hornfelses to such an extent, that reliable petrographic classifications in the field proved extremely tricky (Fig. 6.140).

The volcanogenic series in the deposit area generally strikes at about 20° W, but because of the internal folds, significantly diverging values were measured locally. On the whole, the series dips steeply, mostly to the E, but again, due to the internal folding also between 70° and 90° W.

The productive quartz vein mineralisations occur characteristically at a steep angle to the strike of the volcanogenic series, while dipping almost vertically.

In the westernmost lode (strike approximately 40° W, dip 90°), which actually consists of three parallel quartz veins, four simple pits can be distinguished today. In the second pit from the NW there is a rather friable ladder, by which one all the same can descend into the mine. Two quartz generations can be separated there. The early one has red stains and a porous texture, maybe originating from pyrite decomposition. The later one is more massive and has mineralised with chalcopyrite, bornite and covellite. Even in the hydrothermally much altered hostrocks, one can observe the typical limonite pseudomorphs substituting pyrite. These areas however, had not been mined. A dependable petrographic classification of these strongly altered wallrocks

Fig. 6.142 Stone cist and processing devices (washing tables) near an Early Arab Period settlement in Abirkateib

is no longer possible and can only be derived from the overall geologic context.

Whereas the westernmost trench is predominantly recent, while revealing a hopelessly chaotic extraction method, it seems that the eastern ones near the great tailing heap had already been extracted in antiquity (New Kingdom), and certainly in the Early Arab Period. Mining predominantly followed the rich-ore-zones, which too resulted to a chaotic and cavernous mine structure. The vein located to the W of the tailing strikes 20° W and dips almost vertically, but with a slight westward leaning. Next to the cavernous pits it also has a modern, though derelict shaft. No primary sulphides were found in the vein quartz samples, even though they stem from a marginal zone with significant limonite stains. The central generation consists of solid white quartz, which in spite of missing sulphide mineralisations was nevertheless mined over considerable distances. It must therefore have been at least partially auriferous.

The easternmost vein leading to the slopes just E of the great tailing heap, which too was mined in the chaotic manner as described above, received only little development by modern mining. It strikes NNE-SSW and dips at 80° W.

The overall impression of this site is that of a completely disordered exploitation. It seems obvious that for its implementation not much importance was given to previous planning, as exemplified by the utterly inadequate location of the tailing heap that even ended up by covering some of the factory buildings.

6.4.24 Settlement S of Abirkateib

Geographic position:	20°34′02″ N, 34°54′59″ E

This settlement, which was only seen in GoogleEarth, consists of peculiar-looking hut clusters. According to their architecture, they most likely date to the New Kingdom. They evidently are associated to wadiworkings, whose remains can be barely recognised in the image.

In terms of geology the site is located on the western edge of a small mafic intrusive stock

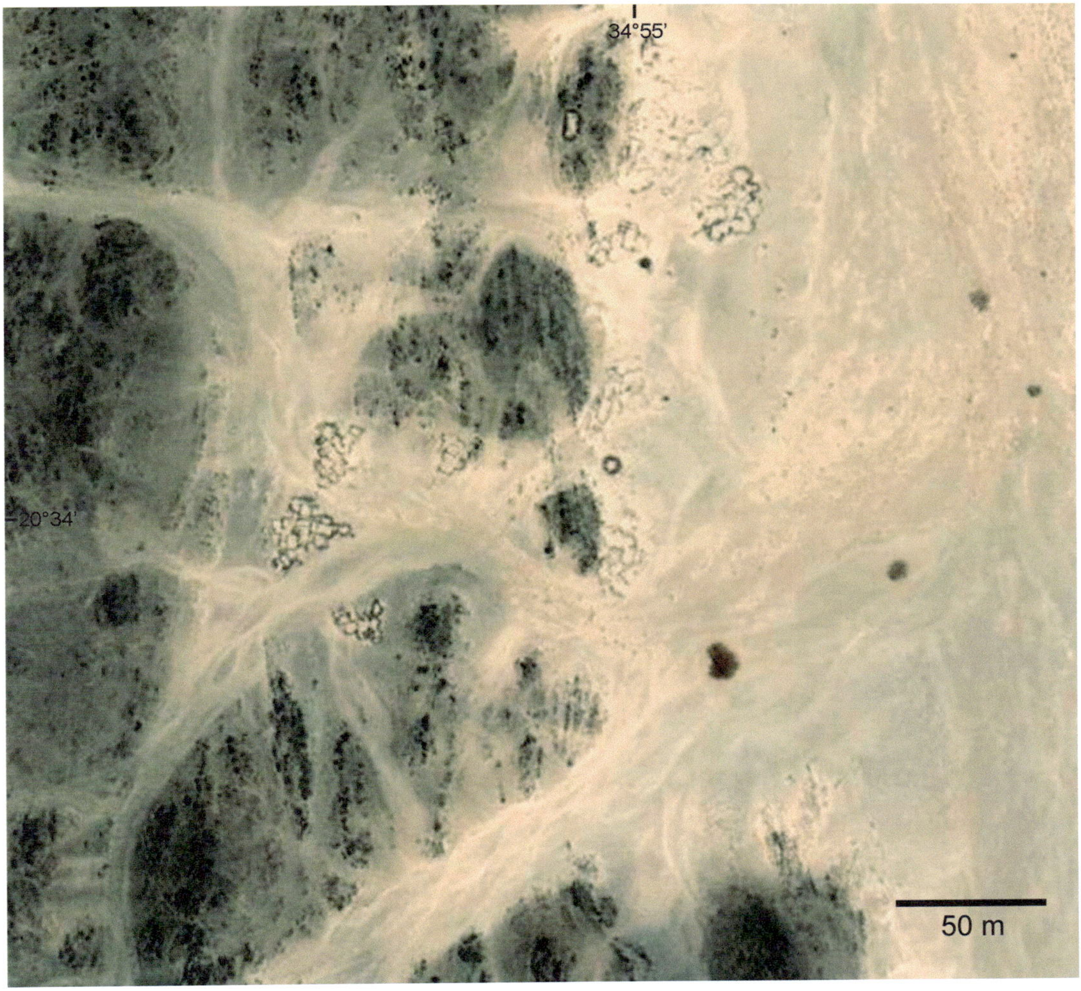

Fig. 6.143 Clustering huts near wadiworkings S of Abirkateib, discovered in a Google-Earth image, probably dating to the New Kingdom (modified Google-Earth image)

within the Nefirdeib series, and at least from that aspect is similar to that at Abirkateib (Fig. 6.143).

6.4.25 Negeim

Geographic position:	
Major mine from the British mining period:	19°27′16″ N, 34°12′16″ E
Eastern adit:	19°27′25″ N, 34°12′20″ E

Next to the main mine (Fig. 6.144) are four large ruins left standing after the Anglo-Egyptian condominium. About 300 m W are the miners' homes. Traces from ancient mining are no longer visible here. In a storage room adjoining one of the houses we found the remains of wooden water- or whisky casks (Fig. 6.145). Old iron cages also scatter about, and shards from large water amphorae (Sir), purple glass shards from English water bottles, ones from beer bottles supplement the legacy from this period.

A map compiled by the Robertson Research International Ltd. (Fletcher et al. 1984b; Boswell 1984a) indicates the locations of the British mining facilities and private dwellings. The Arab huts had erroneously been marked as being pharaonic.

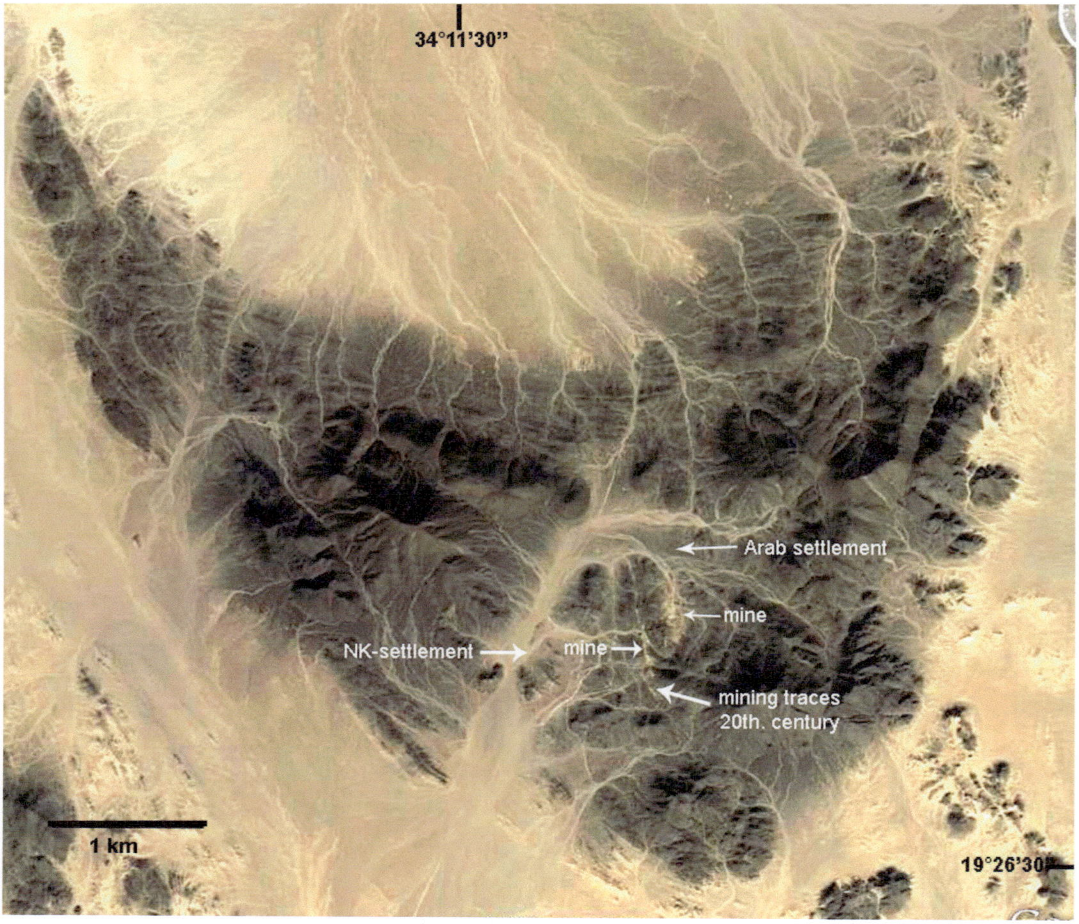

Fig. 6.144 Gold mining district at Negeim (modified Google-Earth image)

In a valley NE to the main mine between 50 and 60, loosely distributed Early Arab huts were counted in the northern vicinity of the adit referred to as the "Eastern Reef". They had been built in the shell-facing technique and display the usual finds of round mills and anvil stones (19°27′35″ N, 34°12′18″ E). Piles of quartz located next to the huts indicate that collected chunks had apparently been prepared here (Fig. 6.146).

The settlement was established around several quartz outcrops, which probably motivated the miners to settle here. In addition, the main quartz vein had certainly been mined as well.

SW of the settlement is the mouth of the Eastern Reef adit, which was probably enlarged in the early twentieth century. In front of it is an aisle for mine carts on both sides of which the overburden was dumped. The tunnel leads about 25 m into the mountain. The year 1983 given at the main mine refers to the recent prospecting activities by the Roberson Research Company.

In a wadi branching-off to the SW between the settlement and the main mine are several work platforms where quartz had been pounded (Fig. 6.147). They were identified by typical, rounded anvil stones with the characteristic central impact depression.

Fig. 6.145 Wooden (whisky?) casks remaining after the British mining phase at Negeim

Fig. 6.146 Early Arab Period mining hut built in the characteristic double-shell technique with in-between fillings of pebbles and clay. Small ore heaps are located right next to the hut

Fig. 6.147 Early Arab Period gold processing site with countless, small pounding stones used for crushing quartz ores

About 300 m from here halfway up the SE slope of the wadi is a quartz outcrop. Our inspection revealed that it had been recently visited by the Robertson Research Company. In the surroundings are few huts with the usual inventory from the Early Arab Period, consisting of countless round mills, anvil stones and slabs. The architectural remains are however in a very poor state of preservation.

About 400–500 m NW to the main mine we found a New Kingdom settlement at (19°27′16″ N, 34°11′49″ E). It consists of about 12–15 rectangular houses consisting of one to two, interconnecting rooms. The rooms are about 12 m^2 large and surrounded by loosely masoned walls. The settlement is not easily found. It lies on both sides of the mouth of a side wadi directly opposite the mine. The terrain is speckled with small stone outcrops, which may in fact be mistaken for ruins.

The New Kingdom oval mills inside the houses have flat and sometimes grooved depressions (Fig. 6.148). The grinding stones are sometimes worn on both sides, as commonly observed on Egyptian examples from the New Kingdom. Small globes or cones from of quartz had served as pounders in the same way as cylindrical stone pestles in Egypt (Fig. 6.149).

Because of the reduced size of this New Kingdom settlement, it is thought that it had been connected to a nearby mine rather than to wadiworkings.

Mining was resumed in 1905 and continued until 1907. Short references appear in the report by A. Llewellyn (1903), who at the time was superintendent engineer for the Sudan Gold Field Ltd., when the Negeim deposit was included to the so-called "Berber Concession" in October 1904 (Boswell 1984a, b).

Initial exploration work began in 1905 at Western Reef with the lowering of ten, approximately 3 m deep prospecting pits over a distance of 305 m. It turned out that thickness variations at the reef ranged between 0.45 and 1.5 m. Panning analyses established gold grades between 6 and 54 g Au/t (average values: 0.8 m thickness and 24 g Au/t).

Fig. 6.148 New Kingdom house ruin at Negeim, built in the simple dry-wall technique. The slight use wear at the numerous ore mills suggests a short occupation period at this site

Fig. 6.149 Globular and conical pounding stones from quartz and quartzite for crushing quartz ores at Negeim

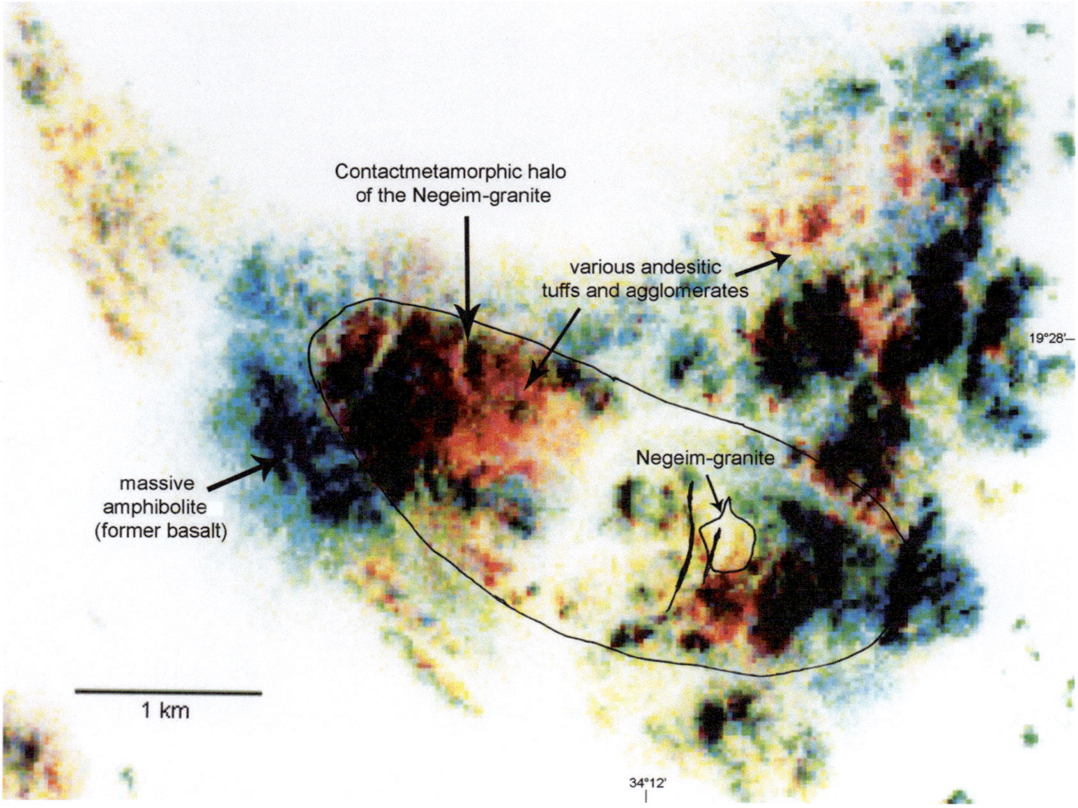

Fig. 6.150 Lithologically processed satellite image of the Negeim area (TM 173/45, channels 7-4-1)

Two ventilation shafts were lowered into the mine while an extra adit was driven into the reef, producing gold grades of 31 g/t over a distance of 18 m at a vein thickness of 1 m. A stretch measuring 6 m with thicknesses around 0.8 m even produced 46 g Au/t. With increasing depth, the shafts were reported to document a widening of the vein. Ventilation shaft 2 is said to have encountered a 0.5 m thick quartz vein containing 64 g Au/t.

Inspite of the lucrative yields, the mine was eventually shut down, mainly for logistical reasons and especially due to a failing water supply.

The exploited quartz vein mineralisations occur in two separate but closely adjacent zones (Fig. 6.150). The so-called Western Reef can be traced over a distance of about 950 m at the surface. Its mean strike is N-S with a dip between 65° and 85° E. Its width varies a lot, probably as a result from intensive folding with marginal crenulations in a slightly schistose wallrock, which actually might suggest a later deformation phase. In its southern part, the second vein system (Eastern Reef) strikes approximately N-S and shifts to a NE-SW strike further to the N. Its dip was recorded to vary between 65° and 85° E, as the thicknesses ranged between 0.4 and 6.3 m (Boswell 1984a, b).

The Western Reef runs through a complex sequence of folded, massive andesite and andesite tuff. The eastern one is initially also located within andesite lava and tuffs, but for the most part in a small granitic to granodioritic intrusion. Accordingly, it is less auriferous than the one in the W.

The wider surroundings of the deposit are also predominantly composed of basalt, andesite and subordinately rhyolite which alternate with agglomerates and conglomerates. There is nevertheless a clear preponderance for volcanogenic rocks, especially as the sediments are for the most regarded as their erosion-products. By contrast, beyond the actual deposit area there are also marbles.

In a more general geologic framework, Negeim seems to be at the southern end of a large syncline as the southern extension of the Hamisana zone, probably near the border to the Oyo-Onib segment.

The deposit of Negeim itself is associated to a dome structure within the southern end of the syncline that was caused by a large granite intrusion. Its uppermost outcrops including a number of small aplite dikes occur in the immediate vicinity of the deposit area. They can be identified indirectly by intense hornfels formations within the volcano-sedimentary wallrock around the quartz vein mineralisation and its geologic environment.

The granite intrusion certainly provided the necessary energy for the formation of the hydrothermal deposit, as the auriferous quartz vein mineralisation evidently is post-intrusive.

Ancient mining seems to have focused on thin pyrite-leading quartz vein mineralisations. Their traces however, seem to have been widely eradicated by modern mining at the beginning of the twentieth century. Apart from a couple of collapsed trench pits, there are thus no evident indications left as to mining activities under the New Kingdom. At the Eastern Reef on the other hand, Early Arab mining attempts may be regarded as affirmative, although gold production probably had footed in the first place on wadiworkings.

The fact that the recorded New Kingdom settlement was located at some distance away from the gold mine shouldn't be seen as a stringent indication to the site's gold mining activities in wadiworkings. As commonly observed for this period, it may rather be connected to a nearby, though no longer determinable water source.

6.5 Group Butena – Wadi Terfawi

6.5.1 Abu Dalala

Geographic position: 21°46′48″ N, 34°29′19″ E

Abu Dalala is a relatively insignificant Early Arab site at the mouth of Wadi Abu Dalala in the Butena plain. Here only a small trench along with one or two houses was observed in the vicinity.

6.5.2 Butena-North

Geographic position: 21°46′20″ N, 34°29′21″ E

This deposit occurs about 1 km S of Abu Dalala within the so-called Butena granite. This too, is a minor site of four, yet relatively well-preserved houses from the Early Arab Period. On a mountain ridge between this site and the previous one we found some painted shards belonging to the "Late Christian" category according to Adams (1986).

Approximately 2 km S of a distinctive mountain range NW of Butena, about 20 isolated, Early Arab huts with double-shell walls are found within the granite outcrops of a flat, sandy backdrop. They are built from the melon-sized granite nodules that lie scattered in plentyful numbers at the surface. One house has a prominent corner entrance highlighted by two vertical stone pillars. It has two interconnected rooms through a lateral passageway. Each house is associated to a small private prayer platform.

A small extraction trench had been lowered into a quartz vein striking 100° E (dip undetermined). The vein occurs in a fine-granitoid rock, which according to the lithologically processed satellite image, apparently represents the hybrid margins of the Butena granite. It had apparently intruded a predominantly rhyolitic to andesitic and intensely folded sequence.

The relatively small quartz vein contains the ore minerals pyrite, chalcopyrite, and noticeable malachite.

No mined quartz veins were found at Abu Dalala 2 and according to the arrangement of the houses, only wadiworkings had been operated here.

6.5.3 Butena

Geographic position:	21°45′53″ N, 34° 29′44″ EN

On a hill are a small trench, about 10–15 huts from the Early Arab Period, and two circular round mills, of which one was still in the stage of being worn-in. This corresponds approximately with the description by Llewellyn (1903), who reports two to three of these mills.

The limited findings concerning the deposit of Butena are somewhat surprising, but because of its fundamental importance it is discussed in the following.

The area of Butena occurs as an isolated geologic peninsula in a sand desert revealing only few basement outcrops. The actual geologic setting therefore is only understood through careful mapping based on a field survey (Fig. 6.151).

The geologic setting pertaining to the ore deposits of this area remains nevertheless a mystery. Probably four different-sized granitoid intrusions intrude a volcanogenic backdrop, consisting mostly of rhyolite and andesite and petrographic variations such as porphyries, tuffs and agglomerates. The intrusion to the NW is exposed by erosion, while the two central ones, which presumably are interconnected at a lower level, are only verifiable in their roof sections where they yet contain greater areas of mappable intruded roof material in all stages of assimilation.

The mentioned volcanogenic rocks occur in a major, E-W striking, anticlinal structure in whose central part the two inner intrusive bodies had penetrated, whereas the two smaller, outer ones had emplaced at the northwestern and southeastern extremities. Regardless of the large anticline, the volcanogenic series, however, are internally still quite complex as they follow the chief structure, while folding in smaller anticlines and synclines. In terms of petrography, the northern flank of the main structure is dominated by rhyolite, whereas the southern flank is almost exclusively andesitic.

According to our experience, the conditions for the formation of vein quartz mineralisations had been rather propitious here. Similar geologic settings are encountered at Gebeit in the Red Sea Hills and Atalla in Egypt. But after closer inspection two major differences remain:

In the Butena complex a close spatial relationship is missing to the mafic and ultramafic series and their (meta)-sedimentary products, especially the greywackes and conglomerates.

Secondly, there are no prominent shear zone systems in whose environment comparably productive gold mineralisations could have formed.

6.5.4 Wadi Tabak

Geographic position:	21°41′54″ N, 34°19′54″ E

We were unable to visit the site at Wadi Tabak. We therefore need to rely on the brief information given by Castiglioni and Vercoutter (1998). The authors noted amongst others, several dozens of buildings with round ground plans buried under the sand and few millstones originating from round mills. Some of the items are dated by W.Y. Adams to the Islamic Middle Ages between AD 850 and AD 1000. Washing tables, ore extraction in quarry trenches occur in the surrounding hills several dozens of metres above the wadi floor.

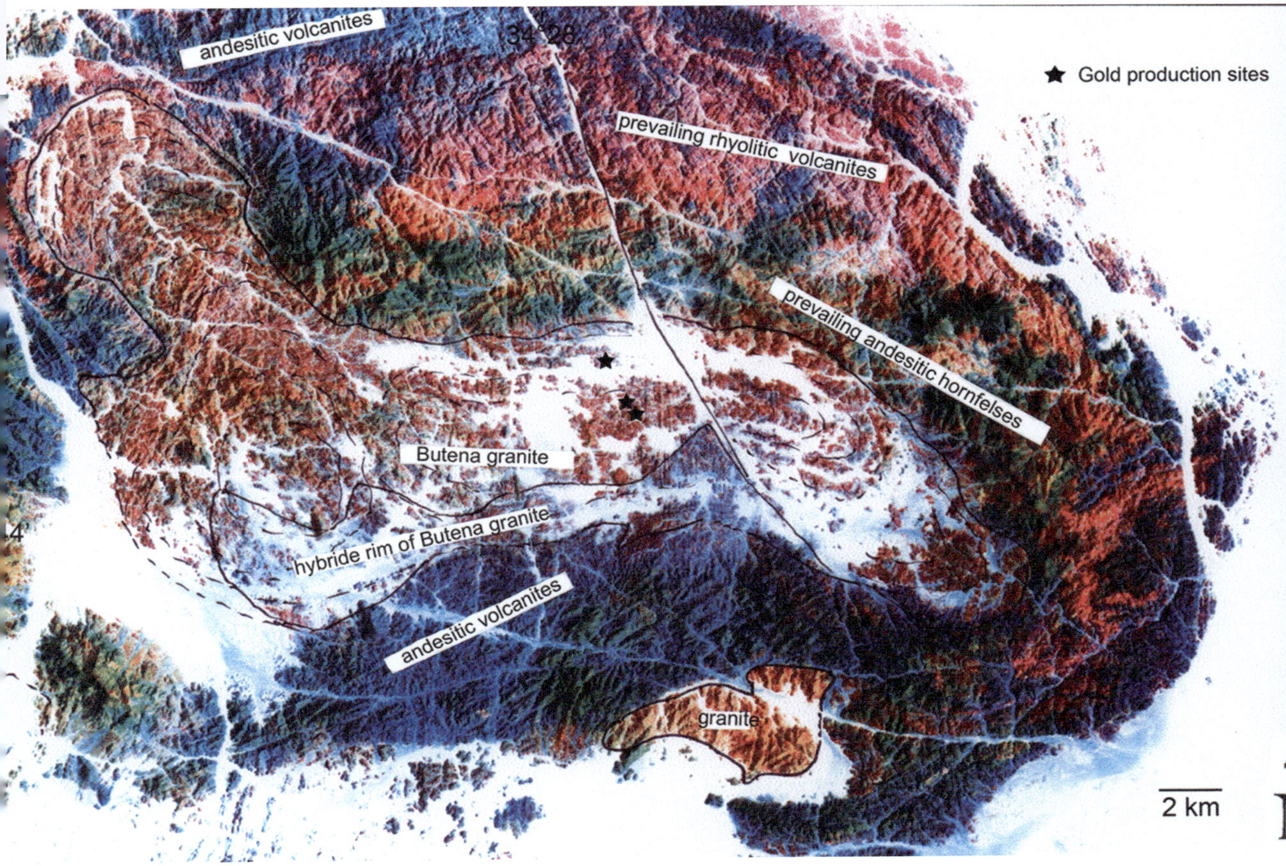

Fig. 6.151 Lithologically processed satellite image of the Butena deposit area. The mineralisation occurs exceptionally inside the granite instead of the hybrid margins (TM 173/45, channels 7-4-1)

6.5.4.1 Shashuateb 1

Geographic position:	21°41′45″ N, 34°13′04″ E
Trench at Shashuteb 1:	21°41′41″ N, 34°13′17″ E

This well-preserved Arab Period settlement consists of scattered, relatively large buildings with several rooms and also single- and two-room huts. A now predominantly sanded-in trench runs through the settlement area from E to W, whose mined part is interrupted at two locations. At least 15 washing tables, of which one has been fully preserved, including the drainage system, are scattered over the entire settlement area and surrounded by conspicuously red tailings (Fig. 6.152).

The houses are not built in the otherwise usual way of agglutinating rooms, but for the most consist of two clearly defined, rectangular spaces

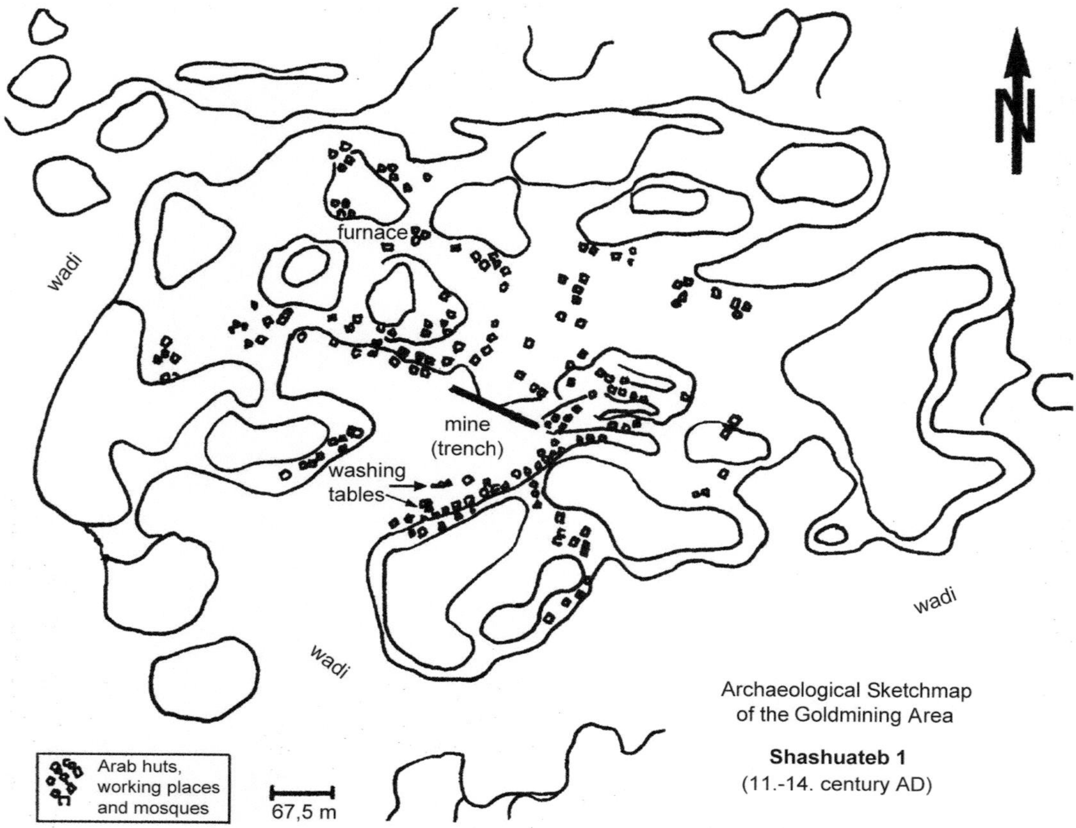

Fig. 6.152 Archaeological sketch map of the Shashuateb district (R. Klemm)

(Fig. 6.153). Their lengths vary between 5 and 6 m, while the rooms themselves are interconnected by inside passageways. Outside, the houses reveal flattened work platforms with accumulations of hammer stones, flat pounding slabs, but relatively few mills. Close to the houses and the work platforms are many small stone cists between 1 and 1.5 m long and 0.6–0.7 m wide. The cists are also found next to the washing tables (Fig. 6.154). In the southern part of the site is a large space reserved for prayers, but otherwise there are also small prayer places next to the houses.

A pit trench had been lowered into the ground at the northern edge of the wadi. The ubiquitous pottery consists of amphora shards and a painted, Nubian ware of the "Late Christian" group (Adams 1986). We also found local Nubian pottery with incised, geometric (rhombic) designs from heavily sand tempered clay.

To the N is a structure we identified as a domed furnace.

Fig. 6.153 Early Arab Period gold mining settlement at Shashuateb1. The mined trench is distinguished in the background by flanking spoil heaps

Fig. 6.154 Well-preserved Arab washing table at Shashuateb 1, including its water collecting basin and backflow channel

Fig. 6.155 The settlement at Shashuateb 2, consisting of singular huts at the wadi edge. Colluvial workings are made out in the foreground through contrasting colours on the slope surface

6.5.4.2 Shashuateb 2

Geographic Position:	21°41′18″ N, 34°11′43″ E

Shashuteb 2 is located about 2,5 km W of Shashuateb 1. The site consists of 20–30, loosely distributed, Early Arab huts. Llewellyn (1903) by contrast, gives a total number of 200.

Further W, on a ridge, is a number of scooped trenches that run down the slope to the wadi. All along its northern slope, the easternmost hill closest to the settlement reveals work platforms and evident traces from colluvium workings, i.e. areas in which auriferous quartz chunks had been collected and selected for processing along the slopes. As reported by Llewellyn (1903), apart from some superficial trial pits and trenches, no genuine underground mines were found. The trenches are easily identified in the mountains by associated grey heaps that contrast to the natural, beige-brown gullies (Fig. 6.155).

Both occurrences at Shashuateb are located at the geologic margins of a massive granite intrusion

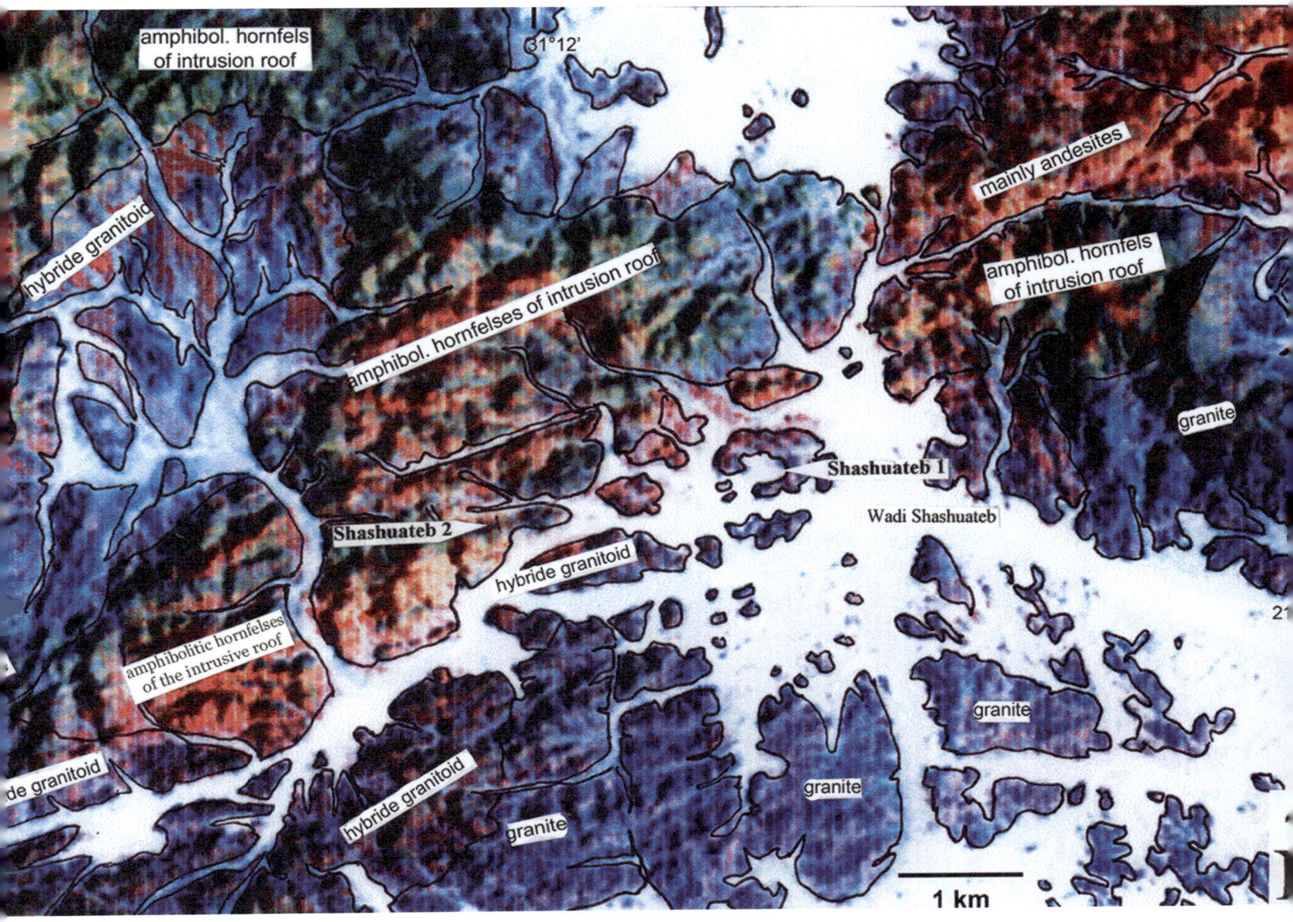

Fig. 6.156 Lithologically processed satellite image of the area around Shashuateb (TM 173/45, channels 7-4-1)

in predominantly andesite volcanics that subordinately alternate with rhyolitic series (Fig. 6.156). Shashuateb 1 is still within the actual intrusive complex, where by assimilation of the andesite roof rocks, the granitic magma had altered to granodiorite and occasionally to diorite. At Shashuateb 2 on the other hand, the trenches are in andesite rock, which had altered so intensely to hornfelses that its hand specimens and thin sections tend to appear like diorite.

The trenches are therefore located within the typically hybrid roof of a granite intrusion.

6.5.5 Kabeseit

Geographic position:	21°07′87″ N, 33°56′88″ E

As we were unable to travel to this large site, we here render the description by Castiglioni and Vercoutter (1998). They report numerous dry-walled buildings with round and rectangular ground-plans. Some walls are rather well-preserved and reach heights of about 1 m. Many buildings are covered under heavy sand drifts. The work platforms are easily discernible and covered with stone hammers. There are also ruins of a mosque, numerous round mills some of which being quite large. Some are complete others are broken. The pottery has been reported to date to the Early Arab Period between the eighth and tenth centuries AD (according to W.Y Adams). Furthermore, large surfaces of ore extraction in trenches were noticed especially at the summit of the hill, above the mining settlement.

6.5.5.1 Wadi Terfawi 1

Geographic Position:	21° 00′40″ N, 34°02′38″ E

On the NW slope of Wadi Terfawi is a settlement with a well-organised structure similar to those discussed at Derahib, Omar Kabash, Bir Kiaw, and Uar in Nubia, as well as Eleiga in Egypt. It consists of an approximately 100 × 50 m large and rectangular building complex with numerous individual rooms in an arrangement along either side of a central road. Unfortunately though, half of the site has been washed away on the SE side. Its original contours are nevertheless easily reconstructable. In accordance with the observations made at the other type-sites, this site too, must be regarded as an Early Arab Period redistributive supply centre for the gold mining community in the surroundings, including warehouses and dwelling quarters (Fig. 6.157).

A now lost well most certainly once furnished the settlement with water. A still functioning well, located 4 km upstream (21°04′47″ N, 34°04′38″ E) reveals that the wadi's subsoil is well-hydrated.

Some additional huts are located in a tributary wadi branching-off to the N. The findings of broken, red granite round mills (Castiglioni and Vercoutter 1998) confirm the Early Arab occupation and gives a direct reference to gold production, which seems to have been based exclusively on wadiworkings.

6.5.5.2 Wadi Terfawi 2 and 3

Geographic position:	21°00′16″ N, 34°02′38″ E and 21°00′00″ N, 34°02′21″ E

There are two settlements in Wadi Terfawi that date to the New Kingdom. They are only located about 700 m apart from each other, site 2 being on the eastern and site 3 on the western side of a wadi mouth.

Both sites display rectangular houses, which partly had been built so close to each other that clusters had formed in the same way as observed at the site S of Abirkateib (cf. Fig. 6.143). Oval stone mills at the surface date both sites to the New Kingdom. As seen in the remaining part of Wadi Terfawi, both settlements had been established for the exploitation of wadiworkings.

6.5.5.3 Wadi Terfawi 4

Geographic position:	20°59′06″ N, 34°02′12″ E

In the confluence area of two side wadis are the ruins of an approximately 30 m long building, which consists of several room units and a large, almost square room in the centre measuring 8 × 8.3 m. It seems as if a settlement similar to that at Terfawi 1 had been planned here, but which eventually had been given up.

About 100 m NE, near the wadi edge, are the rectangular foundation walls of a large building measuring 20 × 5.5 m with two entrance ways.

Approximately 20 more huts scatter along the surrounding slopes. The round mills

Fig. 6.157 Large Early Arab Period settlement in the upper stretches of Wadi Terfawi. The black lines indicate its approximate extent (modified Google-Earth image)

observed there, as well as the remains of a prayer site point unmistakably to the Early Arab Period.

A small mine about 650 m NE in the mountains, seems to represent the traces from prospecting activities rather that a genuine mine.

6.5.5.4 Wadi Terfawi 5

Geographic position: 20°58′55″ N, 34°01′55″ E

This is an unusual, widely scattered settlement to the NW of the wadi. It consists of about 35 small, partly round and rectangular huts spread far apart from each other over several valleys and hills. A small prayer site seems to indicate a date to the Early Arab Period.

6.5.6 Early Arab Settlements

Geographic position: 20°54′59″ N, 35°58′10″ E

Castiglioni and Vercoutter (1998) note a small mining village with sporadic stone buildings, a mosque and medieval pottery of the "Aswan type".

Another, small Early Arab settlement consisting of about ten huts is located in a well-secluded side wadi at position 20°54′18″ N, 34°11′53″ E. Apparently, short-term wadiworkings had been carried out here.

6.6 Group: Wadi Tawil to Eida Arib

6.6.1 Wadi Tawil

Geographic position:	21°51′20″ N, 33°44′35″ E to 21°52′35″ N, 33°45′02″ E

This hitherto completely unknown, but vast settlement area was unfortunately recorded only in a systematic survey in high resolution satellite images in Google Earth.

Wadi Tawil branches-off from Wadi Ismat Omar in a right angle to the N. After about 4 km, a series of large settlements appear in the confluence areas of small tributary valleys, consisting of partly large, multi-room houses associated to areas of well-discernible wadiworkings. Many huts group closely together to form house clusters.

At least three Islamic cemeteries seem to indicate an occupation in the Early Arab Period. The arrangement and shapes of some ground plans seem, however, also to hint to the New Kingdom.

6.6.2 Bir Tawil

Geographic position:	21°51′38″ N, 33°47′40″ E

This small group of huts is mentioned by Castiglioni and Vercoutter (1998) only as a possible mining settlement, because of the lack of associated artefacts. Even the recorded pottery furnished no suggestions as to the site's date.

6.6.3 Ismat Omar

Geographic position:	21°50′22″ N, 33°41′47″ E

An Early Arab mining site of about 30 huts, partially covered by sand is located in a fairly isolated position on the western slopes of a tributary wadi N of Wadi Ismat Omar. An approximately 110 m long trench dug into a NE-SW striking quartz vein, 150 m SE of the settlement, had probably been the motivation for the establishment of the settlement. The large number of recorded round mills as well as all other surface finds point to an exploitation phase exclusively in the Early Arab Period (Fig. 6.158).

6.6.4 Abu Baraga

Geographic position:	21°33′05″ N, 33°37′16″ E

We did not visit this mining site. Castiglioni and Vercoutter (1998) count altogether about 30 singular and clustering huts with circular ground plans. Based on the round millstones, the site was dated to the Early Arab Period.

6.6.5 Small Mining Settlement

Geographic position:	21°32′12″ N, 33°47′05″ E

We discovered this site in a Google Earth image. It is situated in a wadi loop turning S and seems to have been badly damaged by flood waters. The preserved part of the ruins is restricted to an alluvial elevation in the middle of the wadi. The image disclosed no hints as to the date of the site.

6.6.6 Abu Bard

Geographic position:	
Abu Bard 1:	21°31′02″ N, 33°45′24″ E
Abu Bard 2:	21°30′26″ N, 33°45′07″ E

6.6.6.1 Abu Bard 1

We did not visit this site. We therefore refer to the information given by Castiglioni and Vercoutter (1998), who found several dozens of buildings with circular ground plans spread over a vast area. Numerous, large round mills both, complete

Fig. 6.158 Early Arab Period gold mining settlement and mine at Ismat Omar (modified Google-Earth image)

as well as broken from differently coloured granite were recorded together with washing tables. Partly painted ceramic fragments date the site to the Early Arab Period.

6.6.6.2 Abu Bard 2

Abu Bard 2 is located 1.25 km S of Abu Bard 1, at the SE entrance to the same wadi. The settlement comprises approximately 20, round and rectangular huts between 3 and 5 m in diameter. According to the arrangement and the shapes of the architecture, the settlement seems to date to the Early Arab Period. It had probably specialised on wadiworkings.

6.6.7 Abaraga

Geographic position:	21°31′17″ N, 33°35′54″ E

We didn't visit this site either, and therefore again, we submit the observations to Castiglioni and Vercoutter (1998), who noted dozens of mostly singular buildings with circular ground plans spread over a wide area of low, undulating hills. Some are completely buried below the sand. Wheel-made pottery from Aswan was dated to 700–850 AD (W.Y. Adams). No grinding stones and only isolated fragments of quartz were found. Mine trenches are almost entirely covered by sand.

Fig. 6.159 The open, vegetation-covered landscape in Wadi Gagait. Large numbers of well-preserved, Early Arab Period houses are found along its northern mountainous margins

6.6.8 Wadi Gagait

Geographic position:	21°27′08″ N, 33°43′34″ E

Over 30, well-preserved houses and huts are found at this location amidst wadiworkings (Fig. 6.159). They concentrate along the northern edge of Wadi Gagait and around a small elevation near the northern mouth to a side wadi. Most are rectangular and have two rooms, an annex, and a small space for praying outside. There are also round huts with one room. Llewellyn (1903) who also visited the site already noted the difference between the round and rectangular architecture. Often, the houses have a small terrace each facing the wadi. Small stone cists can also be observed between them. The entrance areas often face the slope side, and round work platforms can barely be made out both in- and outside the large rectangular houses.

Among the pottery is the so-called painted, Nubian, Late Christian ware (Adams 1986), which is contemporary to the Early Arab Period. Otherwise, we found numerous fragments from ostrich eggs.

The site is set within a landscape of metavolcanics characterised by chloritised andesites and rhyolites. They occur in lit-par-lit structures together with a fine-grained, slightly porphyritic granite on the outermost margin of a contact metamorphic aureole around the granite stock of Maitib, about 4.5 km to the E (Fig. 6.160).

Another New Kingdom settlement, although heavily diminished by wadi erosion, is situated about 1,200 m SE, on the southern side of the wadi, at position 21°26′45″ N, 33°44′07″ E. One of the few remaining house ruins preserved on the slope is 6.2 m long and 4.2 m wide.

Since no mining traces were found in nearby quartz veins, it is assumed that the settlement had been specialised for the most in wadiworkings.

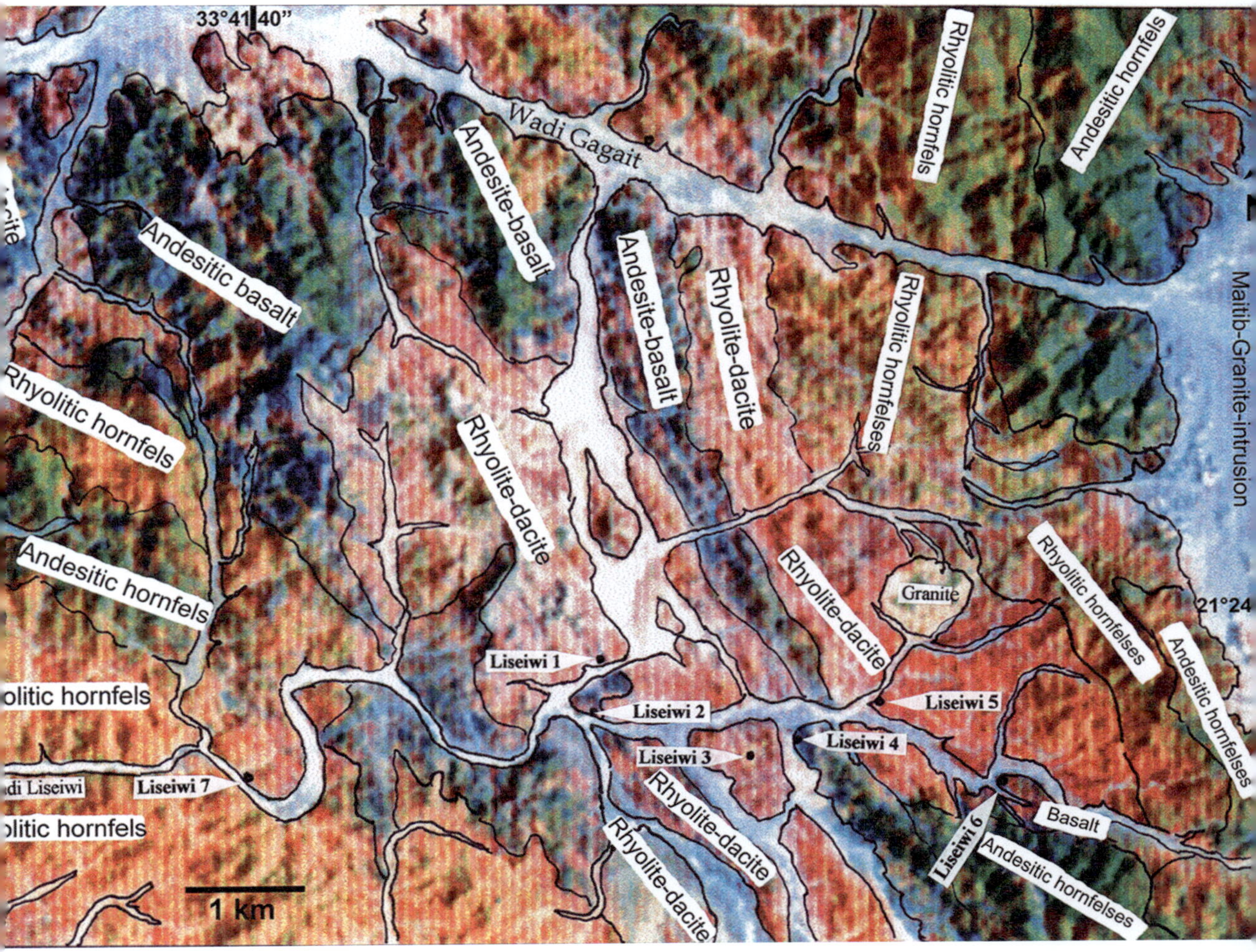

Fig. 6.160 Lithologically processed satellite image of the area to the S of Wadi Gagait and Liseiwi (TM 173/45, channels 7-4-1)

6.6.9 Liseiwi 1

Geographic position:	21°24′45″ N, 33°43′18″ E

This settlement comprises about ten rectangular houses and some small, round work platforms leaning onto a hill slope. The walls from schistose slabs are preserved in heights to about 60–70 cm. Stone cists are also perceptible between the buildings. A relatively well-preserved washing table is located at the outskirt of the settlement (Fig. 6.161).

At this site one initially recognises ruins dating to the Early Arab Period, although these soon turn out to be built over a New Kingdom settlement, of which the flimsy remains are preserved by its stone tools and some pottery. Accordingly, one notices for instance New Kingdom oval mills arranged in small stone circles (Fig. 6.162) and integrated into later wall masonry. Furthermore, innumerable New Kingdom mills and grinding stones are found flushed away over the wadi floor together with mortars and flat pounding slabs from this early phase. The few diagnostic pottery shards have dark vertically running bands and may possibly date to the rule Thutmose III (Holthoer 1977). An analysis of the pottery ware may establish whether it represents an Egyptian import or a local product.

Simple Nubian pottery with incised geometric patterns is found in relatively large amounts at the site. This settlement had probably been involved in the exploitation of wadiworkings.

Fig. 6.161 Early Arab Period settlement remains in Wadi Liseiwi

Fig. 6.162 Wadi Liseiwi 1. Stone circle containing New Kingdom mills hinting to an earlier occupation

Approximately 100 m further W are the ruins of some Early Arab houses containing round mills and anvil stones.

A 6–8 m long mine trench is located in the hills to the N of the wadi, about 650 m SW of the settlement.

Among the petrographic features near the site are amygdaloidal and epidotisised andesite/basalt (Fig. 6.160) with fine quartz bands, limonite linings, as well as fine-grained porphyritic dacite. All quartz samples from the Early Arab occupation display limonite, occasionally with hematite cleft layers.

6.6.10 Liseiwi 2

Geographic position: 21°24′29″ N, 33°43′21″ E

About 20, heavily damaged Early Arabic houses were counted here. They revealed slag residues and pottery fragments of the "Late Christian" type (Adams 1986). Due to the widespread destruction, no other surface finds were recorded. The settlement, as it seems, had operated wadiworkings.

6.6.11 Liseiwi 3

Geographic position: 21°04′22″ N, 33°44′11″ E

Some poorly preserved Early Arab houses are located at the entrance to a southern side wadi. They revealed green glass shards and red as well as bright-coloured fragments from amphorae with incised, wavy lines. Nearby one recognises wadi- and extensive colluvium workings in the adjacent, low-lying mountains to the S. At small quartz outcrops in the wadi beds one furthermore distinguishes mining trenches. After having crossed a broad wadi covered with quartz chunks with colluvium workings to the E of the wadi, one arrives at a site of about ten houses built on a slope. Its sparse finds seem to date to the New Kingdom, although the site seems to indicate that it had been occupied only for a short time. At the foot of a mountain to the W, below the remnants of an elevated Alam, there are two more house ruins.

In addition to the colluvium workings, some poorly exposed trenches striking 50° E and dipping 10° E are found in the dacite rock (Fig. 6.160). The quartz mineralisation seems to have formed inside a vein gap which had opened relatively suddenly, as it has several cavities with idiomorphic quartz crystals and pyrite, the latter of which also occurring in the wallrock. Some gold was detected in a sample.

6.6.12 Liseiwi 4

Geographic position: 21°24′25″ N, 33°44′22″ E

In the quartz-covered, southern wadi and in the smaller side wadis of Wadi Liseiwi traces from intensive wadi- and colluvium workings are observable. On both its sides we counted about 50 house ruins with intermittent, sometimes barely recognisable, round work platforms. Compared to the total number of houses, it was interesting to note the relatively small amount of flat, plate-sized slabs from the New Kingdom, exhibiting only some grinding traces on only one side.

On a terrace in the middle of the wadi is a burial ground comprising about 15 graves. About 100 m further E is a shallow trench and an area with widespread wadiworkings between low hills marked by pointed rock outcrops. Singular houses are especially visible on the hills S side.

Two quartz veins had been exploited in a predominantly andesite rock foliated in a NNW-SSE direction. The southern vein strikes 70° E and dips 65° W, whereas a northern one strikes 80° E.

The entries listed as Liseiwi 1–4 are identical with the ones on a map published by the UNESCO (United Nations 1986).

6.6.13 Liseiwi 5

Geographic position: 21°24′35″ N, 33°44′41″ E

This location is defined by a 75 m long, NE oriented trench. Remains from intensive workings are found along all surrounding wadis and hills (Fig. 6.163)

Fig. 6.163 Wadi- and colluvial workings and trench pit at Liseiwi 5 (modified Google-Earth image)

6.6.14 Liseiwi 6

Geographic position:	21°24′13″ N, 33°45′16″ E

This position refers to a fairly large settlement on either side of NW-SE oriented wadi Liseiwi. The wadi marks the border between dacite rocks in the N and basalt in the S. The settlement consists of large, singular houses with no apparent tools. This might be connected the circumstance that its occupation had probably lasted only for a short period, entailing the removal of all operational tools for their uses at other mining site (Fig. 6.164).

The environment's geology is partly dominated by brecciated and much chloritised basalt (Fig. 6.160) as well as dacite appearing already at the northern side of the wadi. A quartz vein, which had been excavated in several trenches and pits, strikes 50° E and dips 60° SE.

6.6.15 Liseiwi 7

Geographic position:	21°24′24″ N, 33°41′31″ E
Mine:	21°24′35″ N, 33°41′33″ E

The southern edge of this Early Arab settlement is located in a valley bend of Wadi Liseiwi at 21°24′24″ N, 33°41′36″ E. From there, the site

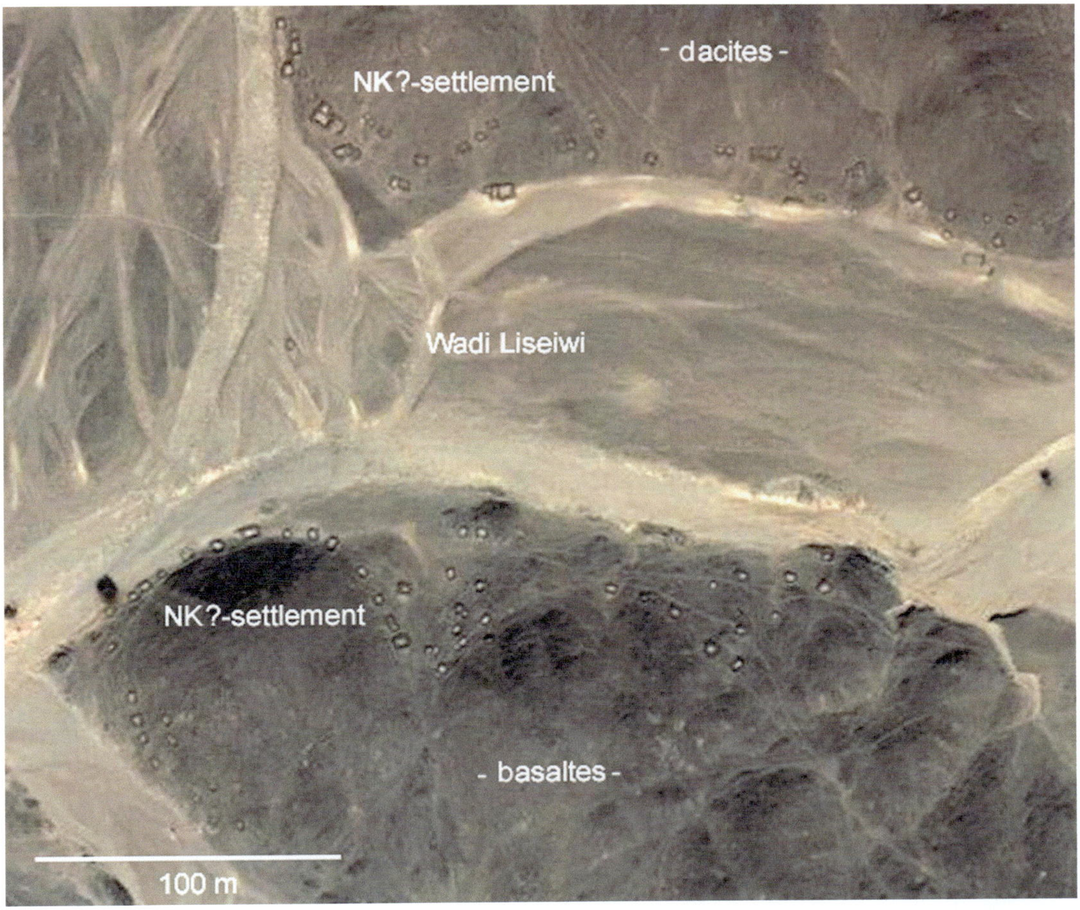

Fig. 6.164 Wadi Liseiwi 6 consisting of two, possibly New Kingdom settlements (modified Google-Earth image)

extends into a northern tributary wadi to about 21°24′32″ N, 33°41′20″ E (Fig. 6.165). A well-preserved washing table surrounded by a tall, horseshoe-shaped tailing (Fig. 6.166) and large, usually rectangular, and detached house ruins with two rooms, distinguishes the appearance of this site. The painted pottery found here is similar to that from Shashuateb and Abu Dalala and therefore date the site to the Early Arab Period (Adams 1986). Near the washing table are fragments from intentionally broken round mills, a widespread phenomenon in the region.

A quartz vein had been mined in a wide trench over about 95 m in an 80° E strike. In its eastern part it is displaced 20 m to the S where it strikes at an average angle of 45° E. Even though this section can be followed for about 45 m, it had been exploited over a distance of only 10 m.

The geologic environment of the deposit is dominated by acid volcanic hornfelses superseded by andesite approximately 1 km to the N. Through contact metamorphism generated by a granite stock about 5 km further S, more or less all rocks in the area are hornfelsic, which however, especially in close vicinity to the vein mineralisation, is blurred because of intense alteration. We were nevertheless able to record a shear foliation striking 10° W.

The large settlement in the main wadi had no doubt developed around wadiworkings. In the northern side wadi, however, the activities had concentrated on the extraction of the quartz lode,

Fig. 6.165 Settlement and extraction areas at Liseiwi 7 (modified Google-Earth image)

as inferred from the visible, although somewhat buried trenches there. We were able to measure its strike only, which at the main trench is 80° E and 50° E at a smaller one, just some metres further S.

All occurrences at Liseiwi are located in a sequence of metabasalts and andesites that partly reveal well-preserved amygdaloidal structures. Furthermore, there are dacites and rhyolites with an average NNW-SSE strike representing a large anticlinal structure, into which the Maitib granite had intruded. Contact metamorphism within these wallrock series thereby decreases gradually within a marginal zone of 10 km. In addition to the usual hornfels formations near the granite itself, this had also led to a noteworthy chloritisation and to some extent epidotisation of the volcanogenic rocks. This is observable everywhere in the aureole of the contact zone and in particular within the mafic volcanics.

The observed accumulation of auriferous quartz veins and the therewith connected concentration of alluvial and colluvial gold production fits in well with the general observations on the accretion of gold deposits around the margins of large intrusions. Noteworthy as well, is that all gold processing sites in Wadi Liseiwi are limited to zones that are decreasingly affected by contact metamorphism and especially to the rhyolite-dacite rock sequences (Fig. 6.160). Apparently, two basic factors had motivated the mineralistaion.

Fig. 6.166 Well-preserved Early Arab Period washing table at Liseiwi 7, surrounded by tailing sands

One had been the more acid nature of the volcanics, which may have favoured the formation of massive, marine sulphide concentrations (cf. the primary ore formation at Ariab). The other had been a certain distance from granite batholiths underlying the entire region, through which the necessary temperature brackets for gold precipitation had been attained.

6.6.16 Liseiwi 8

Geographic position:	21°23′17″ N, 33°37′41″ E

Behind a prominent valley loop facing S in Wadi Liseiwi a mine is located just N of an Early Arab settlement, which consists of "few huts and two or three washing tables" (Llewellyn 1903). The settlement consists of about 15 huts on the eastern side (main settlement) and a house complex with several rooms on the western side with a small, but yet well-preserved domed oven. The slope above the main settlement shows traces from colluvium workings. On the ridge are two old Romibs (pie-shaped graves), of which one to the S contains at least seven burials, which are visible as indentations at the surface and filled with large stones. The northern Romib has no more than two burials. The settlement too, includes three old burial cairns, but some recent Bedouin graves are also visible. The findings at the surface do not permit to establish whether the settlement had initially been connected with gold production or whether it had developed as a mere consequence of a nearby water source.

Gold production nevertheless had taken place here in the Early Arab Period, as evidenced by its even though scarce diagnostic finds.

The settlement and its mines are located in a NNE-SSW striking zone, which is up to 7 km wide. It consists of andesites with some, yet well-recognisable amygdaloidal structures and especially hornblende-porphyric basalts, which had probably been deposited in a terrestrial environment (without pillow-formations). Elongated

Fig. 6.167 Burial complex in Wadi Dom delimited by a circular wall

fault systems cross the zone approximately from NE to SW and have no apparent, genetic connection to a group of quartz veins with an 80° E strike. These had been mined in a main trench in the northern part of the settlement and slightly further S in several pits. During their mineralisation, the quartz veins appear to have been exposed to a swift cleft opening, as the cavities in the mostly coarse, white quartz partly exhibit well-developed, idiomorphic quartz crystal lawns.

The only visible traces from the primary ore paragenesis are preserved in malachite stains in the quartz.

6.6.17 Wadi Dom

In the following discussion, the site sequence no longer strictly follows a N to S order, but rather adheres to individual areas or valleys, but then again ordered from N to S.

6.6.17.1 Dom 1

Geographic position: 21°17′29″ N, 33°41′39″ E

This is an Early Arab Period settlement of approximately 20 houses built in the shell-facing technique on both sides of Wadi Dom, but mainly along the eastern slope and its small side wadis. No finds were made in or around the settlement.

At position 21°17′34″ N, 33°41′20″ E one makes out a stone circle of about 25 m in diameter, which encloses tumuli tombs (Fig. 6.167).

6.6.17.2 Dom 2

Geographic position: 21°16′42″ N, 33°40′57″ E

This Early Arab settlement consists of about 30 houses near a trench that had probably been lowered to no avail, because instead of pyrite, barren and carbonated layers were found inside the quartz.

Fig. 6.168 Wadi Dom, named after the numerous dom palm trees

Several grinding slabs and a flat, circular round mill are located directly at the trench (strike 25° E). A washing table is seen somewhat further away.

6.6.17.3 Dom 3

Geographic position: 21°16′35″ N, 33°41′44″ E

This large, New Kingdom settlement reaches about 100 m deep into an eastern side wadi of Wadi Dom. At the foot of a mountain are two, well-preserved, rectangular houses (outside dimensions: 10 × 5 m), each with two rooms. Westwards near its confluence with Wadi Dom the valley is speckled on both its sides with severely damaged and partly washed-away house ruins with New Kingdom mills and mill fragments. Faintly visible hut ruins spread up the northern slope, whereas on the S side, they extend into the flat plain and continue even into Wadi Dom.

6.6.17.4 Dom 4

Geographic position: 21°16′12″ N, 33°41′02″ E

At this position in the wadi, which is named after its numerous dom palms (Fig. 6.168) is a group of six rectangular huts in the northern, hilly part of a flat plain. Five more house ruins are located in the plain's southern part. In between, there are several, bright-coloured and stone-lined work platforms. Between both areas are a small trench and a washing table as well as numerous round mills together with associated rotor discs. One of them had been perforated from both sides without that the holes had met in the middle. This resulted to the central hole taking the shape of an oval, which meant that its axis was off centre. Consequently, it had never served. We also found a fragment of a furrowed, green glass bottle ("Samsam" flask), as well as some flat "prospector" grinding slabs.

Fig. 6.169 Sanded-in well and spoil heaps at Wadi Dom

About 250 m NE of the settlement Dom 4 is a further, approximately 70 m long trench in the vicinity of ten huts and two washing tables.

6.6.17.5 Dom 5

Geographic position:	21°16′07″ N, 33°41′40″ E

This place is situated in Wadi Dom at an intersecting wadi. About 30 houses scatter loosely on the eastern slope. To the N there is a small trench mine. On the western side of the wadi, one recognises about five round work platforms.

6.6.17.6 Dom 6

Geographic position:	21°15′45″ N, 33°42′44″ E

This site probably corresponds to the one Llewellyn (1903) denoted as "Wadi Domi", even though his coordinates are not consistent with ours.

Wadiworkings had been carried out on a large scale throughout the wadi, both to the E and W. A settlement and a mine are located at the mountain foot.

Approximately 100 m further W into the plain is an open (Fig. 6.169), artificial depression probably representing a former water hole. The surrounding hut ruins have been severely reduced by water erosion. However, two large tailings have survived. Round mills, anvil stones and prospector slabs were observed in large numbers, although not to the extent one might have expected to judge by the numbers of wadiworkings. An area with colluvium workings is noticeable to the NE.

The washing tables next to the tailings are remarkably flat. Like the huts, they had been assembled with the spiky, rhyolitic rocks from the adjacent mountain.

Sites Dom 1, 2, 4, and 6 exclusively date to the Early Arab Period.

All gold mining sites in Wadi Dom are geologically located at the northern extension of the Wadi Naba anticline. It is made up of acid volcanics, especially dacite, rhyolite and quartz porphyry next to subordinate remnants of overlying andesite series.

According to their position in this saddle and their sheared adjustment to the large Tanasheb syncline, the rocks recurrently display shear foliations with a median strike at 35° W. The trenches partly follow this orientation, though without necessarily adhering to the mineralised quartz vein. Especially with regard to the approximately 70 m long prospecting trench at Dom 4, one gets the impression that it had deliberately been traced in an acute angle to the foliation and perpendicularly to the quartz mineralisation, which doesn't really make much sense.

It seems nevertheless evident that the mining activities in Wadi Dom, both during the New Kingdom as well as the different phases of the Arab Period, had primarily been restricted to wadiworkings and that the different mining trenches merely reflect the outcome of inexpert prospecting efforts to locate adequate, auriferous quartz veins for exploitation in underground mines, though without any apparent success.

6.6.18 Tilat Abda

Tilat Abda designates an area marked by extensive wadi- and colluvium workings scattered widely over the hills S of Wadi Tilat Abda, between 21°17′30″ N, 33°39′24″ E and 21°14′40″ N, 33°39′16″ E. Along this stretch one particularly notices the colluvium workings that jut out as red patches on the slopes. Architectural remains are otherwise quite scarce, and tools are rarely found.

The Minex Sudan Company (Roberson Research) has carried out a comprehensive survey of the gold deposits in this region as well as in areas of Wadi Dom, and especially near the Tanasheb occurrence. During our field work we were able to draw on the indications given in the Minex report, so we therefore decided to adhere to the site numbers attributed by Boswell et al. (1981), even though it does not strictly follow the order from N to S.

Castiglioni and Vercoutter (1998) dated Tilat Abda to the eleventh and twelfth centuries AD by aid of the recorded round mills and some of the pottery. However, the partly good state of preservation of the architecture would point to a later date.

Castiglioni and Vercoutter (1998) continue in mentioning some grinding stones from grinding mills, which accordingly would hint to a date in the New Kingdom. We did not find any such tools, but because of the numerous, contemporary remains in the surroundings, the likelihood of an earlier occupation phase in the New Kingdom cannot be entirely ruled out for this site.

6.6.18.1 Tilat Abda 10

Geographic position: 21°16′20″ N, 33°39′24″ E

Site Tilat Abda 10 is an extensive settlement consisting of rectangular, relatively well-preserved houses. They distribute loosely in a vast area defined by heaps of wadiworkings. A very well-preserved washing table, including its drainage system and at least two comparatively large, rectangular praying areas are located near the houses.

The relatively good state of preservation of the houses and the washing tables, as well as the neatness of the prayer site, and especially the high debris heaps at the wadi- and colluvium workings that appear to be little affected by the yet intensive erosion, may suggest a more recent date for this site (Fig. 6.170).

Some round mills and fragments from rotor stones, but very few anvil stones were found, which might indicate that iron mallets had been used.

6.6.18.2 Tilat Abda 11

Geographic position: 21°15′39″ N, 33°39′16″ E

This site consists exclusively of wadiworkings, some trenches, and round work platforms. No architecture was found.

Behind a ridge to the W are yet more trenches, quartz outcrops, and wadiworkings, but no houses. All sites have in common that none is associated

Fig. 6.170 Arab Period settlement remains spreading far into the hills at Tilat Abda. Red-brown coloured areas in the background indicate colluvial workings

to genuine mines. Only trenches and colluvium workings were observed, revealing just few tools. The recorded washing tables are as a rule well-preserved. Tilat Abda 10 is the main settlement in the area, and the remaining sites consist mostly of only small, round work platforms, and are therefore not discussed separately.

The noticeable predilection for colluvium workings at the sites in Tilat Abda, Wadi Dome, and Naba, as well as the good state of preservation of the architecture may indicate an occupation in a more recent past. In this respect, the time of Mohamed Ali in the early nineteenth century (1821 and later) comes to mind, especially bearing in mind that the area was among others accessed via Wadi Gabgaba and Wadi Abaraga, which had been protected by two major "Ottoman" forts.

The various occurrences in Tilat Abda are located at the edge of a syncline, which is severely marked by tectonic displacement. It essentially consists of rhyolites, andesites and basalts, and is extremely elongated and sheared towards the SW. The Tilat Abda granite had intruded this syncline at the eastern edge and thereby caused a saddle-like uplift of its northern end (Fig. 6.171). Through this intrusion, former rhyolites and ignimbrites had partly transformed to porphyritic structured hornfels, which sporadically though reveal their original identity as ignimbrites.

6.6.18.3 Some Remarks Concerning the Wells

The two wells in the area of Tilat Abda, which have been recorded in the topographic map, are both dry. The same goes for the well at Abaraga and for that at Bir Gagait, where we had decided to pitch our camp. We were unable of even finding the latter. Wells in the Nubian Desert in fact dry-up systematically. During floods, they tend to silt up completely, after which they are neither dug, maintained nor protected by covers, etc. As this area is extremely arid and virtually devoid of vegetation, it thereby becomes increasingly uninhabitable. Subsequently almost, no Bedouins are encountered here anymore.

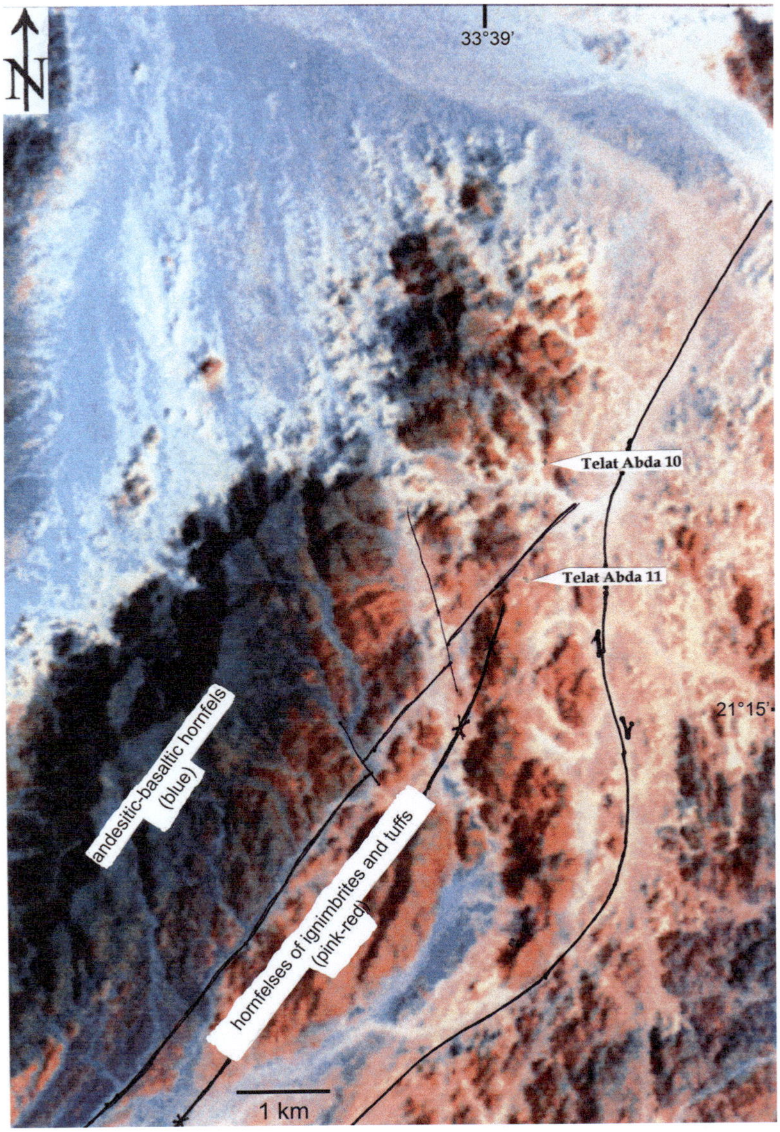

Fig. 6.171 Lithologically processed satellite image of the area around Tilat Abda (TM 173/45, channels 7-4-1)

6.6.19 Tanasheb

Gold exploitation in the area of Tanasheb does not predate the Early Arab Period. In terms of geology, this nevertheless vast area may be regarded as a continuation of the Tilat Abda zone. The distinction between the first seven settlement- and extraction sites is to some extent arbitrary, as the sites themselves are partly not separated from each other by any particular physical feature. The indicated geographic positions serve therefore only as reference values. Sites Tanasheb 8–12 by contrast, are located in the roof area of a granite intrusion and hence in a zone of basaltic volcanics altered by contact metamorphism.

Geographic positions:
TS 1: 21°11′41″ N, 33°37′36″ E
TS 2: 21°12′03″ N, 33°36′55″ E
TS 3: 21°12′04″ N, 33°36′46″ E
TS 4: 21°12′17″ N, 33°36′47″ E
TS 5: 21°12′28″ N, 33°36′55″ E
TS 6: 21°12′27″ N, 33°36′32″ E
TS 7: 21°12′14″N, 33°36′28″ E

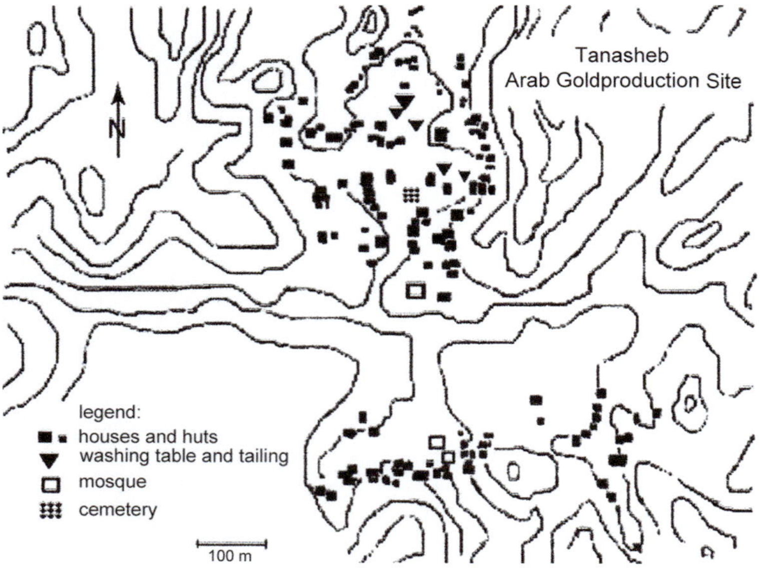

Fig. 6.172 Sketch map of Tanasheb 1, traced after a satellite image (R. Klemm)

Fig. 6.173 Agglutinating round- and rectangular houses at Tanasheb 1

Fig. 6.174 An unusually well-preserved house with rooms in a magazine-like arrangement at Tanasheb 1. A washing table is seen respectively in the front and rear of the picture

The area encompassing the sites Tanasheb1–7 extends N and S of a flat syncline, through which runs a wide wadi (Fig. 6.172).

The house walls are still partly preserved to heights of 70 cm with intact capstones (Fig. 6.173). A 20 m long storage facility consisting of nine remarkably long and narrow room units contains a strikingly high amount of amphora fragments (Fig. 6.174).

The houses (Fig. 6.172) scatter in the plain as agglutinating and singular, both round and rectangular huts. Stone anvils were not found, probably because of the availability of iron mallets.

In the middle of the northern part of the settlement is a cemetery containing 15 graves.

At least five, well-preserved washing tables with strikingly pink tailing residues can be seen in the terrain. At one of the tailings we found a large shell fragment of an ostrich egg. Despite the good state of preservation of the washing tables, they give no indication as to the actual nature of the original surface cover of the inclined surface that captured the gold dust.

All round mills we saw had been intentionally shattered. In one case we even found the stone, which had served to crush an adjacent rotor disc (Fig. 6.175).

Site TS 1 obviously was the central settlement for the entire area. As opposed to the usually observed, small, individual prayer places, it had

Fig. 6.175 Intentionally smashed, Arab Period rotor disc at Tanasheb and the stone slab, which had apparently served for break it

for that purpose a large, rectangular square in its northern part and likewise two smaller ones in its southern quarter.

Ore extraction was carried out in trenches as well as in wadiworkings in the closer and wider vicinity.

The mining area of Tanasheb 1–7 which exclusively dates to the Arab Period is located in a hilly, though nonetheless relatively flat terrain. It is geologically characterised by its location near the centre of a large syncline. To the W, it is cut by a NNE-SSW striking fault. To the S it becomes narrower through tectonic thrust. It therefore represents the southwestern continuation of Tilat Abda and accordingly displays the same geologic units. Visible rhyolites and ignimbrites with well-preserved weld layers form the bottom of this syncline and are superseded by andesite-basalt sequences that do not extend into the deposit area (Fig. 6.176).

The strike values of the mined trenches range between 40° and 45° E. Because of the poor observation conditions at the surface, the dips couldn't be measured. As the veins largely adhere to the orientation of the layers, they should vary around 30° NW.

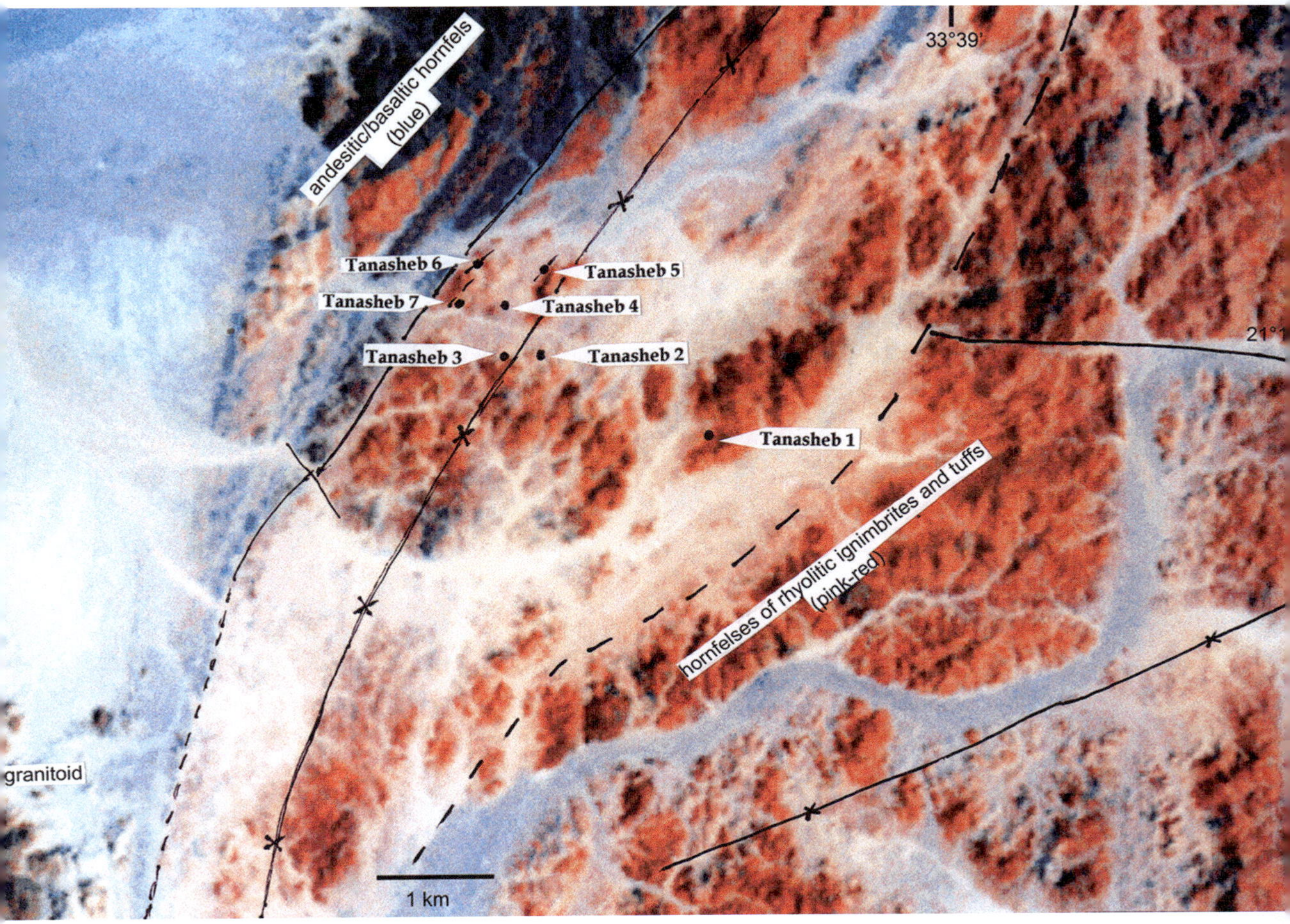

Fig. 6.176 Lithologically processed satellite image of the area around Tanasheb 1–7 (TM 173/45, channels 7-4-1)

6.6.19.1 Tanasheb 8–12

TS 8:	21°10′43″ N, 33°33′18″ E
TS 9:	21°09′13″ N, 33°32′15″ E
TS10 (settlement):	21°09′00″ N, 33°32′17″ E
TS 11:	21°09′00″ N, 33°32′42″ E
TS 12:	21°08′58″ N, 33°32′43″ E

6.6.19.2 TS 8

Eighteen round mills were counted here inside Arab Period hut ruins. Remains from washing tables were also found.

6.6.19.3 TS 9

TS 9 is a small group of Arab Period hut ruins near a vast plain with wadiworkings.

6.6.19.4 TS 10

This widely spread-out settlement (Fig. 6.177) consists of a great number of carefully built huts from spiky porphyry-andesite rocks in a relatively flat terrain. Some are also made from granite (Fig. 6.178). Apart from some small trenches, a larger and elongated, for the most sanded-in trench is visible at the western edge of the settlement (Fig. 6.179). There are more huts built from weathered granite boulders, which reveal only the rough outlines of the individual ruins further NE, beyond a depression.

Some houses have carefully made entrances with well-masoned corner pillars. Tools were found even in the most remote huts. The settlement had therefore probably no central ore processing site. The washing tables by contrast, concentrate near several, large trenches and are generally a little narrower than usually observed (about 70–75 cm wide and 4 m long). They are surrounded by the typical, small, cubic anvil stones, pounding/grinding slabs, and round mills.

The dwellings at TS 10 consist of multi-room, rectangular to rounded houses, or just single room huts. The multi-room houses are relatively small, about 2–3 m wide, but up to 14 m long. Their ground plans correspond exactly to the Arab Period house models reproduced by Llewellyn (1903) who also differentiated between straight and rounded walls. A large prayer site is located in the northern part of the settlement.

The large size of the settlement seems to indicate that gold production in addition to the exploited trenches had essentially been based on wadiworkings.

The archaeological context at Tanasheb 8–12 resembles much that observed at Tilat Abda and Tanasheb 1–7. Noteworthy again, are the relatively good preservation of the architecture, and the partial scarcity of tool finds. Here too, one inevitably gets the impression of a significantly more recent settlement than the Early Arab Period sites discussed further up.

The deposits at Tanasheb 8–12 occur within the same tectonic context as Tilat Abda and Tanasheb 1–7, but this time in a petrographic environment specific to the western limb of the syncline. Tanasheb 8, 10 and 12 are bound to differently hybridised granitoids intruded to partly assimilated amphibolites that have altered their chemical composition to granodiorite and tonalite. Tanasheb 9 and 11 on the other hand, occur in amphibolite (former basalt) altered by contact metamorphism.

Despite the light blue shades in the processed satellite around Tanasheb 10 image (Fig. 6.180), the rocks at the surface at Tanasheb 10 are predominantly represented by substantially modified granitoids through wallrock assimilation. This had partly led to the formation of gabbro-like hornblende rock, which from N to S are crossed by acid dikes and contain rounded fragments of porphyry andesite wallrock.

A quartz vein excavated in a 20 m long trench strikes 60° E through the granitoid bedrock. It is circumscribed by a pyritised alteration zone in the wallrock, and its main generation is white, milky quartz with malachite layers and limonite in clefts. Another quartz generation is grey and contains pyrite as well as cleft hematite. Finally, there is a third one whose fine

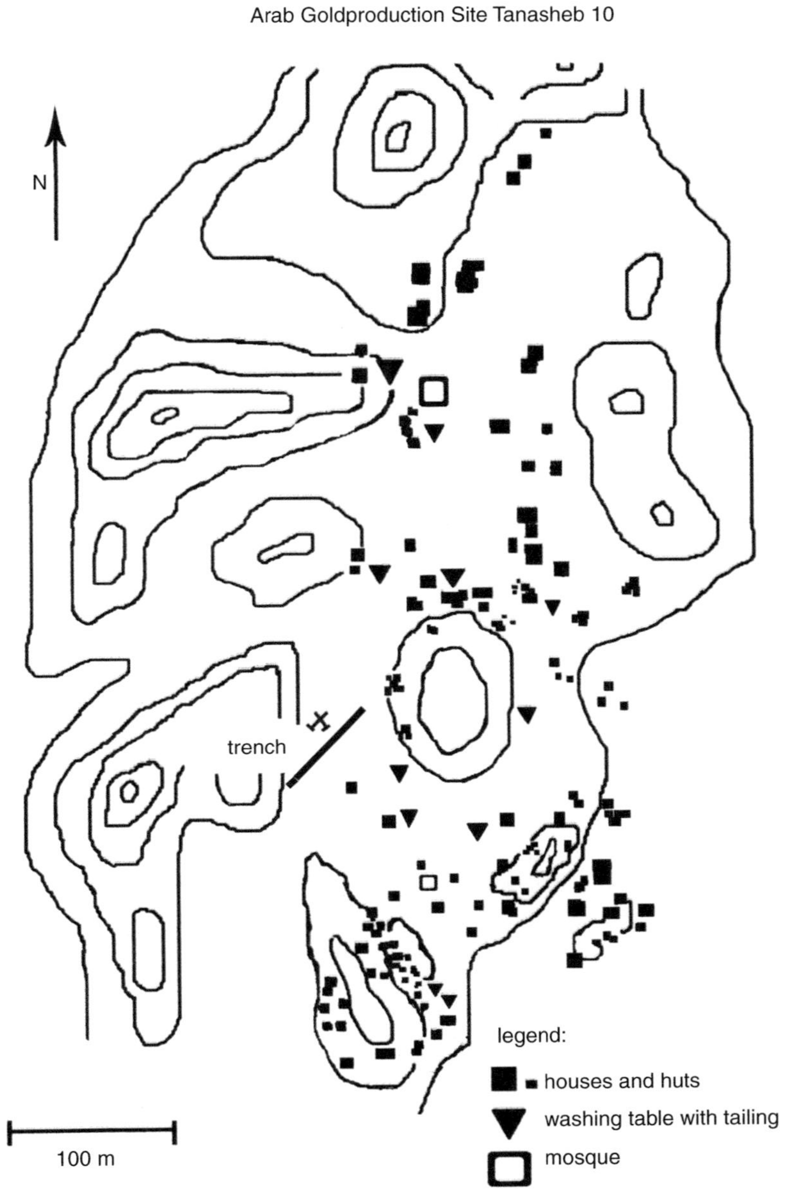

Fig. 6.177 Archaeological sketch map of Tanasheb 10 based on a satellite image (R. Klemm)

fissures contain visible gold in limonite after pyrite and some galena.

6.6.19.5 Tanasheb 11

This site is defined by 12, vaguely distinguishable houses with single-shell walls and some washing tables. Two excavated trenches, whose quartz veins respectively strike at 55° E (northern vein) and 40° E (southern vein) occur in a fine-grained granite. The excavated white quartz contains idiomorphic quartz crystals as well as malachite, galena, and sphalerite.

Fig. 6.178 Widely dispersed and relatively well-preserved huts at Tanasheb 10

Fig. 6.179 Sand-blown trench pit at Tanasheb 10 with flanking spoil heaps

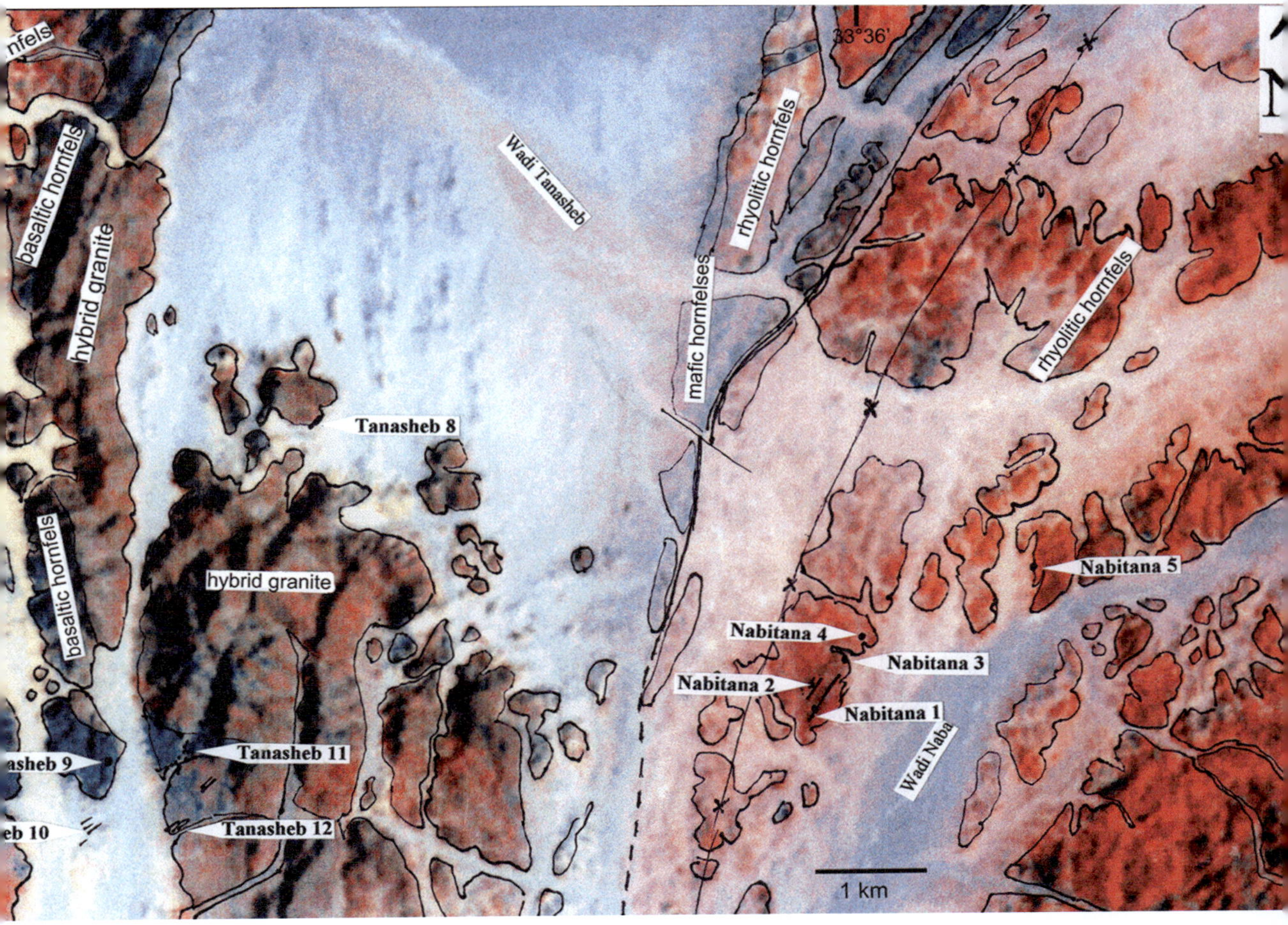

Fig. 6.180 Lithologically processed satellite image of Nabitana 1–5 and Tanasheb 8–12 (TM 173/45, channels 7-4-1)

6.6.19.6 Tanasheb 12

An unexploited vein with a 70° E strike consisting of a strikingly white quartz is located next to some poorly-preserved Arab Period houses in a granite setting. Further S two other veins were, however, mined in trenches. They contain pyrite and galena in white quartz.

6.6.19.7 Tanasheb-West

Geographic position:	21°13′28″ N, 33°29′38″ E

This Arab Period settlement consists of 26 houses and numerous, round work platforms. The houses are round and agglutinate to multi-room complexes. Wadiworkings had apparently been carried out in the surroundings. Tools consist of round mills and some so-called prospecting slabs.

6.6.20 Wadi Naba

The numerous gold producing sites in Wadi Naba are here exceptionally numbered in an order from S to N. They are nonetheless easily identified in the Google-Earth image by their exact geographic coordinates.

6.6.21 Nabitana

Geographic positions:	
Nabitana 1, main mine:	21°09′40″ N, 33°35′57″ E
Nabitana 2, W trenches:	21°09′43″ N, 33°35′47″ E
Nabitana 3, settlement:	21°09′47″ N, 33°35′53″ E
Nabitana 4, two settlements:	21°09′55″ N, 33°35′54″ E

By contrast to the sites discussed above, gold production in the Nabitana area was primarily based on underground mines. A narrow, approximately 200 m long mine stretches from the wadi bottom to the top of a hill. Mining started again in the early twentieth century to which wooden abutment frames still testify (Fig. 6.181). At the foot of the mine there are two, relatively wide washing tables near tailings. Beyond a hill to the N are the scattered remains of an Arab Period settlement and an associated, well-preserved mosque.

The geologic environment of the quartz veins mined at Nabitana represents the southern continuation of the Tilat Abda/Tanasheb syncline (Fig. 6.182). To the W the syncline is noticeably contracted by tectonic thrust. In a more regional context the entire area consists of a large fold system which was compressed from E toward W and ensuing large-scale shearing in a NNE-SSW strike. During the final phase of the fault block tectonics the syncline was in addition crossed in a measurable distance of 55 km by a fault line showing displacements of up to 4 km length.

These had probably been the tectonic conditions, under which the quartz veins at Nabitana had formed. The vein clefts had apparently opened during the last fault tectonics in a NNE-SSW strike and had subsequently mineralised through hydrothermal circulation. The thermal energy source responsible for these convection cells remains however unknown. Lenticular granite apophyses recurrently occur within acid units at Nabitana and its surroundings and may therefore hint to a nearby intrusive body, but without revealing anything as to its actual location.

The quartz vein mineralisations consist of several mother lodes, of which only two, an eastern, and a western one had been exploited.

The eastern lode is well exposed by trenches over a distance of about 200 m. It strikes 10° E and plunges steeply towards W. Its angular discordance to the shear zone foliation striking 30° E is relevant for the determination of its relative age, in other words, it is evidently later and its formation probably coincides with the above mentioned, last extension phase of tectonic fracturing.

The wallrock consists of rhyolitic tuffs with partly well-discernible layering structures and agglomerates of an identical chemistry.

The ore paragenesis composes mostly of limonitised pyrite, malachite and some cleft hematite. The pyrite continued to develop into the wallrock (pyrite pseudo-morphism).

The western lode runs parallel to the eastern one. It had only been mined in trenches and small pits, but intensive colluvium workings were observed in its continuation on a slope. The auriferous quartz was probably restricted to fine, only few centimetre wide veins running paralell to the main, but seemingly barren vein to the W. No noteworthy mineralisation was observed inside them.

6.6.21.1 Nabitana 5

Geographic position:	21°10′12″ N, 33°36′51″ E

This site consists of two, large houses and six to eight round work platforms. A large, but shattered grinding slab, which normally is attributed to prospecting work in the Early Arab Period, was found a little further away, near a washing table in the western part of the site. By contrast to the oval stone mills of the New Kingdom, such stones are usually very flat and display only little use wear at their surfaces. The surrounding area is characterised by small-scale colluvium workings.

6.6.22 Wadi Naba 1

Geographic position:	21°10′24″ N, 33°37′20″ E

A work platform and three large, rectangular houses containing several rooms from the New Kingdom are located in the plain on the N side of

Fig. 6.181 Ancient mine at Nabitana which was reactivated for a short time at the beginning of the twentieth century

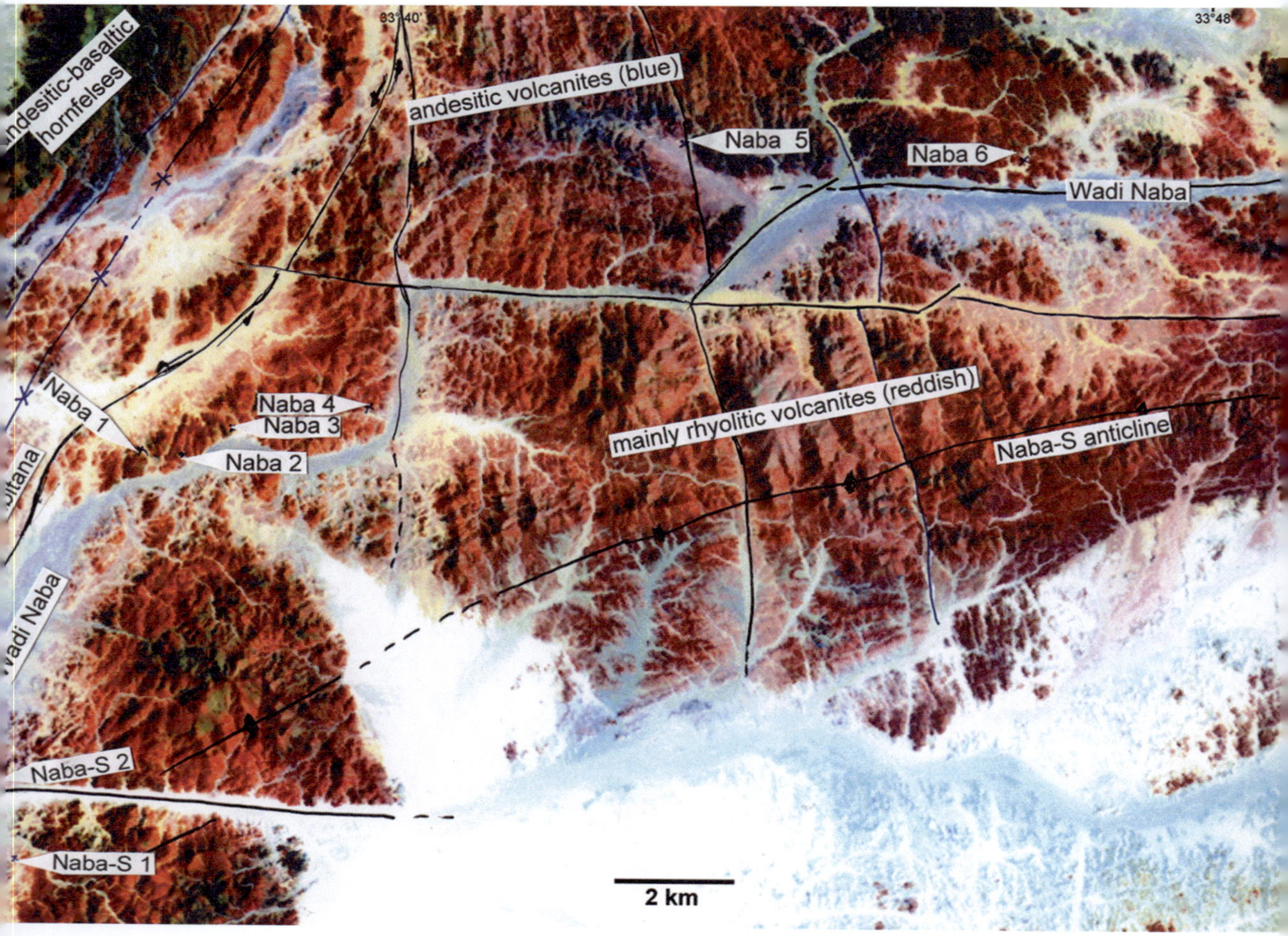

Fig. 6.182 Lithologically processed satellite image of the Wadi Naba area (TM 173/45, channels 7-4-1)

E-W oriented Wadi Naba (Fig. 6.183). The site is just below a quartz outcrop on a hill, and eight additional work platforms scatter over its slope. Sixteen New Kingdom mills were found here. As they are only little worn, they seem to indicate that this was a prospecting site only. Some pottery shards belonging to a brown, burnished ware were also found. The house walls had been built of roundish rock nodes and intermittent pebbles for purposes of stability.

6.6.23 Wadi Naba 2

Geographic position:	21°10′22″ N, 33°37′41″ E

To the E of Naba 1, again on the N side of the wadi, is a spacious, hilly area in which Early Arab round houses with one or two rooms (three have several rooms) widely scatter. To the N are colluvium workings. Otherwise, no traces from mines or trenches were identified near the site, which also includes the quartz outcrops. A series of individual work platforms are seen aligned along the northern edge of the wadi and its recesses.

Some huts and two washing tables are furthermore located in a northern incision between Naba 2 and 3.

6.6.24 Wadi Naba 3

Geographic position:	21°10′34″ N, 33°38′03″ E

A large, contiguous dwelling complex with an integrated washing table, numerous round

Fig. 6.183 New Kingdom settlement at Naba 1 with ore mills displaying strikingly shallow depressions from use wear (*arrows*), thus indicating a short processing phase at this site

mills, and many cubic anvil stones are located in a wadi incision. Further up the small wadi there are two houses and small colluvium workings on either its side but especially on the eastern side, whose surface had been treated in a kind of wide trench. Indicated by red marks halfway up the western slope of the incision, painted shards of the "Late Christian" (Adams 1986) type were found, which might indicate that water vessels had once been standing here for cooling in the wind. At the far end of the wadi is a water retention dam.

A series of huts stretches along the wadi edge to a large plain, where we observed a number of graves. Associated wadi- and colluvium workings emerge in the processed satellite image of the area.

6.6.25 Wadi Naba 4

Geographic position:	21°10′50″ N, 33°39′38″ E

At this position are a washing table surrounded by three, round work platforms, red tailings in different shades, a rotor stone, and small cubic anvil stones. To the E of this facility are a small trench and three round work platforms.

As usual, the flat end of this well-preserved washing table faces the slope in order to assure the water's return. In this case the backflow channel is covered with flat stone tiles. The entire site measures about 10 × 10 m.

Regardless of local differences, sites Naba 1–4 to a large extent group within an internally consistent geologic and petrographic area (Fig. 6.182). In spite of the prominent fault in Wadi Naba, it seems that the deposit area to its S belongs, although slightly displaced, to the anticline extending E of the great Tilat Abda-Tanasheb syncline, whose northern part however, proves difficult to map, because of complex internal folds.

Thereby, the deposit area is located exactly within the flexure transition between the eastern syncline and the flanks of the western anticline. Some of the quartz vein mineralisations follow this NE-SW direction (Naba 4), which is also the case for an unusual quartz carbonate intercalated

Fig. 6.184 New Kingdom settlement and mine complex at Naba 5 (modified Google-Earth image)

between acid volcanite series and exposed over long distances, but which had remained unexploited.

In petrographic terms, the area is marked by fine-grained rhyolites, rhyolite breccias (hyaloclastites?) and quartz porphyry displaying a slight foliation parallel to the layers in an average 40° E strike.

Quartz veins occur at various locations at Naba 1 at a 40° W strike (unexploited), Naba 2 at a N-S strike (unexploited), and to the NE of Naba 3 at 45° W (unexploited, in spite of an elongated exposure). Only the one occurring in a long trench, in a highly altered rhyolite wallrock at 60° E, at Naba 4 exhibits minor traces from mining. This finding is surprising in that the gold prospectors, among which the New Kingdom miners at Naba 1, had evidently failed to locate the auriferous quartz veins. This may imply that the actual productive mineralisations must be quite thin and that economically viable gold mining had already been assured through the gathering of quartz ores in the alluvial and colluvial sediments. This hardly forecasts favourable prospects for the modern gold mining industry in this area.

6.6.26 Wadi Naba 5

Geographic position:	21°13′28″ N, 33°42′30″ E

In a broad side wadi branching-off 9.2 km further up Wadi Naba to the NW is a fairly large site dating exclusively to the New Kingdom (Fig. 6.184).

Fig. 6.185 New Kingdom ruins at Naba 5 with oval stone mills and a central washing table (*arrow*)

A reasonably well-preserved washing table was found at the site. It much resembles the one seen at Nesari 2, which too, is oriented parallel to the slope - in other words, the water used in the washing process was drained away from the table. Later Arab Period specimens on the other hand, are oriented opposite to the slope, thus allowing for an automatic return of the water. Since no drainage channels were observed on the New Kingdom washing tables (Fig. 6.185), the quartz slurry had to be collected at the foot and taken back again in order to repeat the process.

The settlement consists of 30–40, remarkably well-preserved house ruins clinging to a small group of hills and the slopes of the surrounding mountains. A remote house to the SE with several rooms (of an overseer?) displays small pedestals that probably served as bedsteads. It has six

Fig. 6.186 Somewhat remotely located "overseer's home "comprising six rooms, at Naba 5

interior rooms, and its entrance faces the wadi. Its outer dimensions are 9 × 8 m (Fig. 6.186).

The walls of the houses are generally composed of large blocks. In and around the houses one notices typical New Kingdom oval mills, including the associated grinding stones. To the W of the settlement is a mine, which continues after about 80 m further to the W for another 60 m (Fig. 6.187). The excavated quartz vein is only 2–3 cm wide (Fig. 6.188). In addition, at the edge of a mountain there is a large trench leading down its slope.

Approximately 2.5 km further S, hidden away in the hills on the western side of the wadi, are two more, probably Early Arab settlements consisting respectively of 15 houses and huts (21°12′03″ N, 33°42′33″ E) and 12 huts (21°12′00″ N, 33°42′41″ E). However, we were only able to record them in a Google Earth scan.

Fig. 6.187 Trench pit and drift adit in intensively sheared andesite wallrock, at Naba 5

Fig. 6.188 Few centimetres-thick quartz vein at Naba 5 in intensively sheared andesite wallrock

6.6.27 Wadi Naba 6

Geographic position:	21°13′13″ N, 33°45′54″ E

In a small plain to the SE of a group of wadiworkings is a set of six round work platforms, containing no tools. We found a few more in the immediate surroundings, but there was no associated, central settlement. This small number of huts stands in stark contrast to the extensive wadiworkings of the immediate environment. It is therefore possible that a once larger settlement had been completely washed away by the wadi erosion.

Sites Naba 5 and 6 are located in an area that had been much exposed to internal folding and fault tectonics. Both are located within residual andesite layers whose relics overlie the acid series.

In compliance with the multiple tectonic stresses in the area, the andesite rocks reveal different lineation patterns. At Naba 5 one is oriented 20–10° W, whereas another 70–90° W, which naturally had led to complex, internal micro-fold structures.

A gold quartz vein unaffected by tectonic deformation was however mined in a shallow, 2 m deep trench. Following the shear direction it dips perpendicularly at a 10° W strike in an andesite wallrock altered almost to mylonite. The approximately 20 cm wide ore zone itself contains only few centimetres of mineralised pyrite in the centre. The wallrock displays a significant pyrite alteration zone of 20 cm on both sides of the vein.

Wadiworkings grouping around an unexploited a quartz vein (50° W) in an andesite hostrock had been the only operations in the catchment area of Wadi Naba 6.

6.6.28 Mine Nabi 1

Geographic position:	21°06′35″ N, 33°35′58″ E

The partly well-preserved house ruins in this widely stretched settlement are for the most sanded-in (Fig. 6.189). As neither processing

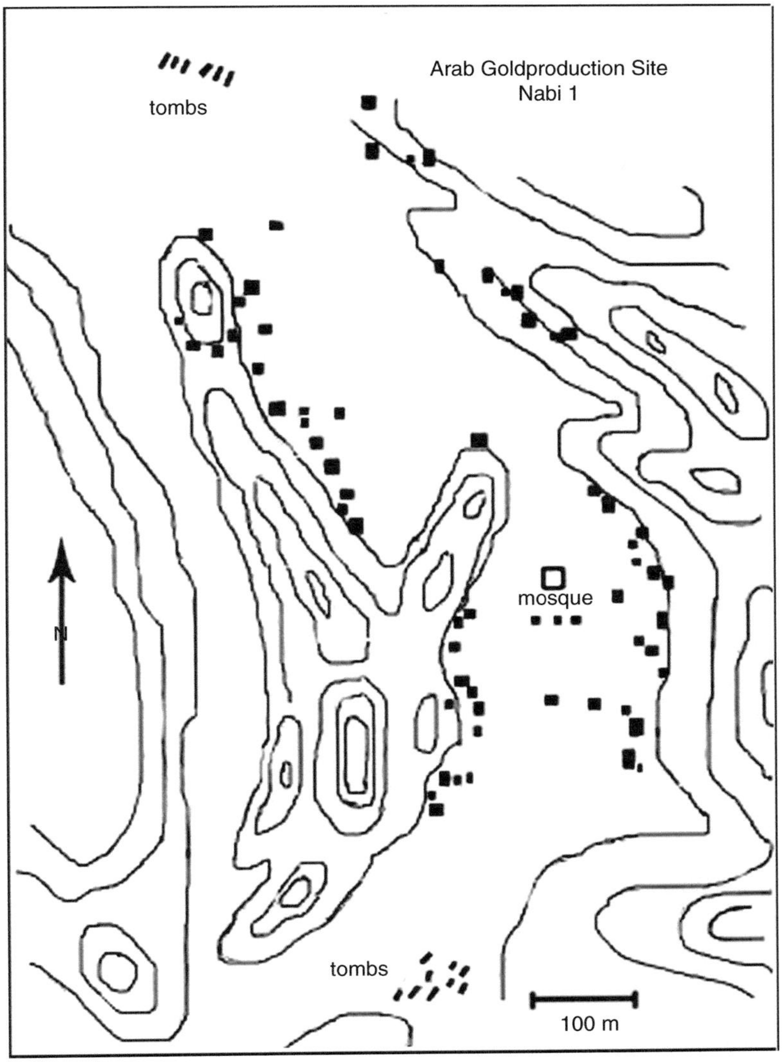

Fig. 6.189 Archaeological sketch map of Nabi 1, based on satellite image (R. Klemm)

tools nor tailings were found here, we assume that ore processing had been operated at some other location. Llewellyn (1903) reported a relatively poor outcome after a hole-drilling campaign to establish several wells in the area. In that context, he also showed a picture of a mine (Llewellyn 1903), which we unfortunately weren't able to find. It is therefore conceivable that the ore processing had actually taken place at this latter site.

Many of the round to angular houses have two rooms, but most have only one. The settlement's most important structure is a comparatively large mosque measuring 8 × 5 m and built in the shell-facing technique with a mihrab pointing directly E. It was accessed through three entrances with respectively one the S, N, and W (Fig. 6.190). Otherwise, all buildings are filled-in with sand and are built of relatively small stones, even though the walls are still preserved to heights reaching 1.5 m. To the S is a cemetery with nine tombs oriented in NNE-SSW. They consist of oval sand heaps that had been covered with stones. Other six tombs are located in the NW.

Next to the houses, there are occasionally small individual prayer sites. As painted pottery

Fig. 6.190 Well-preserved mosque at the Early Arab Period settlement Nabi 1

is completely absent, it is conceivable that the site was entirely populated by Muslims. The vastness of the system indicates that probably anyone could build a dwelling as he pleased. This may represent an important criterion for the late period following Al-Omari.

Because no mine was found in spite of our efforts, we must assume that the site had specialised on processing ores from wadiworkings.

The overall finding remains nonetheless highly unsatisfactory. Llewellyn (1903, Plate XIX) explicitly draws attention to a wide, open mine, which from its appearance may well date to the New Kingdom. We can't think of any explanation why we were unable to find this yet distinct landmark, especially bearing in mind that Llewellyn's picture of the site itself is displayed in the same plate and matches exactly with our observations in the field.

Geologically, the site is located directly S of a vertical fault slip, by which the present rock massif had sunk in relation to the broader northern massif.

In terms of petrography, the units near the site consist of remarkable agglomerates with up to 10 cm large components that overlie fine-grained tuffs and lapilli tuffs with foliation planes at 60° E. Thereby this unit forms a significant tectonic incongruity to the rest of the surrounding rocks.

6.6.29 Nabi 2

Geographic position:	21°07′18″ N, 33°35′47″ E

Beyond Wadi Qar, at the southern inlet of Wadi Naba and into a plain with low hills is a typical settlement from the New Kingdom. However, it is hardly recognisable as such, because all the structures have badly suffered. Numerous New Kingdom stone mills are located widely scattered in the alluvial sediment sands. The entire New Kingdom tool assemblage is represented here, including the flat mortar slabs and their cylindrical pestles. Because of their protected situation, some hillside hut ruins have remained in a relatively

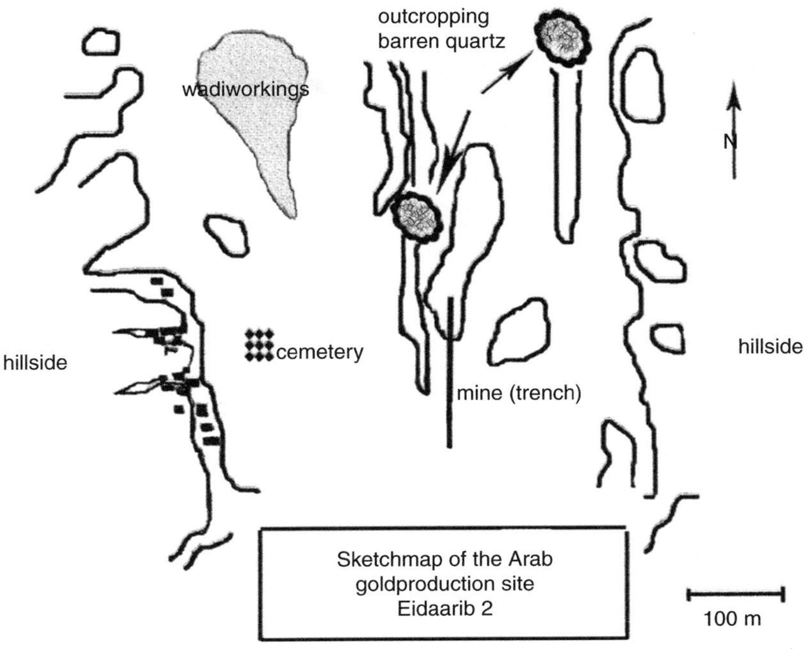

Fig. 6.191 Area sketch map around the Early Arab Period site of Eidaarib 1, based on satellite image (R. Klemm)

good state of preservation. They are built from purple-grey tuff rocks. Wadiworkings had apparently been carried out as well, of which however, wadi floods have wiped out all visible traces.

6.6.30 Eidaarib

6.6.30.1 Eidaarib 1

Geographic position: 21°03′16″ N, 33°34′34″ E

Twelve to 15 hut ruins lie on a flat slope close to wadiworkings. One of them is built entirely of quartz chunks, and is leaning on a quartz outcrop that had furnished the building material. To the W of the settlement is an associated trench with traces from small-scale mining activities.

6.6.30.2 Eidaarib 2

Geographic position: 21° 02′54″N, 33° 34′23″E

According to its description and coordinates, this Early Arabic mining site is evidently not the one Llewellyn (1903) referred to as Idarib. This is why we opted to call it Eidaarib, after the nearby mountain of Gebel Eidaarib. It composes of about eight, large house complexes to the W of a long, silted-up trench that once must have been quite deep, considering the size of the debris heaps next to it (Fig. 6.191). It is located in the low hills to the W of Gebel Eidaarib, one of the dominant mountains in this granite area. Numerous, barren quartz veins that accordingly have remained unexploited are found here. On the other hand, there are distinct traces from wadiworkings. At the N side of the site is a separate house complex in a small ravine with a dam at its upper end. Two more dams are located near the southern sector of the site, where there is another small number of house ruins. A nearby burial ground contains about 15 N-S oriented tombs.

Numerous, deeply worn-out round mills group along together with characteristic small, cube-shaped pounding stones near two washing tables. Broken round mills and pounding stones had also been partly integrated into the house walls, which may suggest a second occupation phase at the beginning of the twentieth century. An iron scraper used for brushing up crushed

Fig. 6.192 Scraper (or hoe) from iron used in the wadiworkings at Eidaarib in the early twentieth century

quartz probably testifies to this late occupation phase (Fig. 6.192).

Eidaarib 1 and Eidaarib 2 are located W of Gebel Eidaarib, in the low hills of an approximately 5 km long and 2 km wide basement elevation at the northern border zone of an extremely dry, sandy desert stretching far to the S and interrupted by only few, short mountainous outcrops.

From a perspective of tectonics, the contingent mountain range at Eidaarib represents the southwestern, intensely sheared foothills of the Naba-anticline (Fig. 6.193).

This zone is essentially composed of hornfelsic dacite andesite, foliated andesite porphyry, and occasional amphibolite (former basalt?). By their proximity to the granite floor of Gebel Eidaarib, they are significantly affected by contact metamorphism. A slight foliation can be discerned near both deposits. The granite floor itself exhibits concurrent foliation with a dominant fracturing.

The area is traversed by numerous quartz vein swarms striking at 10° E and with a chiefly vertical dip. Yet only few had been mined in trenches and pits. Most are filled-in by sand drifts and therefore not available for observation. Only in the so-labeled Trench 1 an approximately 1–1.5 m thick quartz vein displaying limonitic staining was identified.

6.6.31 Wadi Idarib

Geographic position:	21°06′30″ N, 33°25′06″ E

Three hills containing the severely wind-blown remains of a settlement rise above the middle of a flat desert at this position (Fig. 6.194). The round huts, which for the most had been built leaning on rock outcrops on the slopes, have generally one room with the entrance facing the plain. The building material is the local, dark andesite. Altogether there are about 60–80, widely scattered houses in an arrangement suitable for carrying out wadiworkings, as attested to by the evidence in the surrounding plain. There are also some excavated trenches, of which the longest one in the plain measures about 200 m (Fig. 6.195). Round mills, anvil stones and heavily sanded-in washing tables are numerous. The site seems to be relatively unaffected by later disturbances.

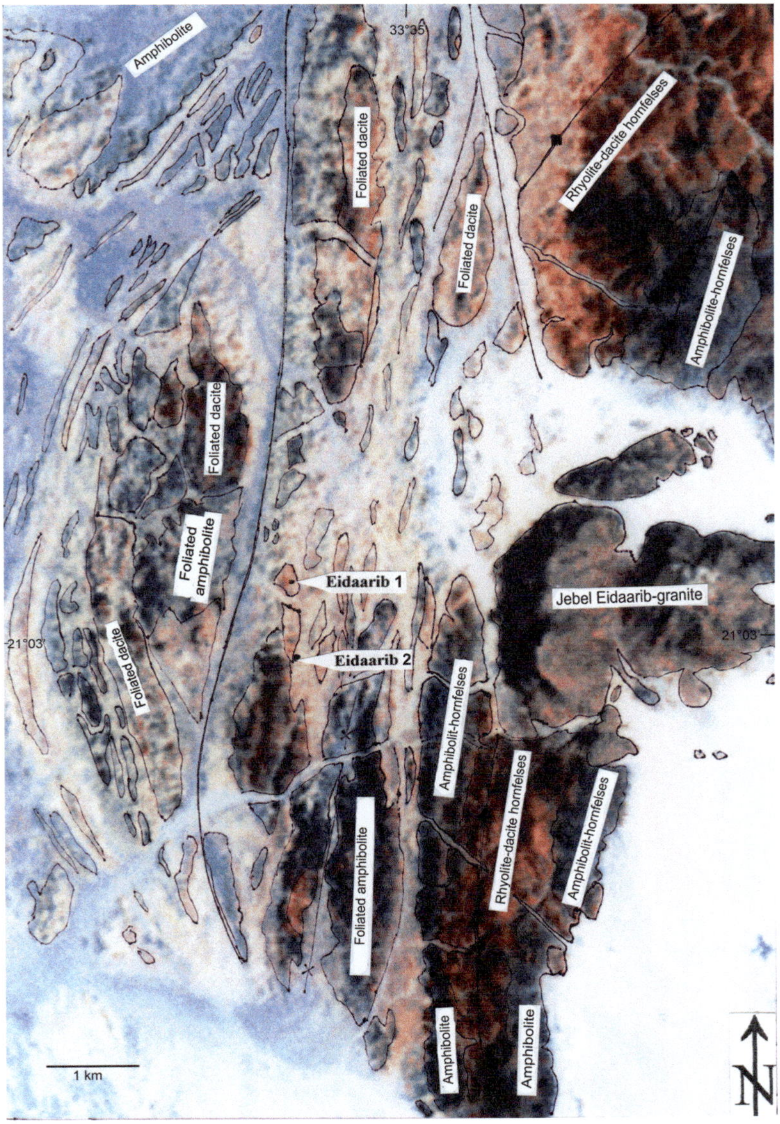

Fig. 6.193 Lithologically processed satellite image of the Eidaarib area (TM 173/45, channels 7-4-1)

According to Llewellyn (1903), a tomb stone had the year 307 Muharram inscribed on it, which would correspond to 918 AD.

Because of the poor observation conditions in this sand-covered desert, the description of the site's geologic setting proves to be somewhat difficult.

The irregular distribution of low hills and small mountain ridges nonetheless reveal that the area predominantly consists of andesite volcanics with tuffs and intermediate layers of unfortunately very poorly exposed metasediments. These had apparently been locally intruded by apophyses from an undeterminable granitoid intrusion

Fig. 6.194 Sand-blown Early Arab Period gold mining site at Idarib, displaying the usual mill assemblage

and sporadically occur as medium-grained granite outcrops.

The volcano-sedimentary series display clear foliations in an orientation between 10° and 30° W and are mostly intensively folded in these directions. The resulting rock is naturally prone to erosion and consistently leveled-out to virtually one surface.

We were all the same able to find a number of mostly blown-in trenches. One was sufficiently enough cleared to reveal its strike at 10° E. It was also open over a distance of about 200 m. The exploited quartz is milky-opaque and exhibits pseudomorphic limonite-pyrite which continues into the wallrock. Llewellyn (1903) was able to identify a modest gold grade of about 1.5 g/t.

Fig. 6.195 Widely dispersed Early Arab Period site at Wadi Idarib and its 200 m-long trench pit (*arrows*). Gebel Kerat Berar is seen in the background

6.7 Group: Mosei – Omar Kabash

6.7.1 Mosei

Geographic position:	21°10′05″ N, 33°07′08″ E

This site, where about 100 huts attest to significant mining activities in the past, was already mentioned by Llewellyn (1903). Even though there are numerous, short quartz veins in the area, gold seems to have been extracted in wadiworkings only. Because of serious water shortages we had to cut short our excursion before visiting the site. Castiglioni and Vercoutter (1998) describe Mosei as consisting of numerous buildings with circular plans and dry stone masonry spread over a wide area. They further noted many oval grinding mills. Undoubtedly, this was a gold mining site from the New Kingdom Period.

6.7.2 Nabi (?) (B5 in Castiglioni and Vercoutter 1998)

Geographic position:	21°07′32″ N, 33°08′07″ E

We were unable to visit the site for the same reasons as given above. It should however be kept in mind that Wadi Nabi with its numerous gold processing sites is only some 46 km E from the position given here and that there is therefore some reason at least to doubt the name of this site. However, the information given by Castiglioni and Vercoutter (1998) has nonetheless to be taken for true. They report buildings from dry stone walls on circular plans scattered over the surrounding hills, and shallow ore mining traces from trenches. In a side wadi a barrage from dry stone masonry probably served to regulate the runoff.

From these indications one may assume that this was an Early Arab Period mining settlement with a simple water supply system.

Fig. 6.196 Ruin of a miner's house at Umm Nabardi from the early twentieth century. The mills integrated to the wall masonry have partly fallen out again

6.7.3 Fort Murrat

Geographic position:	21°04′40″ N, 32°55′45″ E

The Ottoman fort Murrat lies on the northern slope of a sharp bend in Wadi Murrat. According to an inscription on a stela (Castiglioni and Vercoutter 1998), this relatively well-preserved complex was the garrison of an Egyptian-Sudanese regiment and was built in 1897, shortly after the abolition of the Mahdist uprising.

Early Arab pottery, a nearby Islamic cemetery, as well as some hieroglyphic rock inscriptions located directly next to the now silted-up well, however, point to much earlier occupation phases at the site.

6.7.4 Umm Nabardi

Geographic positions:	
Main shaft:	21°07′09″ N, 32°46′11″ E
New Kingdom settlement:	21°07′28″ N, 32°46′17″ E
Modern settlement:	21°06′42″ N, 32°46′53″ E

At this site, though it today is contaminated with recent detritus, one still recognises the no less than 250, unusually large, New Kingdom "elliptical" mills reported by Llewellyn (1903) to lie between discernible heaps from wadiworkings in a depression between the so-called Red Hill and the Main Hill. Some of the mills had also found their way into the stonework of later houses and subsequently fallen out again with the disintegration of the clay mortar and the houses' collapse (Fig. 6.196).

To the E of the site are a number of Early Arab Period houses, and in the NW we ended up after a long search in finding the New Kingdom settlement already alluded to by the typical mills (Fig. 6.197). The settlement is surrounded by wadiworkings, through which a wide track has been traced in a recent mining operation. At least three washing tables are distinguishable. Their inclined surfaces are oriented parallel to the slope, which is usual for such devices in Northern Sudan. The large mills are partly made from very fine porphyry andesite and scatter in large numbers within the barely discernible house ruins (Fig. 6.198). The mill in figure (Fig. 6.198) has a secondary depression at its grinding surface,

Fig. 6.197 Severely damaged New Kingdom settlement at Umm Nabardi, revealing a washing table in the left foreground

Fig. 6.198 New Kingdom mill from (andesite-) porphyry at Umm Nabardi with a later added use wear depression

Fig. 6.199 Round and rectangular house ruins at the mining settlement of Umm Nabardi from the first half of the twentieth century

which is a possible indication for its functional reclaim in a later occupation phase, most probably the Kushite Period.

One of the houses yielded a diagnostic shard dating to the New Kingdom. It belongs to a small pot with a light brown, linear decoration as represented in Holthoer (1977, pl. 53).

A large and fairly ruinous, recent mining estate consists of rectangular clay huts with gravel-tempered walls, as well as a loose arrangement of round stone constructions of schistose rock (Fig. 6.199).

Our studies in Umm Nabardi were restricted to a very short time, due to critical water shortages. We therefore draw on Boswell's report (1984b) for the description and history of the site.

Thereby, modern mining industry had begun in 1901 with an award handed out to the Egypt and Sudan Mining Syndicate for exclusive prospecting in an area measuring 56,980 km^2 between 20–22° N latitude and 32–34°30′ E longitude. With an additional lease granted in 1902, three shafts plus several ventilation shafts were lowered into the ground below Umm Nabardi (West, Central, and East Shaft). The first level consisted of a drift of 600 m, and in the beginning gold grades between 90 and 120 g/t were reported at several occasions. Although comparable values were never again attained before the mine's shutdown in 1919, a total of 3,474 kg of gold at an average grade of 17 g/t were retrieved during the mining operations. Moreover, another 306 kg were recovered from old tailings.

Due to failing supplies in World War I, a layoff of 300 miners in an influenza epidemic in 1918 and expiring reserves as a result of a war-related interruptions in the prospecting work, the mine was closed on August 24th 1919 and its entire hardware sold, including the tracks of a short subsidiary railway line, leading from station no. 6 of the railway Wadi Halfa-Abu Hamed to the mine.

The Umm Nabardi deposit is located S of the Gebel Rafit massif, which forms a basement island in the middle of a vast and extremely dry, sandy desert. The few former wells at Muftah (22.5 km SW), Jerrifat (8.5 km N), Tonaiba (2 km SW),

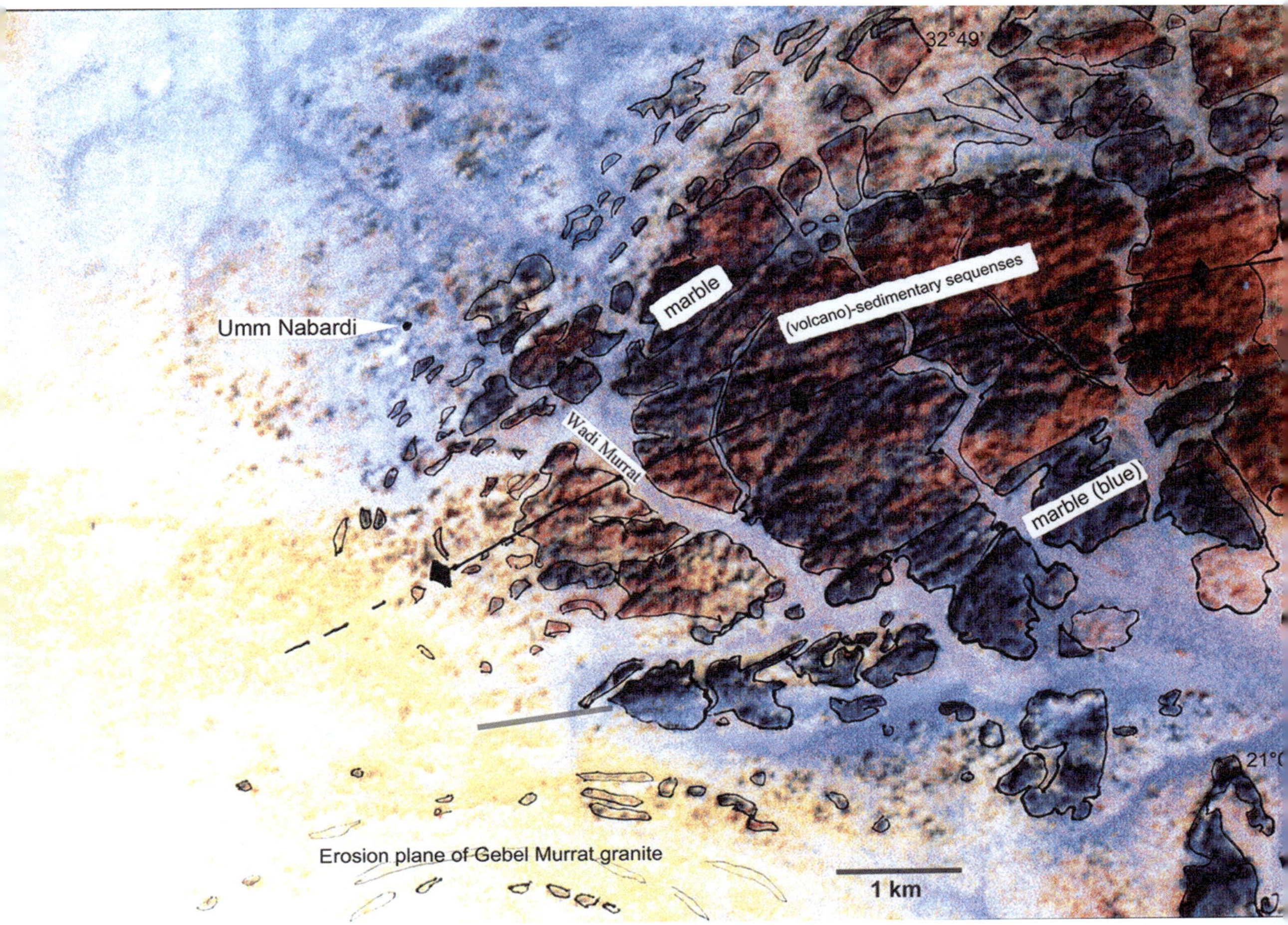

Fig. 6.200 Lithologically processed satellite image of the Umm Nabardi area (TM 174/45, channels 7-4-1)

and Murrat (16 km SE) are today either all sanded-in or dried-up.

Umm Nabardi itself is located at the SW edge of this island and is much covered by sand.

In terms of geology, Umm Nabardi is at the NW edge of a 25 km long and curved saddle oriented E-W. It is predominantly composed of sedimentary sequences with significant portions of acid tuffs and some contact-metamorphic rhyolites. A vast granite intrusion dome measuring about 8 km in diameter has emplaced in its centre, whose large, almost circular erosion surface is exposed in several hills. In its several kilometre-wide contact zone (volcano)-sedimentary series have transformed to hornfelsic rocks, and especially former limestone- and dolomite layers have recrystallised to coarsely-crystalline marbles (Fig. 6.200).

The immediate geologic context of the auriferous quartz mineralisation at Umm Nabardi is in essence represented by tightly folded, NE-SW striking, volcano-sedimentary rock series. In the SE and their central parts just E of the mine they are intruded by a porphyry (quartz) microdiorite.

The individual courses of this series are not very regular, as they often disintegrate causing difficulties from poor exposure in plotting the area in a detailed map. Furthermore, intense shearing and pronounced green schist metamorphism (epidotising, chloritising and partial structural re-crystallisation) had a negative effect on the determination of the original rocks.

Some medium- to coarse-grained, often limonitic (after pyrite) sandstones and metaarkoses, as well as small, intermediary layers of well-stratified pelites and dark-grey dolomite marbles were nonetheless discerned with a relative degree of confidence. Alternating sequences between chlorite-sericite-pelite rocks and pyrite-yielding ash tuffs however proved to be problematical. Additional layers of rhyolite-dacite occasionally transpire in an ignimbritic appearance.

Some of these sequences, which generally compose of former acid tuffs and rhyolites, are quite exceptional, due to intense stains by limonite. One is in fact drawn to think of a seafloor, exhalative environment, which may have led to the formation of SEDEX deposits. The so-called Red Hill in the Umm Nabardi district with its productive vein mineralisations in fact constitutes a striking example for this mineralisation type.

The porphyritic quartz diorites are also secondarily chloritised (Boswell 1984b), and their structure is poorly preserved. This should be viewed as a hydrothermal reaction rather than a consequence of regional metamorphism. At any rate, they were the most likely suppliers of the thermal energy behind the formation of the hydrothermal deposit.

The vein mineralisations at Umm Nabardi divide in a main reef system in the N and another one in the S. The aforementioned Red Hill Reef actually forms the northern one. All strike more or less E-W and dip between 80° and 55° S.

Boswell (1984a) furnishes a detailed description of the various individual veins within the three mother lodes. The ore paragenesis here consists of pyrite, arsenopyrite, to a lesser extent of chalcopyrite and possibly other copper sulphides, and gold.

Noteworthy as well, is the observation that the Red Hill apparently conceals an anticlinal structure into which the vein mineralisation had been folded, as it plunges both to the NW and the SE, and because the mineralisation appears in a boudinaged pinch-and-swell structure. In the S part of the folded lode it bends upwards similar to a syncline, thus supposedly forming an "ore basin" with exceptionally rich gold ores.

The main reef system had been extracted over 800 m and exhibits a series of ancient workings, especially along its surface. Such trenches were also observed at the Red Hill. In both cases, ancient underground mining dates exclusively to the New Kingdom Period, as attested to by countless oval millstones and other gold processing tools found outside the mine. On the other hand, there was no evidence dating to the Arab Period. Unfortunately, the associated processing sites have been devastated to such an extent by the early 20th century mining industry that an adequate reconstruction of their original arrangement seems no longer possible. The only hope would be to carry out salvage excavations at the site.

As for the southern reef, mining does not seem to have taken place there in ancient times.

We were not able to visit Boswell's (1984a) briefly described 'northwestern' and 'southeastern workings' where more, well-preserved settlement and extraction traces are supposed to be found.

Because of our water supply problems we weren't able to inspect the sites at Jebel Mundera, Umm Fit Fit, and Rod el-Ushal. The following descriptions therefore largely foot on the information given by Lewellyn (1903), Boswell (1984b) and Castiglioni and Vercoutter (1998).

6.7.5 Jebel Mundera

Geographic position:	20°54′10″ N, 32°48′17″ E

According to Llewellyn (1903), the occurrence is located about 5 km SW of Gebel Mundera, a protruding granite stock, which is already seen from a distance in the plain. The indications given by Llewellyn clearly suggest occupation phases in the New Kingdom and Early Arab Period.

The area is predominantly composed of a sequence of basaltic and andesite lava and some lapilli tuff folded in a NE orientation. Especially the lapilli tuff contains pyrite. The entire sequence is intruded in the NE by the "younger" granite of Gebel Mundera, which also penetrates the older granodiorite of the basement.

Ancient workings concentrate on four quartz veins between 0.2 and 2 m wide and up to 120 m long, generally striking NE with an 80° dip SE. They had apparently only been mined in shallow trenches and are today filled-in with sand.

6.7.6 Umm Fit Fit

Geographic position: 20°46′15″ N, 32°24°22″ E

For this site, Castiglioni and Vercoutter (1998) noted dozens of round and rectangular, dry-walled buildings spreading-out amid low, undulating hills. Numerous oval millstones are made from diorite originating from a quarry located on the slope of a nearby hill. Some millstones have a deep cavity in their middle.

According to this and the photo published by Castiglioni and Vercoutter (1998, 110), Umm Fit Fit is safely dated to the New Kingdom. The cavities observed in the mills could indicate a secondary reuse in the Kushite Period similar to the finding at Umm Nabardi.

According to Boswell (1984a), the geologic environment at Fit Fit is composed of an ophiolite mélange of predominantly basaltic and sedimentary sequences. The latter consist of sandy schist, chlorite and talc schist, and dolomitic, often brecciated marbles. Especially the chlorite schist often contains pyrite (pseudomorphic) and magnetite.

Even though there are marked traces from wadiworkings, at least two exploited quartz veins with collapsed shafts were identified with respective depths of 11 and 6 m.

6.7.7 Rod El-Ushal

Geographic position: 20° 41′34″N, 32° 37′44″E

Castiglioni and Vercoutter (1998) report numerous round buildings of dry stone masonry spread over a wide area and almost completely covered by sand in a zone of low, undulating hills.

They did not make any finds specific to the site's function or date.

6.7.8 Abu Siha

Geographic position: 20°36′47″ N, 33°01′47″ E

Llewellyn (1903) reports the discovery of "elliptical mills" at this site. At the indicated position in a wadi covered with sand and gravel we found several, mostly broken New Kingdom stone mills, each with a later added groove from secondary reuse, thus drawing a link to the Kushite mill type (Fig. 6.201). Furthermore, there was a fragment of a round rotor stone and a relatively large, grooved stone axe (Fig. 6.202). Abu Siha is therefore particularly interesting, because the highly weathered axe type is otherwise only known from Old/Middle Kingdom contexts. It is nevertheless possible that such tools had continued to remain in use in Nubia for some time. On the other hand, they have so far been limited to sites in the immediate vicinity of the Nile Valley such as at Duweishat, Abu Sari and Sokar. Abu Siha is about 260 km in a straight line from Abu Sari and thereby the most distant site from the Nile known to yield mining tools from the Old/Middle Kingdom. The New Kingdom occupation phase is attested to by the large number of mills that had apparently been functionally reclaimed in a later (Kushite) period. A final occupation had occurred in the Early Arab Period.

The wadi bed has undergone severe remodeling through water erosion, although it is interesting to note that a track runs through the valley referred to by Llewellyn as a "camel track".

Fig. 6.201 New Kingdom mills from Abu Siha partly displaying secondary use wear, probably from the Kushite Period

Fig. 6.202 Much weathered groove mallet from Abu Siha indicative of an early processing period at this site

6.7.9 Omar Kabash 1–19 (Fig. 6.203)

Geographic positions:		
1:	20°17′21″ N, 33°17′37″ E	(large settlement)
2:	20°16′13″ N, 33°18′22″ E	(small group of 10–30 houses)
3:	20°15′36″ N, 33°18′13″ E	"
4:	20°15′24″ N, 33°18′38″ E	"
5:	20°14′14″ N, 33°18′31″ E	"
6:	20°13′45″ N, 33°18′56″ E	"
7:	20°12′07″ N, 33°20′30″ E	"
8:	20°13′39″ N, 33°21′52″ E	(about 70 houses and a cemetery)
9:	20°15′52″ N, 33°20′36″ E	(20–30 houses from blue marble)
9a:	20°16′14″ N, 33°20′16″ E	"
10:	20°18′31″ N, 33°19′33″ E	(small house groups)
11:	20°18′33″ N, 33°20′12″ E	"
12:	20°18′25″ N, 33°20′30″ E	(large house complex near colluvium workings)
13:	20°18′06″ N, 33°20′31″ E	(houses and prayer site)
14:	20°18′02″ N, 33°20′42″ E	(small house groups)
15:	20°17′50″ N. 33°20′56″ E	"
16:	20°18′02″ N, 33°21′07″ E	"
17:	20°17′38″ N, 33°20′30″ E	"
18:	20°15′45″ N, 33°17′14″ E	"
19:	20°14′52″ N, 33°16′09″ E	houses and tombs

6.7.9.1 Omar Kabash 1

Omar Kabash 1 consists of a regularly laid-out settlement to the N of an extensive Early Arab gold mining district. It spreads out both sides of a central street (Fig. 6.204), and is much reminiscent of the sites at Uar, Derahib, and Shashuteb (Bir Kiaw). It therefore probably also served as a supply depository for the surrounding region. Its central street is oriented E-W and is about 200 m long. The rectangular houses on both sides of the street are for the most covered by drift sand and measure about 3 × 4 m, have two rooms, and are oriented in a perpendicular position to the street. In between, there are also markedly narrower rooms, clearly hinting to their probable function as storage rooms. At the southern side of the western access to site is a mosque. Two more are located in the central parts and one towards the S, just outside the settlement.

The function of Omar Kabash 1 as a supply station is primarily supported by the wide central street, which in fact could be accessed with camels to allow for the direct transfer of goods to the depositories and the storage jars. Next to the facility there is a cemetery containing between 10 and 12 graves.

6.7.9.2 Omar Kabash 2–6

In a wider area, mainly to the S of the northern settlement, one comes across small house clusters associated to wadiworkings. They generally date to the Early Arab Period and are located near hilly and mountainous outcrops of steeply dipping, bluish-white marbles in this otherwise flat and quartz-covered landscape. Like Omar Kabash 1, all sites are covered by sand to such an extent that often only the topmost wall layers stick-out of the surface. As far as the eye reaches, the surface is marked by traces from wadiworkings. Repeatedly one comes across houses and house clusters associated to round work platforms, which have partly been included to the map in Fig. 6.203. Due to the substantial sand cover, we weren't able to establish whether gold had not also been extracted in trenches (Fig. 6.205).

The potshards usually consist of light olive-green, slightly mottled, and well-fired amphorae.

Each of these settlements composes of approximately 20–30, relatively large houses built mostly from local schistose rocks. They are generally rectangular and have several rooms; some had been extended with round annexes.

The site Omar Kabash 7 seems to be contemporary with the ones at Omar Kabash 1–6. It again consists of 20–30, more or less well-pre-

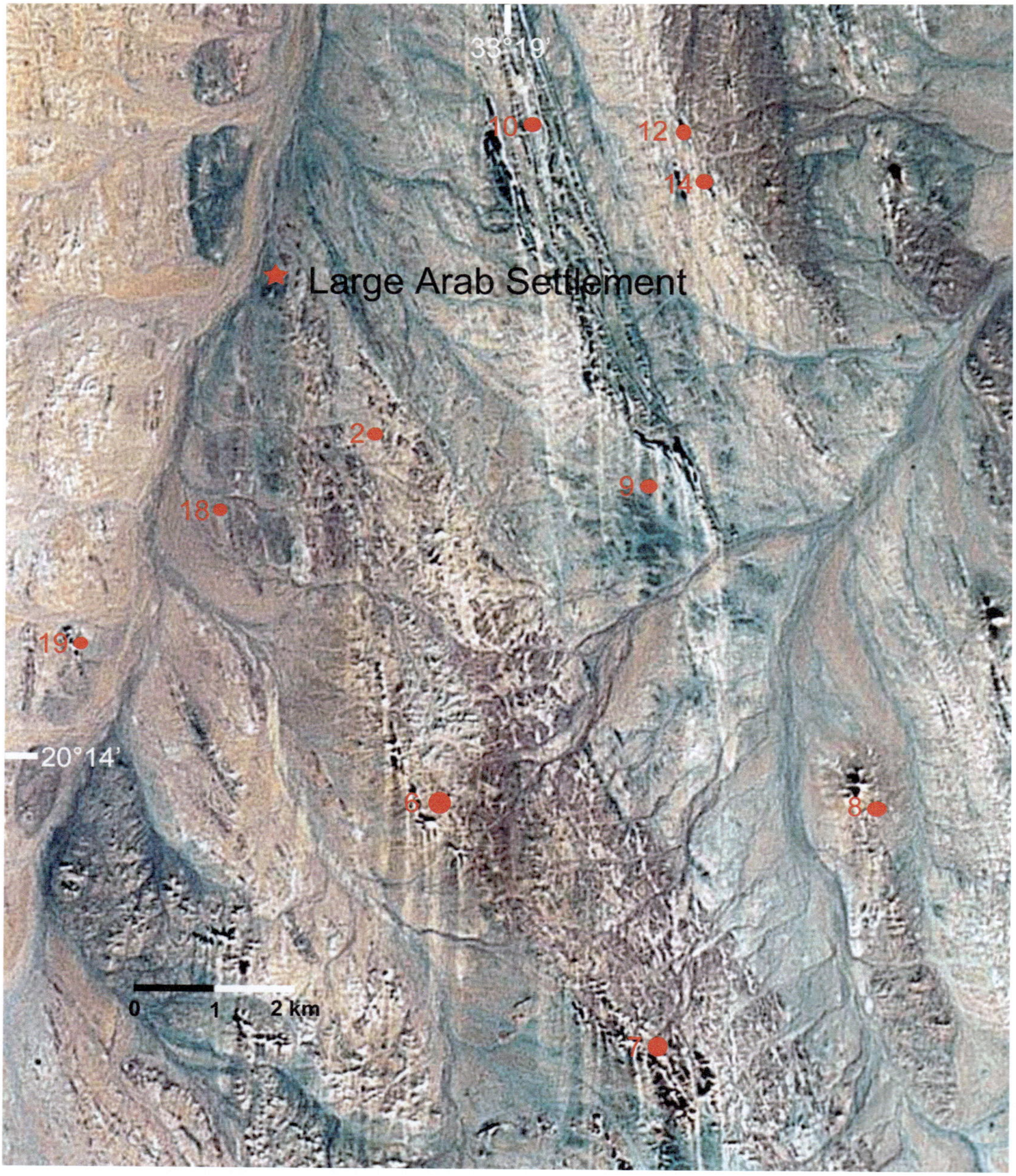

Fig. 6.203 Sand-covered area at Omar Kabash giving the positions of selected gold production sites. The black strips oriented mainly NW-SE represent marble outcrops (modified Google-Earth image)

Fig. 6.204 Partial view over the large Early Arab Period building complex at Omar Kabash and its central alley

Fig. 6.205 Heavy sand drifts in the Omar Kabash plain. Areas marked by wadiworkings are recognised in the background

Fig. 6.206 Sand-blown ruins in the plain of Omar Kabash (Omar Kabash 18)

served, mainly rectangular houses with several rooms. Several work platform groups more or less of the same size distribute in the open terrain where quartz chunks had been collected for further processing. The fact that very little tools are found may be an indication for the short-lived nature of the working sites.

Omar Kabash 8 strongly contrasts with the previous ones in that it constitutes an autonomous site of 50–70 houses and a cemetery with 21 graves.

All houses have more than one room, and some have round mills built into their walls. Therefore, at least two occupation phases can be surmised. We even discovered a stock of unfinished or previously unused mills. In addition, we found dotted pottery shards hitherto unknown to us. There were also shards from brown glass. These differences therefore allocate the site to a settlement group other than that of Omar Kabash 1–7.

Omar Kabash 9 is a site consisting of 20–30 houses built from the local white-blue marble stones that contrast in a spectacular manner to the yellow sand dominating the open landscape in the background. Yet again, pottery and tools are very sparse, but just about everywhere one comes across surface workings.

Omar Kabash 9a is located approximately 1 km N of Omar Kabash 9 along the same blue marble streak. It again forms a cluster of about 20–30 house ruins surrounded by surface workings. We found a collection of shell fragments from ostrich eggs, and our Sudanese staff member, Khazim Mustafa, claims to have seen ostriches in the area in the 1980s. So far, eggshell fragments have only been observed near washing tables at Arab Period sites.

Omar Kabash 10–11 consist of small house clusters along the marble outcrop.

Omar Kabash 12 is a somewhat larger settlement site and is aligned along the wadi opposite a number of hills displaying traces from workings.

Fig. 6.207 Steeply dipping, blue marble layers cropping out at the surface. The extent of weathering and shearing of the volcano-sedimentary rocks between them is so advanced that they no longer are safely determined in the field

Omar Kabash 13 is located further S, on the opposite side and has a distinctive space reserved for prayers.

Omar Kabash 14 is on the E side of the plain and consists of about 15 houses built around a small hill.

Omar Kabash 15 forms a compact cluster of house ruins.

Omar Kabash 16 consists of about 11 houses.

Omar Kabash 17 contains six large houses in the wadi bed and eight to ten more that disperse within a wider radius.

Omar Kabash 18 is a site counting between 15 and 20 houses, of which one is equipped with imposing pillars at its entrance. The ruins measure about 4 × 5 m, have two rooms, and are extended by rectangular, 3 × 4 m large annexes (Fig. 6.206). Again, no tools were recorded, apart from quartz globes that may have served for pounding gold ores.

Omar Kabash 19 consists of three to four, small huts. Graves and unidentifiable, loose rock arrangements scatter over four prominent hills in the surroundings.

Visible geologic features are primarily dominated by tightly folded and steep marble layers of a blue-white colour. In general, they are severely fractured at the surface and otherwise arranged in solid NNW-SSE oriented hill groups (Fig. 6.207). The metavolcanic and metasediment series in-between are also steeply folded and sheared to form spiky and schistose structures. Their petrographic composition is typical for the Western Nefirdeib series and comprises marble-rich rhyolites, acid tuffs, prasinites and fine clastic sediments. These intermediary layers are noticeably less resistant to weathering than the marble and therefore appear in the area in the form of few low hills with severely decomposed outcrops. Due to the overall disintegration of their mineral structures, the original petrography of these rocks is no longer determinable with confidence, neither in the field nor under the microscope.

Fig. 6.208 The Sarras district. Light-coloured zones along the wadi courses represent wadiworkings and E-W oriented, partly mined quartz-veins. The flooded areas are in a dark brown colour (modified Google-Earth image)

6.8 Group: Nile Valley from Sarras to Abu Hashim

6.8.1 Sarras

Geographic position:	21°33′47″ N, 31°04′47″ E

Our search for Sarras ended near the waterline in the flooded wadis along Lake Nubia, where in a narrow gorge about ten rough and small, round huts had been erected on quartz outcrops. They undoubtedly belong to the periphery of the actual mining district, which has fallen victim to the dam. During our survey of the surrounding mountains we nonetheless discovered two trenches and several processed locations in the terrain.

As a result from the district's inaccessibility at the time of our visit, the findings were relatively meagre. The situation was on the other hand more propitious in March 2003 when a high resolution satellite image was available and at several positions individual mine pits were captured (Fig. 6.208).

The deposit area of Sarras is located at the western edge of an elongated granite intrusion in an amphibolite environment. It is highly affected by contact metamorphism, due to the nearness of the granite. This evidently auriferous zone is also replicated on the eastern side of the intrusion, although corresponding mineralisations need yet to be discovered there.

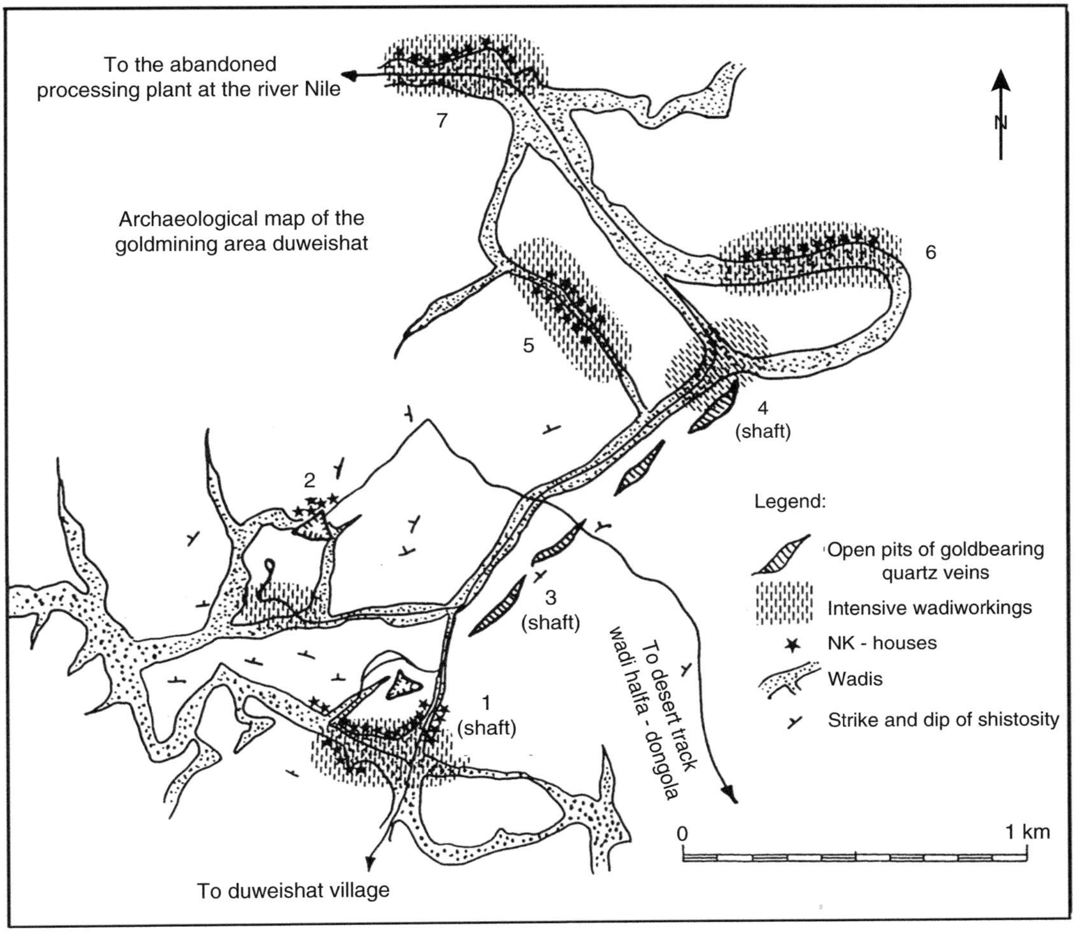

Fig. 6.209 Sketch map of the Duweishat area (modified after Grundmann 1989)

6.8.2 Duweishat

Geographic positions:	
Site 1:	21°21′46″ N, 30°58′29″ E (main mine)
Site 2:	21°22′04″ N, 30°58′21″ E (large Old Kingdom extraction area)
Site 3:	21°21′57″ N, 30°58′46″ E (modern shaft)
Site 4:	21°22′15″ N, 30°59′16″ E (small mines and wadiworkings)

Referring to Grundmann (1989) and Klemm and Klemm (2000), the sketch map of Duweishat (Fig. 6.209) shows the positions of the most important archaeological sites in the area.

The processing plant of the modern mine, which was shut-down by its Italian operators in the 1960s, was located directly at the Nile, where a large tailing was subsequently flooded by the rising waters of the Nile. Only a staggered line of buildings further up the slope remain apparent as impressive ruins. Behind them opens a vast plateau with a spectacular view over the Nile with former hill tops now peaking as islands out from the lake. On virtually all one notices ruins (Fig. 6.210). The Nile level is comparatively high at here and actually still rising according to observations made at Wadi Halfa, where newly established arable land has recently been submerged by the swelling water.

Fig. 6.210 Ruins of the early twentieth century processing site at Duweishat next to the Nile

A pottery registry grid that probably had been laid out during archaeological excavations on the plateau, was found near the relatively well-preserved ruins of recent miners' dwellings. It still included fragments from Kerma bell-beakers that had apparently been recovered from an excavated necropolis consisting of simple, rectangular graves that had been lowered into the schistose subsoil.

6.8.2.1 Site 1

This is a very impressive ancient mine (Fig. 6.211) consisting of several levels. Over 2 m tall abutments from solid, white quartz had been left standing inside its gently sloping hollow (Fig. 6.212).

The debris heap outside the mine contained archaic stone hammers as well as later New Kingdom grinding stones.

An inspection of the 13 houses belonging to the associated settlement from the New Kingdom (21°21′41″ N, 30°58′28″ E), revealed the following findings: Between house 8 and 9 (numbering sequence beginning in the N) are the probable remains of a washing table, as indicated by the accumulation of mills, anvil stones and some tailing sand. In house 9 at least three intact, plus numerous, probably recently shattered grinding stones were found.

Houses with three rooms are often associated with round work platforms and generally group in the S or E of the site. Although the Italian operator had left the premises in a relatively orderly state, grease jerrycans branded with the "Shell"-emblem had been left behind. Their lids had been transformed to gold pans and strainers after recent gold seekers had apparently chosen the lee of the walls as a location for gold panning in their spare time.

To the E of the house alignment several old paths lead up to the mentioned mine which is located behind a ridge.

The abundantly found pottery consists mainly of amphora fragments from the New Kingdom (cf. Holthoer 1977). The mills are made from various types of rocks, among which slabs from the local, bright-coloured dolomite marble.

Fig. 6.211 Ancient and modern mouth openings at shaft 1 at Duweishat as well as remaining pillars at the upper mouth of the removed quartz vein (*arrow*)

Fig. 6.212 Top row of mouth openings along the heavy quartz vein at Duweishat (shaft 1). Quartz pillars have been left standing to prevent the markedly sheared rock ceiling from caving in

Fig. 6.213 Remains of quartz layer removed in the Old/Middle Kingdom Period, just S of shaft 2. The surface of the recumbent wallrock is stained red through oxidation of its original pyrite to limonite. Mining continued in the New Kingdom, but by this time in deep mines, as visible by the diagonal row of adits in the background

Between the collapsed walls of house no. 10 at least ten New Kingdom mills and mill fragments were seen. They are all very large and equipped with a seat at the narrow end. At the SE side of the house a semi-circular wall may have served as a windscreen. Additional remains from architecture reveal that the site had been partly washed away in an erosion gully of the neighbouring valley, which runs from E to W. The site must therefore initially have been much larger.

6.8.2.2 Site 2

A series of adits in a mountain ravine, although enlarged in modern times, apparently date back to New Kingdom times (Fig. 6.213). Near the adits mouths are work platforms with scattered New Kingdom mills and associated grinding stones, as well as large amounts of amphora shards. On the gently rising, eastern slope just opposite the adits mouths one is struck by a large, red-stained surface, covered by bulky lumps of light-coloured quartz. The area continues to the top of the ridge, while revealing badly disturbed, New Kingdom work platforms. Essentially the same is also true for the terrain to the W, just above the adits mouths. Between the quartz chunks inside the red area one finds tools of partly considerable sizes (Fig. 6.214). They identify as spherical pounders and hammers, with rounded surfaces that measure 30 cm and weigh about 5 kg and are made from ultramafic rock and gneiss. Some of them have split through the middle, but still show clear impact marks. They evidently date to a period before the New Kingdom and might stand in some context with the mentioned graves containing the Kerma bell-beakers, which in one way or the other had stood in some connection with the mining activities here.

The red-coloured area in fact represents an alteration zone, which had once been covered by a large sheet of auriferous quartz. During the Middle Kingdom or possibly even earlier, this rather thick and mostly barren quartz vein had been removed and cleared away to the side, leaving behind a recumbent, thin, and auriferous quartz layer which had subsequently been crushed

Fig. 6.214 Old/Middle Kingdom rock pounders retrieved from the area of the removed quartz layer at Duweishat

and pounded on site with the heavy gear just mentioned. Grinding tools had therefore been unnecessary, which also accounts for their total absence within this area.

The large quartz chunks scattering over the surface merely represent the remnants of the removed, barren quartz layer and had therefore not been processed.

After the removal of this thin gold-quartz layer later New Kingdom miners had gone over to extract the continuation of the vein on the other side of the ravine, this time by driving galleries into the mountain. At the resumption of modern, state-supervised mining at this site this was fortunately taken into account and most of the ancient traces have thereby been spared.

The faintly recognisable remains of a washing table probably dating to the New Kingdom are located under the debris, at the ledge to the escarpment.

6.8.2.3 Site 3

This site is defined by several recent, open-cast trenches, although some also seem to be earlier, which is actually recognised by the tarnished colours of the debris heaps. In the vicinities there are also wadiworkings and faintly recognisable washing tables as well as house ruins and work platforms.

6.8.2.4 Site 4

Site 4 is similar to site 1 and site 2. Here as well, a gently dipping quartz vein had been mined in a series of underground mines. It too, had initially been dismantled as a flat lying sheet covering a red-coloured surface, before underground mines had been driven into the veins of deeper lying sections. The surrounding environment is again characterised by extensive wadiworkings.

6.8.2.5 Site 5

Site 5 is located approximately 600 m NW of site 4 (21°22′24″ N, 30°58′56″ E). No less than 12, well-preserved houses yield New Kingdom mining tools, among which some of the oval grinding mills reveal at least one secondary depression. As discussed earlier-on, this may indicate to a later exploitation of the deposit during the Kushite-Meroë Period. The houses are built with flat rock

Fig. 6.215 New Kingdom house and New Kingdom gold mill (*arrow*) with secondary use wear, probably dating to the Kushite Period at Duweishat, site 5

slabs, which partly accounts for their relatively good state of preservation. Yet again, they display semicircular annexes (Fig. 6.215). Further upstream there are more, severely damaged houses. They are associated to wadiworkings on a neighboring wadi terrace, where one also notices a trench.

6.8.2.6 Site 6

Site 6 (21°22′23″ N, 30°59′36″ E) and site 7 (21°22′46″ N, 30°58′48″ E) too, consist of New Kingdom house alignments parallel to the wadi. This time they are less well-preserved and exhibit only few tools.

In all, five settlements dating to the New Kingdom were counted in the Duweishat district. Site 1 is located in a broad, rubble-filled wadi just S of the central mine and was therefore probably associated to this mine. Most of the oval grinding mills were recorded at one location within the confines of the settlement. This can be explained by the proximity to the mine, but at the other sites too, mills are numerously represented and one occasionally also finds them within the masonry of recent buildings.

So far, there has been no evidence for any mining activity later than the Ancient Egyptian/Nubian Periods in the Duweishat district, at least until about 1904, the year when modern mining began at the site (Klemm and Klemm 2000).

Throughout the deposit area the rock sequence of Duweishat is dominated by 10 cm–1 m thick layers, consisting of reddish biotite gneiss with granitoid chemistry and occasionally yielding garnet, amphibolite, and garnet-staurolite gneiss. These layers are sporadically superseded by metasediments, arkoses and marbles, which are viewed as likely eroded remnants from the Nefirdeib nappe of the Gabgaba terrane. These series had partly been subjected to multiple folding, whose extremely complex structures yet have not been fully understood. We nevertheless managed to determine the general arrangement of the folds with a strike of the axial b-surface in a NE-SW orientation.

Apparently, these gneiss sequences no longer belong to the Meso- and Neoproterozoic aggrega-

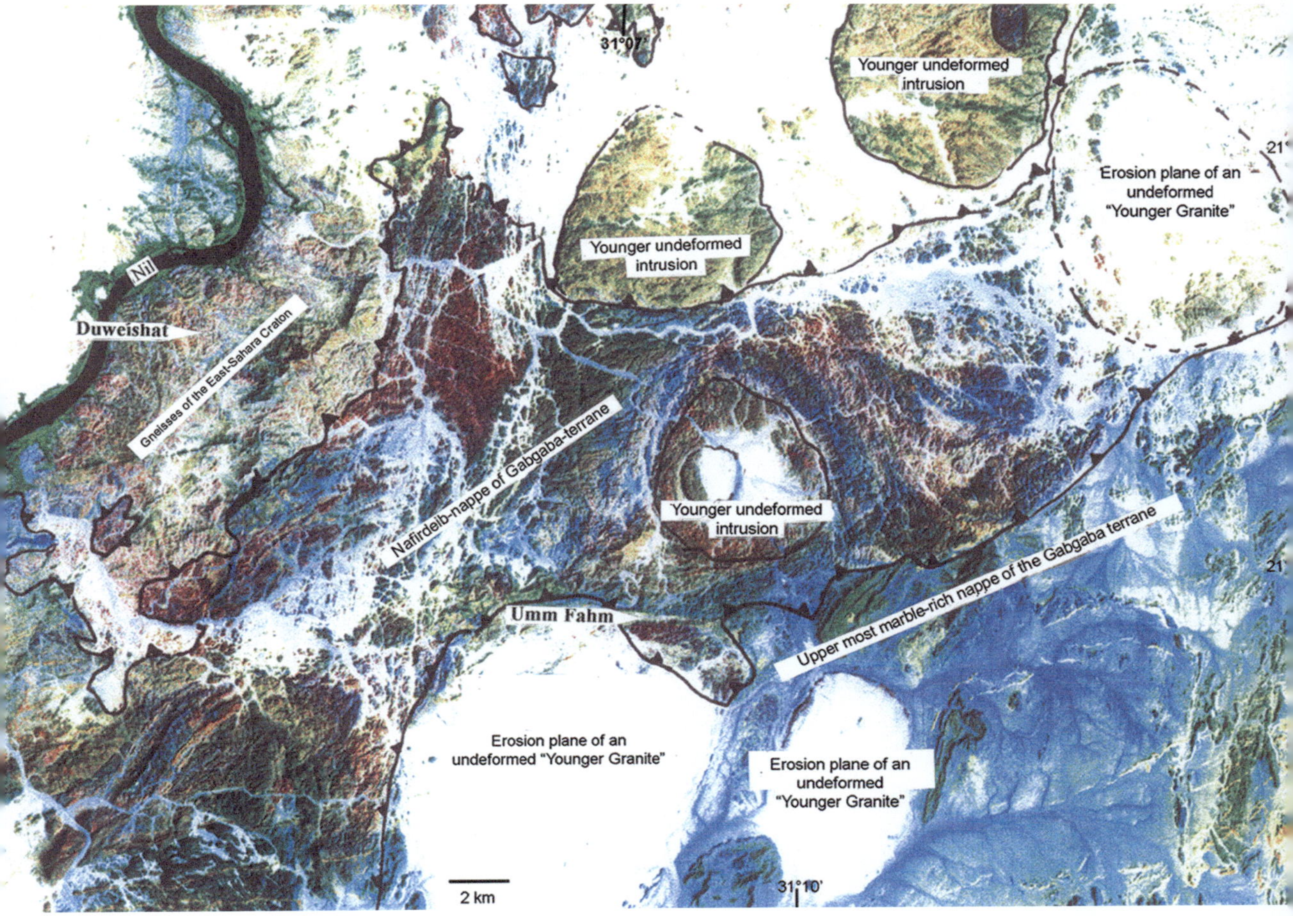

Fig. 6.216 Lithologically processed satellite image of the Duweishat area. It reveals the limit of overthrust by the Nefirdeib nappe above the gneiss of the East Sahara Craton as well as somewhat more to the E, the border of the upper nappe of the Gabgaba terrane (TM 175/45, channels 7-4-1)

tion terranes, that further E determine almost exclusively the geologic framework of NE Sudan. They more seem to belong to the bedrock of the East Sahara Craton, which at this location was overthrust by nappe units of the Gabgaba terrane.

The lithologically processed satellite image shows the outline of the geologic compounds within the wider surroundings of Duweishat (Fig. 6.216). It clearly displays the residues of the marbles overlying the gneisses and in the far W, the overthrusted remains of the Nefirdeib nappe.

The auriferous quartz vein mineralisation follows the general NE-SW strike of the deformation plane of the gneisses. It concentrates within two vein systems. The one to the NW had been mined at a maximum distance of 70 m during the Old/Middle Kingdom and particularly the New Kingdom. During the modern mining phase a shaft (no. 2) had been lowered into it for investigation purposes, though without prompting any meaningful exploitation.

The other main vein system to the SE had also been intensively mined during the New Kingdom Period. Its orientation is indicated by shafts 1, 3, and 4. Here the productive vein in fact displays various widths in a pinch-and-swell structure and had therefore yielded much diverging gold grades (Fig. 6.209).

The most significant extraction area at this south-eastern vein system in the New Kingdom had obviously been operated near shaft 1. The

Fig. 6.217 The early site at Umm Fahm 1 and New Kingdom wadiworking settlements further N (modified Google-Earth image)

exploited veins had mineralised in several generations and display a slightly wider ore paragenesis spectrum than usually observed. It consisted of pyrite, arsenopyrite, chalcopyrite, stibnite, aurostibnite, and gold.

6.8.3 Umm Fahm 1 (Fig. 6.217)

Geographic position (mine and Old Kingdom site):	21°17′09″ N, 31°07′14″ E

During our visit here it turned out that this site had been a former copper mine rather than a gold mine.

A group of 12–15, simple hut ruins with round and rectangular ground plans are aligned at the foot of a hill, just below an approximately 50 m long, horizontal mine oriented parallel to the slope. The extracted minerals consist of a greenish malachite copper ore. At its eastern end this open, though sanded-in trench turns into an underground mine (Fig. 6.218) that reveals a 15 cm wide and well-distinguishable mineral vein. Because of its strikingly green discoloration, it is

Fig. 6.218 Mine at Umm Fahm with clearly distinguishable, green malachite stains along the vein's wallrock (*arrow*)

thought that its exploitation had occurred in a remote period. In this context, we actually found polished, red shards belonging to so-called Medium bowls. There were also numerous, hand-made, and carrot-shaped vessels of a coarse ware, between 25 and 30 cm tall and widely used in the Old Kingdom as simple storage jars.

Because of the fundamentally different technologies involved in copper ore processing, the typical gold processing tools such as mills are missing. On the other hand, we recorded heavy stone hammers in the debris heaps that had been used in the ore extraction process.

According to the findings therefore, the site at Umm Fahm most probably dates to the Old/Middle Kingdom.

The site name Umm Fahm meaning "mother of coal" perhaps refers to slag and charcoal resi-

Fig. 6.219 Unusual rectangular structures (former basins?) on the slope at Umm Fahm that may have played some role in ore processing

dues that may have scattered at the site's surface over lengthy periods. Today, however, nothing of the like is found here.

Directly below the horizontal mine are the huts of the relatively small settlement mentioned above. To compensate for the slope, they had been built on individual, small artificial terraces.

The round work platforms are delimited by low walls that are somewhat taller at the side facing the mountain than that towards the wadi. Rock outcrops at the slope side had been integrated to the walls.

The round huts are disposed in agglutinating groups of two to three units associated to stone-bordered work platforms facing the wadi. They thereby form four clearly distinguishable groups.

An unusual, rectangular structure oriented towards the wadi was also recognised in an aerial photograph of the site. It consists of four, stone-bordered platforms, each platform being clearly subdivided by a small separation wall. At one side the platforms are interconnected by passageways. The entire structure is oriented in a downhill scaling, and the lowest platform has a central exit to the wadi. The function of the structure is unknown though it may have served for separating (copper) ore dust by aid of the wind (Fig. 6.219). From the top (at the level of the mine) one faintly distinguishes similar structures on both its sides.

The Umm Fahm copper mineralisation is located in a hornblende-gneiss, overlying a chlorite-garnet-mica-schist, which in turn is superseded by marbles. Considering the exposed outcrops, it is quite unlikely that this represented a major copper deposit. The great care taken in its exploitation may to some extent reflect the scarcity of this metal in Old Kingdom Egypt.

Especially interesting in terms of geology is the occurrence of a hydrothermal quartz-copper mineralisation in gneiss units that apparently are not part of the so far described sequences in the Gabgagaba-terrane. It rather belongs to the East Sahara Craton, which in the area around Umm Fahm appears in a geologic window below the Nefirdeib series. In the processed satellite image (Fig. 6.220) of the wider periphery around Umm Fahm, these units are described as hornblende gneiss, garnet-chlorite-schist and gneiss.

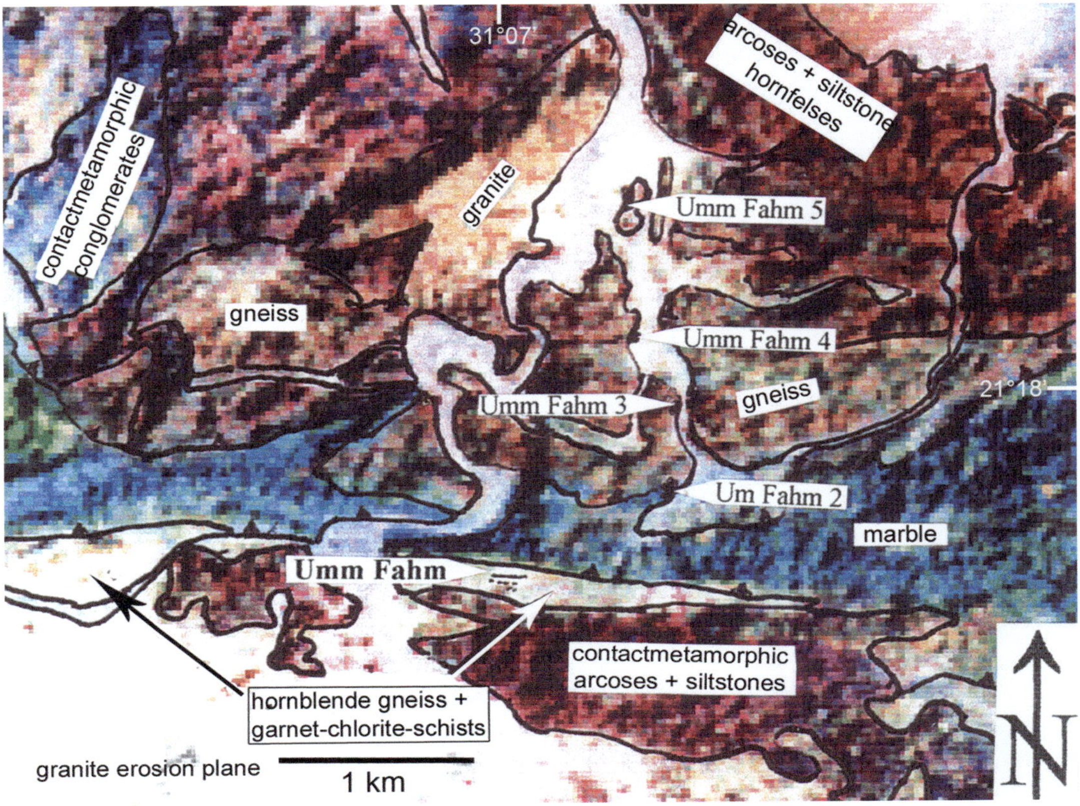

Fig. 6.220 Lithologically processed satellite image of the Umm Fahm area (TM 175/45, channels 7-4-1)

6.8.4 Umm Fahm 2

Geographic position: (New Kingdom settlement):	21°17′29″ N, 31°07′38″ E

At Umm Fahm 2 an alignment of New Kingdom houses stretches along the edge of a wadi pervaded with lime concretions, which makes driving conditions extremely hazardous, even for off-road vehicles. The New Kingdom settlement consists of a heterogeneous blend of rectangular houses with inside passageways and complexes of agglutinating, round buildings. We believe that exclusively wadiworkings had been operated here and that the ores had been brought to the Nile for further processing. Water in return, would then have been transported back to the site, which may actually account for the large amounts of observed amphora shards.

The houses are built from flat stone slabs, which is why in protected areas the walls are up to 1.2–1.3 m tall. Work platforms are occasionally noticed next to the larger houses. The most striking aspect about the settlement is the complete lack of tools. It seems therefore, that they had either been removed after its desertion, or else that the ores had been collected and transported away to the Nile, where processing sites are known to have existed between Farras and Kerma (Vercoutter 1959).

6.8.5 Umm Fahm 3

Geographic position:	21°17′52″ N, 31°07′34″ E

At this location, there are several poorly preserved, round houses with annexes built from weathered granite. The huts are neatly positioned between rock outcrops. Wadiworkings had probably been operated here too, since the entire area is densely covered with quartz chunks.

6.8.6 Umm Fahm 4

Geographic position:	21°18′17″ N, 31°07′25″ E

This site consists of 13 New Kingdom houses without find inventories.

6.8.7 Umm Fahm 5

Geographic position:	21°18′25″ N, 31°07′25″ E

About 100 m further downhill from Umm Fahm 4 are a number of huts that stand out by their unusual entrances.

The settlements at sites 2–5 may probably have been associated to groups of prospecting miners who had checked the gold contents in this quartz-covered district. Forming mobile work gangs, they would have carried their tools with them instead of making new ones for each time they had settled at a new location. Except for Umm Fahm 2, which is one of the largest of these settlements simultaneously associated to yet clearly discernible wadiworkings, Umm Fahm 3–5 consist of 10–13 round huts with each an attendant work platform. Each site is respectively marked by a small Alam on the nearest elevation, higher than 10 m.

6.8.8 Gold Production along the Nile between Ginis and Sesebi

Between Ginis (20°48′29″ N, 30°23′06″ E) and Sesebi (20°06′37″ N, 30°32′32″ E) at the eastern mounts, one repeatedly comes across areas of debris heaps from wadiworkings along the western flanks of the hills, but that reveal no apparent link to nearby gold processing sites. The latter had probably been located directly at the Nile banks, which have now disappeared through shifting of the river course and particularly through recent overbuilding. Our search for ancient gold processing sites at the river banks, which also involved collecting oral information from the local inhabitants, produced no satisfactory results. The noticeable concentration of New Kingdom temples and fortresses between Amara and Sesebi may thereby have some logical connection to a possible gold wealth of this district (Fig. 6.221).

Unfortunately, there is therefore only little additional archaeological evidence that may help to reconstruct the history of ancient gold production between Ginis and Sesebi and beyond. To date, we have been unable of findind any sites related unambiguosly to gold processing in the area.

At the opposite side, along the eastern banks between Ginis and Ager and even further up the Nile towards Tombos, there are a number of medieval fortifications, in whose walls one yet again repeatedly discerns partly well-preserved New Kingdom gold-ore grinding mills. At any rate, these findings clearly point to substantial gold processing activities. Such New Kingdom mills were found in complexes at the Island of Sai, further at Koyeka, Suarda (Fig. 6.222), Fort Iraw, Koya, Ager, Tondi, and Sesebi. The hinterland is characterised by wide areas with wadi- and colluvium workings wherever the geologic preconditions for goldquartz formation are given (Fig. 6.223).

The yields obtained from these wadi- and colluvium workings would have equaled if not surpassed those in the Eastern Desert (Fig. 6.224). In addition, improved supply logistics and unproblematic availability of large work contingents are only some of the significant advantages the proximity to the Nile Valley could present compared to the distant gold mining sites in the Eastern Desert. A hypothesis of a focal point of Ancient Egypt's gold mining industry in this area

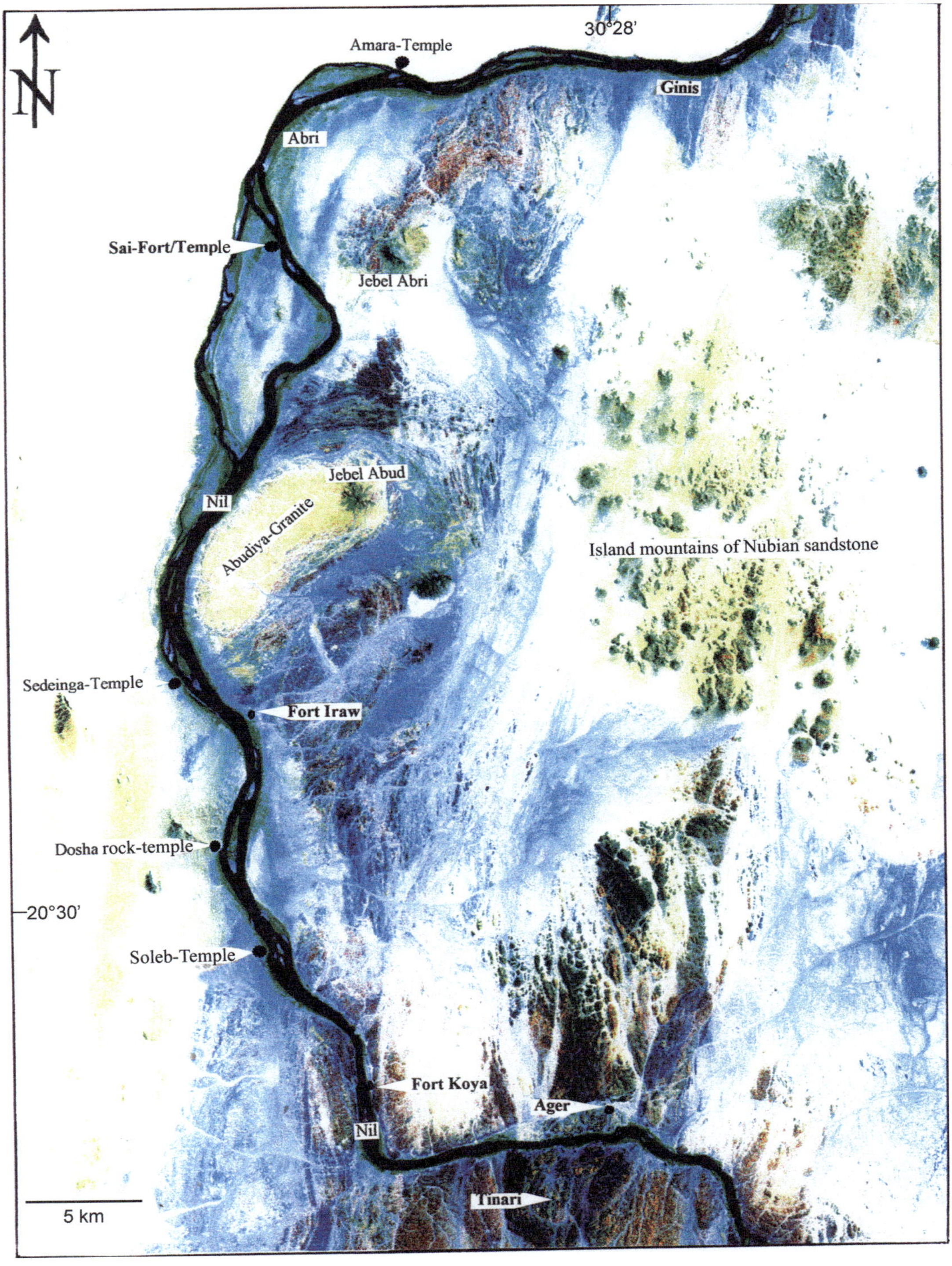

Fig. 6.221 Lithologically processed satellite image of the area between Ginnis and Ager. Blue colours represent predominantly quartz covered surfaces that also coincide with traces from wadiworkings (TM 175/46, channels 7-4-1)

Fig. 6.222 Remains of a medieval fort at Suarda and countless New Kingdom mills contained in the structure's stone work and debris

is moreover supported by an undeniable concentration of New Kingdom temples between Amara and Sesebi, forming a religious power hub on the western Nile banks, which doubtless played an important role in safeguarding the incoming gold.

The situation is to some extent illustrated by a scene depicted in a tomb dedicated to an official called Huy from the time of Tutankhamun (Fig. 2.2), in which Nubians are performing a delivery of large amounts of gold ring ingots and gold dust contained in bags (Davies and Gardiner 1926). Also, a previously mentioned letter addressed to Amenhotep IV (Akhenaten) wherein the Mitanni king, Tushratta, asks for more gold deliveries in arguing that gold had merely to be picked up in Egypt (orig.: "gold in quantities like dust"), may in fact reflect the situation in this vast wadiworking district, close to the Nile Valley

Notwithstanding, firm archaeological proof, comprising New Kingdom ore mills as well as quartz ore heaps brought to light in excavations by the University of Cambridge inside the temple city of Sesebi, now confirm the presence of the gold processing industry in the area (Spence and Rose 2009).

6.8.9 Sai Island

Geographic position:	
New Kingdom temple:	20° 44′14″ N, 30° 19′54″ E
Ottoman fort:	20° 44′11″ N, 30° 19′53″ E

The NE sector of the Nile island of Sai is covered by innumerable, round tombs. There, at the island's eastern banks is a New Kingdom temple and further S, a large Ottoman fort measuring 123 × 83 m, equipped with two impressive towers on its W side.

Numerous New Kingdom mills exhibiting substantial traces from secondary use wear with deep abrasion hollows add significantly to the discussion on ancient gold production in the area. This mill type is in fact found much more frequently near the Nile than in the Nubian Eastern Desert.

The island's aspect is characterised by a dense cover of paleo-rubble and a band of vegetation band around its rims. Traces from former wadiworkings are especially concentrated near the numerous basement outcrops in the island's central parts.

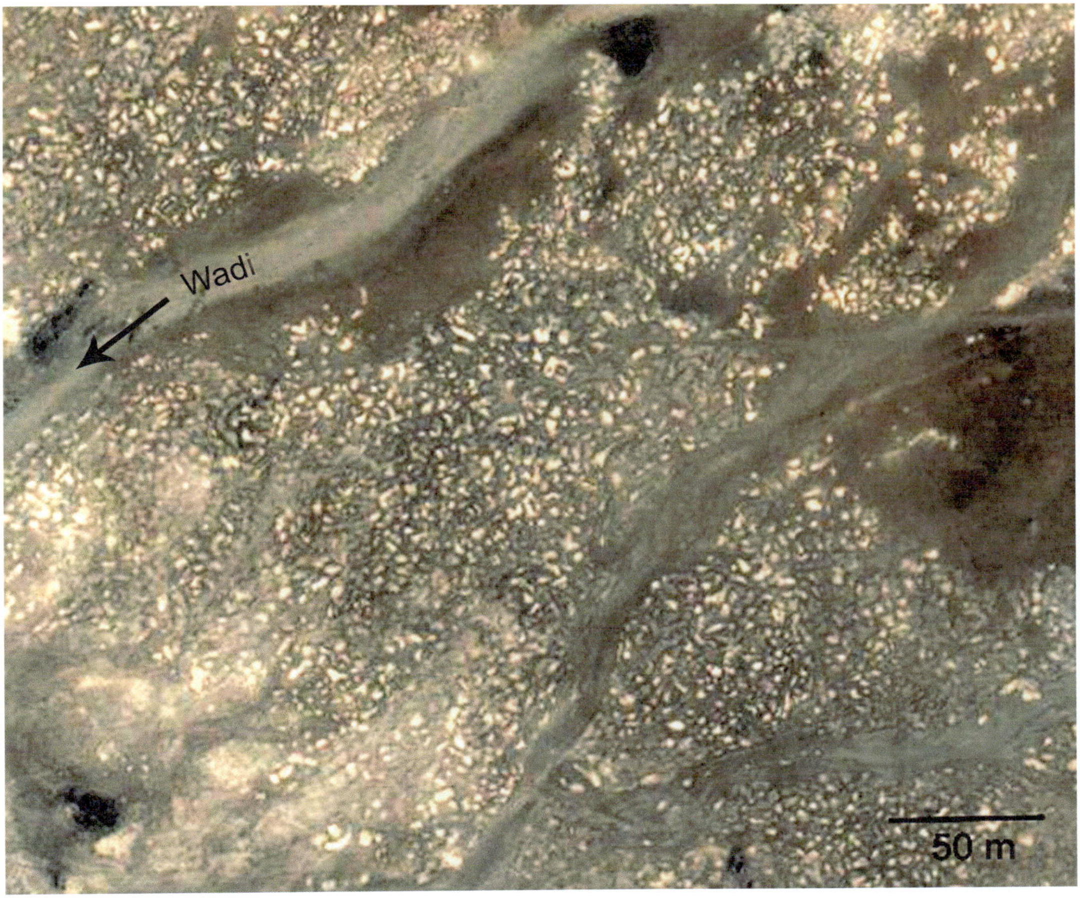

Fig. 6.223 Extensive wadi- and colluvial workings in the north-eastern hinterland of Fort Iraw (modified Google-Earth image)

Fig. 6.224 Looking E over the wide, quartz-covered plain at Ginis which also is marked by traces from wadiworkings

The temple precinct is dated principally to the New Kingdom, although the epigraphic evidence refers to forerunner complexes from the Old Kingdom (Geus 2004).

In the masonry and ruins of the Ottoman fort one distinguishes spolia containing obliterated cartouches of Amenhotep IV (Akhenaten) and other kings of the eighteenth dynasty.

Inside and outside the fort the surface is littered with numerous New Kingdom mills.

The associated sandstone quarries that provided the building material are still visible today near the river banks.

6.8.10 Ager

Geographic position:	20°21′53″ N, 30°39′17″ E

Just off the newly tracked road between Fort Koya and Abu Sari is a New Kingdom gold processing site with hut remains and numerous oval stone mills, anvil stones with multiple impact surfaces, and fist grinders. Since we found no traces from mines in the mountains, as confirmed by the local population, it is probable that quartz sediments had been extracted there and taken to the site for processing. With the implementation of a recent track leading straight through the site, some mills have been moved aside. Over 50 hut ruins are located in the drop between the Nile and the track and may probably be ascribed to the gold processing site.

6.8.11 Abu Sari

Geographic positions:	
Recent complexes:	20°17′00″ N, 30°38′04″ E
Shaft 1:	20°17′02″ N, 30°38′51″ E
Large New Kingdom settlement:	20°17′21″ N, 30°38′08″ E

Arriving at Abu Sari one is first struck by the white glisten of the quartz-covered hills that dominate the landscape here (Fig. 6.225).

Here the scenery offers a view over enormous areas of small, bulging heaps from wadiworkings, and speckled with New Kingdom house ruins and millstones, as well as recently dug shafts. The vastness of the unfolding panorama enhances the impression of the magnitude of the scale at which gold quartz ores had been removed from the Nile Valley's adjacent plains, thereby contributing significantly to the gold production of the New Kingdom.

The white landscape of Abu Sari very much resembles the surface of a glacier, although the temperatures during our visit by far exceeded temperatures around 40 °C. Many of the archaeological remains are severely sand-blown and most of the sparse pottery evidence has been abraded down by the wind.

As opposed to the wadiworking areas closer to the river in the N, where ore grinding mills are completely lacking, here at Abu Sari, about 6 km to the E of the Nile, mills are again found.

After a brief resumption of mining under the Nuba Development Company between 1901 and 1904, Bishop acquired the mining rights for the district in the 1940s where, in addition to his engagement at Gebeit, operations continued until his passing in 1953. During this most recent phase, four deep shafts without branch lines were lowered into the ground, though without striking the sought-after quartz vein. Genuine exploitation can therefore be ruled out for this location. The shafts had been executed near former wadiworkings or close to visible quartz outcrops. Bishop had most likely planned to establish branching drifts to the productive veins, but death caught up on him before he could do so.

The entire area is covered with quartz chunks and crossed by quartz veins oriented in different directions. It hence represents an almost ideal location for wadi-and colluvium workings. The remaining traces from these activities appear in the form of characteristic, small spoil heaps from the New Kingdom. Bishop in fact, became inspired by the old wadiworking method when he decided to develop a modern variant, whose traces in an aerial view are recognised as long, parallel stone lines crossing the countryside. The terrain was thus systematically marked out in strips of around 2.20 m width by aid of

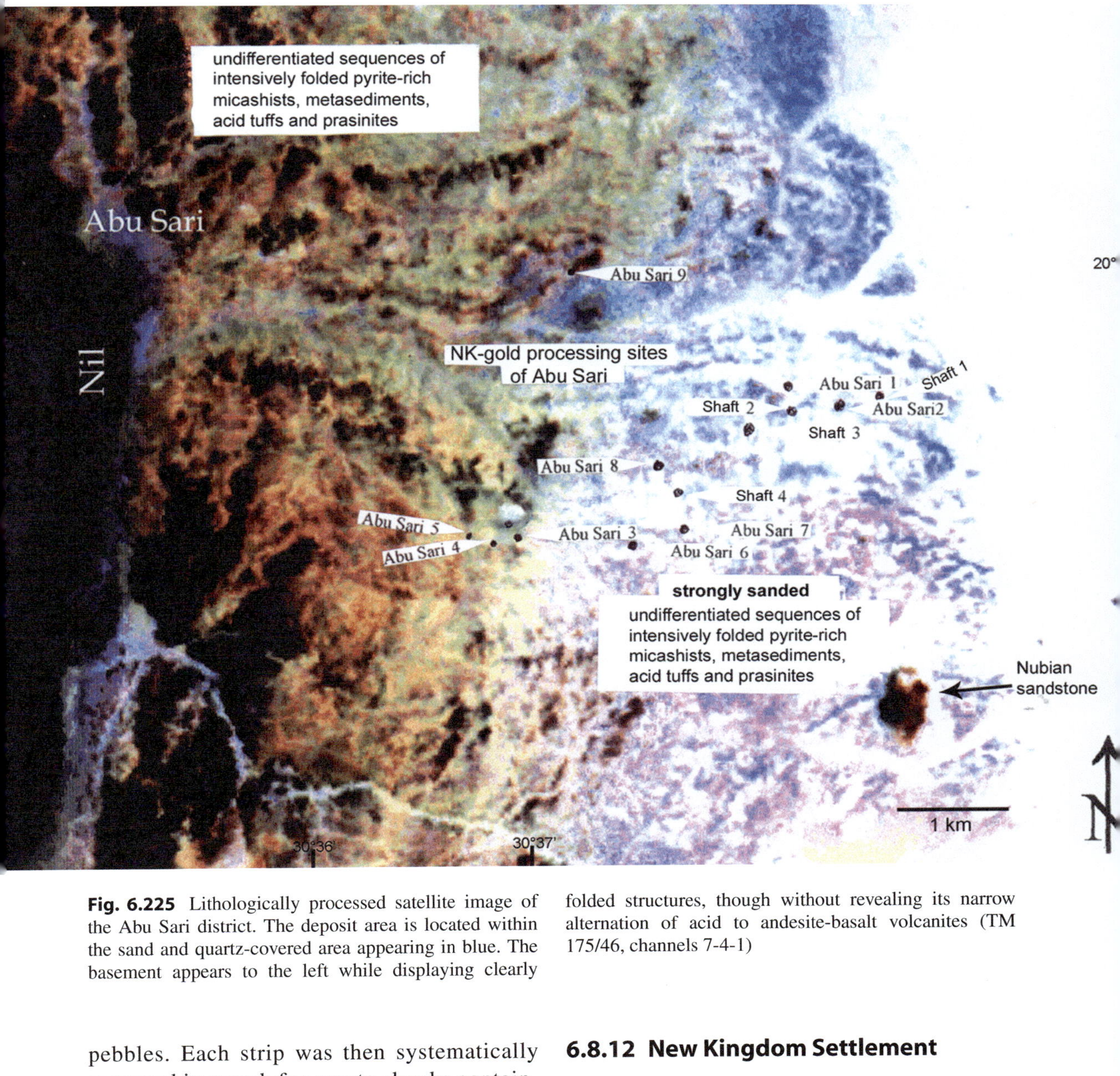

Fig. 6.225 Lithologically processed satellite image of the Abu Sari district. The deposit area is located within the sand and quartz-covered area appearing in blue. The basement appears to the left while displaying clearly folded structures, though without revealing its narrow alternation of acid to andesite-basalt volcanites (TM 175/46, channels 7-4-1)

pebbles. Each strip was then systematically surveyed in search for quartz chunks containing gold ores. Along the strips stone cists containing the discarded material judged unsuitable for further processing were set up (Fig. 6.226). In a sense, these cists correspond to the small spoil heaps sorted out in the ancient wadiworkings.

The auriferous ore chunks on the other hand were transported to the processing plant at Abu Sari, near the Nile.

6.8.12 New Kingdom Settlement

A large, well-ordered New Kingdom settlement with sparse pottery finds is located at this position (Fig. 6.227).

Hardly any stone mills are found in its wide wadiworking area. It seems therefore that the collected quartz ores had not been processed there but rather at the settlement, where they occur in large numbers. Every now and then one comes across few New Kingdom mills and typical fist

Fig. 6.226 Remains from wadiworking operations carried out in the first half of the twentieth century, during which auriferous quartz chunks were systematically retrieved from marked-out strips and barren material was piled up in stone boxes

Fig. 6.227 Unusually large New Kingdom houses at Abu Sari in a good state of preservation

grinders near depleted quartz outcrops, but never more than 12–15 units, which is by far too little compared to the vastness of the area. Some of these mills display only very slight traces from use wear, which may suggest that they only had served for probing purposes.

Furthermore, a whole variety of different-sized rock hammers were recorded, ranging from the simple fist hammer to the much heavier mallet that was manipulated with two hands, as it weighs up to several kilograms. These often consist of amorphous rock lumps clearly identified as tools by their chipped impact surfaces. The rocks used as tools had been gathered on location, and it seems that they been selected and intentionally stocked in piles at certain places. In addition to the ubiquitous quartz rubble, the terrain consists of flatly shaped rocks in b-axis formations. From the air they are easily mistaken for architecture ruins. Compared to the expansion of wadiworkings, the scarcity of architectural remains is quite striking. Abu Sari should therefore also be counts to this wadiworking zone easily reached from Ginnis in the inhabited zone of the Nile Valley.

The numerous stone hammers had therefore probably been used for the rough pounding in order to facilitate the ore transport to the processing sites at the Nile.

6.8.13 **Abu Sari-North** (21°17′40″ N, 30°37′13″ E)

This site is located on a slope about 1.75 km NW of the first New Kingdom settlement. It is relatively well-preserved although very sand-blown and in a relatively remote location from the just discussed sites. Apparently this New Kingdom settlement had specialised on the local wadiworkings only, whose traces still visible (Fig. 6.228).

Since most of the entire mining district is much affected by sand drift, the few exposed areas hardly allow for any satisfactory observations on the surrounding's geologic composition. Further W towards the Nile Valley the sand cover decreases considerably, but the geology is assumed to be more or less the same, which allows for much better observation conditions. Accordingly the area is tightly folded, with flat schistose sequences from carbonate mica-slates, which at times display large amounts of pyrite. There are also metasediments, slightly metamorphic, acid tuffs, and prasinites, as often observed elsewhere in the so-called Nefirdeib series of the Nubian basement. The fold axes strike within a narrow E-W bracket and dip while slightly undulating in both directions. They are therefore not to be mistaken for the more gneissic and metamorphic Duweishat sequences that belong to the original African basement.

6.8.14 **Abu Sari-South** (20°17′00″ N, 30°37′41″ E)

This is a New Kingdom settlement whose houses are preserved in several masonry layers and contain the characteristic find inventory. The mills again are worn down to unusually narrow and slender shapes. The settlement counts between 25 and 30 houses. They measure about 3 × 6 m and have a central, dividing wall and an inner passageway, which fully matches with the New Kingdom house type usually encountered in Egyptian mining areas.

Both here and on the other side, near Tinari, only stone tools had been utilised, as metal tools had been too precious, especially considering the scale of the operations. This partly accounts for the observed high density of stone hammers, whose sizes vary considerably (Fig. 6.229).

6.8.15 **Mines** (20°16′35″ N, 30°36′50″ E)

Two ancient mines located close to each other were found at this location. Both are completely filled-in with sand, but at their margins work traces are still distinguished at three different quartz generations. Between 30 and 40 New Kingdom house ruins scatter loosely over a wide area. The houses are built from the local, flat and elongated stones, but nonetheless are preserved

Fig. 6.228 Thinly spread-out New Kingdom settlement at Abu Sari North (modified Google-Earth image)

in only one masonry layer. The tool assemblage seems to be complete and consists of oval stone mills, grinding stones and pounding slabs with central depressions. They lie strewn along the sides of the houses, which for the most have a rounded shape and consist of one room. Additional mills and hammers measuring between 10 and 25 cm are found near the mines. The different hammer sizes to some degree probably reflect tool specialisation with regard to the processing of different quartz fractions. On the other hand, they also seem to indicate the use of stone tools inside the actual mines, contrary to what had been normal in Egypt by the time of the New Kingdom,

6.8.16 Tinari

Tinari is located on the W bank of the Nile, from where the actual mining district is best reached on foot.

6.8.17 Sokar Mine

Geographic locations:	
Shaft 1:	20°19′23″ N, 30°29′49″ E
Shaft 2:	20°19′23″ N, 30°29′53″ E
Shaft 3:	20°19′17″ N, 30°29′34″ E

Several ruins at Sokar attest to a mining site dating to the first half of the twentieth century.

Fig. 6.229 Stone hammers from Abu Sari-South dating probably to the Old/Middle Kingdom

Light-green glass shards with impressed floral ornaments from this site are similar to ones from Derahib.

Four shafts had been lowered into the ground in that phase. Shaft 1 is located next to an ancient mine (Fig. 6.230). In the debris heap just outside the old mine, we found small, globular tools from green dolerite. They had probably been used as fist hammers with two, rounded pounding surfaces (Fig. 6.231). They are indicative of a local mining industry in a remote period. Extensive wadi- and colluvium workings are found in the surrounding area, especially near a small settlement to the S of the main mine.

About 800 m S in a straight line is an alignment of approximately 15 round huts in a narrow gorge whose diameters vary between 1 and 2 m and containing no artefacts. No traces from mining were detected at the margins of the wadi.

In the wadi that leads over a distance of 3 km to the village of Tinari are the remains of a recent shaft (shaft 3) and a washing table. The people in the village report that their grandmothers had once worked here and that they had removed the remaining mills from the site. Further investigations revealed the relics of three more, faintly discernible washing tables. We found another five to six, round huts partly integrated into the mountain rock near shaft 3. A long trench with small stone hammers scattered at its edges is situated along the foot of this mountain.

Three mining phases have thereby been identified in the district of Tinari- Sokar. The first had been a very ancient, indigenous Nubian one comprising the mine at Sokar itself. According to the sparse pottery finds, the second phase had apparently taken place in the New Kingdom Period and was based on wadiworkings. The most recent one dates to the early twentieth century.

The auriferous quartz veins in the Tinari-Sokar district are embedded in a much deformed anticline with roughly N-S oriented, flat lying fold-axes. The structure consists of a colourful petrographic sequence of fine-grained rhyo-

Fig. 6.230 Mine at Sokar with spoil heap containing tools comparable to ones dating to the Old Kingdom in Egypt

lite, acid tuff, porphyry crystalline tuff, and agglomerates that partly are distinctly schistose and sheared. Through this largely N-S striking shear the anticline structure is extended toward the S. The auriferous quartz veins run along the ac-planes, perpendicularly to this structure.

The ore paragenesis near shaft 1 was seen to consist of pyrite, malachite from chalcopyrite (?) antimonite, tennantite, and gold occurring in two generations: (1) pyrite, arsenopyrite (?), chalcopyrite, and gold. (2) tennantite and gold.

Fig. 6.231 Stone hammers and a grooved mallet (right) found near the Sokar mine as witnesses of early gold mining in Nubia, probably in Old/Middle Kingdom

6.8.18 Tondi

Geographic position:	
Tondi:	20°11′11″ N, 30°32′40″ E
Tondi-S:	20°09′59″ N, 30°32′51″ E

We did not visit these sites but were able to identify them in a Google-Earth satellite image (Fig. 6.232). The characteristic arrangement and geographic location of the houses seem to hint to only one occupation in the New Kingdom. Gold production had probably been based on wadi- and especially colluvium workings.

The geologic setting is very similar to that at Abu Sari and Sokar (Tinari). It consists of tightly folded and somewhat schistose sequences of rhyolites and acid tuffs, and occasional intercalations of metasediments without marble.

6.8.19 Kerma

Geographic position (crucible-heap):	19°36′27″ N, 30° 24′46″ E

Following an indication in the Dunn report (1911) mentioning a site with thousands of tons of crucibles for gold refining and assaying, we decided to stop over to take a closer look at the site. It is located approximately 500 m NE of the so-called Western Defufa (large mudbrick building) and today is protected by a fence. Two enormous heaps of ash and broken vessels from refractory clay, sand, and slags dominate the site. According to a personal communication by Ch. Bonnet, the so-called crucibles are in fact forms for baking bread used in the temple cult. The interpretation by Dunn may thereby be dismissed.

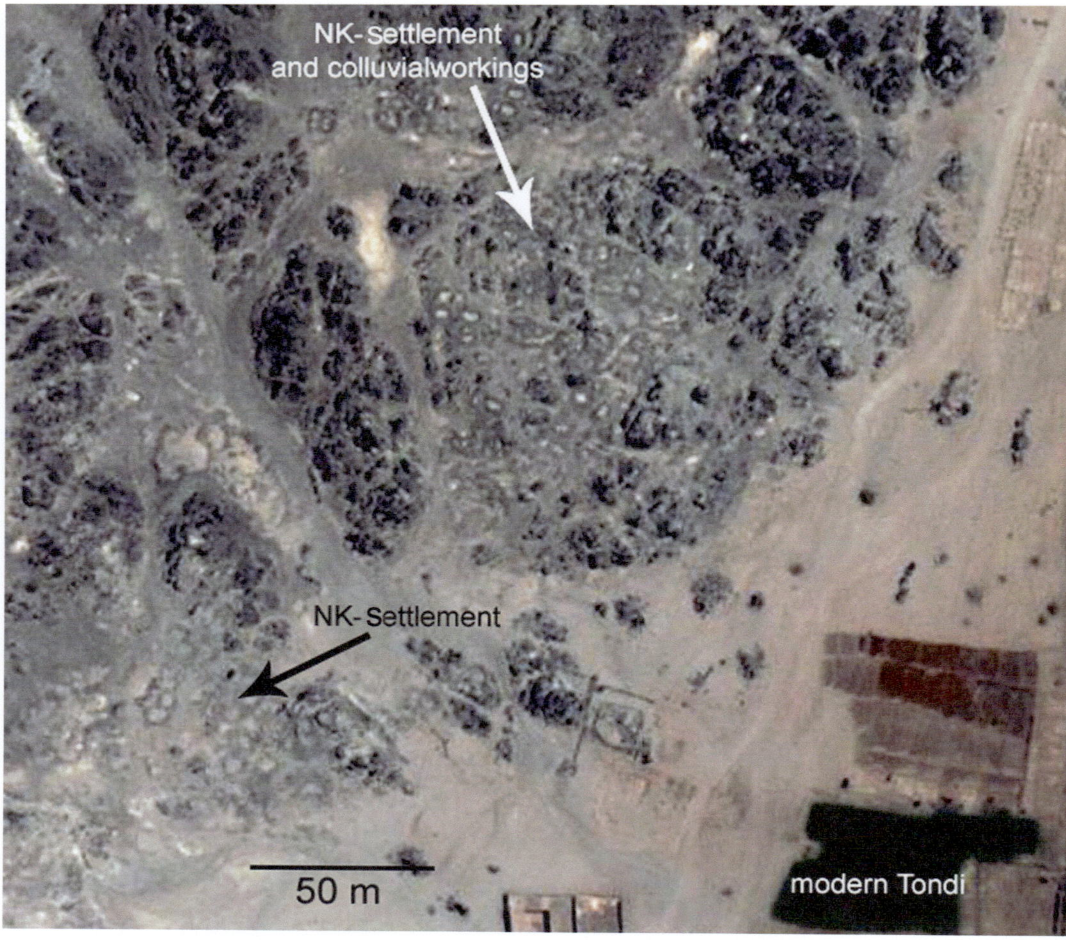

Fig. 6.232 Ancient settlements and wadi- and colluvial workings at Tondi (modified Google-Earth image)

6.8.20 Gold Mining Sites between Gebel Barkal and Abu Hamed

There are no noteworthy gold mining sites close to the Nile Valley in the sector between the Third Cataract and Gebel Barkal, as this area consists of Nubian sandstone only and therefore lacks the necessary basement rocks.

The basement reappears again at Nuri, primarily at the borders to the Bayuda Desert, but also in the desert regions to the NW. Consequently, gold mining sites may be expected along this section of the Nile.

Our expectations were in fact confirmed by the recent discovery of a gold processing site at Hosh el-Gurub by a team from the Chicago Oriental Institute (Harms 2007; Harrell 2010). It consists of a New Kingdom settlement with the characteristic oval stone mill inventory, although it was dated to a period between 2000 and 1500

Fig. 6.233 Ruins of the fortress at Shamkhiya with debris at the foot of the hill

BC based on an adjacent burial ground containing Kerma pottery. On the other hand, systematic gold production probably hadn't begun here before 1500 BC, a phase in which Upper Nubia had come under the control of Egypt.

According to the published figures Harms (2007), the site had possibly even been re-occupied in a later phase under the Kushite rule, as indicated by a number of reclaimed and additionally worn mills.

Similar mills can be observed in the masonry and inside the courtyard of a medieval castle to the N of the village of Shamkhiya, as well as on the SW bank near the island of Tanta, and particularly at several sites on the island of Mograt.

6.8.20.1 Shamkhiya

Geographic position:	19°27′57″ N, 32°59′39″ E

Dr. Sheikh (GRAS) drew our attention to this fort at Shamkhiya with its New Kingdom houses, grinding mills, and its wealthy pottery deposits. The houses had been reported to be located at the edge of the cultivated strip, at the foot of the fort. However, the ruins here are more likely to represent looted graves. The much dilapidated, medieval fort is built on granite rock and directly overlooks the Nile (Fig. 6.233). Inside its walls there are more ruins. Countless oval ore mills had been integrated into both the masonry of the buildings inside the fort and that of the ramparts themselves. They belong to the original, "classic" New Kingdom type, but for the most display a secondary depression at the grinding surface (Fig. 6.234). Other mills exhibit grinding surfaces more worn than the classic New Kingdom mills and are much smaller (Fig. 6.235). Large amounts of light Meroë pottery as well as slags seem to suggest that the site had been occupied over lengthy periods.

A local peculiarity is represented by mills that had been shaped directly into the bedrock granite (Fig. 6.236). The general finding therefore seems to indicate that the site on this massive elevation had functioned as a fortified gold processing site already in the New Kingdom and that it later had been re-occupied under the Kushite rule. The ruinous buildings and the fort had then been rebuilt with the original construction mate-

Fig. 6.234 New Kingdom broken ore mill displaying later use wear, from the stone masonry of the medieval fort at Shamkhiya

Fig. 6.235 Deeply worn, small ore mill in the fort area at Shamkhiya, probably dating to the Kushite Period

Fig. 6.236 Runner stone and New Kingdom type mill in the granite-gneiss bedrock at the Shamkhiya fortress

rial as well as with former mills in the Christian Medieval Period. The findings from Shamkhiya as well as the Nile fortifications into whose walls stone mills had repeatedly been integrated, may suggest that there had been no significant, domestic gold production under the Medieval Christian/Nubian kingdoms, if one excludes unverifiable gold panning activities in the river sediments. This may hint to a difference between the local, Christian/Nubian population living at the Nile and the gold-seeking, Arab population in the Eastern Desert. With the conversion of the Nile Valley to Islam in the mid fourteenth century this latter region was probably depopulated.

Gold panning today is in fact a widespread activity among the village's girls. At the sedimented terrace banks near the mouth of a large wadi coming in from the Bayuda Desert (Fig. 6.237) they claim to recover up to 5 g of gold within about 4 h. This is to some extent substantiated by a noticeable wealth of the village whose white-washed houses are already made out from a distance.

Nearby, we were told of a wadi referred to as Wadi Mahariq ("Valley of mills"), a place where there allegedly are many mills. Our attempts to find it, however, were luckless, in spite of the help of a local guide.

6.8.21 Medieval, Christian Cemeteries

Although not necessarily connected with the present topic, we include the GPS-coordinates of a number of cemeteries as another, local peculiarity of the area between Shamkhiya and Abu Hamed. The cemeteries are all located above the southern banks of the Nile, and regardless of their old age, they continue to be respected by the local population:

Fig. 6.237 Nubian girls from the village at Shamkhiya washing gold at the Nile banks, near the confluence with a long wadi coming in from the Bayuda Desert in the N

"Cemetery 1" (19°31′15″ N, 33°03′53″ E) is located near the village of Kimelab, on a sandy elevation, its graves oriented in an E-W direction (Fig. 6.238).

Another cemetery, which we labeled "Cemetery 2" is found approximately 700 m further E, near Karbab village (19°31′15″ N, 33°04′18″ E). The graves orient 100–110° E. We were told that they are Christian ("masahen") graves.

"Cemetery 3" (19°31′10″ N, 33°05′12″ E) is located 1.6 km further E in the village of Salatib. In contrast to the two previous cemeteries, it consists of ring-shaped tombs.

"Cemetery 4" is located about 800 m further E (19°31′04″ N, 33°05′37″ E) whose graves distinctly display an E-W orientation.

6.8.21.1 Mograt Island

To the S and SW of Abu Hamed several fortifications dating mainly to the Medieval Christian Period are grouped on the 30 km long and up to 4 km wide Nile island of Mograt. During our survey it became apparent that most of the indications to ancient gold mining sites that all seem to point to the New Kingdom, are somehow connected to the medieval forts. At all sites New Kingdom ore mills either scatter on the ground or are included to wall masonries. Also, large areas, especially along the quartz-covered, greyish-white flanks to the N and S of the island's central ridge show evident traces from wadi- and colluvium workings.

6.8.21.2 Karmel Fortress

Geographic position:	29°27′25″ N, 33° 20′13″ E

Today the ancient fortress near the village of Kereseikal forms a derelict heap of debris surrounded by the remains of its former ramparts

Fig. 6.238 Medieval Christian tombs, near Kimelab village at the northern banks of the Nile, Bayuda Desert

Fig. 6.239 Ruins of the medieval Christian fortress at Karmel, on Mograt island. The debris contains ore mill fragments

(Fig. 6.239). We found between five and seven, light-coloured granite fragments from New Kingdom grinding mills, randomly dispersed over the area inside the walls. Another mill was found outside the fortress, at the edge of a cultivated field (Fig. 6.240).

Fig. 6.240 Partly preserved ramparts at Karmel (modified Google-Earth image)

6.8.21.3 Kurduma

Geographic position:	19°27′12″ N, 33°22′14″ E

This site consists for the most of a mudbrick wall comprising layers of greenish-black amphibolite rocks. We also found a fragment of an ore mill contained inside it. The courtyard of the complex was covered with thorny shrubs stopping us from pursuing our examination of the site.

6.8.21.4 Sehan

Geographic position:	19°30′46″ N, 33°16′02″ E

This site consists of an open space at the edge of an extensive wadiworking area, just S of the thoroughfare road at Sehan. We counted approximately ten, large New Kingdom grinding mills (Fig. 6.241). They certainly are no longer at their original location, but still witness past ore processing activities in the area. More mills and unmistakable traces from gold workings are found to the W of Sehan.

At some places in this impressive granite landscape we furthermore came across petroglyphs representing animals (Fig. 6.242).

6.8.21.5 El-Mikeseir

Geographic position:	19° 32′13″ N, 33° 09′12″ E

An almost square fortification whose walls measure 65 × 58 m is located directly at the northern banks of the island. The fort has two gates and semicircular bastions. Numerous New Kingdom

Fig. 6.241 Fragments from large New Kingdom stone mills at Sehan, Mograt island. Agriculture has considerably contributed to the decimation of the archaeological evidence for the area‘s once possibly intensive gold production

Fig. 6.242 Granite rock with animal carvings at Sehan, Mograt island

Fig. 6.243 Virtually square, medieval fortress at Mikeseir, Mograt island, revealing evidence for New Kingdom gold production (*arrow*)

grinding mills were found inside the complex (Fig. 6.243), even though none had been integrated to the stonework.

6.8.21.6 Abu Alalik

Geographic position:	19°32′04″ N, 33°08′31″ E

Numerous New Kingdom grinding mills, partly exhibiting secondary use wear depressions are located at the village square on the northern side of the road track. There were also mill fragments with unusually broad grinding surfaces (Fig. 6.244).

6.8.21.7 Ras El-Gezira

Geographic position:	19°31′38″ N, 33°06′27″ E

This highly interesting site at the westernmost point of Mograt Island consists of a poorly preserved, Medieval Christian fort with just barely distinguishable ramparts (Fig. 6.245). In its surroundings and especially between its western side and the river bank are numerous New Kingdom grinding mills, often displaying, as frequently on this island, unusually wide grinding depressions (Fig. 6.246). Poorly preserved hut remains, or possibly even only work platforms, were also observed. Particular to this site are a number of granite blocks arranged in alignments and inclined eastwards at the fort's eastern front (Fig. 6.247). Their function is unknown but they may have served defensive purposes in the same way as palisades.

6.8.21.8 Abu Kuweib

Geographic position:	19°31′56″ N, 33° 16′00″ E

The mud brick wall remains of the Abu Kuweib fortress are located in a picturesque setting, on granite nodes about 8 km W of Abu Hamed. Yet again, we recorded characteristic oval grinding mills whose initial, flat grinding surfaces had been deepened significantly by secondary use wear. Observed pounding stones consisted mostly of fist-sized quartz globes.

Fig. 6.244 New Kingdom mill with secondary use wear from the village square at Abu Alalik, Mograt island

Fig. 6.245 Largely destroyed remains of Ras Gezira fortress, at the W end of Mograt island. Countless New Kingdom mills once used in gold production are still found in the debris of its collapsed walls

Fig. 6.246 Unusual, possibly defensive stone installations on the E side of Ras Gezira fortress

Fig. 6.247 Large New Kingdom mills with later use wear depressions from the Ras Gezira fortress

The fortresses had possibly been first established during New Kingdom for the protection of the local gold production, as this was deep inside Nubia and therefore foreign territory occupied by Egypt. They are located close to the Nile, as water was essential for ore processing. With the reoccupation of the sites after the New Kingdom, possibly in the Kushite or Meroë Period, gold production could take up again. The geologic preconditions were anyhow given and had been reliably identified due to the knowhow of the local and Egyptian prospectors.

During the refurbishing of most forts in the Medieval Period, ancient mills ended up to serve for building material of new complexes. The numerous Christian Medieval cemeteries are probably to be set within a context the area's regain in importance.

6.8.21.9 Abu Hashim

Geographic position:	18°58′34″N, 33°35′33″ E

This site is situated at the Nile near the railway track S of Abu Hamed. Dunn (1911) already reported ancient wadiworkings and trenches in the area.

We noticed house ruins and individual trenches nearby the village. However, there were no signs of the former presence of an ancient gold producing site. The trenches did not display the diagnostic markers for gold extraction. Most of the house ruins date to the time of the railway construction.

6.9 Bayuda Desert

At the end of the twentieth century, wide areas of the Eastern Bayuda Desert were still being actively exploited according to the traditional gold mining methods in wadi- and colluvium workings, mostly without modern machinery. Striking similarities with ancient methods inspired us to pay the mining communities a visit in the hope to gain extra insight into the ancient working methods by illustrating those we had reconstructed from archeological findings with present-day observations (Fig. 6.248).

We visited there four gold mining sites in which work was carried out according to identical procedures (Fig. 6.249).

The work gangs we were able to follow (Fig. 6.250) comprised various Sudanese ethnic groups. Local Nubians hence cooperated peacefully, side-by-side with tribesman from S Sudan.

The entire area, especially in the wadis, is speckled with metre-high heaps containing residual, discarded material (Fig. 6.251). They correspond to the now usually much flattened heaps left behind in the wadiworkings from the New Kingdom and the Early Arab Period.

In a first step the barren ores are separated at the surface from the possibly auriferous quartz chunks. The gold ore is then collected in plastic bags, while the rest is dumped on piles. The gold contained in the quartz is usually invisible to the naked eye. It is normally indicated to by a grey tint from finely distributed sulphide minerals with tiny gold sequins. The quartz chunks are hewn out of the wadi floor by means of small iron hacks. The large, barren rocks are discarded forming heaps and holes scattering the ground (Fig. 6.252). In some cases deeper trenches develop, as the quartzes are being removed.

The ore chunks are then crushed at the surface, which also was done in the ancient processing method. In this step a blunt iron hammer in stead of a stone pounder is used to pound the ore down to approximately the size of a pea. To prevent the quartz splinters to project, a jute cloth is placed at the point of impact (Fig. 6.253). The obtained quartz gravel is then brought to a diesel powered ore mill where it is transformed to fine powder and filled into sacks.

The sacks are then taken back to the original work crew, where usually women and children

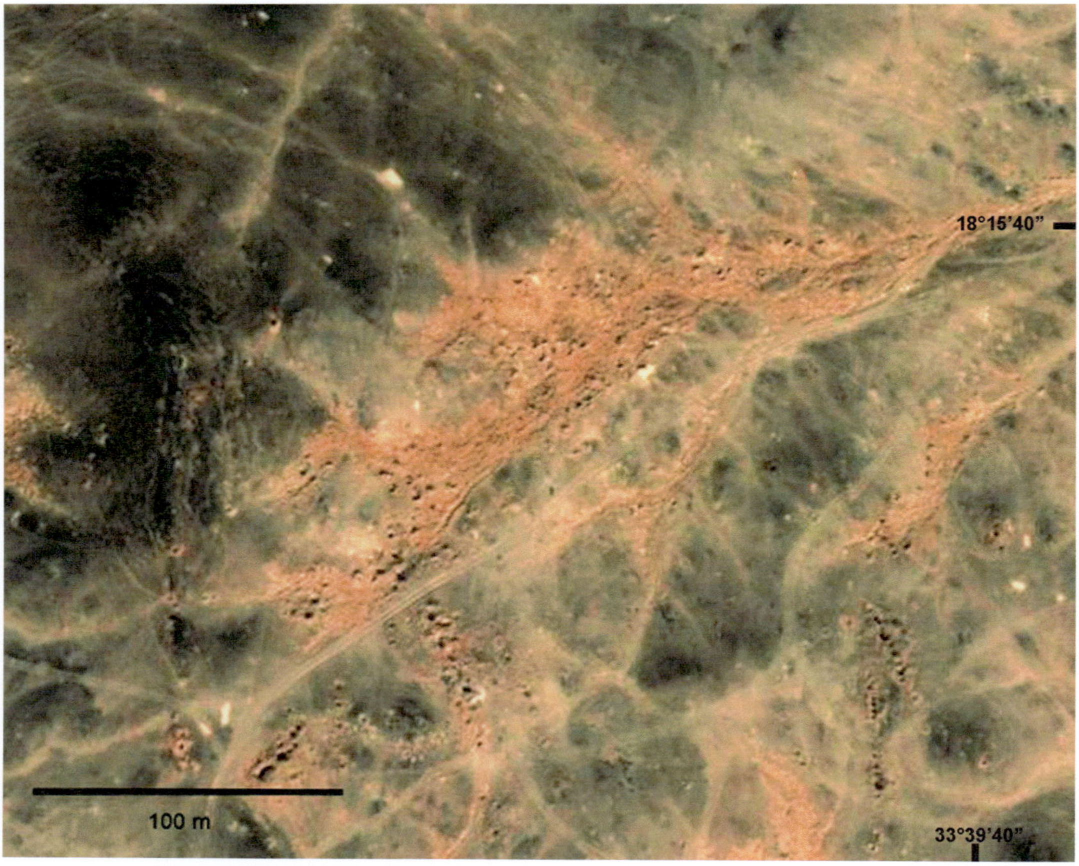

Fig. 6.248 Recent wadiworkings at El-Higagiya in the Bayuda Desert, concentrating clearly within older wadi courses (modified Google-Earth image)

wash and strain the meal in holes filled with water, before washing it out again in sheet metal bowls to obtain evenly grained sand (Figs. 6.254).

This fine sand is then panned in order to separate and remove the lighter earth and silicate particles from the heavier minerals that also include the gold remains left in the pan. This black slurry (Fig. 6.255) consists chiefly of black magnetite (Fe_3O_4) and ilmenite ($FeTiO_3$).

The water is brought from the Nile in barrels on an old truck and is sold on the site to the individual work gangs. Donkeys carrying two plastic cans are also used. The distance to the Nile varies between 8 and 10 km and the donkeys are usually led by women or children.

Instead of separating the gold from the quartz powder in a water-suspension by means of an animal fur attached to an inclined washing table in accordance with the ancient method, a modern, but very dangerous technique is applied to extract the gold from the rest of the heavy minerals:

Pure mercury kept in a plastic bag is now added to the heavy mineral powder and stirred in a bowl together with water, and shaken until all, initially invisible gold sequins contained in the black sand is diluted in the mercury (Fig. 6.256). The gold/mercury amalgam is then separated from the black sand through straining with a finely meshed cloth. It is then heated on a small metal plate over an open fire, which prompts the mercury to evaporate, thence leaving behind a porous mass of pure gold in about pea size (Fig. 6.257), which eventually is transformed to solid gold through evaporation of the mercury. The entire procedure is carried

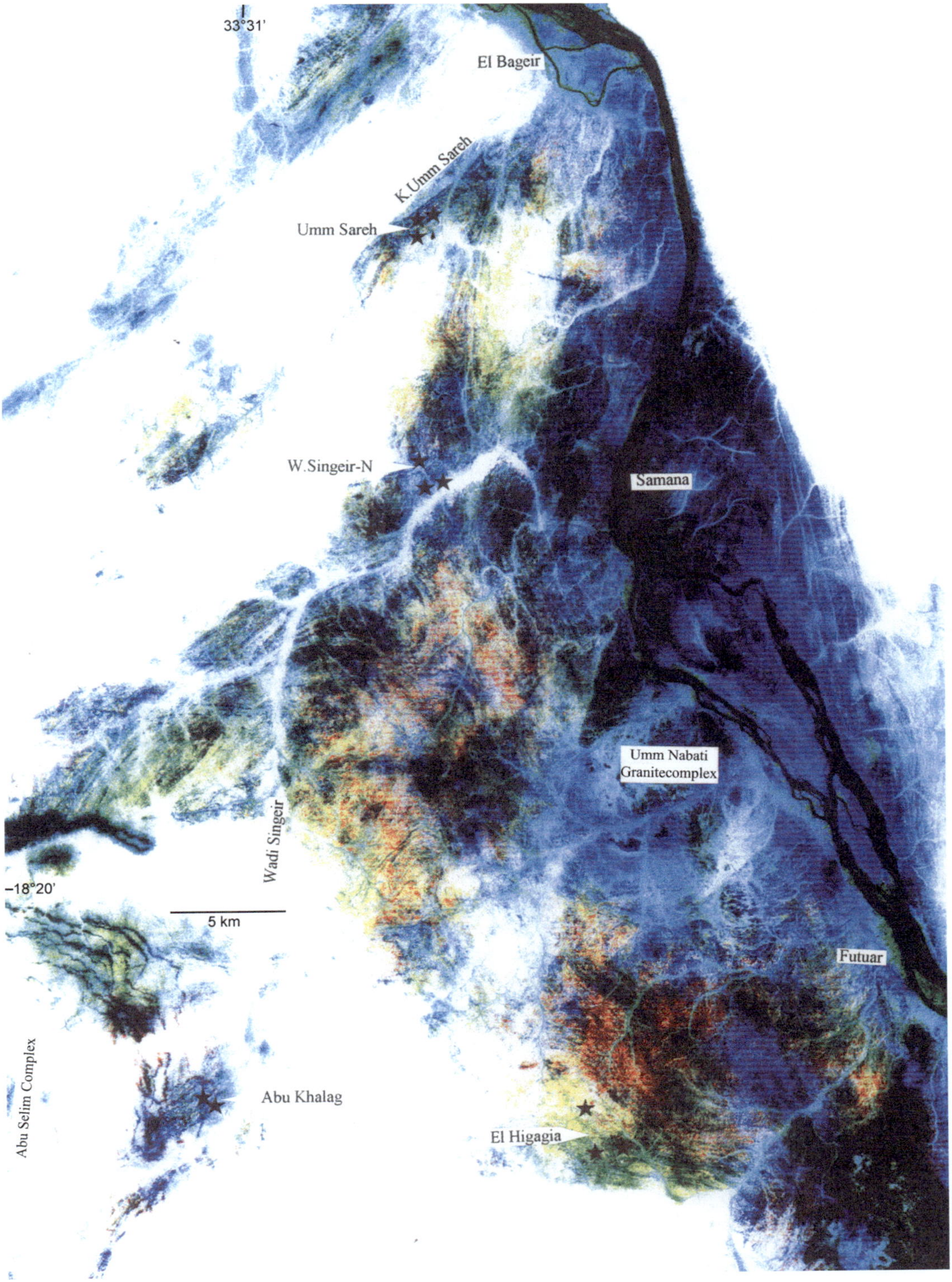

Fig. 6.249 Lithologically processed satellite image of the eastern Bayuda Desert. The rock formations appear in different, dark colours (Gebel Samana), recent sand-drifts are white, and the older ones are blueish (TM 173/47, channels 7-4-1)

Fig. 6.250 Present-day wadiworking activity in the Bayuda Desert, basically relying on almost the same methods as in antiquity

Fig. 6.251 Red spoil heaps from recent wadiworkings at Abu Sugha, Bayuda

Fig. 6.252 Auriferous quartz chunks are being extracted from wadi sediments in pit holes. El-Higagiya, Bayuda

Fig. 6.253 Crushing-down quartz chunks to pea-size before further processing to powder in a diesel-powered ore mill. Hirtawi/Bayuda

Fig. 6.254 Usually women and children are involved in the washing procedure of the finely powdered quartz ore. Kewaifira, Bayuda

Fig. 6.255 Auriferous, black heavy metal fraction (predominantly magnetite and ilmenite), representing the last stage in the concentration process. Umm Sareh/Bayuda

Fig. 6.256 The heavy metals are amalgamised with mercury. Umm Sareh, Bayuda

Fig. 6.257 The gold amalgam is heated on a metal sheet in order to evaporate the mercury from pure gold. Umm Sareh, Bayuda

out without any protection over an open flame, thereby exposing the labourers to the life-threatening fumes.

6.9.1 Umm Sareh

Geographic position:	18°38′25″ N, 33°35′08″ E

The given position is only approximate, as it is intended to indicate merely the area of our survey work along the extensive wadis and slopes of Umm Sareh and its neighbouring gold mining districts of the Eastern Bayuda.

Derelict wadiworking areas are encountered already over widespread areas in Umm Sareh, and only few people live there today.

They live in simple huts, and benefit from basic services consisting of small sale stands. They sleep in rudimentary beds made from woven mats, fastened to wooden frames, which interestingly enough, very much resemble the beds known from Ancient Egypt.

Women and children again participate in the gold production. They usually work at the waterholes under the protecting shade of a small canopy attached to wooden poles.

There are also two primitive, about 5 m deep mines at Umm Sareh (Fig. 6.258). During our stay, they were being exploited without any ground support or protection. The extraction is operated in three shifts of six hours each with teams of five miners. Each work gang keeps the ore it is able to remove during the shift. This may cause problems and lead to disagreements, as the gold grades in the quartz seam may vary considerably. However we were assured on several occasions that such disputes had never led to serious conflicts.

In the district of Umm Sareh auriferous vein deposits are apparently also subdivided into several sectors for which again, individual work teams are designated.

6.9.2 Abu Sugha

Geographic position:	18°30′05″ N, 33°34′ E

Here as well, extensive stretches in the wadi reveal deserted wadiworkings that form elongated, red bands in the landscape.

Water is found here at several locations at depths of only 1.5 m below the surface. Today its transport is done by donkeys in a constant shuttle back and forth to the workings.

6.9.3 El-Higagiya

Geographic position:	18°15′7″ N; 33°39′35″ E

This occurrence is located E of Abu Khalag. Here, six to eight miners work in shallow mines and pits under most archaic conditions. The entire area is all the same covered with quartz rubble and work seems to pay-off nonetheless.

According to the workers, the gold yield adds up to 1–3 g per day and head, which is above the average income in Sudan at the end of the 20th century.

6.9.4 Abu Khalag

Geographic position:	18°11′03″ N, 33°23′ 34″ E

Recent mines are also observable in the area of Abu Khalag. The deposit has recently been depleted and is now deserted. The mines can be made out at the surface as reddish heaps of debris from selected and leached-out ores. To judge by the volume of the churned up masses of rock, hundreds of people must have worked until recently in something possibly similar to a goldrush frenzy, and subsequently abandoned the mined are as soon as they had arrived.

Fig. 6.258 Primitive deep mine at Umm Sareh, Bayuda. Only the use of metal tools distinguishes this mine from the ancient mining methods

7 Sequential and Spatial Distribution of Gold Mining Sites in the Egyptian and Nubian Eastern Deserts

In the previous discussion on the ancient gold mining sites in the Eastern Deserts of Egypt and Nubian Sudan their processing periods are established according to the approximate dates of the recorded artefacts at the site surfaces whenever possible. They show that many deposits had been exploited during two and even more periods.

Figures 5.1 and 6.1 display the spatial distribution of the more than 230 recorded gold mining sites in the surveyed area. They also show that all deposits are found within the Precambrian basement and, apart from a few cases, more specifically within the geologic overthrust series of the Arabian-Nubian Shield. The resulting genetic implications have been amply discussed in previous chapters and for the present purposes will not be addressed again.

The geographic distribution of the sites for each period is particularly interesting. Despite the relatively rough chronological resolution, significant conclusions can be drawn concerning this relatively small but nonetheless important economic sector of the Ancient Egyptian and Nubian economies.

One of the first things the distributional maps reveal is the strikingly low number of sites in the early periods. This is in part due to the accepted circumstance that there had been fewer mining sites in the early days of the industry but also that their traces have been shrouded and even eradicated by later activities. Furthermore, because the present survey project was exclusively based on the recording of surface finds, it is quite likely that a considerable number of ancient sites reveal considerable more evidence from these early periods if they were investigated in systematic archaeological excavations. For the lack of time and especially excavation licenses, it was beyond our possibility to carry out more detailed investigations. We can therefore only hope that the present results will have an inspiring influence on future more detailed archaeological investigations.

The probably most ancient gold mining sites in Egypt are represented by the so-called "Earliest Hunters" group, which was first described by Winkler (1938) and thought to date back to the end of the fifth millennium BC. Analysis carried out on a yet small number of samples antedating the Early Dynastic Period, whose beginning is normally set at around 3000 BC, showed that the artefacts had been made from placer gold based on their purity above 90 %, which by far exceeds the primary averages for Egyptian gold. The nuggets had formed through natural cold-welding and forging of minute gold flakes by continuous movement of rubble in fluvial transport. The slightly more soluble silver contents were gradually leached out by water from the areas near the nuggets' surfaces, which led to a relative increase of the gold's fineness. Due to the poor hydrology in the form of sporadic wadi floods, since the last ten-thousand years nugget formation processes are likely to have stalled with the conclusion of the Pleistocene Period.

Umm Eleiga is the only site we visited that may be considered as attributed safely to this early group (Fig. 7.1, no. 122). As reported this is one of the few areas still marked by a Pleistocene landscape of river valleys crossing a district with

R. Klemm, D. Klemm, *Gold and Gold Mining in Ancient Egypt and Nubia*, Natural Science in Archaeology,
DOI 10.1007/978-3-642-22508-6_7, © Springer-Verlag Berlin Heidelberg 2013

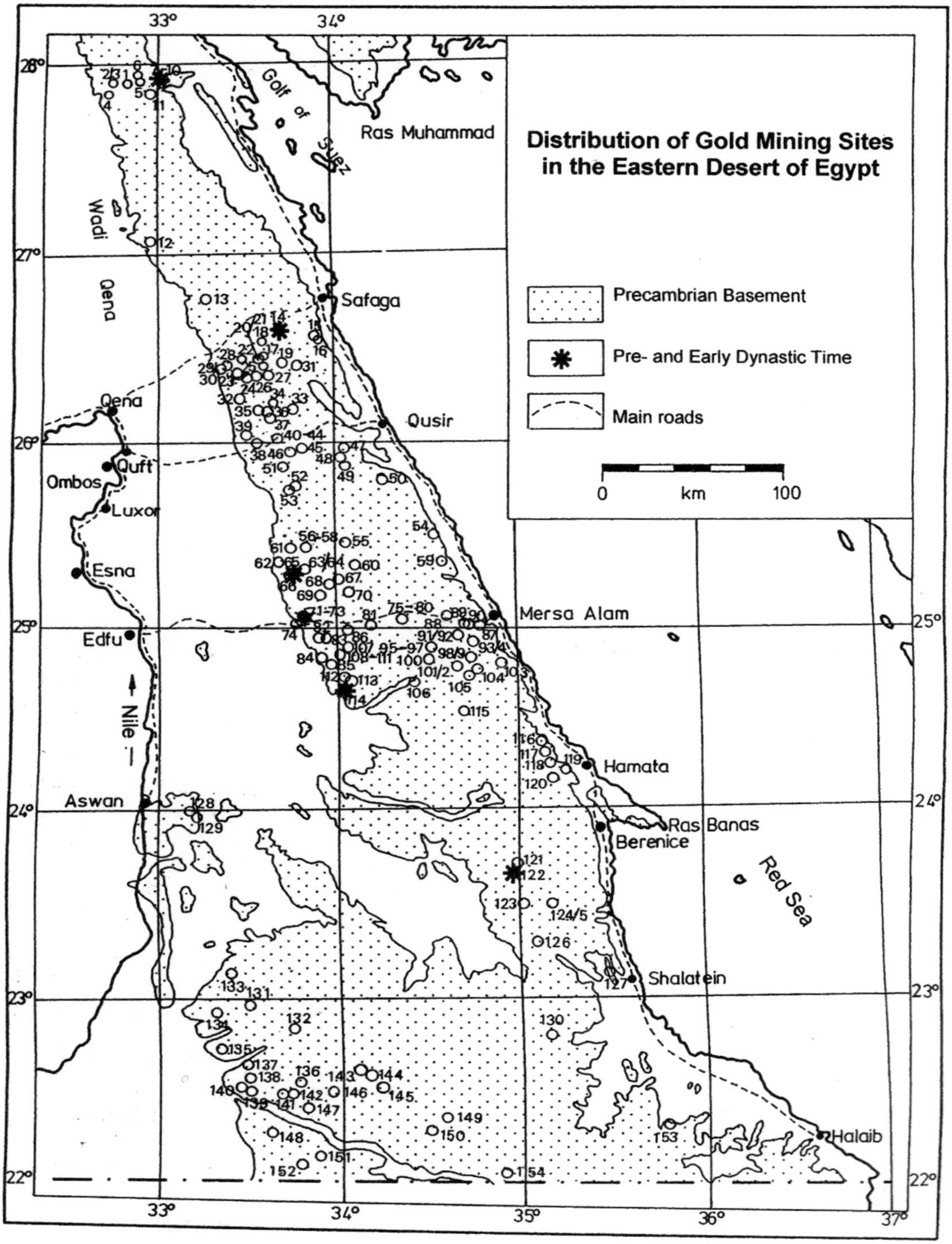

Fig. 7.1 Spatial distribution of Pre- and Early-Dynastic gold mining sites in the Egyptian Eastern Desert (numbering after Klemm and Klemm 1994)

recumbent gold quartz veins that later on, chiefly in the New Kingdom and Early Arab Period, had undergone systematic exploitation.

It may be assumed that during Predynastic times, most of the originally freely available gold nuggets in such wadi grounds had already been removed by early nomadic groups. In later periods therefore, new supplies for this highly sought-after material needed to be found. At the end of the fourth millennium apparently first experiments were made to extract the metal directly from quartz veins with simple stone hammers. This time coincides generally with the first appearance of copper artefacts in the Early Nagade Period (Lucas and Harris 1962). In these early days of metallurgy, at least as far as the Egyptian Eastern Desert is concerned, it is conceivable that there had been a parallel development for both copper and gold mining, which from a perspective of geology only seems reasonable.

In the previous chapters we repeatedly referred to the association of sulphide minerals and gold in hydrothermal quartz vein deposits. There, the sulphide minerals consist mainly of pyrite (FeS_2), chalcopyrite ($CuFeS_2$), chalcocite (CuS_2), arsenopyrite (FeAsS), sphalerite (ZnS) and galena (PbS), to name but a few. Because of oxidative processes at the surface, the sulphide minerals tend to dissolve under formation of sulphuric acid and remove the metal contents. This is also the case for the relatively low soluble copper contents, which however precipitate within the clefts of the parent quartz and the adjacent wallrock in the form of copper carbonates (predominantly bright-green malachite $Cu_2[(OH)_2/CO_3]$.

Copper carbonate is easily recognised by its vivid colour and is relatively simple to transform to metallic copper in a charcoal kiln. These were therefore also the ores the copper prospectors had been looking for, first in the territories of the Eastern Desert, later in the Sinai, and from the New Kingdom onwards also at Timna, today Israel.

The early copper miners no doubt occasionally came across visible primary gold occurrences along the quartz veins they were excavating. They certainly also ended up by eventually recovering this metal separately.

The recognition of the linkage between both metals then led to the malachite linings near the quartz veins also becoming the main criterion for gold prospectors in the following Early Dynastic Period. Consequently, the distinction between ancient gold and copper mines in some cases may be problematic.

In spite of evident malachite linings along the wallrock, the most important early mines in Nubia and Egypt must exclusively have been based on the extraction of gold. The number of identifiable Pre- and Early Dynastic gold mines in the Eastern Desert is still very low (Fig. 7.1), though to which two must be added that are located in the immediate vicinity of the Nile in Lower Nubia (Duweishat, Abu Sari ?).

The basic function of the heavy, two-hand-hammers in the shape of calabashes, as well as granite discs from mining contexts indicate to early extraction activities in the Pre- and Early Dynastic Periods. The two-hand-hammers had apparently served to loosen and especially crush the auriferous quartz rock to a fine powder in situ. In a second stage the released gold dust was then separated mechanically from the rest of the quartz powder, probably in a water suspension. So far, there is no archaeological evidence that may furnish firm indication as to how exactly this was done. The disc-shaped granite slabs, however, seem to have been used mainly for work within the malachite-yielding clefts in the wallrock and less so for processing the hard quartz material from the vein. Only few mines associated to this type of technology are known so far in the Eastern Desert: Wadi el-Urf in the N, Abu Mureiwat and eventually Bokari in the centre, and Higalig in the S.

The mines associated to grooved mallets reveal a slight technological progress in this early phase of incipient deep mining. The grooves of the mallets served to attach a forked wooden shaft. With the shaft, the impact from the blow was increased significantly. As a result, such stone mallets became susceptible to suffer from splitting and chipping. They therefore occur often in large numbers in discarded find contexts near the mines where they had been used. This method too, consisted of the grinding of the mineral ores to a fine powder before separating the gold from the rest of the barren quartz powder in an undocumented method probably involving the use of water.

Early gold mines displaying grooved mallets were recorded at the mines at el-Urf, Umm Balad, Barramiya, and Bokari in the Eastern Desert of Egypt, and at Duweishat, Tinari, Abu Sari and Abu Siha in Nubia. The mines had probably remained in use during the Old Kingdom, which is supported by an inscription in the temple of Djedkare-Isesi (around 2330 BC), in which Nubia for the first time is described as "land of gold" (Grimm 1988). The therein mentioned three Nubian place-names may possibly even partly designate the sites we were able to inspect.

These early underground mines, whose side walls usually are very smooth, essentially represent a further development in the tradition of the original simple pit trenches. The mines therefore probably evolved at ore-rich rock outcrops containing visible gold along the usually narrow quartz veins. With ongoing excavation, the trench transformed to a mine as it penetrated deeper into the ground. In order to follow a few centimetre-thick vein, it was often necessary to remove much of the barren wallrock in order to obtain a corridor wide enough to hold a miner.

Although dwelling huts are recurrently found near these early mines, genuine settlements as known from later periods yet need to be identified with assurance. In this respect, the best observed occupation remains have been observed at the Early Dynastic/Old Kingdom sites at Wadi Dara, Bokari and Abu Mureiwat.

A marked change in the extraction technology of gold and copper (malachite) deposits occurred at the beginning of the Old Kingdom. The former, heavy tools were now being replaced by ones easier to handle, especially because of their light weight and increased ergonomics, as exemplified by hammers with integrated grips that could be manipulated with only one hand. Their average length varies between 18 and 20 cm with a diameter of about 5 cm. They usually consist of dolerite. They are mostly broken-off at an angle at their blow surface. Fist-sized dolerite globes, of which some are flattened on one side, are also relatively common. They had probably served to crush (grind?) the extracted ore to a fine grain.

Stone mortars and matching pestles that may have served in an additional crushing phase were found at only few sites, whose main exploitation periods do not ante-date the Middle Kingdom. Such mortars are usually only preserved in fragments and are so rare that they are unlikely to have played a significant role in ore processing technologies. On the other hand, since only fragments remain, it is also conceivable that complete specimens may over the centuries have been removed and reused by mobile Bedouin groups. The final gold processing method in the Old and Middle Kingdom Periods remains mostly obscure.

All deposits associated with such tools are characterised by the mentioned malachite linings at the wallrock next to the quartz veins. Some had at first been limited to copper extraction and then maybe later (also) to that of gold. This for example was the case at the workings of Wadi Dara, Umm Balad, and el-Urf, but probably also Sagia, Soleimat, and Hamash in the Egyptian Eastern Desert, and most definitely Umm Fahm in Nubia.

Usually, the actual mining started at the ore-yielding surface outcrops of the quartz vein and through continuing excavation work then developed into open trench pits that could reach depths between 10 and 15 m. Most of these mines have nevertheless collapsed in the meantime, hence conceiling their original depths, which in the end can only be established individually by means of excavations. Wadi Dara and Umm Balad 3 form an exception with respect to this aspect of the early mining phase.

The dating of these tools to the Old and Middle Kingdom has been affirmed by few, but nevertheless diagnostic pottery finds.

The distribution of gold (−copper) deposits assigned to the Old and Middle Kingdoms in the Precambrian basement of the Egyptian Eastern Desert is shown in Fig. 7.2. There are clear concentrations in the western part, near the important routes Qena-Safaga and Quft-Quseir on the one hand, and at Edfu-Marsa Alam on the other. A third cluster of gold (and copper) mining sites

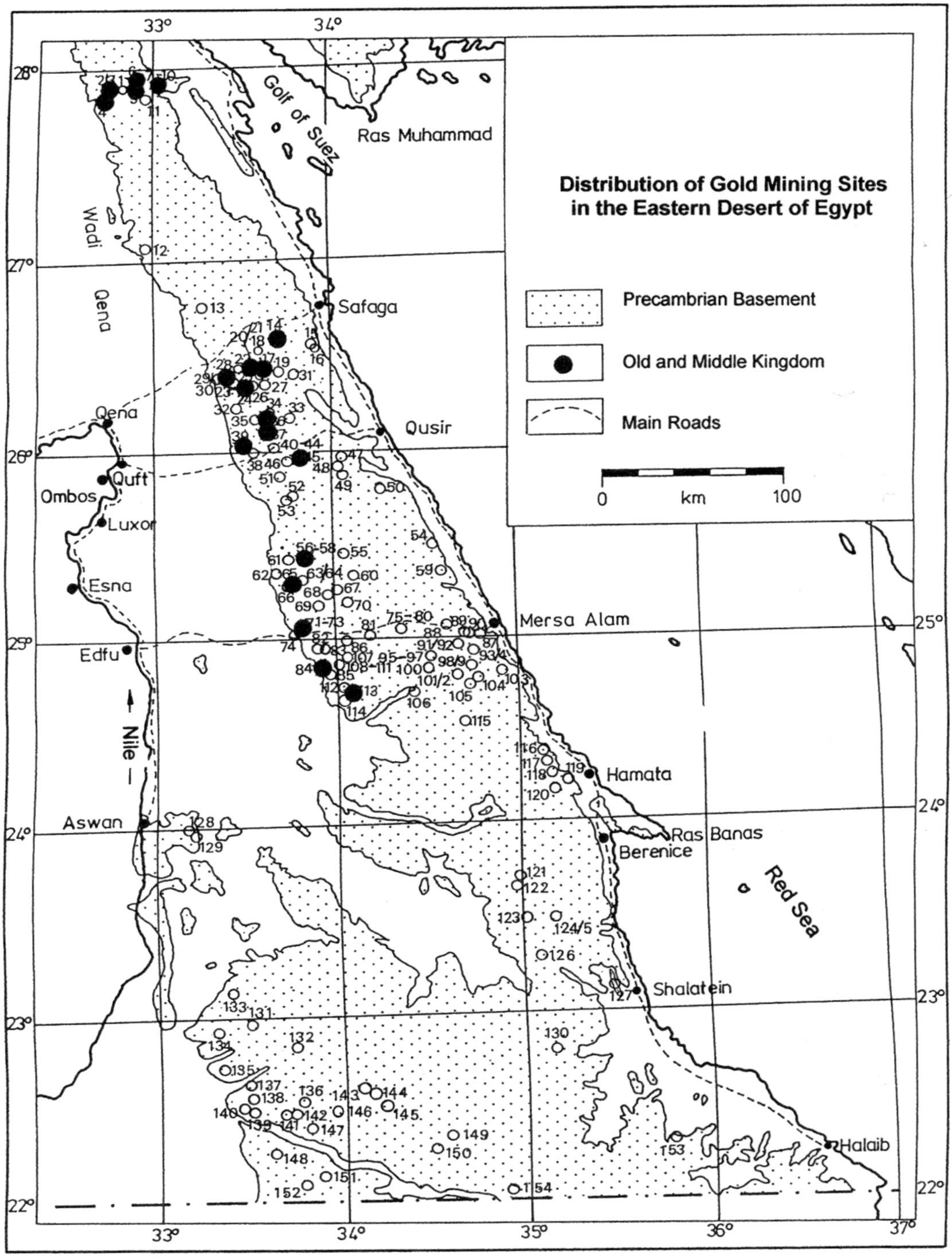

Fig. 7.2 Spatial distribution of Old- and Middle Kingdom gold mining sites in the Egyptian Eastern Desert (numbering after Klemm and Klemm 1994)

in an apparently exclusive location is recognised in the northern part. Its particularity is mainly explained by the fact that in the later periods, except briefly in Early Arab Period, mining was no longer practiced here. This is to some extent surprising, as in its early phases mining industry had developed here in particularly skillful way. The mining tools in this region nevertheless don't reveal any differences to those observed in the southern sites. The specificity of this region seems at first rather connected with the nature of its ore-yielding quartz veins, which at Wadi Dara dip steeply into the mountain, and while displaying exclusively short, open trench pits, can then be pursued in deep mines only. At Umm Balad 3 the ores occur in outcropping, but flat-lying vein systems leading almost horizontally into the mountain, permitting no other exploitation than underground mining. But since the remaining occurrences in this northern region exhibit more or less the same features as those in the S, its particularity may after all be deceptive. It primarily proves that the miners had no particular difficulties in adapting to the technical problems set by the spatial orientation of the veins as long as the metal contents justified their exploitation.

The reclaim of the region by the mining industry in the Early Arab Period had probably a lot to do with a previous depletion of the local deposits, but had probably only lasted for a short while as a side-event of the so-called goldrush affecting the entire Eastern Desert of Egypt and Nubia in this period.

The fundamental question concerning the Old-Middle Kingdom gold/copper mine concentrations in the western Precambrian basement still remains unanswered. In the case that mining operations had primarily been conducted by mobile desert groups, it seems surprising that they apparently hadn't been attracted by the eastern gold deposits, which they inevitably would have noticed during their migrations. The answer to this paradox seems to lie in the ancient prospecting method, which chiefly relied on the presence of malachite linings at the mineralised quartz veins. In the more eastern zones such linings are relatively rare. Even the few rock inscriptions from this period, which only have a distant connection to gold mining, are all located in western areas (e.g. at Dungash).

Issues pertaining to the so-called "fire-setting" method are often raised in discussing early pit trench and underground mining. This technology consists of lighting a fire directly in front of the face of the vein to be worked with the intention of decreasing the rock's tenacity by gushing water over it to create a temperature shock thus slackening the rock. Yet, this method requires much fuel as well as water, of which both are just barely available in arid desert zones.

According to Weisgerber (1989, 1990), the technology was applied in Neolithic and Early Bronze Age mining industry in Europe. Agatharchides in Diodorus Siculus (Oldfather 1935) attests the method to the Ptolemaic Period (~220 BC) in Egypt.

Moreover, Castel and Soukiassian (1989) claim to have identified smoke traces from fire-setting in a Middle Kingdom context at the galena mines of Gebel el-Zeit, although this is still somewhat doubtful as they may originate from a later, domestic context.

In spite of these references we did not observe any conclusive indications whatsoever for the fire-setting technique during our numerous visits of the ancient, opencast and underground gold mines neither in the Eastern Desert of Egypt nor that of Nubia.

A genuine technological revolution exclusively restricted to the gold mining industry occurred at the beginning of the New Kingdom (Fig. 7.3). With the consolidation of state rule towards the middle of the second millennium BC, numerous gold mining sites in Egypt's Eastern Desert had become accessible through the use of new processing equipment. And, in the wake of the sack of the Kerma Kingdom, this was now also true for Nubia. From the early New Kingdom onwards, gold mining not only expanded into Egypt's more eastern Precambrian basement zones, but also into southern areas around Wadi Allaqi, and especially throughout vast regions in Nubia, both close to the Nile, as well as in the open desert zones to the E, extending to the Red Sea Hills (Fig. 7.4).

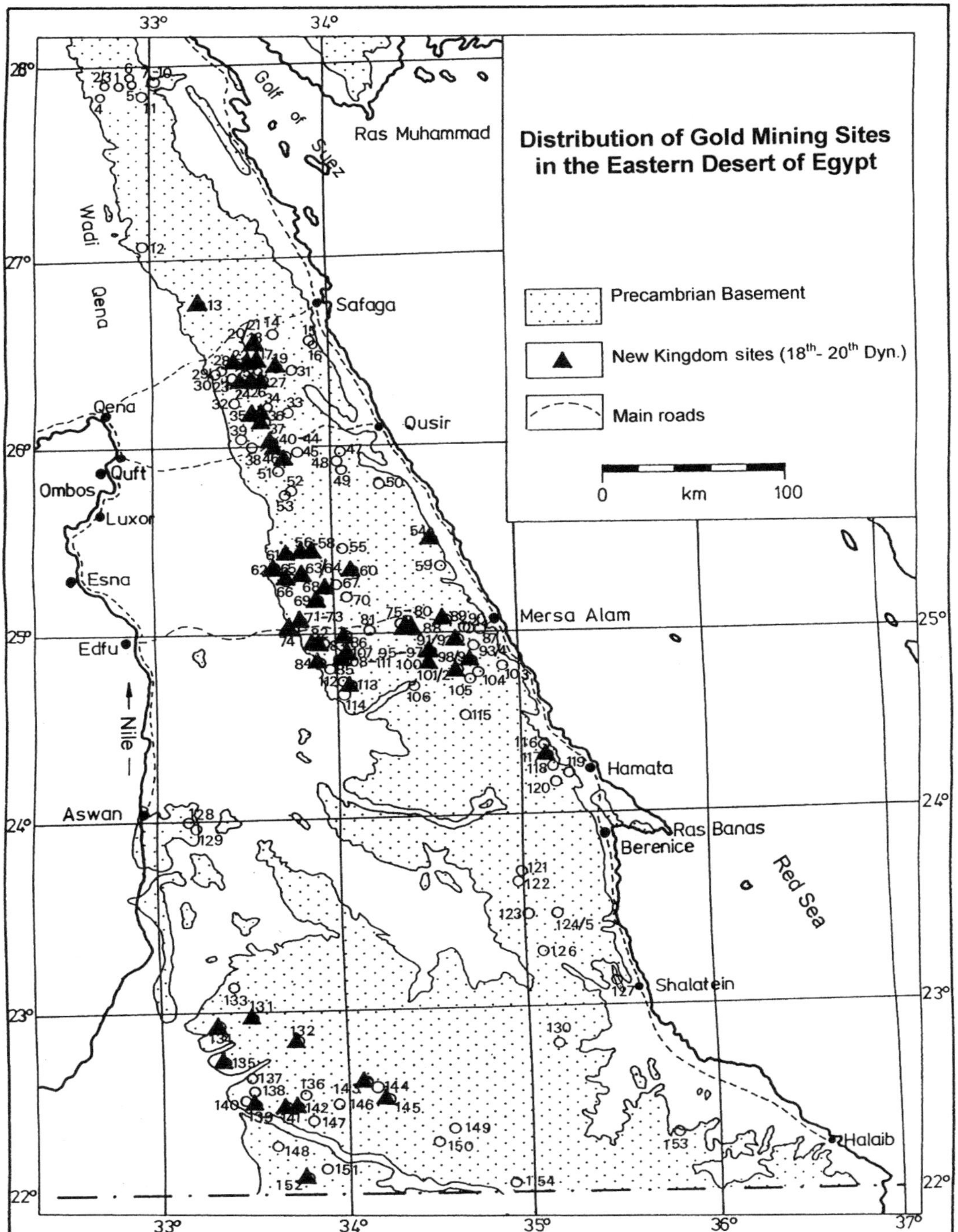

Fig. 7.3 Spatial distribution of New Kingdom gold mining sites in the Egyptian Eastern Desert. Note the abandonment of sites in the northern Eastern Desert and the massive expansion into the southern regions, especially around Wadi Allaqi (numbering after Klemm and Klemm 1994)

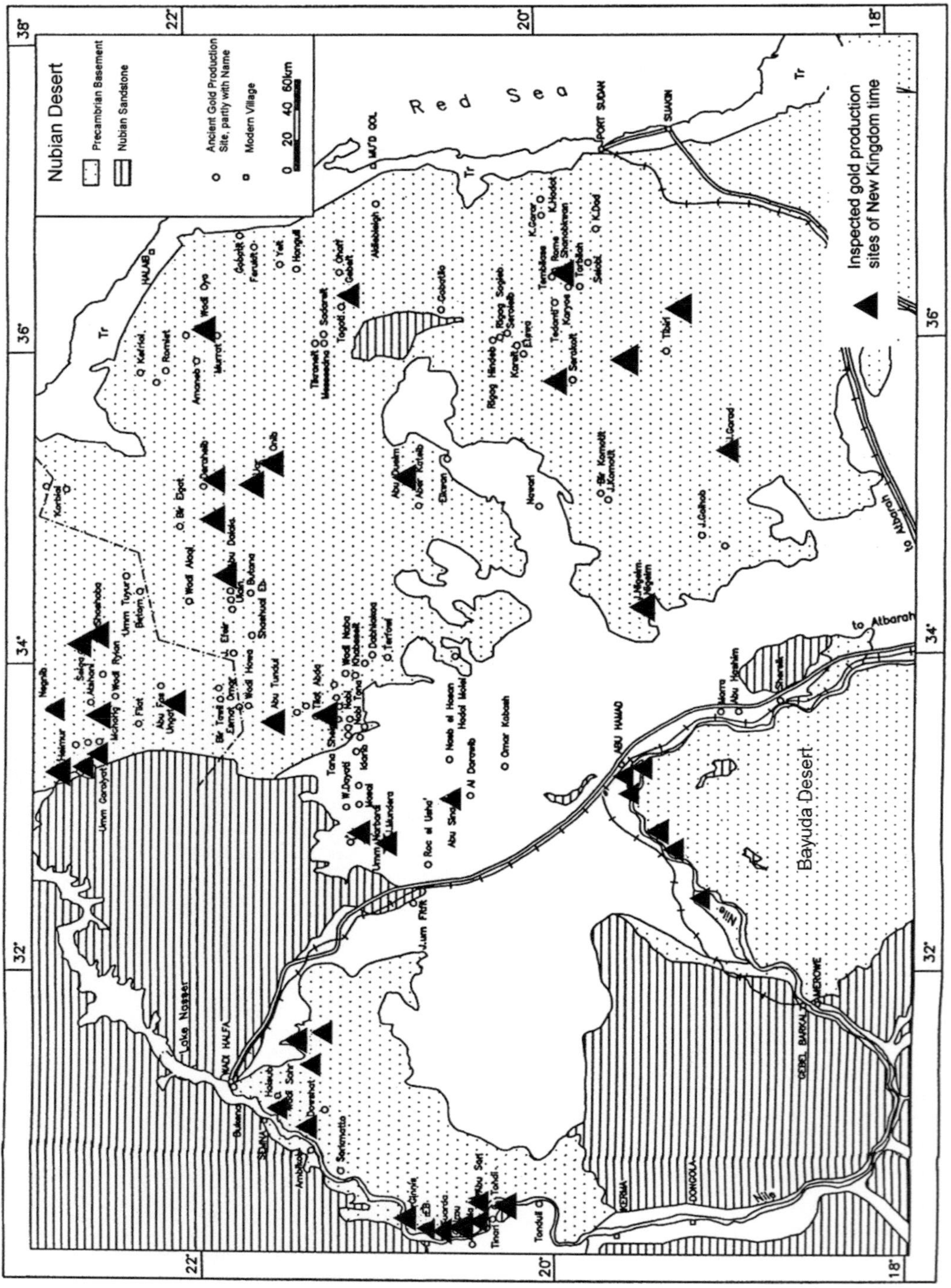

Fig. 7.4 Spatial distribution of gold mining sites in Nubia during the New Kingdom. At most sites gold was mined in wadiworkings, particularly along the Nile (modified after Klemm et al. 2002b)

The gold industry's significant expansion throughout Egypt and into Nubia is also reflected by a palpable increase of wealth and variety of gold artefacts from this period.

Already known mining areas in the eastern desert were re-opened on a large scale and considerably extended. In their surroundings, many new and especially newly discovered deposits were developed for mining. Comprehensive ore prospecting campaigns stretching over wide areas of the Eastern Desert resulted to the development of a great number of highly productive gold mining areas, especially in its eastern zones.

Entirely new gold deposit areas were discovered and developed in and around Wadi Allaqi, the southernmost, until that time for the most neglected area inside Egypt, which was known as Wawat.

Through the destruction of the Kingdom of Kerma and the Egyptian occupation of Nubia, numerous auriferous districts throughout the Precambrian basement between the Nile Valley and the Red Sea Hills had become accessible to the Egyptian ore prospectors. The development of the gold deposits in the Eastern Desert of Nubia was fostered especially from the reign of Thutmose III onwards. Amenhotep III even had a viceroy appointed in Kush for administrative purposes that most probably included matters pertaining to the gold mining industry.

In addition to improvements in the actual mining engineering technologies, which systematically comprised abutments and pillars from ore bodies, thereby significantly increasing security inside the mines, copper and bronze chisels were added to the already existing stone tools.

The most significant innovation, however, was the organised prospecting of hills and wadi floors for auriferous quartz rocks originating from veins that had ended up in the wadis through desert erosion (wadiworkings).

With little exercise it was possible even for the less skilled to distinguish between barren and auriferous quartz fragments within a given wadi sector. Hence, large work gangs could be deployed, who equipped with hoes and baskets, within particular area would simply pick up the ore chunks from the wadi ground. Today such workings are recognised by the flattened remnants from spoil heaps in the wadi beds.

In comparison to the arduous work inside a trench mine, where few workers extracted only small amounts of material, the method of gathering-up ore chunks in the wadis multiplied the efficiency, as virtually unlimited numbers of labourers could be hired, even if the chances of collecting barren material by error was much higher.

The New Kingdom houses and huts differed according to the required number of workers. We generally recorded only few hut ruins of varying sizes and orientations near the mines and containing small numbers of grinding mills and pottery.

As to the organisation of the labour force, the related titles, and functions as stated in the written records, a comprehensive work was carried out by Wilsdorf (1952)

Near the wadiworkings, however, the houses and huts are not only frequently larger and subdivided into several rooms, but are in addition arranged in regular alignments along the wadi edges. This seems to reflect a strict organisation, and the larger sizes of the houses suggest somewhat better living conditions for the labourers in the wadiworkings. As opposed to the settlements near the deep mines, the ones near the large wadiworking areas were furthermore usually associated to nearby wells, not only for ore processing purposes, but also for the supply of the workers.

In addition to the actual mining and ore extraction techniques, fundamental innovations are also observed in the processing technologies. The most noteworthy one was the introduction of a stone mill with a first normally flat and later oval grinding surface and a hand-held runner stone, with which pea-sized ore grains previously prepared on a large anvil stone were ground down to a fine ore meal of only few microns. The mills occur in varying sizes but because of their yet generally considerable weights, are often found left lying in relatively large numbers at their original locations, where they had once been used.

Another decisive advance in the gold processing technology is represented by the introduction of the so-called washing table during the New Kingdom, which has so far remained unreported from the earlier sites. It consists of a dry masonry stone structure with an inclined table surface, whose lower end opens into a small stone basin. From there a return channel leads back to a slightly deeper basin at the other end of the structure. Such devices are well- known from the Ptolemaic and especially the Early Arab Period, but hitherto unreported from New Kingdom gold mining sites in Egypt, where they usually have been damaged beyond recognition.

Thereby the essential phase of gold washing is also for the first time attested to for the New Kingdom. The identification of this installation as a device used for washing gold is partly owed to Agricola (1546) who had described a "hearth" used in the mining industry towards the end of Medieval Period in Germany. Some examples from the Early Arab Period still reveal the residues of a mortar coating on the inclined surface, which of course are no longer preserved on the ones from the New Kingdom more than 2000 years older. The washing tables are often observed amidst surrounding tailing heaps of different sizes consisting of fine, more or less sharp-edged quartz powder often stained red by iron oxides. Usually, the tailings are in better states of preservation at Ptolemaic and Early Arab sites than at the older ones from the New Kingdom, where erosion has had substantially more time to reduce if not completely eradicate them. As the washing table's function was to separate the gold from the quartz, we are now able to restitute almost without interruption the entire chain of operations in the gold processing phase. Agricola (1546) describes how quartz ore powder is poured in a water suspension over a hushing board with wooden ribs covering a "hearth" (= washing table), whereby the heavy minerals are deposited in the grooves while the lighter silica flour remain suspended in the flowing water. In Egyptian gold mining these heavy minerals would have consisted mainly of iron oxides and to a lesser extent of various sulphide minerals, as sulphide zones were generally avoided in New Kingdom mining. The method would therefore hardly have produced a satisfactory result. A much more effective separation would have been obtained through using sheepskins on the inclined surfaces. The sharp-edged gold sequins would thereby have been retained within in the greasy and fine fur of the sheepskin, whereas the other heavy minerals and the quartz would have been flushed into the basin.

For the lack of archaeological evidence it is not clear which of the two separation methods had been applied. Neither the wooden lattice nor the sheepskin would have survived even in this arid zone. The sheepskin method, however, does have the advantage of being more efficient. In this method the sheepskin is removed from the washing table as soon the adhesive lanoline retaining the gold sequins has decreased beyond a certain degree. It is then set on fire in order to retrieve the gold from the remaining ashes. Whatever the method applied, no direct archaeological evidence concerning this process may be expected from the field. Sheepskins were at any rate available, as witnessed by the bones observed in several waste piles. As already stated, we also view the legend of the Golden Fleece to advocate the use of sheepskins in the gold separation process.

The maps showing the distribution of the New Kingdom gold mining sites in the Egyptian Eastern Desert, Nubia and the Nile Valley first of all underscores their preponderance over the entire Eastern Desert.

We already voiced some thoughts as to the beginning phases of this upsurge in the gold mining sector during the New Kingdom Period. It was outlined that the political preconditions in Nubia had been set in the early eighteenth dynasty with its conquest by the Egyptian King Thutmose I at around 1500 BC. His army went deep into Nubia until the fifth cataract, safeguarding the access to the goldmines. From then a high Egyptian official was based in Nubia to control the gold production. Sparse pottery evidence from deep inside the Nubian Eastern Desert dating to the time of Thutmose III was found at gold mining sites specialised on wadiworkings, deep mines were nonetheless found at Umm Nabardi, Shikryai, and Onib that together with Oyo, Gebeit, and Shanobkwan reach right to the Red Sea Hills.

These widely dispersed operations within occupied and in logistical terms difficult desert territory had only been possible within a context of pacification of the local populations. By its local, Nubian characteristics, the observed pottery remains lead to suggest that at major part of the labourers working at these sites had belonged to such local populations using oval ore mills introduced by the Egyptian miners. It is therefore questionable whether they had worked here by their free will.

The near absence of cemeteries in the immediate surroundings of the gold mining sites in both, the Nubian and Egyptian Eastern Desert is quite striking. Although understandably virtually no burial was found in the wadiworking areas, practically the same applies for the settlements near the deep mines. One might be tempted to argue that this stood in some connection with the Egyptian custom to repatriate the deceased, although this seems hardly plausible for such distant territories as Nubia. Possibly the scarcity of graves in the vicinities of New Kingdom gold mining sites can to some degree be explained by the application of relatively high safety standards, reflecting some sense of responsibility by those in charge. Texts by the kings Sethos I and Ramesses II mentioning work gangs charged with digging wells for the miners on their way to the goldmines may be put in this context (Schott 1961; Zibelius-Chen 1994).

Such conjectured conditions are only conceivable under strict and highly organised governance. In Nubia such conditions were given only until the end of the eighteenth dynasty (about 1300 BC), and in the Nile Valley they seem to have been in a continuous decline until the middle of the nineteenth dynasty (about 1200 BC.) However, evidence shows that Ramesses II had nevertheless been able to maintain the gold mining industry in Wadi Allaqi and that its downfall only becomes palpable with the reigns of his successors.

In Egypt's Eastern Desert, however, gold continued to be mined well into the time of the twentieth dynasty, and declined after the reign of Ramesses III. This at least is suggested by donation lists from the mortuary temple at Medinet Habu (Hölscher 1957) and details given in the Papyrus Harris (Schädel 1936), although the therein specified enormous amounts of gold are not very credible.

There is no convincing evidence pointing to gold mining in the Egyptian Eastern Desert between the twentieth dynasty and the beginning of the Ptolemaic Period. Apparently, the fading central power was in spite of some illustrious rulers no longer capable of asserting its authority to uphold logistics and food supply and guarantee protection of the routes and the workforce in remote areas of the Eastern Desert. This is perplexing in that the quarries delivering the raw materials for the heartland's sarcophagi, statues, and shrines were still functioning as late as the thirtieth dynasty in Wadi Hammamat and even far-flung desert sites such as Rod el-Gamra (Harrell 2002).

While in much of the Egyptian Eastern Desert gold mining activities seemed to have come to an almost complete standstill in the late New Kingdom, the evidence points to an active gold mining industry in post-New Kingdom Nubia (Fig. 7.5). The geographic distribution of sites largely reflects countless and later reclaimed New Kingdom mills recognised by deep moulds in the centre of the grinding surfaces. Their use wear suggest a reuse in a somewhat altered technique, although this seems to adhere to the original function of the Egyptian models. This later phase is probably dated to the Kushite (Meroë) Period. The scanty evidence from the pottery recovered at the site surfaces unfortunately fall short in confirming a clear-cut "Kushite" horizon. A pending elucidation of this matter will require systematic excavations, which so far have been wanting altogether. Confusing in this case is also why similar traces are completely missing in the Egyptian Eastern Desert under the twenty-fifth, Kushite dynasty. A possible explanation is that the Kushite kings ordered their gold from Nubia. Relief blocks attributed to King Piye (740–713 BC) in the temple of Mut in Luxor shows five cargo ships coming from Nubia, one of them was loaded with gold (Benson and Gourlay 1899).

A distinct resumption of gold mining activities in the Egyptian Eastern Desert doesn't become apparent before the Ptolemaic Period. This industry, however, kept its focus mainly on the central regions of the Eastern Desert (Fig. 7.6) and concentrated on deposits already under exploitation during the New Kingdom.

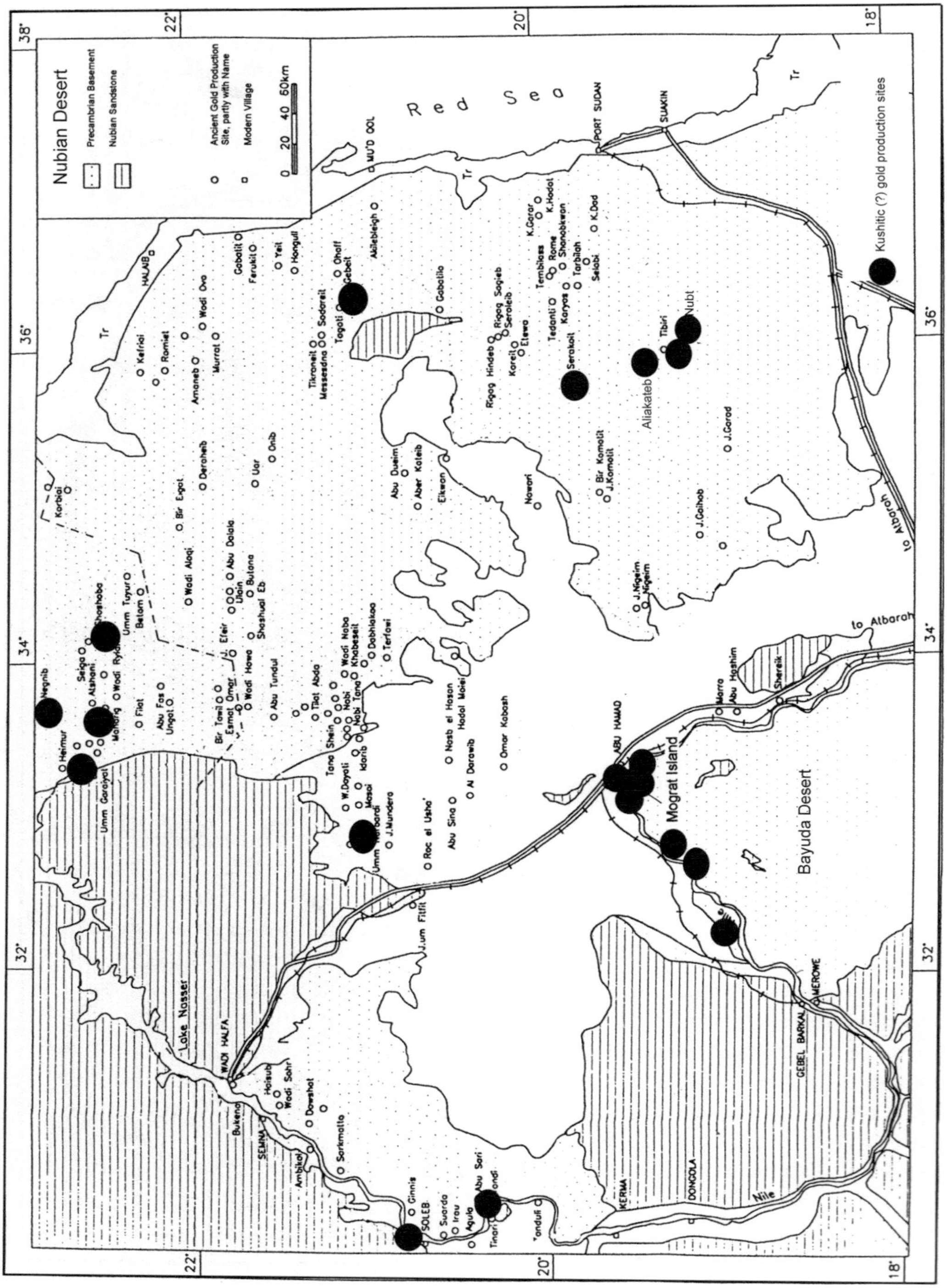

Fig. 7.5 Spatial distribution of the conjectured Kushite gold mining sites in the Nubian Desert. The relatively low number is partly due to the little evidence in general from this period (modified after Klemm et al. 2002b)

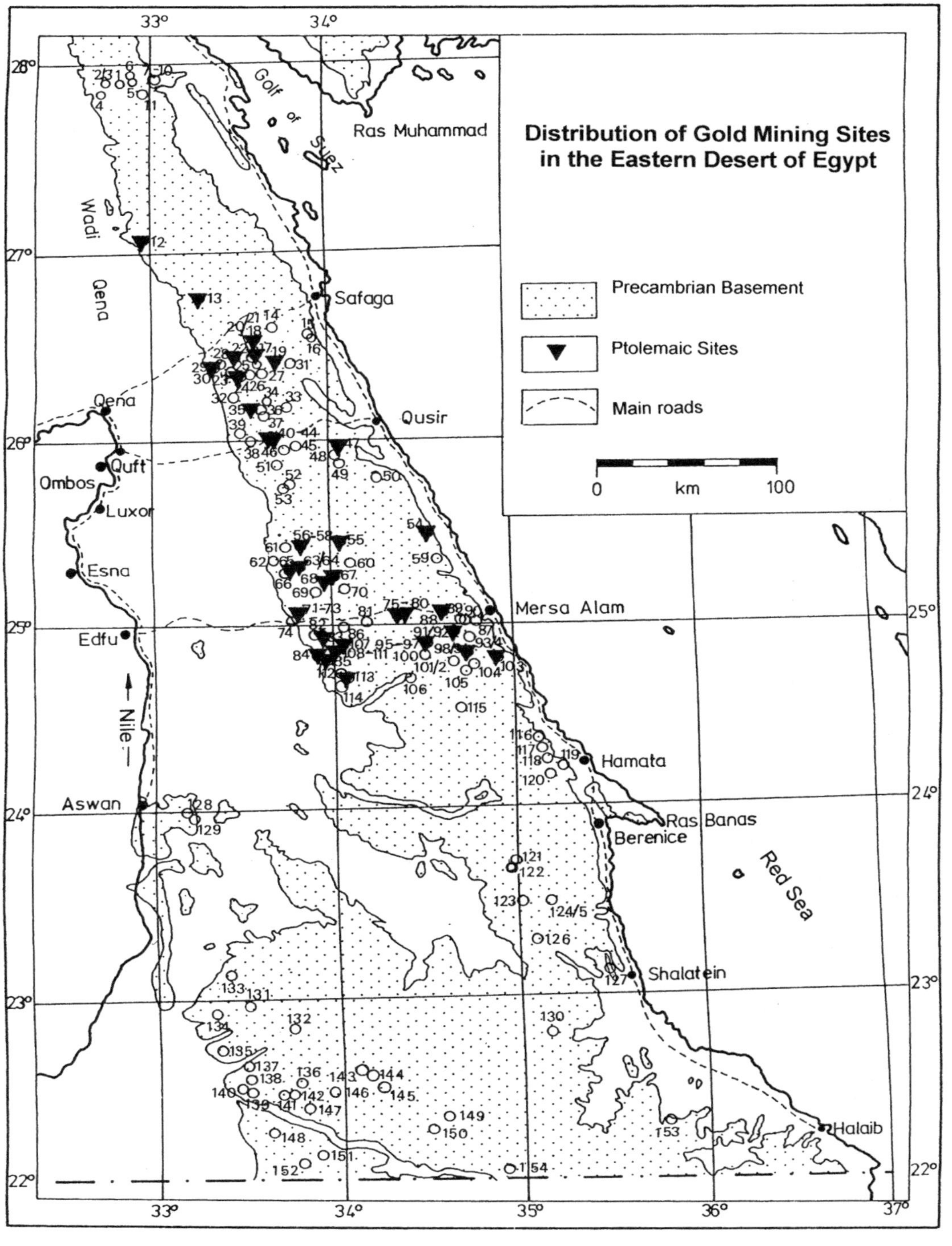

Fig. 7.6 Spatial distribution of the gold mining sites in the Egyptian Eastern Desert during of the Ptolemaic Period. Note the marked limitation of sites to the central Eastern Desert and their total lack in Wadi Allaqi, not to mention Nubia (numbering after Klemm and Klemm 1994)

Thanks to the experience from Greek and Macedonian mining technology, the Ptolemaic Period brought some significant innovations to the Egyptian gold mining industry. Of particular interest in this context was the introduction of dome-like roof structures, especially inside the stopes of flat-lying quartz veins. Losses in mining were thereby reduced significantly compared to the pillar-technique of the New Kingdom.

A very important innovation was the introduction of a completely new mill type. It consisted of a usually heavy stone block with a 60–80 cm long and around 25 cm wide, concave grinding surface. Its apron-shaped runner stone too, was heavy and equipped with two handle grips. With this mill quartz ores previously crushed to pea-size were now ground-down significantly more efficiently and above all, to a finer grain, which consequently led to a higher output of tiny gold particles.

Until the end of the New Kingdom, mining of auriferous quartz veins was restricted to the removal of oxidised sulphide ores in depths, which already on grounds of ventilation didn't exceed 35 m. The reason for this was the intimate intergrowth of sulphide minerals with gold, which even on a sub-microscopic level has the tendency to incorporate into sulphide minerals such as pyrite and arsenopyrite. These gold contents had only become exploitable to the miners in the New Kingdom through near-surface oxidation processes of such sulphide minerals.

The introduction of a ring-shaped, hydrodynamic separation device ("lavérie hélicoïdale" after Conophagos 1988) thereby represented yet another essential innovation in the ore processing techniques. It permitted an efficient separation of gold as well as sulphide minerals and therefore an isolation of the latter in view of subsequent metallurgical roasting processes through which the gold contents were extracted.

Large rectangular basins as reported from Gidami, Bir Semna, Barramiya, and Samut (buried) may probably have served as large water retentions and therefore only have had an indirect connection to gold processing.

Excluding the three northern sites at Ghozza, Abu Shehat, and Fatira there is a noteworthy concentration of Ptolemaic gold mining sites within the central areas of the Egyptian Eastern Desert, which to some degree is surprising, in that well-protected routes also led through the southern parts of the desert to the Ptolemaic harbour of Berenice. The numerous New Kingdom mining sites in Wadi Allaqi and its surroundings too, had apparently been neglected under the Ptolemaic rule, which lastly disputes the account of Agatharchides (according to Diodorus Siculus) denouncing the atrocious work conditions at the gold mining sites in Wadi Allaqi. This positioning however, may probably be due to a recurrent erroneous correlation of the "southern border to Ethiopia" with Wadi Allaqi.

The probable reason for the clustering of Ptolemaic gold mining sites in the central Eastern Desert was supposedly nonetheless connected to an increasing insecurity within the more southern areas. Here one notes the strengthening of local nomadic tribes, who under the later name of the Blemmyae became an increasing concern to the Roman administration.

With the Roman occupation of Egypt in the period following 30 BC, a new mill type was introduced. As a rule, it consisted of two superimposed stone discs. The lower, basal stone is usually between 15 and 20 cm thick while its diameter ranges between 60 and 80 cm. Its grinding surface was worn in with an upper, well-rounded rotor stone over a surface between 40 and 70 cm in diameter. By aid of a central axis and an eccentric handle bar attached to a perforation, the runner stone is set-off into rotary motion. This milling technology, which is known to have developed in the mining industry of Celtic Europe, was subsequently transferred to Egypt by the Romans, although its diffusion within the local gold mining industry at first was very restricted (Fig. 7.7).

However, at most sites in the Egyptian Eastern Desert and at all sites in Nubia, whose find contexts revealed a connection of this mill type to gold processing, such activities are dated by pottery finds and/or the presence of indisputable Islamic prayer sites to the Early Arab Period. Initially, this finding was unexpected, as the Roman occupation forces in Egypt were not renowned for their reticence when it came to

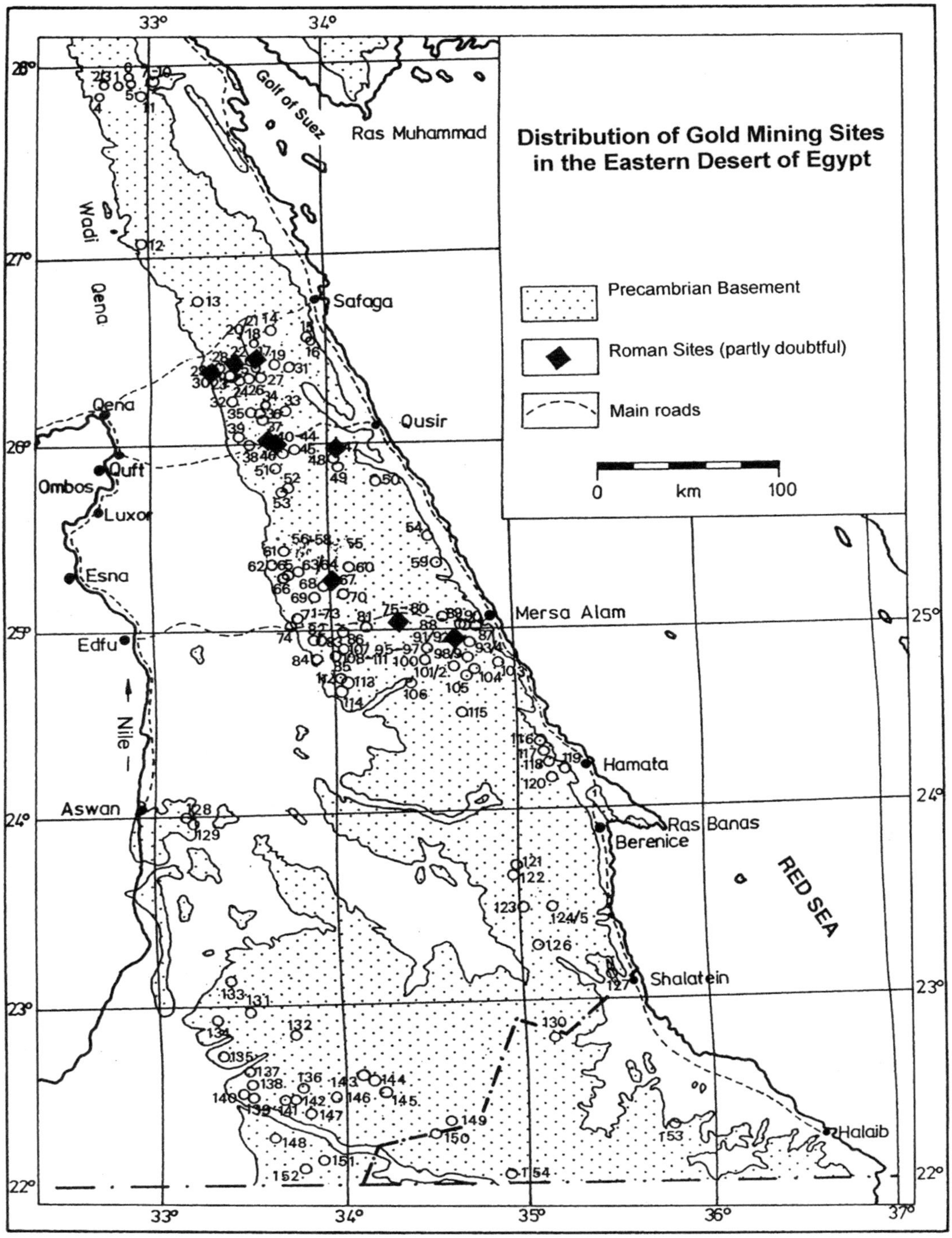

Fig. 7.7 Spatial distribution of the Roman-Byzantine gold mining sites in the Egyptian Eastern Desert. The extremely low number may result from instability caused by attacks from belligerent Blemmyes-tribes (numbering after Klemm and Klemm 1994)

recruiting forced labour. In addition, numerous and in some cases, gigantic stone quarries were operated under the Roman and later Byzantine rules, predominantly in northern parts of the Eastern Desert where access routes had been elaborately secured with forts (Klemm and Klemm 2008b). This indeed, reflects to some degree a palpable instability in these regions. An adequate securing of relatively small mining settlements spread over vast and thinly populated territories would have been too demanding on logistics, so that viable and dependable gold mining in the Roman-Byzantine Period was probably only conceivable with nearby, already fortified desert routes. The mining districts located in the southern parts of the Eastern Desert as well as in Nubia thereby fully escaped Roman control. The disruption in the eastern desert was most likely instigated by the steadily increasing belligerence of nomads from the Blemmyes tribe, whose raids were even reported to have reached the Nile Valley.

Towards the middle of the Roman Empire, the already flawed gold mining industry apparently fell into further decline and just barely survived until the Early Arab Period on a much reduced scale at Bir Umm el-Fawakhir, at the road connecting Kuft and Quseir (Meyer et al. 2000).

It was only with the rule of A. Ibn Tulun (835–884 AD) that gold mining began to expand again in the Egyptian Eastern Desert. After caliph al-Mu´tasim (833–842 AD) had lifted annuity payments to the local Arabs (Hasan 1967), the latter partially came into financial trouble. The loss of income forced some to settle down by marrying into Egyptian families. Others decided to move into Sudan, where by acquiring lands they came into armed conflict with the Bedja tribes and especially with the Nubian rulers. The Bedja were finally defeated by the Arab commander Abdullah ibn al-Jahm. But even before the actual reign of al-Mu´tasim there had been a unilateral decision (831 AD), by which the Bedja among other had been instructed to open to the Arabs their lands between "Quseir and Qubban", in other words the entire SE desert. Subsequently, the gold deposits, which had thus far been disregarded by the Bedouin tribes, came increasingly under the control of the newly arrived Arab population. In a following episode marked by the pursuit for gold, which literally developed to a goldrush when a large influx of Arabs disheartened by conditions in Egypt, reached the gold mining districts, and thus provoked the first Bedouin raids onto the mining communities in 854 AD (Hasan 1967). In the aftermath of these and the following, countless clashes, the Arab communities in the Egyptian and Nubian Eastern Deserts were compelled to settle in remote side valleys, in hidden locations away from the main, thoroughfare wadis.

The Persian historian Al-Tabari (quoted by Hasan 1967, p. 51) recounts a field campaign by Mohammed Abdallah al-Qummi in 846, who following a breach of contract, attacked with an alleged army of 20,000 warriors the Bedja under Ali Baba, who as well, had a "great army". The Bedouins evidently managed to demoralise their opponents in an attack-and-withdrawal strategy within their familiar territory. The conflict ended more or less in a stalemate, as Ali Baba was summoned to Baghdad to appear before caliph al-Mutawakkil (847–861), who however, awarded him generously with gifts.

The Arab immigration into the Egyptian Eastern Desert and Nubia continued regardless of that to increase significantly in the aftermath.

The account on an Arab looting campaign under Abd al-Hamid al-Omari is probably the best documented incidence of an individual campaign to the "land of mines" in Nubia as reported by the Egyptian historian al-Maqrizi (1364–1442 AD).

According to Al-Maqrizi, this alleged descendant of caliph Omar I moved with 20,000 men to the "land of the gold mines" in 855 AD. He had previously received the consent by the then ruler of Egypt, Ahmed Ibn Tulun (835–884 AD), who wanted him out of the country. In Wadi Allaqi he began with the exploitation of gold deposits but soon came into conflict with the local Rabi'a and Bedja tribes, after which he moved further to the S (probably to Umm Nabardi). Due to water shortages he was forced to stay close to the Nile but when the Christian Makurian Nubians denied him access to the river several massacres were committed. In the looting campaign that followed,

Al-Omari and his men left a trail of devastation behind them throughout Nubia, until he was eventually assassinated in 869 AD. His head was brought to Cairo to Ibn-Tulun, who incidentally also had the bearer beheaded.

The 60,000 camels Al-Omari was reported to have at his disposal during his gold-campaigns is of course preposterously exaggerated, and under no circumstance stands in relation to the potential amounts the gold deposits would have been able to deliver.

Al-Omari's engagements in the southern Egyptian and Nubian Eastern Desert nonetheless assume a firmly established authority capable of supplying and protecting the gold mining districts that spread over vast regions. It is partly reflected by the archaeological remains of the seemingly master-planned supply stations at Umm Eleiga in southern Egypt, and Derahib, Uar, Bir Kiaw, Wadi Terfawi, Omar Kabash, and probably Betam in the Nubian Eastern Desert. So far, however, the archaeological and epigraphic evidence do not permit to speculate whether immigrant Arabs continued the gold mining operations in Wadi Allaqi and Nubia on a larger scale after the time of Al-Omari.

The amounts of gold harvested by Al-Omari all the same had raised the attention for the region's gold resources among Arab leaders in the following periods. Until the early years of the Fatimid Caliphate attempts were made to revive the numerous mines and wadiworkings in the Egyptian Eastern Desert, and possibly also in Wadi Allaqi, as corroborated by the distribution map (Fig. 7.8). Such attempts would only have had some chance to succeed through pacification of the local nomadic tribes by converting them to Islam. State power in those days was however sapped by internal quarrels and turmoil and above all, the stifling consequences from the crusades, which finally lead to a serious downfall of the gold mining industry in the Egyptian Eastern Desert in the eleventh and twelfth centuries.

As it were, gold mining activities altogether came to a general standstill in Egypt at this time, and it was not until the late nineteenth century that it was resuscitated during a short period lasting until the middle of the twentieth century. This episode was marked by a number of Anglo-Egyptian initiatives to resume gold mining by concentrating their activities almost exclusively on already known deposits exploited as far back as the New Kingdom, although with erratic success. Unfortunately, much of the hitherto well-preserved traces were destroyed through the application of highly intrusive mining techniques, involving modern machinery and infrastructure with a dishabilitating effect on archaeological research. The last mine to close down was the one at Wadi el-Sid, shortly after Nasser's nationalisation program.

Within the framework of a Soviet-Egyptian cooperation program in the1960s and 1970s, an extensive gold prospecting project was launched in the Egyptian Eastern Desert, including modern geologic and geochemical methods. Although numerous anomalies were identified, settlement and mining traces showed that they had already been exploited in antiquity whenever they revealed economically interesting. This disturbing finding definitely validated the prospecting capacities, especially those of the New Kingdom, which by no standards would need to envy modern methods. In recent years, the Egyptian-Australian company, Centamine Egypt Ltd. has been carrying out extensive survey work in the Egyptian Eastern Desert and is currently advertising for share offers. Active mining, however, has started only at Sukkari, unfortunately destroying the ancient traces at this site.

In Sudan (Figs. 7.9 and 7.10) the history of gold mining has undergone a different development since the time of al-Omari. Here at first, apparently seasonally migrating labourers had been coming in from the Arabian Peninsula and by finding occupation mainly in wadiworkings rather than in deep mines, gradually made progress deeper into the more western parts of the Nubian Desert.

The widely spread gold mining sites in the Nubian Desert which consisted primarily of wadiworkings seem mostly to have been exploited in peaceful coexistence with local beduins ever after the period of contention between the nomadic tribes and the Arab invasions from the

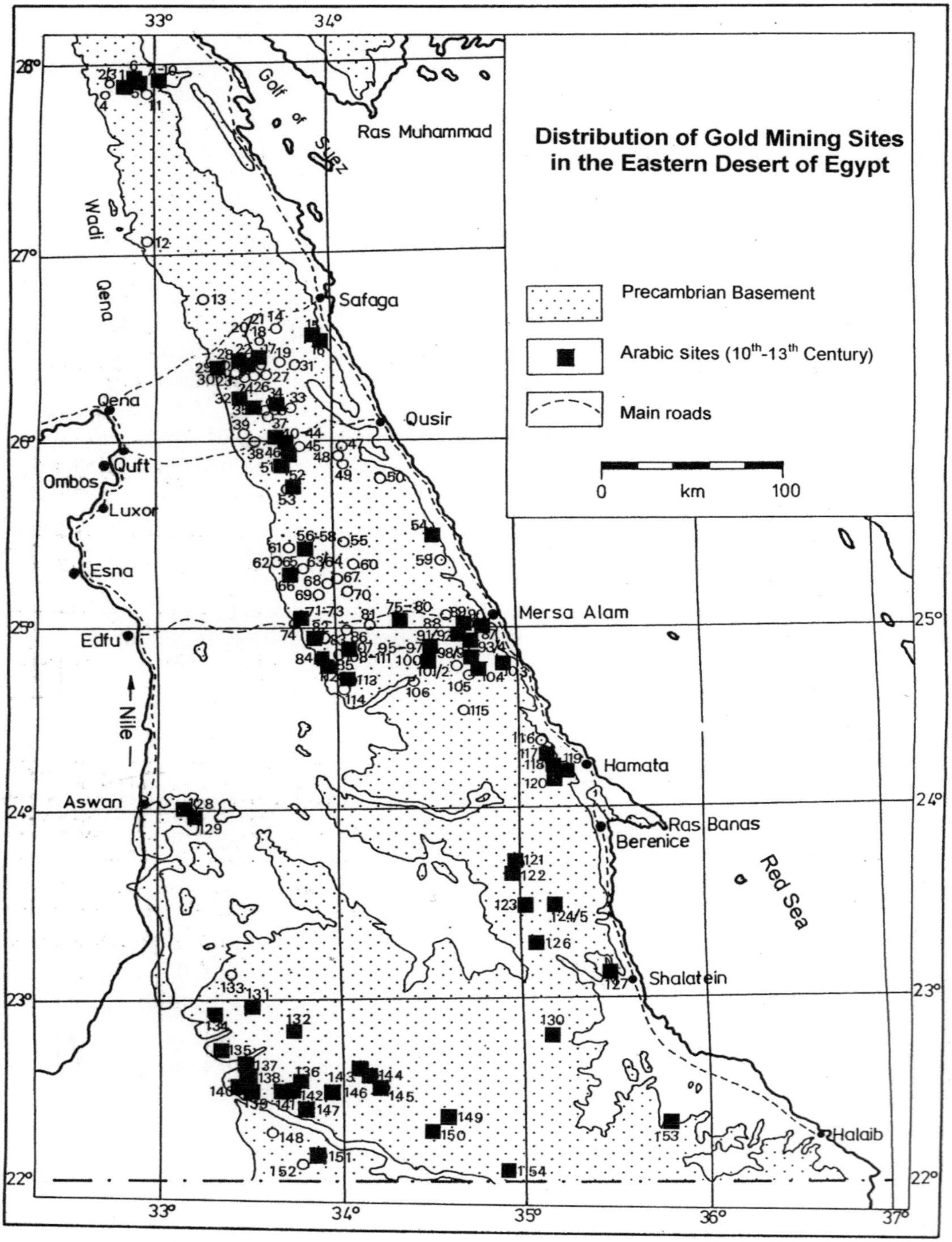

Fig. 7.8 Spatial distribution of Early Arab Period gold mining sites in the Egyptian Eastern Desert. The large number is reflected by intensive wadiworking activities. The conversion of nomadic tribes to Islam probably engendered a pacification of most parts of the desert (numbering after Klemm and Klemm 1994)

Time	Historical period	Relative rate of gold-production	Eastern desert of egypt	Nubian desert and Nubian River Nile valley
3000 BC	Predynastic	?	Sporadic gold findings in wadi beds. First systematic gold mining with large two hand stone hammers.	No archaeological evidence for mining.
2500 BC	Old Kingdom		Sporadic gold mining with ergonomically shaped one-hand stone hammers limited to north and central Eastern Desert.	Sporadic gold mining in close proximity to river Nile (Duweishat) with grooved stone axes.
2000 BC	First.Intermed Period; Middle Kingdom; Hyksos Period		Sporadic gold mining with ergonomically shaped one-hand stone hammers and mortar milling limited to north and central Eastern Desert.	No archaeological evidence for gold mining by Kerma Culture. Gold washing along the Nile (?) (Ameni's report at Beni Hassan) but no archaeological evidence for it
1500 BC	New Kingdom		Systematic gold prospection, intensive mining in underground and wadi workings (oval rubber stone mills, inclined washing tables, bronze chisels, well organised housing areas).	
1000 BC	Late Period; Kushite Kingdom		No archaeological evidence for gold mining.	
500 BC	Ptolemaic Period		Intensive gold mining reusing New Kingdom underground and wadi workings (two-lugged rubber stones on concave stone mills, circular heavy mineral concentrators, inclined washing tables).	Reorganisation of some New Kingdom gold production sites and reuse of New Kingdom tools, espicially of oval rubber stone mills. Isolated findings of Meroitic (?) stone mills at Mograt Island.
0	Roman Period; Byzantine Period; Arab Conquest of Egypt		Scarce gold mining close to protected desert roads (round mills, inclined washing tables).	
500 AD; 641 AD	A.Ibn Tulun			During Christian Kingdoms in Nubia no archaeological evidence for gold mining
1000 AD; 1200 AD	Salah ed Din		Intensive gold mining reusing old underground operations, but also new wadi workings (round mills, inclined washing tables, (?) panning). Gradual abandonment of gold mining.	El-Omari's expedition starts intensive gold rush, reuse of New Kingdom mines and wadi workings around well-organised settlements (round mills, inclined washing tables). Intensive gold mining mainly of wadi workings by Arabs penetrating from Arabian Peninsula into Nubia (round mills, inclined washing tables, (?) panning).
1350 AD	Collapse of Christian Kingdoms in Nubia			Gradual abandonment of gold mining.

Fig. 7.9 Outline of the development of gold mining in the Egyptian and Nubian Eastern Deserts from 4000 BC until 1350 AD (modified after Klemm et al. 2002b)

North. This stability seems to be reflected by the more widely dispersed and for the most open settlements that offered no noteworthy protection against sudden attacks.

Today, similar conditions prevail in the Bayuda Desert, where mining is carried out on a seasonal basis (from October to March) according to the traditional wadiworking method as well as in shallow sub-surface mines, comprising labourers from virtually any part of Sudan.

The distribution map in Fig. 7.10, however, reveals that the gold mining industry's western advance into the Nubian Desert did not infringe on a zone between 20 and 50 km to the E of today's railway track leading from Abu Hamed to Wadi Halfa. As reflected by the numerous defensive structures and fortifications mainly along the eastern banks of the Nile, the Nubian kingdoms of Nobatia and Makuria had therefore probably succeeded in temporarily halting the western progress of the mostly Muslim gold miners. But they eventually converted to Islam towards the middle of the fourteenth century (Hasan 1967), and as a result the strenuous gold

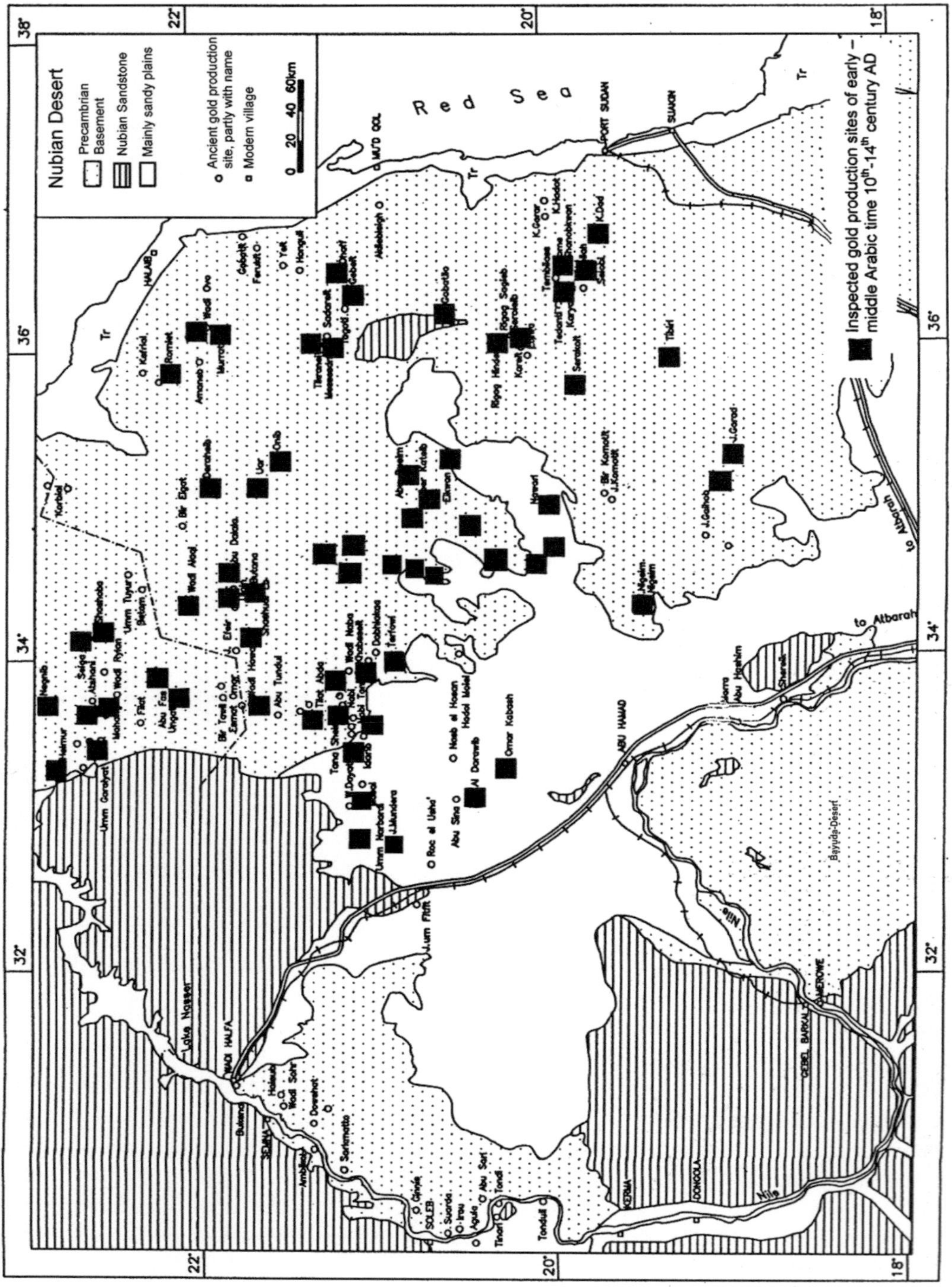

Fig. 7.10 Spatial distribution of the Early Arab Period gold mining sites in the Nubian Eastern Desert. Their limitation to eastern regions with an evident distance to the Nile may possibly be linked to an opposition by the Christian Kingdoms of Nobatia and Makuria until the middle of the fourteenth century (modified after Klemm et al. 2002b)

mining activities in the Nubian Desert came to an end as Arab groups gradually mixed with the Christian population in the Nile Valley.

We were not able to estimate the extent to which gold prospecting and mining activities had taken place in the Nubian Desert at the time of Mohamed Ali's campaign to Sudan, between 1820 and 1823. Some traces from iron hoes, remarkably well-preserved hut remains, as well as settlement patterns diverging from what we were used to see at Early Arab sites were nevertheless observed at Tilat Abda, Tanasheb and Eida Arib. These sites therefore seem to date to more recent periods and possibly even to this episode.

Gold mining resuscitated between the early and the mid twentieth century, as a rule on a modest and private scale, as seen at the deposits at Oyo, Gebeit, Onib, Abir Kateib, Uar, Umm Nabardi, Sarras, Duweishat and Abu Sari. All had already been exploited in the New Kingdom, but only few proved to be satisfactorily profitable (Gebeit, Abir Kateib, and Duweishat).

A large prospecting campaign was undertaken in the 1970s by the Sudanese-Soviet Technoexport cooperation, after which followed a United Nations campaign in the 1980s. Neither of them, however, led to any significant gold mining operations. Additional prospecting in the Nubian Desert was carried out later in that decade, this time under the management of the British Robertson Research Company (Minex Sudan), after which a short mining period followed at the deposit of Gebeit. Recently, Chinese groups have launched intensive prospection work in the Nubian Desert with the aim of reviving old mine deposits, although concrete results have yet to be published. Finally, there is a highly successful gold mining cooperation project in Ariab with the French mining company BRGM and the Sudanese Geological Authorities (GRAS), which however is located in genetic type deposit inaccessible to ancient prospectors as these ores are detected and exploited exclusively by means of highly advanced prospection methods and application of sophisticated metallurgical processing.

Appendices

Appendix A List of Ancient Gold Production Sites in Egypt from N to S (Fig. 5.1)

Compiled after Hume (1936), EGSMA (1986) and own field work.

Wadi Dara
Umm Balad
Wadi el-Urf I
Ghozza
Fatira
Abu Shehat
Wadi Abiyad
Abu Mureiwat
Wadi Gasus
Bir Semna
Semna
Wadi Markh
Wadi Bahlog
Kab Amiri
Gidami
Abu Gaharish
Abu Gerida
Hamama
Aradiya-East
Wadi Sagia
Abu Had
El Rebshi
Wadi Atalla el-Mur
Atalla
Umm Esh el-Zarqa
Wadi Sodmein
Hammamat
Umm Had-South
Bir Umm el-Fawakhir and El-Sid
Umm Soleimat
Wadi Karim
Kab el-Abiad
Terfawi
Wadi Zeidun
Sharm el-Bahari
El Nur
Umm Rus
Wadi Raheiya
Sigdit and Wadi Miyah
Daghbag
Talet Gadalla
Abu Muawad
El Hisinat
Abu Dabab
Umm Samra
Bokari
Abu Qareiya
Umm Salatit
Beza
Umm Salim
Umm Ashayir
El Suwayqat
Barramiya
Atud
Marsa Allam
Wadi Umm Khariga
Umm Quli
Wadi Faraon
Sukkari
Umm Tundub
Ambaud
Kurduman
Zabahiya
Urf el-Fahid
Wadi Umm Rashid
Marwat (Abu Mureiwa)
Umm Hugab
Dungash

DOI 10.1007/978-3-642-22508-6, © Springer-Verlag Berlin Heidelberg 2013

Wadi Dalalil
Samut
Bir Samut
Wadi Mueilha
Hangaliya West
Hangaliya
Umm Ud
Wadi Ghadir
Allawi
Dweig
Lewewi
Hamash
Higalig
Wadi Geili
Wadi Dendekan
Qualan
Abu Rahaiya
Bitan
Umm Eleiga
Hutit
Urga Ryan
Umm Kaliba (Tundeba)
Anbat
Korbiai
Wadi el-Hudi
Wadi Allaqi
Hairiri
Ashira-East
Umm Ashira
Neguib
Heimur
Nile-Valley Block A
Ahmed Village
Umm Garaiyat
Atshani
Marahig (Marahib)
Wadi Murra
Wadi Rylan
Filat
Seiga
Shoshoba
Umm Tuyur
Betam
Abu Fas
Ungat
Romit
Umm Egat

Appendix B List of Ancient Gold Production Sites in Nubia/Sudan (Fig. 6.1)

From N to S within each group.
I. Group: Red Sea Hills
Shishaiteb (or Shishutaib)
Shishaiteb-South
Oyo
Mikraff
Hangul
Ohaff
Gebeit
Gebeit Sharq (East Gebeit)
Tagoti and Tigranit
Walati
Gumarob
Khor Massesana
Gabatilo
Nafirdeib (Lakobdog)
Rigag Sageib (Rigag Hindeb or Ragaghundab)
Shikryai (or Serakoit)
Romi
Sheiteb
Shanobkwan
Selobit
Miradaab
Khor Dat and Miteb
Abidoidib
Romeib
II. Group: Wadi Amur – Ariab
Tibiri
Aliakateb
Wadi Amur
Hadanaib
Igariri
Hashai
Nubt
Ariab
Mishalliet Gurad
Ganait
Derbeikwan
III. Group: East of the Hamisana suture
Kamoli
Wadi Rak and Bir Rak
Camel 1
Camel 2

Uar (Alaar)
Uar, New Kingdom complex
Hufra
Onib
Onib New Kingdom settlements
Onib Minor
Tabon North
Tabon-East
Tabon-South
Egait- North
Egait-South
Adarmo-East
Bir Kiaw
Listi
Shashuteb
Eikwan 1
Wadi Eikwan 1–4
Eikwan-South
Abu Mereim Salalob
Techol
Navarai

IV. Group: West of the Hamisana suture

Derahib
Miri
Wadi Ward Miriyam1–6
Wadi Nafiryam
Nafir 1–6
Nafir-Derahib
Khor Nesari 1–5
Hamisana-North
Hamisana mine
Kazim-City
Wadi Gabaideb 1–5
Caravanserai
Khor Adarmo
Khor Adarmo-West
Khor Adarmo-Fort
Wadi Eleij area
Sukai
Abu Dueim
Karaibitar 1–2
Abirkateib
Negeim

V. Group Butena – Wadi Terfawi

Abu Dalala
Butena North
Butena
Wadi Tabak
Shashuateb 1–2
Kabeseit
Wadi Terfawi 1–5

VI. Group: Wadi Tawil to Eida Arib

Wadi Tawil
Bir Tawil
Ismat Omar
Abu Baraga
Abaraga
Abu Bard
Wadi Gagait
Liseiwi 1–8
Wadi Dom 1–6
Tilat Abda
Tilat Abda 10
Tilat Abda 11
Tanasheb
Tanasheb-West
Wadi Naba
Nabitana 1–5
Wadi Naba 1–6
Mine Nabi 1
Nabi 2
Eidaarib 1–2
Wadi Idarib

VII Group: Mosei – Omar Kabash

Mosei
Nabi (?)
Fort Murrat
Umm Nabardi
Jebel Mundera
Umm Fit Fit
Rod El Ushal
Abu Siha
Omar Kabash 1–19

VIII. Group: Nile Valley from Sarras to Abu Hashim

Sarras
Duweishat
Umm Fahm 1–5
Island Sai
Ager
Abu Sari
Tinari
Sokar mine
Tondi

Kerma
Shamkhiya
Mograt Island
Karmel Fort
Kurduma
Sehan
El Mikeseir
Abu Alalik
Ras el-Gezira
Abu Kuweib
Abu Hashim
IX. Bayuda Desert
Umm Sareh
Abu Sugha
El Higagiya
Abu Khalag

Appendix C Chronology

Chronology of Egypt

Predynastic period	5500–3100 BC
Early dynastic period	
First and second dynasties	3100–2686 BC
Old Kingdom (OK)	
Third to sixth dynasties	2686–2181 BC
First intermediate period	
Seventh to eleventh dynasties	2181–2055 BC
Middle Kingdom (MK)	
Eleventh to twelfth dynasties	2055–1795 BC
Second intermediate period	
Thirteenth to seventeenth dynasties	1795–1550 BC
New Kingdom (NK)	
Eighteenth dynasty	1550–1295 BC
Nineteenth to twentieth dynasties, ramesside period	1295–1069 BC
Third intermediate period	
Twenty first to twenty fourth dynasties	1069–715 BC
Late period	
Twenty fifth to thirtieth dynasties	747–332 BC
Ptolemaic period	332–30 BC
Roman period	30 BC–AD 395
Byzantine period	AD 395–640
Early Arab period	AD 640 – ~1200

After Klemm and Klemm (2008a, b)

Chronology of lower Nubia/Sudan (All dates before 690 BC are approximate)

A-group	ca. 3700–2800 BC
C-group	ca. 2300–1600 BC (partly under Egyptian control)
Kingdom of Kush	
Kerma Ancien	2500–2050 BC
Kerma Moyen	2050–1750 BC
Kerma Classique	1750–1500 BC
New Kingdom	1500–1000 BC (under Egyptian control to the fourth cataract)
Kingdom of Kush	1000–500 BC
Napatan phase	Ninth to fourth centuries
Meroitic phase	Fourth century BC to fourth century AD
X-group	Fourth to sixth centuries AD
Christian period	542–1332 AD
Nobadia Kingdom	Sixth-end of seventh centuries AD
Makuria Kingdom (upper Nubia)	Sixth century AD-1323/1365
Islamic period	AD 1323-present
Early Arab gold production in Eastern Desert	~AD 890 [Al Omari]- ~1350

Modified after Welsby and Anderson (2004)

Glossary

Alluvium Sub-recent filling of a valley floor (in Egypt mostly of a wadi).

Amazonit Pale green Orthoclase (potassic feldspar).

Amphibolite Metamorphic rock consisting mainly of hornblende and to a lesser extent of feldspar.

Amratien Prehistoric period between ca. 4500 and 3500 BC. Named after the site at Amra in Upper Egypt.

Amygdule A gas cavity in a volcanic rock, commonly filled with secondary minerals (amygdaloidal)

Anatexis Melting of pre-existing rock by an intruding magma, changing its chemical composition.

Andesite Fine-grained volcanic rock consisting of hornblende and feldspar.

ANS Arabian-Nubian Shield: geological nappe units mainly of island-arc composition with ophiolites and back-arc sediments.

Aplite Intrusive, mostly fine-grained dike of granitic composition.

Apophysis An intrusive magma tongue into host rocks.

Arkose Feldspar enriched sandstone.

Arsenopyrite FeAsS (Iron arsenic sulphide).

Assimilation Mixing and digestion of solid or molten wall rocks in a liquid Magma.

Back-arc Region of a submarine basin associated with island-arcs and subduction zone.

Bekhen-stone Metagreywacke occurring in Wadi Hammamat/Eastern Desert of Egypt. Used in Ancient Egypt mainly for sarcophagi but also for fine sculptures etc. The expression is the ancient Egyptian word for this kind of stone.

BIF Banded Iron Formation.

Black-Smoker Hydrothermal vent occurring mainly along ocean ridges and forming occasionally mineral deposits.

Blemmyes Nomadic tribe in the Eastern Desert of Southern Egypt and Northern Sudan, attacking Ptolemaic and Roman conquerors in these regions in the first centuries BC and AD.

Chalcopyrite $CuFeS_2$ (Copper-iron sulphide).

Conglomerate Rock consisting of individual rounded clasts within a finer-grained matrix that have become cemented together.

Craton An old and stable part of the continental lithosphere, which often has survived cycles of subsequent tectonics.

Cumulate An igneous rock formed by accumulation of crystals settled-out from magma by gravity.

Cyanide leaching Leaching of gold with cyanide-salt. An extremely toxic metallurgical process.

Dacite Fine-grained extrusive (volcanic) rock often semi-glassy with quartz, feldspar and other silica minerals.

Dike/Dyke Tabular igneous intrusion cutting across the bedding or foliation of country rocks.

Dokhan-volcanics Late Proterozoic volcanics, mainly acid (silica-rich). Named in Egypt after Gebel Dokhan, the source area of "imperial porphyry".

Dolerite Crystalline hard dike rock type of a diorite/gabbro composition.

Dunite Ultramafic rock consisting mainly of olivine and some pyroxene.

Dynasty Era referred to in Ancient Egypt defined by ancestry in royal families.

EGSMA Egyptian Geological Survey and Mining Authority.

Fanglomerate Conglomerate like rock, badly sorted, consisting of large and smaller components.

Fatimids The Fatimid Caliphate ruled over Egypt from 909 to 1171 AD.

Fluid Volatile, water-rich liquid transporting metals, salts and silica trough open systems in the Earth crust and precipitating them by forming mineral deposits.

Galena PbS (lead sulphide)

Gabbro Magmatic rock consisting mainly of calcium-rich plagioclase and pyroxene, rarely hornblende.

Goethite FeO(OH).

Gossan Upper part of a ferrous, weathered former sulphide deposit, occasionally enriched in gold.

Gneiss A foliated, feldspathic rock formed by regional metamorphism. mainly with a parallel texture of its components.

GRAS Geological Research Authority of Sudan

Greywacke Sandstone with fragments of quartz, feldspar, mica, and lithic components (immature sandstone).

Grooved stone hammer Stone hammer with a grooved rim originally fixed on a shaft and used as a mallet.

Green-schist facies Physico-chemical conditions of rather low temperature (~250–300 °C) and pressure, under which sedimentary or magmatic rocks change through metamorphism into chlorite and epidote-yielding rocks.

Hybrid Magma changed its composition by assimilation of different (country) rocks.

Hybridisation Assimilation of different (country) rocks by an intruding magma and modifying its composition.

Hydrothermal Pertaining to a hot water system circulating in the Earth crust and containing various salts and dissolved minerals ± sulphide minerals and occasionally very low gold concentrations.

Ignimbrite Hot, mainly acid ash flow, partly welded.

Isotope One of two or more species of the same chemical element, only differing by the number of neutrons.

Lamprophyre Dark dike rock

Lherzolite Ultramafic rock consisting mainly of olivine and clinopyroxene.

Limonite Collective term for ironoxide/hydroxides like goethite and lepidocrocite [FeO(OH)].

Magma Rock melt.

Magmatism The term magmatism covers all geological processes concerning magma. This includes all parts of formation, uprising in the Earth crust and crystallisation. Magmatism covers volcanism, where the magma erupts at the surface, but also plutonism when it crystallises within depth. Magmatism results to the formation of magmatic rocks

Malachite $Cu_2[(OH)_2/CO_3]$ typical green copper carbonate.

Meidum-bowl Typical ceramic vessel from the Early Dynastic to Old Kingdom Periods in Egypt (~3000–2000 BC). Named after its type-site Meidum in Middle Egypt.

Meta- The prefix "meta-" in front of a rock name, means the original rock type was modified during a metamorphic process.

Metamorphism Transformation of a rock as a consequence of significant variations of pressure, temperature and chemical parameters of its original conditions of formation.

Migmatite Rock composed of igneous and metamorphic materials.

Molasse Mainly marine but also continental formation of conglomerates, sandstones, shales and marls, formed in the fore-basin of an uprising orogen.

MORB Mid-Ocean-Ridge-Basalt

Nappe Sheet-like, allochtonous geological unit moved between few and more than 100 km over an other geological unit.

Open pit, Trench Open cast (open cut) mining operation, in contrast to underground mining.

Ophiolite A section of oceanic crust and upper mantle uplifted and subducted on a continent.

Orogen Mountain belt formed as a mobile belt as result of folding, thrusting and magmatic intrusions and subsequent uprising within the continental crust.

Pan African orogeny Late Precambrian to Early Paleozoic orogeny (between ~850 and 550 ma). In our case, mainly the accretion and overthrusting of various terranes over the Eastern African craton and resulting implications.

Paragenesis Equilibrium assemblage of mineral phases mainly used in description of ore deposits.

Phanerozoic Era of Earth history since about 540 years BP.

Pillow basalt Submarine basalt forming pillow like structures.

Pit (hole) Open vertical diggings in mining activities.

Plagiogranite Granitoid rocks with low potassium content occurring mainly as dikes in ophiolite assemblages.

Pleistocene Geological epoch from about 1.8 to 0.01 million years before present marked by repeated glaciations.

Pluton An igneous complex intruded into the Earth crust.

Prospection Search for economically viable deposits or minerals.

Proterozoic Geologic aeon between 2,500 and 550 million years BP.

Pseudomorph The outward crystal form from another mineral.

Pyrite FeS_2 (Iron sulphide).

Pyroxene Group of mainly dark greenish rock forming silica minerals generally containing Mg, Ca and Fe.

Rhyolite Acid volcanic rock of granitic chemistry.

Romib Local expression of cake-shaped tombs in southern Egyptian and Nubian desert with diametres of 4-8 metres.

Round mill Also referred to as rotation mill or quern. It consists of a lower stone base with a round grinding surface and an upper rotating disc referred to as rotor stone. It was used to grind crushed quartz pebbles down to a powder fraction from which gold flitters were separated.

Salinity Total amount of salts in a water liquid (seawater, hydrotherm etc.)

Sericitisation Formation of fine-grained muscovite (=sericite) as result of hydrothermal alteration or low grade metamorphism.

Serpentinite An ultramafic (silica poor) rock formed by regional metamorphism of rocks from the oceanic mantle.

Shear zone Tabular zone of crushed or brecciated rock as result of crustal shear movements, often mineralised by quartz, carbonate and/or ore forming solutions.

Sheeted dike Dike formed in ophiolite assemblage as result of seafloor spreading.

Siltstone Very fine-grained sandstone.

Sphalerite ZnS (Zinc sulphide).

Sub-aeric Formed under open air conditions.

Subduction zone Down- going zone of oceanic crust sliding beneath either a continental plate or another oceanic plate.

Suture zone In our case the area where two terranes have joined together through collision through nappe movement.

Tailing Spoil heap of milled and washed ore, normally in close to stone mills and washing tables.

Barren Rock devoid of gold or other sought-after, mineral content.

Tectonics Study generally concerned with the structures of the lithosphere and particularly with the forces and movements that have operated to form these structures.

Terrane A crustal fragment of oceanic and/or continental material accreted on an other plate (or craton).

Trench In the present context an opencast mine along a quartz vein.

Ultramafic rock Extremely silica-poor rock, like dunite, harzburgite, lherzolite and (metamorphic) serpentinite.

Ounce Equivalent to 31.1 g – International troy ounce.

Ventilation Natural or artificial ventilation in underground mines

Volatile State of vaporized fluid (substance).

Wadi Dry river bed in desert regions.

Wadiworking Collecting systematically gold-ore (quartz-) fragments in a wadi ground with subsequent crushing, milling and washing.

Wall rock Rock adjacent to a vein, commonly altered by hydrothermal reactions with vein forming fluids.

Washing table Inclined construction to concentrate on its surface the gold by flowing water dispersion.

Xenolith A foreign rock inclusion in an igneous rock.

References

Abd El-Nabi HH, Frisch W (2002) Origin of Wadi Heimur-Abu Swayel gneiss belt, South-Eastern Desert, Egypt: petrological and geochronological constraints. Precambrian Res 113:307–322

Abd El-Nabi HH et al (1997) Results of prospecting for of mineral resources of Wadis Hammad, El-Atrash, El-Ghozah and Fatira El-Beida Area. Internal report EGSMA, Technoexport, Cairo

Abd El-Nabi HH, Fritsch W, Hegner E (2002) Evolution of the Pan-African Wadi Heimur metamorphic sole, eastern desert, Egypt. J Metamorph Geol 18(6): 639–651

Abdeen MM, Greiling RO (2005) A quantitative structural study of late Pan-African compressional deformation in the Central Eastern Desert (Egypt) during Gondwana assembly. Gondwana Res 8:457–471

Abdeen MM, Dardir AA, Greiling RO (1991) Hammamat pebbles in felsite at Wadi Queih, north of Quseir Eastern Desert of Egypt. Ann Geol Surv Egypt 17:73–75

Abdeen MM, Sadek MF, Greiling RO (2008) Thrusting and multiple folding in the Neoproterozoic Pan-African basement of Wadi Hodein area, South-Eastern Desert, Egypt. J Afr Earth Sci 52:21–29

Abdel-Karim ML et al (1996) The ophiolite mélange of Wadi Dungash and Arayis, Eastern Desert of Egypt: petrogenesis and tectonic evolution. Acta Mineralogica-Petrographica, Szeged, XXXVII, 129–14

Abdel-Khalek ML (1978) Tectonic evolution of the basement rocks in the southern and central Eastern Desert of Egypt. Precambrian Res 6(1):1–37

Abdel-Monem AA, Hurley PM (1979) U-Pb dating of zircons from psammitic gneisses, Wadi Abu Rusheid, Wadi Sikait area, Egypt. Inst Appl Geol Jeddah Bull 3(2):165–170

Abdelsalam MG (1993) Structural evolution of the Oko shear zone, Sudan. In: Thorweihe U, Scandelmeier H (eds) Geoscientific research in Northeast Africa. A.A. Balkema, Rotterdam/Brookfield, pp 121–125

Abdelsalam MG, Abdeen MM, Dowaidar HM, Stern RJ, Abdelghaffar AA (2003) Structural evolution of the Neoproterozoic western Allaqi-Heiani suture, Southeastern Egypt. Precambrian Res 124:87–104

Abu Zeid KM (1987) Evaluation of reports on the Sukari gold ore deposit, Eastern Desert of Egypt. Minex minerals Egypt report

Adams WY (1986) Ceramic industries of medieval Nubia. Kentucky Univ Press, Lexington

Agricola G (1546) De re metallica libri XII, Buch 8. Übers. C. Schiffer, VDI-Verl. Düssedorf, 1978

Akaad MM, Noweir AM (1969) Lithostratigraphy of the Hammamat Um Seleimat district, Eastern Desert, Egypt. Nature 223(5203):284–285

Akaad MM, Noweir AM (1978) Geology and lithostratigraphy of the Arabian desert orogenic belt of Egypt between lat. 25°35´and 26°30´. Precambrian Res 6:A6

Akawy A (2007) Geometry and texture of quartz veins in Wadi Atalla area, Central Eastern Desert, Egypt. J Afr Earth Sci 47:73–87

Al-Boghdady AA, Van den Kerkhof AM, Khalaf IM (2001) Composition and origin of fluid inclusions in gold-bearing quartz veins at Wadi Sodmein area, central Eastern Desert, Egypt. In: Geology of the Arab World-proceedings of the 5th international conference in the geology of the Arab World, Feb 2000. Nat Libr Arch Egypt no. 13556, pp 305–312

Al-Maqrizi after: Urbain Bouriant, 1895-1900, Description topographique et historique de l'Égypte, Paris

Alford CJ (1901) Gold mining in Egypt. London

Alfy EM (1933) Project and estimate for Sukari mine. EGSMA report 4/1933

Al-Shanti AMS, Mitchell AHG (1976) Late Precambrian subduction and collision in the Al Amar-Idsas region, Arabian shield, Kingdom of Saudi Arabia. Tectonophysics 30:T 41–T 47

Armanious LK, Barsoum M, Ismail IM (1971) Geological sketch map of Abu Marwat area, 1:5000. Intern report EGSMA/Technoexport, Cairo

Aufrère S (1990) L'Univers minéral dans la pensée égyptienne II. Inst Franç Arch Orient Le Caire pp 353–354

Bakheit AK, Mattheis G (1993) Gold-productive volcanogenic sulphide mineralization in the Ariab Belt, Red Sea Hills/Sudan: evidences for late proterozoic seafloor hydrothermal systems. In: Thorweihe U, Schandelmeier H (eds) Geoscientific research in Northeast Africa. Balkema, Rotterdam, pp 533–539

Ball J (1912) The geography and geology of Southeastern Egypt. Geol. Surv. Cairo, 304 pp

Ballet P (1987) Essai de classification de coupes Type Maidum-Bowl. Cahiers de la ceramique Egyptienne, tome 1, Le Caire, pp 1–16

Barbarin B (1990) Granitoids: main petrogenetic classifications in relation to origin and tectonic setting. Geol J 25:227–238

Benson M, Gourlay J (1899) The temple of Mut in Asher. Murray, London, p 370

Bentor YK (1985) The crusal evolution of the Arabia-Nubian massif with special reference to the Sinai Peninsula. Precambrian Res 28:1–74

Berndt ME, Buttram TE, Early D III, Seyfried WE Jr (1994) The stability of gold polysulfide complexes in aqueous sulfide solutions 100 to 150 ° C and 100 bar. Geochim Cosmochim Acta 58:587–594

Berthe D et al (1979) Orthogneiss, mylonite and non coaxial deformation of granites: the example of the South Armorican Shear Zone. J Struct Geol 1(1):31–42

Beyth M, Stern RJ, Altherr R, Kröner A (1984) The late Precambrian Timna igneous complex, southern Israel: evidence for comagmatic-type sanukitoid monzodiorite and alkali granite magma. Lithos 31:103–124

Bitak M (1968) Studien zur Chronolgie der Nubischen C-Gruppe. Österr Akad Wiss Denkschriften Bd. 97, Wien

Bloss JE (1966) Relics of ancient gold miners. Sudan Notes and Records 20:313–315, Khartoum

Boswell PM (1984a) Detailed exploration of the Jebel Negeim gold prospect, Southern Nubian Desert, Sudan. Robertson research Intern Report 3979

Boswell PM (1984b) Detailed and reconnaissance gold exploration in the Om Nabardi region of the Nubian Desert, Sudan. Robertson Research Intern. Report 3980

Boswell PM, Harrison SJ, Bichan HR, (1981) Detailed and reconnaissance gold exploration in the Wadi Gabgaba Region of the Eastern Nubian Desert, Sudan. Robertson Research Intern. Ltd. Report No. 3663

Botros NS (2002a) Alteration zones: Are they a good target for gold deposits in Egypt? Arab Gulf J Sci Res 20(4):209–218

Botros NS (2002b) Metallogeny of gold in relation to the evolution of the Nubian Shield in Egypt. Ore Geol Rev 19(3–4):137–164

Botros NS (2003) On the relationship of auriferous talc deposits hosted in volcanic rocks and massive sulphide deposits. Ore Geol Rev 23(3–4):223–257

Botros NS (2004) A new classification of the gold deposits of Egypt. Ore Geol Rev 25(1–2):1–37

Botros NS, Wetait MA (1997) A possible porphyry copper mineralization in south Um Monqul, Eastern Desert, Egypt. Egypt J Geol 41/1:175–196

Boyle DM, Hall DJ, Samani KZ, Ahmed AH (1986) Reconnaissance prospecting in the Khor Arba'at area, Red Sea Hills Province, Democratic Republic of Sudan. Robertson Research Intern. Ltd. Report 4135

Bridges PS, Boswell PM, Bedawi M, El Khalek IA, El Moneim KA (1986) Geological investigations at Barramiya Mine, Egypt. Internal report No.4146, Robertson Research Intern. Ltd

Brown PE, Lamb WM (1986) Mixing of H2O-CO2 in fluid inclusions. Geochim Cosmochim Acta 50:847–894

Čaba Z (1974) The rock inscriptions of lower Nubia, Appendix, Nr. A_1, Prag

Cahen L, Snelling NJ, Delhal J, Vail JR (1984) The geochronology and evolution of Africa. Carendon Press, Oxford, 512 pp

Castel G, Soukiassian G (1989) Gebel El-Zeit I, Les Mines de Galène (Egypte, IIe millénaire av. J.-C.) Fouill. Inst. Franç.d'Arch.Orient. (FIFAO) 35, Le Caire, p.42.

Castel G, Matthieu B et al (1992) Les Mines de Cuivre du Ouadi Dara. Rapport préliminaire sur les travaux de la saison 1991. Bull Inst Franç d'Arch Orient (BIFAO) 92:51–65, Le Caire

Castel G, Matthieu B et al (1995) Wadi Dara copper mines. In: Proceedings of the first international conference on mining & metallurgy and conservation of metallic artifacts, Cairo, pp 15–27

Castel G, Köhler EC, Matthieu B, Pouit G (1998) Les Mines du Ouadi Um Balad Desért Oriental. Bull. Inst Franç d'Arch Orient. (BIFAO) 98:57–87, Le Caire

Castiglioni AA, Negro G (1992) La scoperta di Berenice Pancrisia e l'esplorazione archeologica del Deserto orientale nubiano. Atti del VI Congresso Internationale di Egittologia Torino, pp 79–84

Castiglioni AA, Vercoutter J (1998) Das Goldland der Pharaonen. Die Entdeckung von Berenike Pancrisia. Verlag Ph. von Zabern, Mainz, 192 pp

Cauuét B (1991) L'exploitation de l'or en Limousin, des Gaulois aux Gallo-Romains. Annales du Midi, Toulouse t.103, n.194:149–181

Černy J (1947) Graffiti at the Wadi el-Allaki (Wadi Beiga). J Egypt Arch 33:52–57, London

Childe VG (1943) Rotary querns on the Continent and in the Merditerranean Basin. Antiquity 17. New York, pp 19–26

Clifford TN (1970) The structural framework of Africa. In: Clifford TN, Gass IG (eds) African magmatism and tectonics. Oliver and Boyd, Edinburgh, pp 1–19

Conophagos CE (1988) Le Laurium antique et la technique Grèque de la production de l'argent. Ekdotike Hellados S. A, Athen, 458 pp

Coward MP, Ries AC (eds) (1986) Collision tectonics, geological society special. Publication no. 19, pp 67–81

Davies N de G, Gardiner AH (1926) The Tomb of Huy, Viceroy of Nubia in the Reign of Tut'ankhamun (no. 40). The Theban Tomb series 4, London

Davies NM (1958) Picture Writing in Ancient Egypt. Oxford Univ.Press, London, pl X

De Wall H, Greiling RO, Sadek MF (2001) Post-collisional shortening in the late Pan-African Hamisana high strain zone, SE Egypt: field and magnetic fabric evidence. Precambrian Res 107:179–194

Dewey JF (1988) Extensional collapse of orogens. Tectonics 7:1123–1139

Dixon TH (1981) Age and chemical characteristics of some pre-Pan-African rocks in the Egyptian shield. Precambrian Res 14:119–133

Drenkhahn R (1976) Die Handwerker und ihre Tätigkeiten im Alten Ägypten, Äg. Abh. 31, Wiesbaden, 169 pp

Dunn SC (1911) Notes on the mineral deposits of the Anglo-Egyptian Sudan, Khartoum-Edinburgh, 70 pp

Durr S, Zimmer M (1987) Pan-African crustal evolution in the Nubian segment of northeast Africa. Amer Geophys Union, Spec. Publ. 17

EGSMA (1986) Gold in Egypt: a commodity package, compiled by Shokry Z. Gabra, Geol. survey of Egypt, Cairo

Eisbacher GH (1991) Einführung in die Tektonik. Enke-Verl, Stuttgart, 374 pp

El Gaby S (1983) Architecture of the Egyptian basement complex. In: Proceedings of fifth International basement Tectonics, Cairo, 1983

El Gaby S, El-Nady OM, Khudeir AA (1984) Tectonic evolution of the basement complex in the CED of Egypt. Geol Rdsch 73:1019–1036

El Ramly MF and Akaad MK, (1960) The basement complex in the CED of Egypt between lat. 24° 30´and 25° 40´. Geol Surv Egypt, paper Mo. 8, 33 pp

El Sharkawy MA, El Bayoumi RM (1979) The ophiolites of Wadi Ghadir, Eastern Desert, Egypt. Ann Geol Surv Egypt 9:125–135

El Shimi KAM, Soliman AA (2002) Gold Mineralization associated with the Banded Iron Formations in the Central Eastern Desert of Egypt: first record, pp 281–300

El Tokhi M, El Muslem A (2002) Fluid inclusions in the gold bearing quartz veins at Umm Rus area, Eastern Desert, Egypt. Chinese J Geochem 21:131–139

El-Bayoumi RMA (1984) Ophiolithes and melange complex of Wadi Ghadir, Eastern Desert, Egypt. Bull Fac Earth Sci King Abdul Aziz Univ. Jeddah 6:324–329

El-Bayoumi RMA, Greiling R (1984) Tectonic evolution of a Pan African plate margin in Southeastern Egypt. In: Klerkx J, Michot J (eds) African Geology. Tervuren, pp 47–56

El-Bouseily AM, El-Dahhar MA, Arlan AI (1985) Ore-microscopic and geochemical characteristics of gold-bearing sulfide minerals, El Sid gold mine, Eastern Desert, Egypt. Miner Deposita 20:194–200

El-Bouseily AM, Arlan AI, Ghoneim MF, Harraz HH (1986) Mercury dispersion patterns around El Sid-Fawakhir gold mine, Eastern Desert, Egypt. J Afr Earth Sci 5(5):465–469

Eldougdoug A (1990a) Gold anomalies in the late proterozoic felsig-mafic volcano-sedimentary sequences and associated rocks, Gebel Abu Marawat area, Eastern Desert, Egypt. Bull Fac Sci Cairo Univ 58:533–548

Eldougdoug A (1990b) Fossil placer: a potential source for gold in Egypt, Egypt. J Geol 34(2):233–246

El-Gaby S, List FK, Tehrani R (1988) Geology, evolution and metallogenesis of the Pan African belt in Egypt. In: El-Gaby S, Greiling RO (eds) The Pan African belt of Northeast Africa and adjacent areas. Vieweg, Braunschweig, pp 17–63

El-Gaby S, List FK, Tehrani R (1990) The basement complex of the Eastern Desert and Sinai. In: Said R (ed) The geology of Egypt. A.A. Balkema, Rotterdam/ Brookfield, pp 175–184

Eliwa H, Breitkreuz Ch, Khalaf I, El GK (2010) Depositional styles of early ediacaran terrestrial volcanosedimentary succession in Gebel El Urf area, North Eastern Desert, Egypt. J Afr Earth Sci 57:328–344

El-Ramly MF (1972) A new geological map for the basement rocks in the Eastern and South-Western Deserts of Egypt. 1:1.000.000. Ann Geol Surv Egypt 2:1–18

El-Ramly MF, Greiling R, Kröner A, Rashwan AA (1984) On the tectonic evolution of the Wadi Hafafit area and environs, Eastern Desert of Egypt. Bull Fac Sci King Abdulaziz Univ 6:113–126

Emery WB (1931) Preliminary report of the work of the archaeological survey of Nubia 1930–1931. ASAE 31:71–74

Emery WB (1963) Preliminary report on the excavations at Buhen. Kush 11:116–120

Emery WB, Kirwan LP (1935) The excavations and survey between Wadi Es-Sebua and Adindan 1929–1931. IFAO 1935, Le Caire

Fakhry A (1952) The inscriptions of the amethyst quarries at Wadi el-Hudi. Gov. Press, Cairo

Fakhry A (1974) Geochemical studies at the Fawakhir gold mine, Eastern Desert, Egypt 2, infrared absorption spectroscopy as a means of characterizing the primary dispersion haloes of gold deposits. J Geochem Explor 3(3):265–275

Ferrar HT (1913) Note on the Turin papyrus mine plan. Cairo Sci J 7(82):247–251

Fleck RJ, Greenwood WR, Hadley DG, Anderson RE, Schmidt DL (1980) Rubidium-strontium geochronology and plate-tectonic evolution of the southern part of the Arabian shield. U.S. Geol. Surv. Prof. Paper1131, 38 pp

Fletcher RJ, Miles R, Newall P, Ahmed AH, Samani K (1984a) Minex Sudan, Gold exploration in the Red Sea Hills area, September 1983-February 1984. Robertson Research Intern Ltd. Report 3975

Fletcher RJ, Miles R, Newall P, Ahmed AH, Samani K (1984b) Progress report on exploration of the Jebel Negeim area, Feb-Apr 1984

Floyer EA (1893) Étude sur le Nord-Etbai entre le Nil et la Mer Rouge. Le Caire, 157 ff

Fowler TJ (2001) Pan-African granite emplacement mechanisms in the Eastern Desert Egypt. J Afr Earth Sci 32(1):61–86

Francis M et al (1972) On the results of prospecting for gold and rare metals by the Atut party in the area of Atut, Sukari, Abu Rusheid deposits in 1971–72, Central Eastern Desert, ARE. Intern report EGSMA, Cairo

Fullagar PD (1980) Pan-African age granites of Northeastern Africa: new or reworked sialic materials? In: Salem JJ, Busrewil MI (eds) Geology of Libya. Second symposium on the geology of Libya 3. Academic, London, pp 1051–1058

Fullagar PD, Greenberg JK (1978) Egyptian younger granites: a single period of plutonism? Precambrian Res 6:22–33

Gabra SZ (1986) Gold in Egypt: a commodity package. EGSMA, Cairo

Gad S, Kusky T (2006) Lithological mapping in the Eastern Desert of Egypt, the Barramiya area, using landsat thematic mapper (TM). J Afr Earth Sci 44:196–202

Gardiner A (1914) The map of the gold mines in a Rammesside Papyrus at Turin. Cairo Sci J 8(89):41–46

Garson MS, Shalaby IM (1976) Precambrian-lower Paleozoic plate tectonics and metallogenesis in the Red Sea region. Geol Assoc Canada, Sp. Paper No.14, pp 573–596

Gass IG (1982) Upper Proterozoic (Pan-African) calc-alkaline magmatism in North-Eastern Africa and Arabia. In: Thorpe RS (ed) Andesites - orogenic andesites and related rocks. Wiley, Chichester, pp 591–609

Geçkinli AE, Özbal H, Craddock PT, Meeks ND (2000) Examination of the Sardis gold and the replication experiments. In: Remage A, Craddock P (eds) King Croesus' gold, excavations at Sardis and the history of gold refining. British Mus. Press, London, pp 184–199

Geus F (2004) In: Welsby DA, Anderson JR (eds) Ancient Treasures. An exhibition of recent discoveries from the Sudan National Museum. British Museum Press, London, pp 114–121

Ghazala HH (1993) Geological and structural interpretation of airborne surveys and its significance for mineralization, South Eastern Desert, Egypt. J Afr Earth Sci 16(3):273–285

Gillespie JG, Dixon TH (1983) Lead isotope systematic of some igneous rocks from the Egyptian shield. Precambrian Res 20:63–77

Gippsland-Limited (2005) Quaterly activity internet report, July 2005, Seiga drilling results

Glidden H (1954) Khor Nubt Tombstones. Kush 11, Khartoum pp 63–65

Gnoli R (1989) Marmi Antichi. De Luca, Roma

Gold SMW (2009) SMW Gold announces assay results for gold deposits in Egypt. www. SMW Gold.com/ SMW-Gold/news

Goyon G (1949) Le Papyrus de Turin, dit "Des Mines d'Or" et le Wadi Hammamat. ASAE 49:337–392, Cairo

Graefe E (1999) Über die Goldmenge des Alten Ägypten und die Beraubung der thebanischen Königsgräber. Zeitschr Äg Sprache Altertumskd 126:19–40

Green FW (1909) Notes on some Inscriptions in the Etbai District, PSBA 31, London, pp 320–321

Greenwood HR, Hadley DG, Anderson RE, Fleck RJ, Schmidt DL (1976) Late Proterozoic cratonization in Southwestern Saudi-Arabia. Trans Roy Soc London 280(A):517–527

Greenwood HR, Hadley DG, Anderson RE, Schmidt DL (1980) Rubidium-strontium geochronology and plate-tectonic evolution of the southern part of the Arabian shield. US Geol Surv Prof Pap 1131:38

Greiling RO (ed) (1988) The Pan African belt of Northeast Africa and adjacent areas. Vieweg-Verl, Wiesbaden, pp 9–94

Greiling RO, Kröner A, El-Ramly MF (1984) Structural interference patterns and their origin in the Pan-African basement of the south-eastern Desert of Egypt. In: Greiling RO, Kröner A (eds.) Precambr.Tectonics Illustrated, 401–413

Greiling RO, Kröner A, El Ramly MF, Rashwan AA (1988) Structural relationships between the Southern and Central Parts of the Eastern Desert of Egypt: details of a fold and thrust belt. In: El-Gaby S, Greiling RO (eds) The Pan African belt of Northeast Africa and adjacent areas. Vieweg-Verl, Wiesbaden, pp 17–63

Greiling RO, El Ramly MF, Rashwan AA, Kamal El Din GM (1993) Towards a comprehensive structural synthesis of the (Proterozoic) Arabian-Nubian Shield in E. Egypt. In: Thorweihe U, Scandelmeier H (eds) Geoscientific research in Northeast Africa. A. A. Balkema, Rotterdam/Brookfield, pp 15–19

Greiling RO, Abdeen MM, Dardir AA, El Akaal H, El Ramly MF, Kamal El Din GM, Osman AF, Rashwan AA, Rice AHN, Sadek MF (1994) A structural synthesis of the Proterozoic Arabian-Nubian Shield in Egypt. Geol Rdsch 83(3):484–501

Grimal N (1991) Travaux de L'IFAO, Desert Oriental. Bull Inst Franc Arch Orient 91:299–306, Le Caire

Grimal N (1994) Travaux de L'IFAO Desert Oriental. Bull Inst Franc Arch Orient 94:423–434, Le Caire

Grimm A (1988) Ta-Nebu "Goldland" und "Nubien". Gött Misz 106:23–28

Grundmann G (1989) Goldlagerstätten in der NW-Nubischen Wüste (Nordsudan). Mitt Öster Miner Ges 134:47–62

Guéraud O (1942) Ostraca Grecs et Latins de l'Wadi Fawakhir. Bull Inst Franc Arch Orient 41:141–196, Le Caire

Gundlach R (1977) "Goldgewinnung" and "Goldminen". Lex. Ägyptol. II, Harrassowitz, Wiesbaden

Gundlach R (1977) "Expeditionen", Lex. Ägyptol. II, Harrassowitz, Wiesbaden

Habib MS, Ahmed AA, El-Nady OM (1985) Two Orogenesis in the Meatiq area of the CED, Egypt. Precambrian Res 30:83–111

Hall DJ, Agar R, Jakubiak Z, Adel Khalek I (1987) Progress report on the exploration of the El Sid concession, Minex Minerals Egypt

Hamimi Z, Wetait MA, Botros SS (1994) On the geology and deformational history of the basement rocks of Wadi Ummm Balad, Wadi El Urf Area, Northeastern Desert, Egypt. Egypt J Geol 38/2:613–624

Harms W (2007) Discoveries in Sudan reveal economic organization of an ancient African state-the kingdom of Kush. Univ. Chicago Chronical 26, No. 19

Harraz HZ (1995) Primary geochemical haloes, El Sid gold mine, Eastern Desert, Egypt. J Afr Earth Sci 20(1):61–71

Harraz HZ (1999) Wall rock alteration, Atut gold mine, Egypt: processes and P-T-X_{CO2} conditions of metamorphism. J Afr Earth Sci 28(3):527–551

Harraz HZ (2000) A genetic model for mesothermal Au deposits: evidence from fluid inclusions and stable isotopic studies at El Sid Gold mine, Eastern Desert, Egypt. J Afr Earth Sci 30(2):267–282

Harraz HZ (2002) Fluid inclusions in the mesozonal gold deposit at Atud mine, Eastern Desert, Egypt. J Afr Earth Sci 35(3):347–363 (17)

Harraz HZ, El-Dahhar MA (1993) Nature and composition of gold-forming fluids at Umm Rus area, Eastern Desert, Egypt: evidence from fluid inclusions in vein material. J Afr Earth Sci 16(3):341–353

Harrell JA (2002) Pharaonic stone quarries in the Eastern Desert of Egypt and Nubia. In: Friedmann R (ed) Egypt and Nubia-gifts of the Desert. Brit. Mus. Press, London, pp 232–244

Harrell JA (2010) Archaeological geology of Hosh el-Guruf, Fourth Nile Cataract, Sudan: Gadansk Archaeological Museum African Reports, vol 7. pp 71–84

Harrell JA, Brown VM (1992) The oldest surviving topographical map from ancient Egypt (Turin Papyri 1879, 1899 and 1969). J Am Res Centre Egypt 29:81–105

Harris NBW (1982) The petrogenesis of alkaline intrusives from Arabia and northwest Africa and their implications for within-plate magmatism. Tectonophysics 83:243–258

Harrison SJ, Gaskell JL, Bichan HR (1980) A history of Gebeit. Progress report on reconnaissance gold exploration in the Nubian Desert, Sudan, March 1980. Report no. 3564, Robertson Research Int. Ltd

Hasan YF (1967) The Arabs and the Sudan from the seventh to the early sixteenth century. University Press, Edinburgh, 298 pp

Hassan MA, Hashad AH (1990) Precambrian of Egypt. In: Said R (ed) The geology of Egypt. A. A. Balkema, Rotterdam/Brookfield, pp 201–245

Hassan MM, Soliman MM, Azzaz SA, Attawiya MY (1990) Geological studies on gold mineralization at Sukkari, Um Ud, and Samut, Eastern Desert, Egypt. Ann Geol Surv Egypt 16:89–95

Hassan H, Helmy H, Kaindl R (1999) Mineralogy and fluid inclusion studies of the Au-Cu quartz veins in the Hamash area, South-Eastern Desert, Egypt. Miner Petrol 65:69–86

Hassan H, Kaindl R, Fritz H, Loizenbauer J (2004) The Sukari gold mine, Eastern Desert/Egypt: structural setting, mineralogy and fluid inclusion study. Mineral Deposita 39(4):495–511

Hautumm W (1981) Studien zu Amphoren der spätrömischen und frühbyzantinischen Zeit. Fulda, 129 pp

Hayes WC (1972) Late Roman pottery. London, p 387

Helck W (1955–61) Urkunden der 18. Dyn., Nr. 568, Berlin

Helck W (1965) Materialien zur Wirtschaftsgeschichte des Neuen Reiches, Bd. IV, Mainz, p 7

Helck W (1975) Wirtschaftsgeschichte des Alten Ägyptens im 3. und 2. Jahrtausend v. Chr. Handb. d. Orientalistik I, Köln

Helmy HM, Kaindl R, Fritz H, Loizenbauer J (2004) The Sukari gold mine, Eastern Desert, Egypt: structural setting, mineralogy and fluid inclusion study. Mineral Deposita 39(4):495–511

Hilmy ME, Osman A (1989) Remobilization of gold from a chalcopyrite-pyrite mineralization Hamash gold mine, Southeastern Desert, Egypt. Mineral Deposita 24(4):244–249

Hölbl G (2004) Altägypten im Römischen Reich 2: Die Tempel des römischen Nubien. Antike Welt, Zabern-Verl, Mainz, 160 pp

Hölscher U (1957) Univ. of Chicago Orient. Inst. Publ., vol. 83: Medinet Habu, vol. 5, The Temple proper, Part I: pl. 328. Chicago

Holthoer R (1977) New Kingdom Pharaonic Sites. The Pottery Scand Joint Exped. to Sudanese Nubia, Uppsala, 5.1

Hossam AH, Khalil IK, Nagdi MFA (2008) Alteration patterns related to hydrothermal gold mineralization in meta-andesites at Dungash area; Eastern Desert, Egypt. Resour Geol 5:19–30

Hume WF (1934) Geology of Egypt, vol. 2, The fundamental Precambrian rocks of Egypt and the Sudan; part I: the metamorphic rocks. Egypt Geol Surv Departm, Cairo, pp 1–300

Hume WF (1935) Geology of Egypt, vol. 2, part II, The fundamental Precambrian rocks of Egypt and the Sudan: the later plutonic and intrusive rocks. Egypt Geol Surv Departm Cairo 301–688

Hume WF (1936) Geology of Egypt, vol. 3, Minerals of economic value associated with the intrusive Pre-Cambrian igneous rocks and ancient sediments, section 2, Gold Mining in Egypt. Egypt Geol Surv Departm, Cairo, pp 689–818

Hussein AAA, Carter GS, Searle DL (1978) Is the basement complex of Egypt complex? Precambrian Res 6:A27–A28, abstr

Hussein AAA, Ali MM, El-Ramly MF (1982) A proposed new classification of the granites of Egypt. J Volc Geoth Res 14:187–198

Jackson NJ, Ramsay CR (1980) Time-space relationship of upper Precambrian volcanic and sedimentary units in the central Arabian shield. J Geol Soc London 137:617–628

Jakubiak Z (1987) An appraisal of gold potential in the Sukari and Atut areas of the Eastern Desert, Egypt. Intern. report, Greenwich Resources-Robertson Research Intern. Ltd

Jakubiak Z (1989) An assessment of the gold potential of the Wadi Allaqi Area in South-Eastern Egypt. Intern. report, Greenwich Resources PLC

Jenkins RS (1925) Note on ancient gold mine workings in Upper Egypt. Intern. Report Geol. Survey Egypt, Cairo

Kamal El-Din GM (1986) Geology of Wadi Umm Had area, Eastern Desert, Egypt. M. Sc. Thesis. Fac. Sci., Qena, Assiut Univ., Egypt

Khalil KI, Helba HA, Mücke A (2003) Genesis of the gold mineralzation at the Dungash gold mine area, Eastern Desert, Egypt: a mineralogical-microchemical study. J Afr Earth Sci 37:111–122

Kitchen KA (1968) Ramesside Inscriptions II. Oxford, 353 ff

Kitchen KA (1983) Ramesside inscriptions I, vol 6. Blackwell, Oxford

Klemm R (1995) Umm Eleiga, ein prädynastischer Goldfundplatz in der Ostwüste Ägyptens. Münchner Ägyptol Untersuch 4:247–260

Klemm R, Eichler E (1998) Neue Expeditionsinschriften aus der Ostwüste Oberägyptens. MDAIK 54:237–266
Klemm R, Klemm DD (1988) Pharaonischer Goldbergbau im Wadi Sid und der Turiner Minenpapyrus. Beihefte Stud. Altägypt. Kultur, Bd 2:73–87
Klemm R, Klemm DD (1993) Steine und Steinbrüche im Alten Ägypten. Springer, Berlin/Heidelberg/etc, 465 pp
Klemm R, Klemm DD (1994) Chronologischer Abriss der antiken Goldgewinnung in der Ostwüste Ägyptens. MDAIK 50:189–222
Klemm D, Klemm R (1997) Antiker Goldbergbau in der ägyptisch-sudanesischen Ostwüste. Nürnberger Blätter zur Archäologie, Heft 13:149–166
Klemm R, Klemm DD (2000) Antike Goldlagerstätten am Nil im Nördlichen Sudan. Geo-archäologische Nachlese zu früheren Untersuchungen von G. Morteani und Mitarbeitern. Münchener Geologische Hefte A 28:285–298
Klemm R, Klemm D (2005) Fatira (Abu Zabal) and Ghozza-Two New Kingdom and Ptolemaic gold working sites in the Eastern Desert of Egypt. Studies in Hon Ali Radwan, Khaled Daoud (ed), CASAE 34/II, Cairo, pp 95–111
Klemm D, Klemm R (2008a) Mining in ancient Egypt and Nubia. Encyclopaedia of the History of Science, Technology, and Medicine in Non-western Cultures, part 13. Springer, Netherlands, pp 1685–1698
Klemm R, Klemm DD (2008b) Stones and quarries in ancient Egypt. British Museum Press, London, 354 pp
Klemm R, Klemm DD, El Hawari AB (1995) Evolution of Methods for Prospection, Mining and Processing of Gold in Egypt. In: Proceedings of the First Intern. Conf. on Mining & Metallurgy and Conservation of Metallic Artifacts. Cairo, pp 341–354
Klemm DD, Klemm R, Murr A (2001) Gold of the Pharaohs-6000 years of gold mining in Egypt and Nubia. J Afr Earth Sci 33:643–659
Klemm R, Klemm D, Murr A (2002a) Geo-archäologischer Survey im Wadi el-Hudi. Hildesheimer Ägyptolog. Beitr. 48:53–66
Klemm D, Klemm R, Murr A (2002b) Ancient gold mining in the Eastern Desert of Egypt and the Nubian Desert of Sudan. In: Friedman R (ed) Egypt and Nubia-Gifts of the Desert. Brit. Mus. Press, London, pp 215–232
Klerkx J, Deutsch S (1977) Resultats préliminaires obtenus par la méthode Rb/Sr sur l'age des formations Précambriennes de la région d´Uweinat (Libye). Musee Royal d´Afrique Centrale, Tervuren (Belgium). Dept Geol Min Rap Ann 83–94
Knudtzon JA (ed) (1964) Die El-Amarna-Tafeln, 1. Teil, Nr. 29. Vorderasiat. Bibl., Aalen, 145 ff
Koshin G, Bassyuni F (1968) The mineral Resources of the UAR. Part1: metallic minerals, Internal Report EGSMA-Technoexport, Contract 1247, Cairo
Kretschmer U, Scott SD (1976) Phase relations involving arsenopyrite in the system Fe-As-S and their application. Can Mineral 14:364–368
Kröner A (1979) Pan-African plate tectonics and its repercussions on the crust of Northeast Africa. Geol Rdsch 68:565–583, Stuttgart
Kröner A (1985) Ophiolites and the evolution of tectonic boundaries in the Late Proterozoic Arabian-Nubian shield of Northeast Africa and Arabia. Precambrian Res 27:277–300
Kröner A, Greiling R (eds) (1984) Precambrian Tectonics Illustrated. E. Schweitzerbart´scheVerlagsbuchhandlu ng, Stuttgart, pp 401–412
Kröner A, Greiling R, Reischmann T, Hussein IM, Stern RJ, Kruger J, Zimmer M (1987) Pan African crustal evolution in the Nubian segment of Northeast Africa. Am Geophys Union Spec Publ 17:235–257
Kröner A, Reischmann T, Wust H-J, Rashwan AA (1988) Is there any Pre-Pan-African (>950) basement in the Eastern Desert of Egypt? In: El-Gaby S, Greiling RO (eds) The Pan African belt of Northeast Africa and adjacent areas. Vieweg, Wiesbaden, pp 95–119
Kröner A, Todt W, Hussein IM, Mansur M and Rashwan AA (1992) Dating of late Proterozoic ophiolithes in Egypt and Sudan using single grain zircon evaporation technique. Precamb Res 59, 15-32
Kröner A, Krüger J, Rashwan AAA (1994) Age and tectonic setting of granitoid gneisses in the eastern Desert of Egypt and South-West Sinai. Geol Rdsch 83(3):502–513, Stuttgart
Kusky TM, Ramadan TM (2002) Structural control on neoproterozoic mineralizations in the South Eastern Desert, Egypt: an integrated field, Landsat TM and SIR-C/X SAR approach. J Afr Earth Sci 35(1): 107–121
Küster D, Hengst M, Pilot J (1993) Evolution of Pan-African granitoid magmatism along the southwestern Nakasib suture zone, Red Sea Hills, Sudan. In: Thorweihe U, Schandelmeier H (eds) Geoscientific research in Northeast Africa. A. A. Balkema, Rotterdam/Brookfield, pp 111–115
Langwieder G (1994) Die Geologie des Gebietes von Wadi El Sid/Bir Umm El Fawakhir (Ostwüste von Ägypten). Unpubl. Masterthesis, LMU-University, Munich, 122 pp
Llewellyn A (1903) Report on a mining concession in the Egyptian Sudan. London, 78 pp
Loizenbauer J, Neumayr P (1996) Structural control on the formation of the Fawakhir gold mine, Eastern Desert. In: Naim GM, Gaber M (eds) Egypt-Tectonic and fluid inclusion evidence, vol 34. Special Public. Geol. Survey of Egypt pp 477–488
Loizenbauer J, Neumayr P, Fritz H (2002) Genesis of the Fawakhir gold deposit in the Eastern Desert of Egypt. The result of late Pan-African transgressional tectonics. In: 19th Colloquium of African Geology, Morocco, 2002
Lucas A, Harris J (1962) Ancient Egyptian materials and industries. E. Arnold Publ, London
Lyssenko AK, Zarkov VM, Riad AM, Morkos SL, Abu El-Hassan AS, Zaki ME (1975) Geological map of the Barramiya gold ore field. Techonoexport, Contract 78 442, Cairo
Mansour MS, Bassyuni FA, Al-Far DM (1956) Geology of Umm Salatit, El-Hisinat district, Barramiya East Sheet. Geol Survey Egypt, Cairo

Markowitz Y, Lacovara P, Hachfield P (1997) Jewellery fragments from the tomb of Nefertari in the Museum of fine Arts in Boston. In: Goring E, Reeves CN, Ruffle J (eds) Chief of seers Egyptian studies in memory of Cyril aldred. KPI, London, pp 220–228

Meredith D (1952) The Roman remains in the Eastern Desert of Egypt. J Egypt Archaeol 38:94–111

Meyer C (1995) A Byzantine gold-mining town in the Eastern Desert of Egypt: Bir Umm Fawakhir, 1992–93. J Roman Arch 8:192–224, Ann Arbor

Meyer C, Heidorn L, Kaegi W, Wilfong T (2000) A Byzantine Gold-Mining Town: The Bir Umm Fawakhir Survey Project 1993. Oriental Institute Communications 28, Chicago, 92 pp

Meyer C, Bryan E, Omar M, Smither R (2003) Ancient gold extraction at Bir Umm Fawakhir. J Am Res Center Egypt 40:13–53

Miller MM, Dixon TH (1992) Late Proterozoic evolution of the northern part of the Hamisana zone, Northeast Sudan. Constraints of Pan-African accretionary tectonics. J Geol Soc 149:743–750, London

Morgan P (1990) Egypt in the framework of global tectonics. In: Said R (ed) The geology of Egypt. A. A. Balkema, Rotterdam/Brookfield, pp 91–112

Moussa EMM, Stern R, Manton WI, Ali KA (2008) Shrimp zircon dating and Sm/Nd isotopic investigation of Neoproterozoic granitoids, Eastern Destert, Egypt. Precambrian Res 160(3–4):341–356

Mumin AH, Fleet ME, Chryssoulis SL (1994) Gold mineralisation in As-rich mesothermal gold ores of the Bogosu-Prestea mining district of the Ashanti gold belt, Ghana: remobilisation of "invisible" gold, mineral. Deposita 29:445–460

Murr A (1994) Geologische Kartierung der Goldlagerstättenbezirke Fatira und Dungash in der Ostwüste Ägyptens, unpubl. Masterthesis, LMU-University, Munich

Murr A (1999) Genese der Goldlagerstättenbezirke Fatira, Gidami, Atalla und Hangaliya in der Ägyptischen Ostwüste, Münchner Geol. Hefte A 27, 203 pp

Murray GW (1925) The roman roads and stations in the Eastern Desert of Egypt. J Egypt Arch 11:138–150, London

Murray GW (1942) The gold-mine of the Turin papyrus. Bul Inst Eg 24:81–86

Neubauer WH (1958) Reconnaissance report on the Barramiya gold mine. Quseir Phosphate Co, Cairo

Newall PS, Ahmed AH, Samini KZ (1985) Report on Gold Reconnaissance Work in Sudan carried out in Block 36° 00′ E – 36° 20 E, 18° 30′ N- 19° 30′ N between October-December 1984. Robertson Research Intern. Ltd, Project 4043

Nicholson PT, Shaw J (2000) Ancient Egyptian Materials and Technology. Cambridge University Press, Cambridge

Notton JHF (1974) Ancient Egyptian gold refining – a reproduction of early techniques. Gold Bull 7/2: 50–56

Obeid MA (2006) The Pan-African arc-related volcanism of the Wadi Hodein area, South Eastern Desert, Egypt: petrological and geochemical constraints. J Afr Earth Sci 44/3:383–395

Ogden JM (1976) The so-called platinum inclusions in Egyptian goldwork. J Egypt Arch 62:138–144

Ogden JM (2000) Metals in chapter 6 of ancient Egyptian Materials and Technology. In: Nicholson PT, Shaw J (eds). Cambridge University Press, Cambridge

Oldfather CH (Translator) (1935) Diodorus siculus. Library of history: Loeb Classical Library. Harvard University Press, Cambridge

Osman AM, Dardir AA (1989) On the mineralogy and geochemistry of gold bearing quartz veins in the Central Eastern Desert of Egypt and their altered wall rocks. Ann Geol Surv Egypt 16:17–25

Osman AF, Greiling RO, Warr LN, Hilmy ME, Ragab AI, El Ramly MF (1993) Distinction of different syn- and late orogenic Pan-African sedimentery sequences by metamorphic grade and illite crystallinity (W. Zeidun, Central Eastern Desert, Egypt). In: Thorweihe U, Schandelmeier H (eds) Geoscientific research in Northeast Africa. A. A. Balkema, Rotterdam/Brookfield, pp 21–25

Osman AM, Piestrzynski A (1997) Gold from Atalla Gold Mine Area, Central Eastern Desert, Egypt. In: Papunen H (ed) Mineral deposits: research and exploration. Turku, pp 265–268

Osman AM, Pieczonka J, Piestrzynski A (2002) Relationship between gold and tellurides in El Sid gold deposit, Eastern Desert, Egypt. Mineral Pol 33(2):15–26

Petrie WMF (1901) The Royal Tombs of the First Dynasty, vol. II. Egypt Expl. Fund 21, London, pl. 1

Piotrovskiy BB (1964) Ancient Egyptian inscriptions in the Dakka and Wadi Alláki region (in Russ.). Drevnjaja Nubija 1961–1962, Moskva

Plinius dÄ Historia Naturalis, vol 5, XXXVII

Pudlo D, Franz G (1993) Late Panafrican dyke rocks in the Bir Safsaf complex, Western Desert, S-Egypt. In: Thorweihe U, Schandelmeier H (eds) Geoscientific research in Northeast Africa. A. A. Balkema, Rotterdam/Brookfield, pp 67–71

Quibell JE (1898): El Kab. BSAE 3, London, p 7

Rabie H (1972) The financial system of Egypt A.H. 564-741/A.D. 1169–1341, London-New York-Toronto

Ramadan TM, Kotny A (2004) Mineralogical and structural characterization of alteration zones detected by orbital remote sensing at Shalatein district, SE Desert, Egypt. J Afr Earth Sci 40(1–2):89–99

Ramage A, Craddock P (eds) (2000) King Croesus' Gold. Excavations at Sardis and the History of Gold Refining. British Mus. Press, London

Reichmann T, Kröner A (1994) Late Proterozoic island arc volcanics from Gebeit, Red Sea hills, North-East Sudan. Geol Rundsch 83:547–563, Stuttgart

Ren D, Abdelsalam MG (2006) Tracing along-strike structural continuity in the Neoproterozoic Allaqi-Heiani suture, Southern Egypt using principal component analysis (PCA), fast Fourier transform (FFT) and redundant wavelet transform (RWT) of ASTER data. J Afr Earth Sci 44:181–195

Riad AM et al (1977) Results for prospecting for rare metals, non-ferrous metals and gold in the area of Wadi

Dara, El-Urf, Malha, Abu Marwa, Hammad Bali in 1976–1977. EGSMA-Technoexport Contract 78442, Cairo,125 pp
Rice AHN, Sadek MF, Rashwan AA (1993a) Igneous and structural relations in the Pan-African Hammamat Group, Igla Basin, Egypt. In: Thorweihe U, Schandelmeier H (eds) Geoscientific research in Northeast Africa. A. A. Balkema, Rotterdam/ Brookfield, pp 35–39
Rice AHN, Osman AF, Abdeen MM, Sadek MF, Ragab AI (1993b) Preliminary comparison of six late- to post- Pan-African molasse basins, E. Desert, Egypt. In: Thorweihe U, Schandelmeier H (eds) Geoscientific research in Northeast Africa. A. A. Balkema, Rotterdam/Brookfield, pp 41–45
Richter A, Schandelmeier H (1990) Precambrian basement inliers of Western Desert; geology, petrology and structural evolution. In: Said R (ed) The Geology of Egypt. A. A. Balkema, Rotterdam/Brookfield, pp 185–200
Ries AC, Shackleton RH, Graham RH, Fitches WR (1983) Pan-African structures, ophiolites and melanges in the Eastern Desert of Egypt: a traverse at 26°N. J Geol Soc London 140:75–95
Ries AC, Dawoud AS, Schackleton RM (1985) Geochronology, geochemistry and tectonics of the NE Bayuda Desert, N. Sudan: implications for the western margin of the late Proterozoic fold belt in NE Africa. Precambrian Res 30:43–62
Roubet C (1989) The grinding stones of site E-78-3 Wadi Kubbaniye. In: Wendorf F (ed) The prehistory of Wadi Kubbaniye, vol III. Southern Methodist Univ. Press, Dallas, pp 473–489
Saad NA, El-Bouseil AM, Khalil KI (1996) Alteration patterns in the Umm Rus gold mine area, Eastern Desert, Egypt. Acta Mineral-Petrogr 37:51–74
Sabet AH et al (1976) On gold mineralization in the Eastern Desert of Egypt. Ann EGSMA VI:201–212
Sanders GER, Owen TRH (1951) Notes on ancient villages at Khor Nubt and Khor Omek, Sudan Notes and Records 32, Khartoum, 326–331
Säve-Söderbergh T, (1941) Ägypten und Nubien, Lund
Schädel HD (1936) Die Listen des großen Papyrus Harris. Leipziger Äg. Stud. 6, Glückstadt
Schäfer H (1902) Ein Bruchstück altägyptischer Annalen. Abh. Preuß. Akad. Wiss., phil.-hist. Klasse, Nr.1
Schandelmeier H, Darpyshire DPF, Harms U, Richter A (1988) The East Sahara craton: evidence for Pre-Pan-African crust in NE Africa West of Nile. In: El-Gaby S, Greiling RO (eds) The Pan African belt of Northeast Africa and adjacent areas. Vieweg-Verl, Wiesbaden
Schmid F (2001) Geologische Kartierung im Gebiet Wadi Eleij und Geologisch-tektonische Untersuchungen zu lokalen und regionalen Deformationsstrukturen. Unpubl. Masterthesis, LMU-University, Munich
Schott S (1961) Kanais: Der Tempel Sethos I. im Wadi Mia. NAWG, Jahrg. 1961, Nr.6, Göttingen
Schweinfurth G (1903) Die Wiederaufnahme des alten Goldminenbetriebes in Ägypten und dem Sudan. ASAE 4, Cairo, 268 ff
Schweinfurth G, Ermann A (1885) Alte Baureste und hieroglyphische Inschriften im Wadi Gasus. Berlin, pp 11–14
Seipel W (ed) (2001) Gold der Pharaonen. Ausstellungskatalog des Kunsthistorischen Museums Wien, 143 pp
Sethe K (1964) Dodekaschoinos. Das Zwölfmeilenland an der Grenze von Ägypten und Nubien, UGAÄ 2/3. Leipzig 1901 (Göttingen 1964)
Shablowsky R (1973) Geological sketch map of Wadi Umm Huqab, Rashid area scale 1:25000. EGSMA-Technoexport report, Cairo
Shackleton RM (1988) Contrasting structural relationships of Proterozoic ophiolithes in Northeast and Eastern Africa. In: El-Gaby S, Greiling RO (eds) The Pan African belt of Northeast Africa and adjacent areas. Vieweg, Wiesbaden, pp 183–193
Shackleton RM (1994) Review of Late Proterozoic sutures, ophiolitic melanges and tectonics of eastern Egypt and North-East Sudan. Geol Rundsch 83:537–546
Shackleton RM, Ries AC, Graham RH, Fitches WR (1980) Late Precambrian ophiolitic melange in the Eastern Desert of Egypt. Nature 285:472–474
Shalaby JM, Stumpfl E, Helmy HM, El Mahalaw MM, Kamel M (2004) Silver and silver bearing Minerals at the Um Samiuki volcanic massive sulphite deposit, Eastern Desert, Egypt. Mineral Deposita 39 (5–6)
Shaw I (2002) Life on the Edge: Gemstones, Politics and Stress in the Deserts of Egypt and Nubia. In: Friedman R (ed) Egypt and Nubia-Gifts of the Desert .Brit. Mus. Press, London, pp 244–251
Shazly EM, Adel Hady MA, El Nasharty FA (1988) Gold exploration by remote sensing techniques in Wadi El Allaqi areas, Egypt. Photogrammetria 42(6–8):303–310
Sidebotham E St (1991) Römische Straßen in der ägyptischen Wüste. Antike Welt 22. Jg., Nr.3, Mainz
Sidebotham E St (ed) (2008) The Red land: The illustrated archaeology of Egypt's Eastern Desert. Amer. Univ Cairo, Cairo, 424 pp
Spence K, Rose P (2009) New fieldworks at Sesebi. Egyptian Archaeol 35:21–24
Stacey JS, Hedge CE (1984) Geochronologic and isotopic evidence for early Proterozoic continental crust in the eastern Arabian shield. Geology 12:310–313
Stern RJ (1981) Petrogenesis and tectonic setting of late Precambrian ensimatic volcanic rocks, CED of Egypt. Precambrian Res 16:195–230
Stern RJ, Hedge CE (1985) Geochronologic and isotopic constraints on late Precambrian crustal evolution in the Eastern Desert of Egypt. Am J Sci 285:97–127
Stern RJ, Sellers G, Gottfried D (1988) Bimodal dike swarms in the North Eastern desert of Egypt: significance for the origin of late Precambrian a-type granites in Northern Afro-Arabia. In: El-Gaby S, Greiling RO (eds) The Pan African belt of Northeast Africa and adjacent areas. Vieweg, Wiesbaden, pp 147–179
Stern RJ, Kröner A, Bender R, Reischmann T, Dawoud AS (1994) Precambrian basement around Wadi Halfa, Sudan: a new perspective on the evolution of the East Saharan Craton. Geol Rundsch 83:564–577

Stern RJ, Johnson PR, Kröner A, Yibas B (2004) Neoproterozoic ophiolites of the Arabian-Nubian Shield. Developm Praecambr Geol 13:95–128
Stoeser DB, Camp VE (1985) Pan-African microplate accretion of the Arabian Shield. Geol Soc Am Bull 96:817–826
Stoeser DB, Stacey JS (1988) Evolution, U-Pb geochronology and isotope geology of the Pan-African Nabitah orogenic belt of the Saudi Arabian Shield. In: El-Gaby S, Greiling RO (eds) The Pan African belt of Northeast Africa and adjacent areas. Vieweg, Wiesbaden, pp 17–63
Störk L (1977) "Gold", Lex. Ägyptol, vol II. Harrassowitz, Wiesbaden
Sultan M, Tucker RD, El Alfy Z, Attia R, Ragab AG (1994) U-Pb (zircon) ages for the gneissic terrane west of the Nile, Southern Egypt. Geol Rdsch 83(3):514–522
Takla MA, Eldougdoug A, Gad MA, Ramsy AH, Eltabbal HK (1997) Gold-bearing quartz veins in ultramafic and mafic rocks, Hutit and Um Tenedba, South Eastern Desert, Egypt. Ann Geol Surv Egypt XX:411–432
Tawab MMA (1991) Gold exploration in Egypt from pharaonic to modern times. Zentralbl Geol Paläont Teil I 1991/11:2721–2733
Tawab MMA, Castel G, Pouit G, (1990) Archéo-Géologie dans des anciennes mines de cuivre et d'or des régions el-Urf/Mongul-Sud et Dara-Ouest. Bull. Inst. Franç. d'Arch. Orient. 90, Le Caire, 359–364
Taylor WEG, Hamed El Kazzaz YA, Rashwan AA (1993) An outline of the tectonic framework for the Pan-African orogeny in the vicinity of Wadi Um Relan area, South Eastern Desert, Egypt. In: Thorweihe U, Schandelmeier H (eds) Geoscientific research in Northeast Africa. A.A. Balkema, Rotterdam/Brookfield, pp 31–34
Thomas E (1913) The ancient mine plan of the Turin Papyrus. Cairo Sci J 7(82):158–160
Tichatschke Chr (2001) Geologische Kartierung im Gebiet des Wadi Eleij und Geochemische Untersuchungen an Vulkaniten und Plutoniten des Arbeitsgebietes. Unpubl. Masterthesis, LMU-University, Munich
United Nations, Dptm. of Technical Cooperation and Development (1986) Mineral Exploration of the Egypt/Sudan Area of Integration.- Compilation Report, Geology and Mineral Potential: Preparatory Assistance (DP/UN/RAB-80-014/2), New York
Updegraff RT (1982) A study of the Blemmyes. Ann Arbor (Diss.1978)
Vail JR (1988) Tectonics and evolution of the Proterozoic basement of Northerastern Africa. In: El-Gaby S, Greiling RO (eds) The Pan African belt of Northeast Africa and adjacent areas. Vieweg, Wiesbaden, pp 195–226
Vercoutter J (1959) The Gold of Kush. Two Goldwashing Stations at Faras-East. Kush 7, Khartoum, 120–153
Vercoutter J (1993) Or et Politique. Hommage à Jean Leclant, Bibl.d'Étude, 106/2, Cairo, 403–410
Wallbrecher E, Fritz H, Khudeir AA, Farahad F (1993) Kinematics of Panafrican thrusting and extension in Egypt. In: Thorweihe U, Schandelmeier H (eds) Geoscientific research in Northeast Africa. A. A. Balkema, Rotterdam/Brookfield, pp 27–30
Weigall AEP (1913) Travels in the Upper Egyptian Deserts. Blackwood, London, 193 pp
Weisgerber G (1989) Montanarchäologie – Grundzüge einer systematischen Bergbaukunde für Vor- und Frühgeschichte und Antike, Teil 1. Der Anschnitt 41, H. 6, 190–204.
Weisgerber G (1990) Montanarchäologie – Grundzüge einer systematischen Bergbaukunde für Vor- und Frühgeschichte und Antike, Teil 2. Der Anschnitt 42, H. 1, 2–18.
Welsby DA, Anderson JR (eds) (2004) Sudan- Ancient Treasures. The British Museum Press, London, 336 pp
Wetait MA (1997) Petrology, geochemistry and tectonic evolution of the Pan-African rocks of Wadi Umm Balad-Wadi El Urf area, Northeastern Desert, Egypt. Egypt J Geol 41/2B:431–455
Wetait MA, Botros NS (1997) Baryte mineralization in the south of Um Monqul area, Eastern Destert, Egypt. J Afr Earth Sci 25(3):485–489
Whiteman AJ (1971) The geology of the Sudan republic. Clarendon Press, London, 290 pp
Willis KM, Stern RJ, Clauer N (1988) Age and geochemistry of late Precambrian sediments of the Hammamat series from the NE Desert of Egypt. Precambrian Res 42:173–187
Wilsdorf H (1952) Berg-und Hüttenleute im Altertum bis zum Ausgang der Römischen Republik, ihre wirtschaftliche und soziale Lage. Freiberger Forschungshefte, Reihe D, Heft 1, Berlin
Winkler HA (1938) Rockdrawings of Southern Upper Egypt II. Arch. Survey Egypt, London
Woelk D (1966) Agatharchides von Knidos. Über das Rote Meer. Übers. u. Kommentar, Bamberg
Yoyotte J (1975) Les sementyou et l'exploration des régions minières à l'Ancien Empire. Bull. Soc. Franç. d'Egyptologie 73, Paris
Zalata A et al (1972) The results of prospection work for gold and rare metals by the Barramiay Party in the area of Barramiya in 1971–1972.-EXMA-Techoexport int. Report Cairo
Zibelius-Chen K (1994) Die Kubanstele Ramses II. und die nubischen Goldregionen, Hommages à Jean Leclant, IFAO, Le Caire, N 2, pp 411–417
Zoheir BA (2004) Gold mineralization in the Um El Tuyor area, South Eastern Desert, Egypt: geologic context, characteristics and genesis. Inaug. Dissertation. LMU-University Munich, 159 pp
Zoheir BA, Klemm DD (2007) The tectono-metamorphic evolution of the central part of the Neoproterozoic Allaqi- Heiani suture. South Eastern Desert of Egypt. Gondwana Res 12(3):289–304
Zoheir BA, Moritz R (2007) Fluid evolution and gold deposition conditions of the El-Sid Gold deposit, Egypt. Eur Curr Res Fluid Incl 258, Abstr
Zoheir BA, Robert M (2007) Fluid evolution and gold deposition conditions of the El Sid gold deposit, Egypt. Eur Curr Res Fluid Incl Bern, Abstr 258

Zoheir BA, Helba H, Khalil KI, Bondnar R (2007) H_2O-CO_2-CH_4-bearing fluid inclusions in Quartz: insights into the origin and evolution of two different hydrothermal Au deposits from the Egyptian Eastern Desert. Abstr. Session Econ Geol

Zoheir BA, Qaoud NN (2008a) Hydrothermal alteration geochemistry of the Betam gold deposit, south Eastern Desert, Egypt: mass-volume-mineralogical changes and stable isotope systematics. Appl Earth Sci: IMM Trans Sect B 117(2):55–76

Zoheir BA, El-Shazly AK, Helba H, Khalil KI, Bodnar RJ (2008b) Origin and evolution of the Umm Egat and Dungash orogenic gold deposits, Egyptian Eastern Desert: evidence from fluid inclusions in quartz. Econ Geol 103:405–424

Zoheir BA, Akawy A, Imbarak H (2008c) Role of fluid mixing and wallrock sulfidation in gold mineralization at the Semna Mine area, central Eastern Desert of Egypt: evidence from hydrothermal alteration, fluid inclusions and stable isotope data. Ore Geol Rev 34(4):580–596

Zoheir BA, Akawy A (2009) Genesis of the Abu Marawat gold deposit, central Eastern Desert of Egypt. J Afr Earth Sci 57:306–320

Index

R. Klemm, D. Klemm, *Gold and Gold Mining in Ancient Egypt and Nubia*, Natural Science in Archaeology,
DOI 10.1007/978-3-642-22508-6, © Springer-Verlag Berlin Heidelberg 2013

H

I

J

K

L

M

MIX
Papier aus verantwortungsvollen Quellen
Paper from responsible sources
FSC® C105338

If you have any concerns about our products,
you can contact us on
ProductSafety@springernature.com

In case Publisher is established outside the EU,
the EU authorized representative is:
Springer Nature Customer Service Center GmbH
Europaplatz 3, 69115 Heidelberg, Germany

Printed by Libri Plureos GmbH
in Hamburg, Germany